SAUNDERS GOLDEN SERIES

Donald J. Burton and Joseph I. Routh

University of Iowa, Iowa City, Iowa

Essentials of Organic and Biochemistry

1974

W. B. Saunders Company · Philadelphia · London · Toronto

W. B. Saunders Company: West Washington Square
Philadelphia, PA 19105

12 Dyott Street
London, WC1A 1DB

833 Oxford Street
Toronto, Ontario M8Z 5T9, Canada

Library of Congress Cataloging in Publication Data

Burton, Donald Joseph, 1934–

Essentials of organic and biochemistry.

(Saunders golden series)

1. Chemistry, Organic.

2. Biological chemistry.

I. Routh, Joseph Isaac, 1910– joint author.

II. Title.

QD251.2.B85 547 74-9429

ISBN 0-7216-2210-0

Essentials of Organic Chemistry ISBN 0-7216-2210-0

Last digit is the print number: 9 8 7 6 5 4 3 2 1

PREFACE

Scientific knowledge in the past 10 to 15 years has grown so rapidly that it has been exceedingly difficult to keep science courses abreast of the times. Prior to this period of rapid development, the beginning college course in chemistry, for example, consisted mainly of inorganic and theoretical chemistry with an introduction to qualitative analysis. So many interesting events took place in the fields of organic chemistry and biochemistry that an attempt was made to integrate the principles of inorganic, organic, and biochemistry in beginning college chemistry courses. Several textbooks, including *Essentials of General, Organic and Biochemistry* by the present authors and Darrell Eyman, were written for students not planning to become chemistry majors. Several courses covering material in these three fields were designed for students in nursing and allied health fields, as well as for those interested in a general survey of chemistry as part of a broad background in preparation for other professions.

More recently, several textbooks have appeared that emphasize the chemistry of the environment and the multiple problems of pollution. The development of science and the occurrence of textbooks of this type, coupled with the increased popularity of science articles in the news, has stimulated a quest for scientific knowledge in the present generation of students. The healthy growth of student participation in college and university committees and administration has resulted in the demand for more realistic and relevant chemistry courses in the curriculum.

The trend at present, therefore, is toward courses that meet the needs of students, whether they are in liberal arts, medical technology, nursing, or other health-related fields. A major movement in this direction is a demand for a one semester course in general chemistry followed by a one semester course in organic and biochemistry.

The present text is designed to introduce the principles of organic and biochemistry to students who have had one or more semesters of general chemistry. The authors are convinced that it is no longer sufficient to present only a general description of the fundamentals of organic and biochemistry. They have, therefore, attempted to include an explanation of why chemical reactions follow certain pathways and to explore the mechanisms of important reactions in these two fields of chemistry. In the first section, the chemistry of organic functional groups and the chemical mechanisms of the reactions they undergo are emphasized. The chemical nature and properties of insecticides, pheromones, plastics, natural polymers, and compounds of biochemical interest are introduced in this section. This material is followed by a brief but fairly complete section on biochemistry. In addition to an explanation of the chemistry and metabolism of the major foodstuffs, this section contains a description of the cell, its enzymes and coenzymes, and energy-releasing mechanisms. The chemistry of heredity and DNA and RNA and the biochemistry of drugs and drug action are the subjects of other chapters. A laboratory manual *Experiments in Organic and Biochemistry* has also been prepared to accompany the text.

The authors are indebted to the many students they have had the opportunity to teach in beginning chemistry courses. They would also like to thank their colleagues, especially Darrell Eyman, with whom they shared the teaching of these students, for valuable suggestions and criticisms. Finally, the authors are most grateful to their publisher for thoughtful advice and continued interest in the preparation of the manuscript for publication.

CONTENTS

REVIEW OF FUNDAMENTAL CONCEPTS

CHAPTER 1

The *objectives* of this chapter are to enable the student to:

1. Understand the basic concepts of atomic and molecular orbitals.
2. Write the electronic configurations and Lewis symbols for the representative elements and their ions.
3. Explain the formation of an ionic compound and the nature of ionic bonds.
4. Explain the formation of covalent bonds and how the forces of interaction arise in a covalent bond.
5. Recognize the trends of electronegativity in the periodic table and their usefulness in predicting ionic and polar covalent bonds.
6. Understand atomic orbital hybridization and its relationship to molecular geometry.
7. Draw contributing forms of the resonance hybrid of molecules not adequately described by a single Lewis formula.
8. Define hydrogen bonding and explain how hydrogen bonding imparts some abnormal properties to molecules.
9. Define solute, solvent, solution, and electrolyte.
10. Understand the concept of K_w, K_a, and K_b.
11. Calculate pH, pOH, and pK of an aqueous solution.
12. Describe a titration and understand the significance of a titration curve.
13. Calculate $[H^+]$ and $[OH^-]$ from K_w, pH, or pOH data.
14. Understand the significance of a buffer solution and its role in biochemistry.

BONDING

Two types of bonds are important in chemistry: the ionic bond and the covalent bond. The best way to understand ionic and covalent bonds is in terms of **orbitals.** An **atomic orbital** can be defined as a region in space centered around the nucleus of the atom. The orbital can be thought of as a surface that corresponds to the area in which the electrons in that orbital are most likely to be found. Each atomic orbital may contain zero, one, or two electrons, but not more than two. All orbitals are not identical—they differ in shape, size, and energy.

The shapes of the atomic orbitals of hydrogen are shown in Figure 1–1. Note that

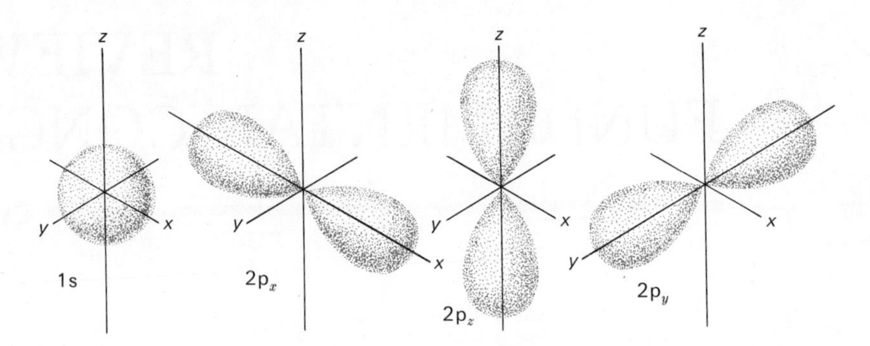

FIGURE 1–1 The shapes of the 1s- and 2p-atomic orbitals of hydrogen.

the **s-orbital** is spherical in shape, indicating that the electrons in this orbital may be in any orientation with respect to the nucleus, whereas the **p-orbitals** consist of two lobes of electron density on either side of the nucleus and are directional in shape.

The electrons of an atom are not arbitrarily assigned to any orbital, but are arranged around the nucleus in spherical shells of increasing radius. The lowest energy shell, given the designation 1, is called the K shell. The next highest energy shell, designated 2, is called the L shell; the third shell is M; and the fourth shell is N. Within any given shell (K, L, M, N) the electrons occupy certain orbitals, called s, p, d, and f, which are indicated in order of increasing energy. Therefore, a 2s-orbital is lower in energy than a 2p-orbital, a 3p-orbital is lower in energy than a 3d-orbital, and so on. The energy levels and sublevels and their relative energies are shown in Figure 1–2.

As shown in Figure 1–2, not every shell contains s, p, d, or f orbitals. For example, the K shell may contain only 1s electrons, the L shell only 2s and 2p electrons, the M shell only 3s, 3p, and 3d electrons, and so on. Note also that in the M and N shells the orbitals overlap and the 4s-orbitals are lower in energy than the 3d-orbitals. In addition, since the orbitals become larger the further away they are from the nucleus, the maximum number of electrons that are theoretically possible in any shell increases as depicted in Figure 1–2.

In order to correctly assign the schematic tabulation of the electrons in any given atom (called its electronic configuration), two restrictions must be kept in mind. First, the **Pauli exclusion principle** states that any orbital may contain no more than two electrons and these electrons must have opposite spins (indicated by ↑ and ↓). Secondly, **Hund's Rule** states that a second electron does not occupy an orbital until each orbital of equivalent energy contains at least one electron. Consequently, as a corollary to this rule, we can state that electrons occupy an orbital only if **all** orbitals of lower energy are filled. Therefore, an electron will occupy a 1s-orbital in preference to a 2s-orbital, a 2s-orbital in preference to a 2p-orbital, and so on.

A shorthand method of expressing electronic configuration is by indicating the shell number and the number of electrons in each type of orbital as shown below:

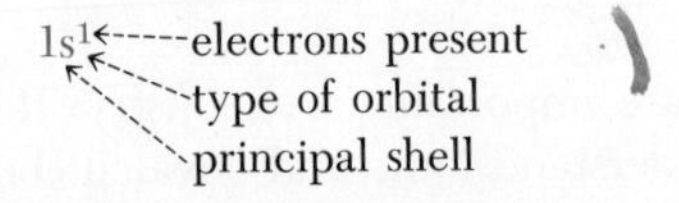

The notation above is for the hydrogen atom which has only one electron in an s-orbital in the K shell. The electronic configuration for helium which has two electrons is $1s^2$. Carbon which has six electrons has the electronic configuration $1s^2 2s^2 2p^2$. The electronic configurations for the first ten elements are shown in Table 1–1 Note that in the electronic configuration notation, the sum of the right superscripts always equals the atomic number

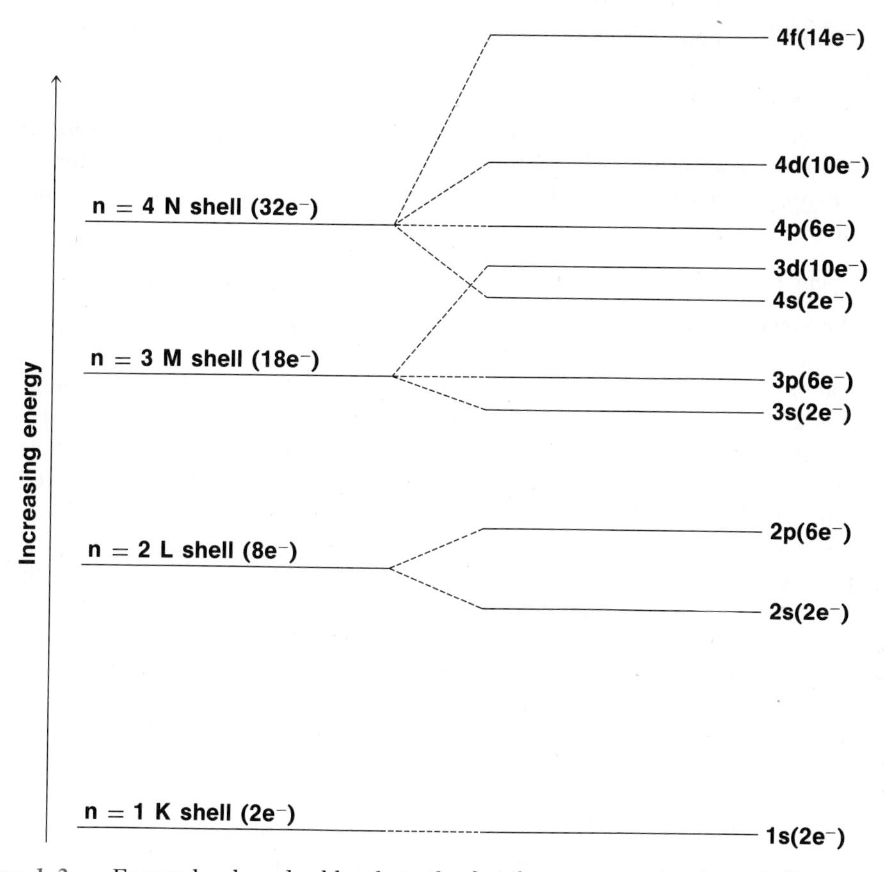

FIGURE 1-2 Energy levels and sublevels in the first four quantum levels or shells.

of the atoms. For ions the electronic configuration is indicated by simply increasing or decreasing the sum of the right superscripts to indicate the number of electrons added or lost as shown below:

$$Li^{+} = 1s^{2}$$
$$F^{-} = 1s^{2}2s^{2}2p^{6}$$

TABLE 1-1 ELECTRONIC CONFIGURATIONS OF THE FIRST TEN ELEMENTS

	1s	2s		2p	
H	(↑)				
He	(↑↓)				
Li	(↑↓)	(↑)			
Be	(↑↓)	(↑↓)			
B	(↑↓)	(↑↓)	(↑)		
C	(↑↓)	(↑↓)	(↑)	(↑)	
N	(↑↓)	(↑↓)	(↑)	(↑)	(↑)
O	(↑↓)	(↑↓)	(↑↓)	(↑)	(↑)
F	(↑↓)	(↑↓)	(↑↓)	(↑↓)	(↑)
Ne	(↑↓)	(↑↓)	(↑↓)	(↑↓)	(↑↓)

It should be pointed out that atoms (and also ions) whose outer shell of electrons has reached a maximum are particularly stable and unreactive. Therefore, the helium atom in which the K shell electrons have reached the maximum of two is one of the so-called inert gases. Similarly, neon which has the maximum eight electrons allowed in the L shell is also unreactive. Atoms or ions which have completed outer shells of electrons are said to have attained the inert or rare gas electronic configuration and desire to neither gain nor lose any additional electrons.

Valence Electrons

The electronic configuration of an element determines its chemical properties, as well as many of its physical properties. The electrons most important in determining chemical properties are those found in the outermost quantum level, or shell. These electrons are called the **valence electrons.** The periodically repeated occurrence of elements with similar properties is due to the periodically repeated occurrence of elements with the same number of valence electrons. For example, in all of the elements in Period 2, starting with lithium, the first shell is filled with two electrons. Lithium has one electron in the second shell; beryllium, two; boron, three; carbon, four; and so on to neon, in which the second shell is completed with eight electrons. The element with the next highest atomic number would naturally start a new period under lithium, since its first two shells are completely filled with electrons, and it has one electron in the third shell. Following sodium, magnesium has its first two shells completely filled with electrons and two extra electrons in the third shell; aluminum has three electrons in the third shell; and so on to argon, which has the third shell filled with eight electrons. The element with the next highest atomic weight would then start a new row, or series, being placed under lithium and sodium. This is potassium, with the first two shells completely filled, with eight electrons in the third shell and one electron in the fourth shell. Since the number of electrons necessary to fill a given shell (starting with shell number one) is 2, 8, 18, and 32, it is not surprising that 2, 8, 18, or 32 elements are needed to complete a horizontal row, or period, in the periodic table. The relationships just discussed may be seen in the following tabulation:

Electrons in	*Li*	*Be*	*B*	*C*	*N*	*O*	*F*	*Ne*
First shell	2	2	2	2	2	2	2	2
Second shell	1	2	3	4	5	6	7	8

	Na	*Mg*	*Al*	*Si*	*P*	*S*	*Cl*	*Ar*
First shell	2	2	2	2	2	2	2	2
Second shell	8	8	8	8	8	8	8	8
Third shell	1	2	3	4	5	6	7	8

	K	*etc.*
First shell	2	etc.
Second shell	8	etc.
Third shell	8	etc.
Fourth shell	1	etc.

Another symbolism for indicating the electronic configuration of an atom is very helpful in predicting chemical properties. The **Lewis symbol** is written as the letter(s) denoting the element, surrounded by dots symbolizing the number of valence electrons. The letter(s) of the symbol represent the nucleus and all electrons in inner closed shells, and the dots represent the valence electrons. Examples of Lewis symbols are:

Li·   Be: ·B: :Ċ·   :N·   :Ö·   :F·   :Ne:

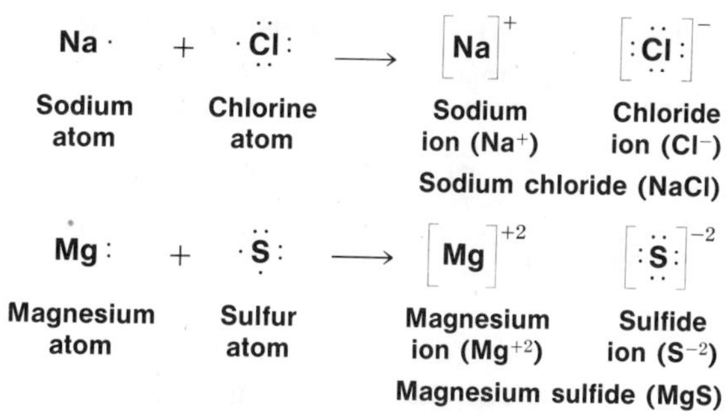

Na·	+	·Cl:	⟶	$[Na]^{+}$	$[:Cl:]^{-}$
Sodium atom		Chlorine atom		Sodium ion (Na^{+})	Chloride ion (Cl^{-})
				Sodium chloride (NaCl)	
Mg:	+	·S:	⟶	$[Mg]^{+2}$	$[:S:]^{-2}$
Magnesium atom		Sulfur atom		Magnesium ion (Mg^{+2})	Sulfide ion (S^{-2})
				Magnesium sulfide (MgS)	

FIGURE 1–3 The process of electron transfer commonly occurs in the formation of inorganic salts.

The Lewis symbols for elements in one vertical column, or group, in the periodic table differ only in letters, since each has the same number of valence electrons. Thus, the Lewis symbols for nitrogen and phosphorus, both in Group VA, are $:\dot{N}\cdot$ and $:\dot{P}\cdot$.

IONIC BONDS

The nature of ionic bonding may be illustrated by considering the formation of sodium chloride. In the overall reaction, it appears that sodium atoms react with chlorine atoms to form the compound sodium chloride. As seen in Figure 1–3, however, an electron is transferred from the outer shell of the sodium atom to the outer shell of the chlorine atom. This loss of an electron from sodium occurs readily and requires a relatively small amount of energy, which is called the ionization potential. When this electron is presented to the chlorine atom, the chlorine atom readily accepts it, with the release of energy, which is called the electron affinity. The sodium ions and the chloride ions that are formed in the process are much more stable than the atoms, and resemble the inert gases neon and argon, respectively, in their valence electron configuration. The ions differ from the inert gas atoms in that they are no longer neutral but bear positive (Na) and negative (Cl) charges. These opposite charges attract each other and are responsible for the strong ionic bond between sodium ions and chloride ions in the compound sodium chloride. Compounds in which the atoms are held together by ionic bonds are called ionic compounds.

A compound formed by the transfer of two electrons to yield an outer shell of eight valence electrons in each ion is magnesium sulfide. A relatively small amount of energy is required to remove the two outer electrons of magnesium. They are accepted by the outer shell of the sulfur atom with the release of energy. This electron transfer is illustrated in Figure 1–3. The ionic bond between magnesium and sulfide ions results from the strong electrostatic attraction between the oppositely charged ions.

In simple chemical reactions, atoms are commonly represented by the symbol alone. However, ions alone, or in compounds, are represented as the symbol bearing either positive or negative charges. For example, note ions such as Na^{+} and Mg^{++} and compounds such as $Na^{+}Cl^{-}$ and $Mg^{++}S^{--}$.

The number of electrons transferred in forming an ion corresponds to the combining capacity of an atom. The bonding resulting from electron transfer is called electrovalence, or ionic bonding. It can readily be seen that the combining capacity is equal to the number of electrons gained or lost by an atom when it is converted into an ion.

COVALENT BONDS

About 1916, it was suggested that two atoms may combine by sharing valence electrons. The process of joining atoms to form molecules by the sharing of electrons

H· + ·H ⟶ H:H or H—H

Hydrogen atom | Hydrogen atom | Hydrogen molecule

:C̈l:C̈l: or |C̄l—C̄l| | :N:::N: or |N≡N|

Chlorine molecule | Nitrogen molecule

FIGURE 1-4 The sharing of electrons, or covalence, illustrated by hydrogen, chlorine, and nitrogen molecules.

is called covalence. As an example, hydrogen gas consists of molecules that contain two hydrogen atoms held together by a force resulting from the sharing of a pair of electrons, as shown in Figure 1-4. This figure illustrates the use of Lewis symbols for molecules with covalent bonds. In the symbol for the molecule, a pair of dots placed between the atoms represents a bond.

Hydrogen chlorine, nitrogen, and other diatomic gaseous elements show similar behavior in sharing electrons to the extent that their valence shells are filled. In the chlorine molecule, each chlorine atom with seven valence electrons is able to attain a complete shell by sharing one pair of electrons with the other chlorine atom. In the nitrogen molecule, a filled shell is attained only if three pairs of electrons are shared. These examples are represented schematically in Figure 1-4 with Lewis formulas.

It is stated that the hydrogen and chlorine molecules are held together by a single bond, but that the nitrogen molecule has a triple bond. Double bonds are also known to exist in many molecules.

The rationalization of the existence of a force between atoms which share a pair of electrons is aided by the following hypothetical experiment. In this experiment, diagrammatically represented in Figure 1-5 a, two hydrogen atoms are allowed to approach one another until their 1s-atomic orbitals overlap and mutually occupy the space between the nuclei. The result of the overlap is an **electron cloud** associated with the molecule. An electron cloud associated with more than one nucleus is referred to as a molecular orbital. In the molecular orbital formed here by the combination of atomic

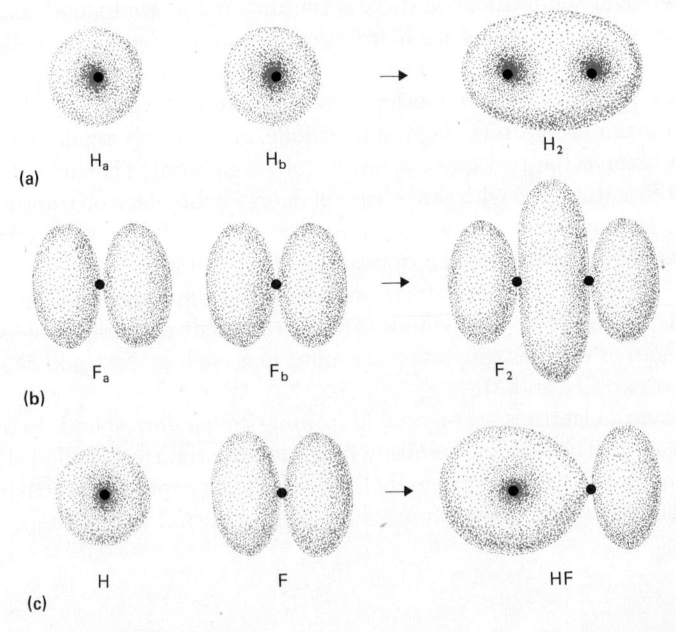

FIGURE 1-5 (*a*) Hypothetical approach of the electron clouds of two hydrogen atoms and the resulting electron cloud of the hydrogen molecule. (*b*) A similar approach of the p-orbitals of two fluorine atoms. (*c*) The approach of a 1s-orbital of a hydrogen and a 2p-orbital of a fluorine.

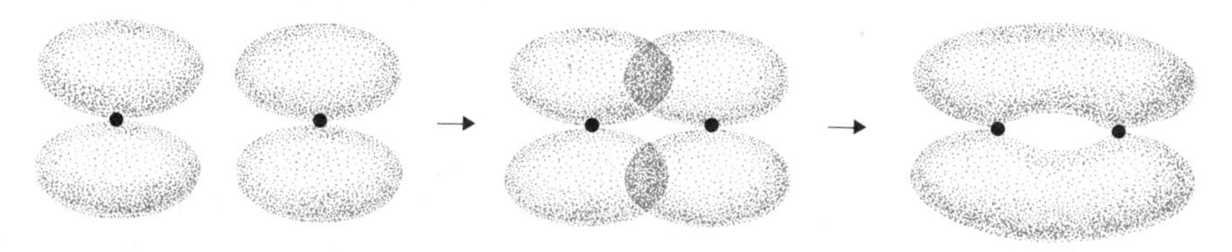

FIGURE 1-6 Side-by-side overlap of atomic p-orbitals to give a pi-bond.

orbitals, there is a high probability of finding the electrons in the region between the nuclei.

Before the atoms approach one another, the only forces which exist are between the electron and the nucleus of a given atom. When the atoms are brought close together, four new forces arise. Two of these tend to destabilize the molecule relative to the separated atoms. These are the repulsions of the nuclei for one another and the repulsions of the electrons for one another. These repulsions are expected, since it is known that similarly charged objects repel one another. One of the attractive forces arises because the electron originating from atom H_a is attracted to the nucleus of atom H_b. The same force holds for the electron on atom H_b interacting with the nucleus of atom H_a. The second attractive force arises from the interaction of the opposed small magnetic fields associated with the oppositely spinning electrons. The summation of the contributions of these four forces determines the total force of interaction between the two atoms and, consequently, the net stabilization of the molecule relative to the separated atoms.

The results of a similar experiment involving the approach of fluorine atoms is shown in Figure 1–5 b, where it is seen that atomic 2p-orbitals on the approaching atoms overlap to form a region between the nuclei where the probability of finding electrons is quite high. Figure 1–5 c shows a third type of experiment in which a 1s-orbital on hydrogen overlaps a 2p-orbital on fluorine to form a covalent bond.

In each of the three hypothetical experiments shown in Figure 1–5, the overlap of the electron clouds of approaching atoms results in an increased electron density along the internuclear axis. A covalent bond, resulting from increased electron density along the axis connecting the nuclei of adjacent atoms, is called a **sigma-bond** (σ-bond).

A covalent bond between adjacent atoms can also result from the overlap of p-orbitals in a side-by-side fashion, as shown in Figure 1–6. Overlap of electron clouds in this manner results in increased electron density above and below the internuclear axis. A bond of this type is called a **pi-bond** (π-bond).

POLAR BONDS

"Pure" covalent bonds, as are found in homonuclear diatomic molecules, and "pure" ionic bonds, as are found in salts like NaCl, represent the extremes of bonding. In the former, the electron pair is shared equally between atoms (Fig. 1–7), whereas in the latter there is no sharing of electrons, but rather a complete transfer of an electron from one

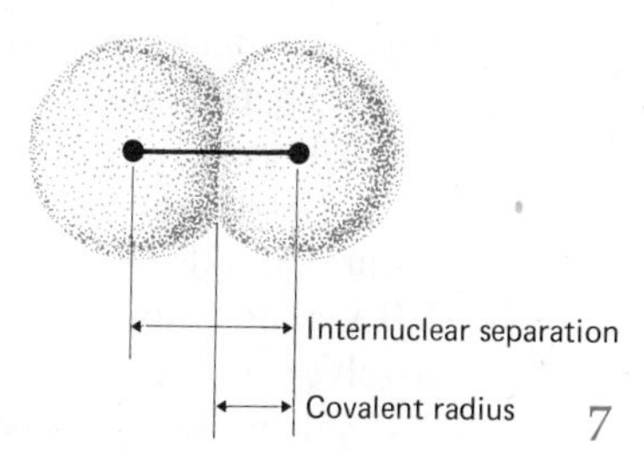

FIGURE 1-7 Covalent radius of a homonuclear diatomic molecule.

atom to another. Bonding intermediate between "pure" covalent and "pure" ionic is very common. Since electrons may not always be shared equally between atoms, but yet are not necessarily completely transferred from one atom to another, there exists the possibility of unequal charge distribution in a covalent bond.

For example, in the compound hydrogen chloride, HCl, the shared pair of electrons is attracted more by the chlorine end of the molecule than by the hydrogen end. The result is an unequal charge distribution, with the chlorine end more negative and the hydrogen end of the molecule more positive. Hydrogen chloride may be represented as follows:

$$\mathrm{HCl} \quad \text{or} \quad \mathrm{H:\ddot{\underset{..}{Cl}}:} \quad \text{or} \quad \mathrm{H^{+}:\ddot{\underset{..}{Cl}}:^{-}} \quad \text{or} \quad (+\ \ -)$$

Even though the molecule is electrically neutral, the center of the positive charge does not coincide with the center of the negative charge. The molecule is called a dipole and is said to possess a dipole moment. Dipole moments are often designated by the symbol ⟼. In this symbol, the pointed end corresponds to the negative end of the dipole; the length is related to the magnitude; and the orientation of the dipole is indicated by the orientation of the symbol. When placed in an electrical field, such molecules will line up with their negative ends facing the positive plate (electrode) and their positive ends facing the negative plate (electrode). The dipole character of these compounds gives rise to the names polar bonds and polar covalent compounds.

Compounds whose atoms equally share a pair of electrons will have the center of their positive charge coinciding with the center of their negative charge. These compounds do not exhibit dipole characteristics and are called nonpolar molecules with nonpolar bonds. Hydrogen molecules (H_2) and chlorine molecules (Cl_2) are examples of nonpolar covalent compounds. In general, if a molecule is composed of two of the same kind of atoms, the bond between them will be nonpolar and the molecule will be nonpolar. If two different atoms make up the molecule, the bond is polar and the molecule is polar.

Electronegativity

It has already been stated that in the compound hydrogen chloride the chlorine atom has a greater attraction for the electron pair than does the hydrogen atom. The attraction of an atom for shared electrons depends on the amount of energy required in the transfer of electrons from or to its outer electron shell. The attraction for valence electrons varies from element to element and is called its electronegativity. The relative electronegativity values that have been determined for some of the common elements are represented in a partial periodic table as shown in Table 1–2. These values are related to the ability of the atoms to attract shared electrons and thus increase the negative charge of their end of a molecule.

The **electronegativity** of an element is related to its ionization potential and its electron affinity. Elements which have high ionization potentials and high electron affinities exhibit high electronegativities. Inspection of Table 1–2 will reveal that elements with the highest electronegativities are found in the upper right corner of the periodic table. There is also a tendency toward increasing values across a period to the right and toward decreasing values moving down the elements of a group. If two elements with greatly different electronegativities combine, the bond will be highly polar or ionic in nature. Metals in Groups IA and IIA combining with the nonmetals in Groups VIA and VIIA almost always form ionic compounds with ionic bonds. When the combination involves two elements with similar values, the bonds are usually covalent. If the compound is covalent, the atoms with the greatest difference in electronegativity will form the more

TABLE 1–2 Relative Electronegativity Values of the Common Elements

IA	IIA	IIIB	IVB	VB	VIB	VIIB	VIII			IB	IIB	IIIA	IVA	VA	VIA	VIIA	O
1 **H** 2.1																	2 **He** 0
3 **Li** 1.0	4 **Be** 1.5											5 **B** 2.0	6 **C** 2.5	7 **N** 3.0	8 **O** 3.5	9 **F** 4.0	10 **Ne** 0
11 **Na** 0.9	12 **Mg** 1.2											13 **Al** 1.5	14 **Si** 1.8	15 **P** 2.1	16 **S** 2.5	17 **Cl** 3.0	18 **Ar** 0
19 **K** 0.8	20 **Ca** 1.0	21 **Sc** 1.3	22 **Ti** 1.5	23 **V** 1.6	24 **Cr** 1.6	25 **Mn** 1.5	26 **Fe** 1.8	27 **Co** 1.8	28 **Ni** 1.8	29 **Cu** 1.9	30 **Zn** 1.6	31 **Ga** 1.6	32 **Ge** 1.8	33 **As** 2.0	34 **Se** 2.4	35 **Br** 2.8	36 **Kr** 0
37 **Rb** 0.8	38 **Sr** 1.0	39 **Y** 1.2	40 **Zr** 1.4	41 **Nb** 1.6	42 **Mo** 1.8	43 **Tc** 1.9	44 **Ru** 2.2	45 **Rh** 2.2	46 **Pd** 2.2	47 **Ag** 1.9	48 **Cd** 1.7	49 **In** 1.7	50 **Sn** 1.8	51 **Sb** 1.9	52 **Te** 2.1	53 **I** 2.5	54 **Xe** 0
55 **Cs** 0.7	56 **Ba** 0.9	57-71 — 1.1-1.2	72 **Hf** 1.3	73 **Ta** 1.5	74 **W** 1.7	75 **Re** 1.9	76 **Os** 2.2	77 **Ir** 2.2	78 **Pt** 2.2	79 **Au** 2.4	80 **Hg** 1.9	81 **Tl** 1.8	82 **Pb** 1.8	83 **Bi** 1.9	84 **Po** 2.0	85 **At** 2.2	86 **Rn** 0
87 **Fr** 0.7	88 **Ra** 0.9	89- 1.1-															

polar bonds. Two like atoms with no difference in their electronegativity values will obviously form nonpolar covalent bonds.

MOLECULAR GEOMETRY

The shape of a molecule is determined by the distances between bonded atoms in the molecule, called bond lengths, and the angles between bonds, called bond angles. In simple diatomic molecules, such as H_2 and CO, the bond length completely describes the geometry of the molecule. In triatomic and higher polyatomic molecules, more than bond lengths must be stated to completely describe the geometry of the molecule. In such molecules the geometry is determined by the nature of the central atom, such as oxygen in H_2O and carbon in CCl_4.

The predominant geometry found for molecules with four atoms bonded to a central atom is tetrahedral, as shown in Figure 1–8. In this type of molecule, the central atom is located at the center of the tetrahedron and the bonded atoms are found at the apices. This geometry occurs when four bonding pairs of electrons seek the maximum distance of separation in order to minimize the repulsions. The only other geometry observed for pentatomic molecules is square planar, in which the bonded atoms are located at the corners of a square with the central element located in the center of the square. This geometry is limited to transition metal compounds.

The shapes of complex molecules can usually be predicted by considering the principles for simple molecules. Shapes of specific simple and complex molecules will be discussed in later chapters.

ATOMIC ORBITAL HYBRIDIZATION

The shape of a molecule can be enlightening as to the bonding interactions among the constituent atoms. In the cases of molecules having all single bonds, the relative orientations of the atoms establish the relative orientations of the sigma-bonds. For example, in methane, CH_4, previously stated to have tetrahedral geometry, four sigma-bonds, bonding a carbon atom to each of four hydrogen atoms, are pointed toward the apices of a tetrahedron. The assumption of interbond angles of 109.5° is consistent with the experimental observation that each of the four hydrogen atoms bonded to carbon has the same chemical properties, and thus must be involved in identical bonding. These experimental observations are inconsistent with a bonding model which involves the use of pure atomic orbitals by carbon.

If the carbon atom in methane uses four pure atomic orbitals to bond to each of four hydrogen atoms, the molecule would not be tetrahedral. This is because the four orbitals of the valence shell of carbon, one s-orbital and three p-orbitals, are not equivalent. The use of pure atomic orbitals would result in three sigma-bonds which are mutually

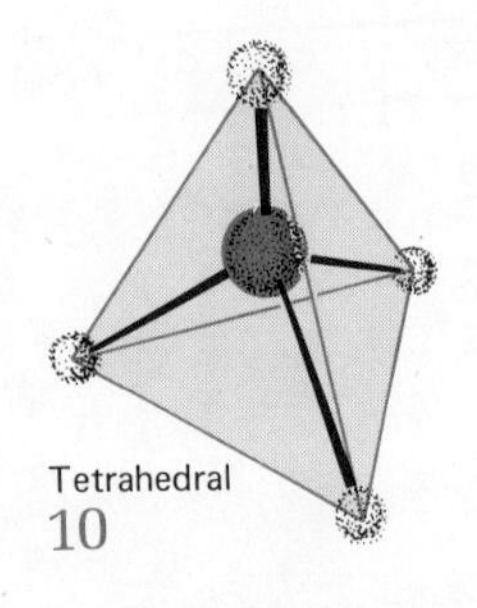

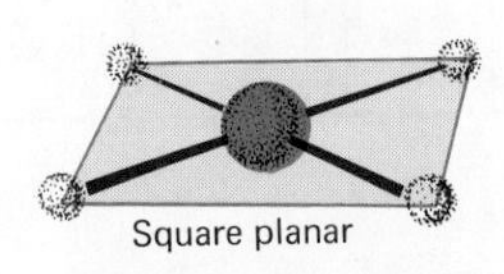

FIGURE 1–8 Geometrics of molecules with four atoms bonded to a central atom.

FIGURE 1-9 An sp^3-hybrid orbital.

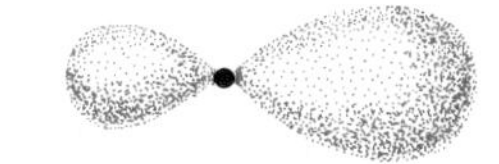

perpendicular, owing to the overlap of the three 2p-orbitals of carbon with three 1s-orbitals of hydrogen. The fourth sigma-bond would arise by overlap of the carbon 2s-orbital with a hydrogen 1s-orbital. Clearly this bonding model would not produce a tetrahedral arrangement of four equivalent bonds, as is the case in methane.

A set of four equivalent sigma-bonds in methane is explained by considering the combination of the carbon 2s-orbital and the three carbon 2p-orbitals as generating a set of four equivalent hybrid orbitals. These four hybrid orbitals, designated sp^3-orbitals, are not like s-orbitals or p-orbitals. Each appears similar to a p-orbital with the two lobes being of different sizes (Fig. 1–9). The relative orientations of the four hybrid sp^3-orbitals are such that the larger lobe of each points toward the apex of a tetrahedron (Fig. 1–10).

For simple molecules having a total of three bonded atoms and nonbonding electron pairs around the central atom, the concept of atomic orbital hybridization can be used to explain the observed geometry. The bonded atoms or nonbonded electron pairs found at the corners of an equilateral triangle are bonded to a set of three equivalent sp^2-hybrid orbitals on the central atom. This set of hybrid orbitals, arising from the combination of an s-orbital and two p-orbitals, results in bond angles of 120° (Fig. 1–10). In this bonding scheme, one p-orbital on the central element which is not included in the hybridization assumes an orientation perpendicular to the plane of the three hybrid orbitals.

Molecules having a total of two bonded atoms and nonbonding electron pairs on the central atom are linear. This geometry is explained by considering a set of two equivalent sp-hybrid orbitals on the central element and two unhybridized p-orbitals oriented perpendicular to the sp-hybrid orbitals and perpendicular to one another. The use of sp-hybrid orbitals by a central element results in a bond angle of 180° (Fig. 1–10).

MULTIPLE BONDS

As indicated earlier, in many molecules more than one pair of electrons is shared between two atoms. In such multiple-bonded molecules, there must be pi-bonds involved, since only one sigma-bond can form between any two atoms. In the double-bonded molecule carbon dioxide, CO_2, each carbon-oxygen interaction involves one sigma-bond and one pi-bond, as shown in Figure 1–11. The sigma-bond arises by overlap of the electron clouds of an sp^2-hybrid orbital on oxygen with an sp-hybrid orbital on carbon, whereas the pi-bond involves overlap of p-orbitals from each atom.

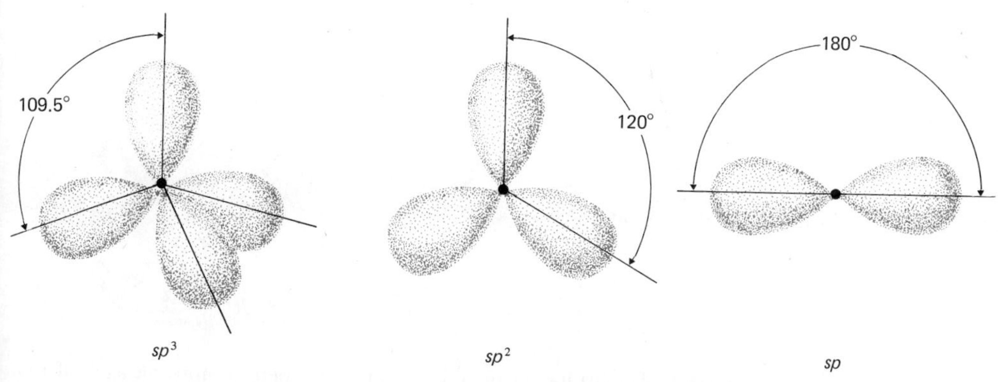

FIGURE 1-10 The relative orientations of equivalent hybrid orbitals.

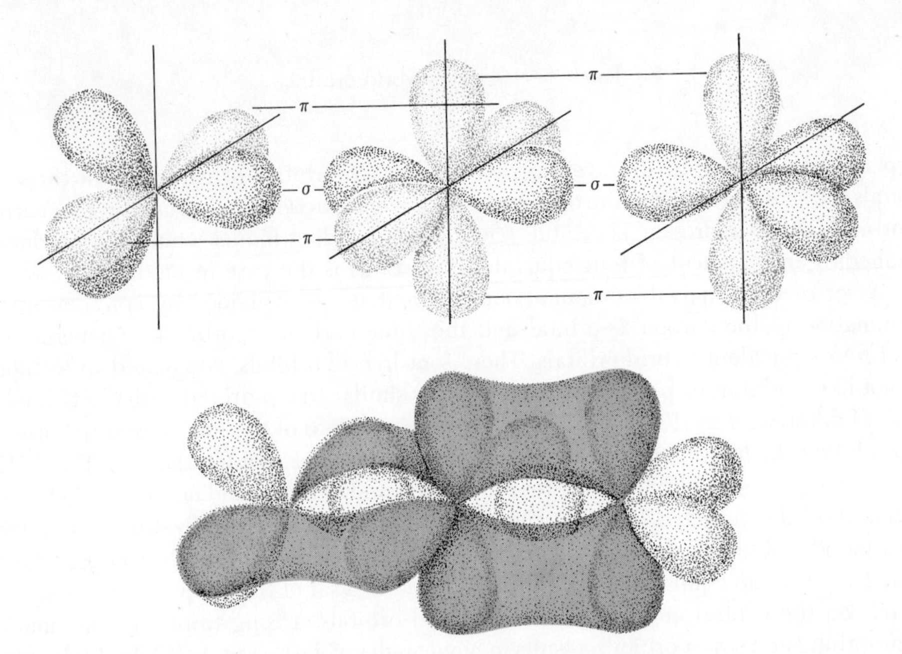

FIGURE 1-11 The approach of hybridized (shaded) and atomic (open) orbitals on carbon and oxygen to form sigma- and pi-bonds. The nonbonding electrons shown in the lower Lewis formula occupy hybrid sp^2-orbitals on oxygen.

In the triple-bonded nitrogen molecule, N_2, the sigma-bond can be ascribed to the overlap of the electron clouds of two sp-hybrid orbitals. The two pi-bonds arise by the overlap of two p-orbitals from each atom. This triple-bonded system is represented in Figure 1-12.

For many molecular species, it is not possible to define a single Lewis formula which is consistent with experimental facts. For example, the molecule SO_3 is planar triangular and has all sulfur-oxygen bonds of equal length, but the Lewis formula requires one sulfur-oxygen double bond. The following three Lewis formulas are equivalent in that each has one double bond.

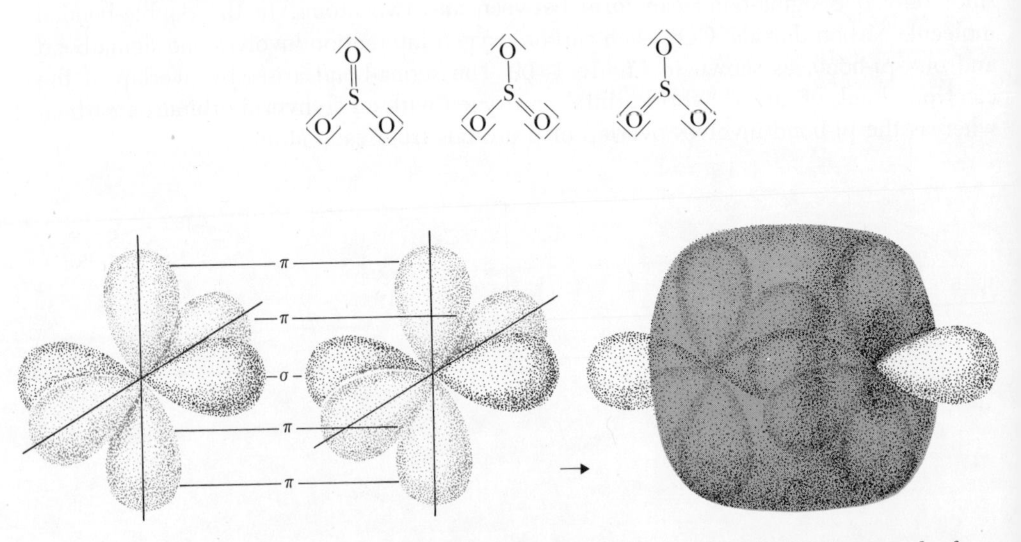

FIGURE 1-12 The approach of hybridized (shaded) and atomic (open) orbitals on each of two nitrogen atoms to form a sigma-bond and two pi-bonds.

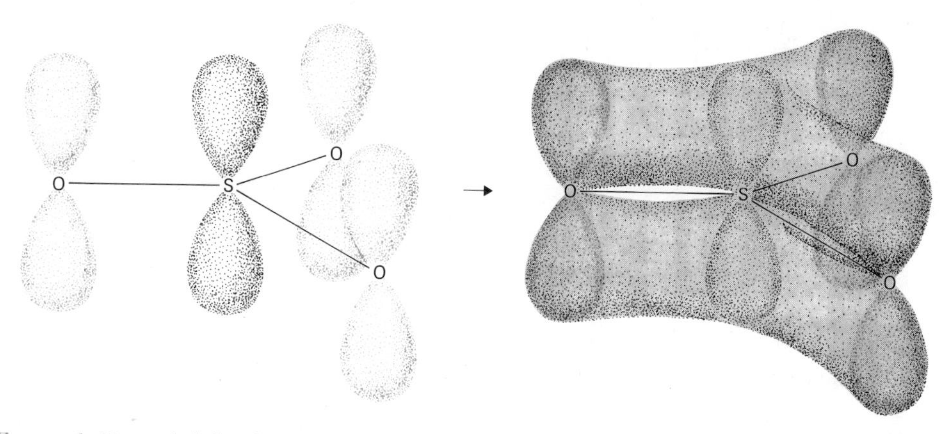

FIGURE 1-13 A delocalized π-bond in SO_3 arising from the overlap of four p-orbitals.

The real bonding in SO_3 can be described as a resonance hybrid of these three contributing forms and others. Any one contributing form inadequately describes the bonding, but SO_3 has characteristics of each of the contributors. Each bond is stronger than a single bond but weaker than a double bond.

When considering the bonding in molecules which are inadequately described by a single Lewis formula, the following principles are used:

1. Resonance contributors are limited to those Lewis formulas which differ only in the arrangement of electrons without shifting any atoms.
2. The hybrid of the resonance contributors represents the actual bonding of the molecule.
3. The resonance hybrid is more stable (lower in energy) than any of its contributing forms.
4. An increased stabilization is observed for molecules described as resonance hybrids. This stabilization, called the resonance energy, is greatest when the contributing forms have equal energy.
5. Contributing forms which have the lowest energies (are most stable) make the greatest influence on the nature of the hybrid and on the resonance energy.
6. Separation of unlike charges to two identical atoms gives a high energy (less stable) resonance form with minimal contribution to resonance energy.

It is also possible to describe the bonding in SO_3 and other molecules, inadequately described by a single Lewis formula, by using sigma-bonds and delocalized pi-bonds. Delocalized pi-bonds involve molecular orbitals which entail overlap of p-orbitals from more than two nuclei in a pi-fashion. This is schematically represented for SO_3 in Figure 1–13. In this figure it is seen that p-orbitals on sulfur and each of the oxygen atoms, all perpendicular to the plane of the molecule, overlap to give a pi-orbital spread over the four atoms. This orbital can contain only two electrons and results in one bond spread over three sulfur-oxygen interactions.

WATER

Water is very important to living organisms. As a substance essential to our existence, water ranks next to oxygen in importance. The human body can survive several weeks without food, but only a few days without water. Water is the medium in which biological reactions take place; it is a means of transporting nutrients in both plants and animals. The digestion of food, the circulation, the elimination of waste materials, and the regula-

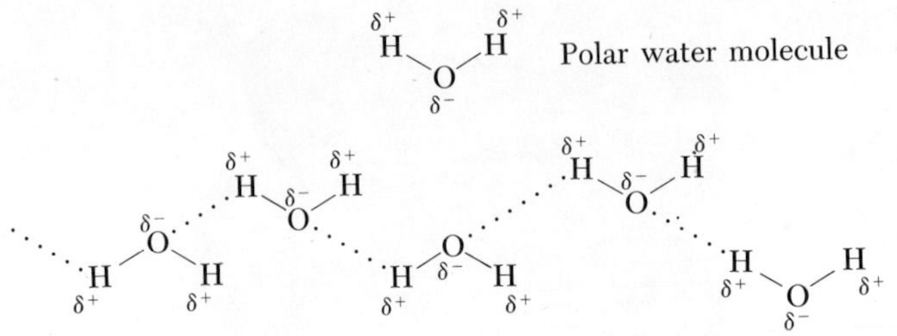

Polar water molecule

FIGURE 1-14 Hydrogen bonding in water.

tion of acid-base balance and body temperature, as well as other vital functions, depend on an adequate supply of water.

Pure water has no odor, taste, or color. It freezes at 0°C (32°F) and boils at 100°C (212°F). These values are abnormally high for a compound with such a low molecular weight. For example, hydrogen sulfide has a molecular weight of 34 and boils at −60°C, and methane, which has a molecular weight of 16, boils at −162°C. The abnormal physical properties of water are attributable to the fact that pure water is not a collection of isolated molecules, but consists of clusters of water molecules held together by high intermolecular attractive forces. These intermolecular forces are called **hydrogen bonds,** and arise because of the very strong attractions between the nonbonding electrons on the oxygen and the partially positively charged hydrogen atoms on the adjacent polar water molecules as shown in Figure 1-14.

The efficiency of hydrogen bonding decreases markedly when hydrogen is bonded to larger, less electronegative atoms and is generally important only when hydrogen is bonded to atoms of Groups V, VI, and VII, such as nitrogen (N), oxygen (O), and fluorine (F). Organic molecules, such as alcohols and carboxylic acids, in which oxygen is bonded to hydrogen (see Chapters 6 and 10), also show similar abnormal physical properties owing to hydrogen bonding. Complex biochemical molecules, such as DNA (see Chapter 16), also involve hydrogen bonding. Although hydrogen bonding results in the formation of weak bonds relative to the bond strength of a normal sigma-bond, these types of bonds play an important part in both organic and biochemistry, as will be seen in later chapters of the text.

If a substance such as salt or sugar is dissolved in water, the resulting homogeneous mixture is called a **solution.** When dissolution in water of a compound such as salt occurs, the forces holding the sodium ions and chloride ions together are overcome by the attraction of the water molecules for the ions as depicted in Figure 1-15 (water molecules associated with the sodium and chloride ions are represented by the symbol ↦). In a similar fashion, water will also attract other polar molecules. Solutions of polar compounds in water are called **aqueous solutions.** In a solution, the substance that is dissolved is called the **solute,** whereas the substance in which the solute is dissolved is called the

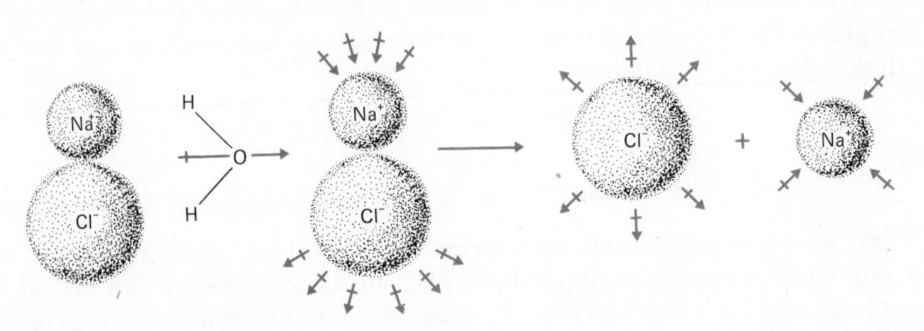

FIGURE 1-15 Sodium chloride going into solution illustrates the effect of water on ionic compounds.

solvent. In solutions in which it is not obvious which component is the solute or the solvent, it is common practice to call the solvent the substance present in the greatest amount.

Although water is a common solvent for polar solutes, such as sugar and salts, it is not satisfactory for nonpolar solutes, such as fats or oils, because the energy required to separate solvent molecules is not compensated for by a weak solute-solvent interaction. Such nonpolar solutes are more readily soluble in nonpolar solvents like ether, carbon tetrachloride, or gasoline. In general, it can be stated that ionic and polar solutes are most soluble in polar solvents (such as water), and nonpolar solutes are most soluble in nonpolar solvents.

Ionization

Ionization is the process whereby compounds split into positively and negatively charged ions when dissolved in a solvent. Compounds that ionize when dissolved are called **electrolytes,** since a solution of ions conducts an electrical current. Some simple generalizations can be made about electrolytes in water as follows:

1. When an electrolyte is dissolved in water, it either dissociates into ions or reacts with water (hydrolyzes) to generate ions.
2. The sum of the positive charges that result from the dissociation of the electrolyte is equal to the sum of the negative charges.
3. Nonelectrolytes that fail to conduct an electric current when in solution do **not** dissociate to form ions.
4. Ions possess properties different from the corresponding uncharged atoms or molecules, and are responsible not only for the electrical properties but also for the chemical properties of a solution.

Strong electrolytes are ionic compounds and hydrolyzable covalent polar compounds that dissociate completely into ions in dilute solutions. **Weak electrolytes** are substances that dissociate only slightly into ions when in solution and exist essentially as undissociated molecules. Most salts, strong bases, and strong acids, such as HCl, H_2SO_4, and HNO_3, are classed as strong electrolytes. Examples of weak electrolytes in water are the base ammonia (NH_3) and acids such as H_2CO_3, H_3BO_3, and acetic acid (CH_3COOH).

It is important to remember that ionization processes are always reversible; therefore, the ionization of an electrolyte is an equilibrium reaction. For strong electrolytes, the equilibrium lies to the right (the ionized side of the reaction), whereas for weak electrolytes the equilibrium lies to the left (the undissociated side). For example, when the polar covalent gas, hydrogen chloride, is added to water, it undergoes hydrolysis to give the **hydronium ion,** H_3O^+, and the chloride ion, Cl^-, as indicated in the following equation.*

$$HCl + H_2O \rightleftharpoons H_3O^+ + Cl^-$$

In this example the ionization equilibrium lies far to the right and the dissociation into hydronium ion and chloride ion is about 95 per cent complete.

In contrast to hydrogen chloride, when acetic acid is dissolved in water, dissociation occurs only to about 1 to 2 per cent, and the equilibrium lies far to the left of the equation as shown below:

$$CH_3COOH + H_2O \leftrightharpoons H_3O^+ + CH_3COO^-$$

---

*In aqueous solutions discrete H^+ ions never exist, but are always hydrated.

TABLE 1-3 Dissociation of Typical Electrolytes

	Dissociation into Ions (per cent)
Hydrochloric acid	95.0
Nitric acid	92.0
Sulfuric acid	61.0
Acetic acid	1.3
Carbonic acid	0.17
Boric acid	0.01
Sodium hydroxide	91.0
Potassium hydroxide	91.0
Ammonium hydroxide	1.3
Most salts	70–100

Consequently, only small amounts of hydronium ion are formed, and acetic acid behaves as a weak acid compared with hydrochloric acid in which a large concentration of hydronium ion is formed. Therefore, in general, acids that dissociate almost completely into hydronium ions at ordinary concentrations are called **strong acids,** and those that dissociate to a small extent into hydronium ions are called **weak acids.** Similar arguments can be made for **strong vs. weak bases,** depending on the concentration of hydroxide ion formed on dissociation. In Table 1–3 is given the degree of dissociation for some typical electrolytes.

The above classification of acids and bases is convenient for aqueous solutions. Other definitions or classifications of acids and bases for nonaqueous solution will be discussed in Chapter 2.

Ionization of Water

Even in the absence of solutes, an ionic equilibrium exists in water. This equilibrium involves the slight dissociation of water which is represented in the following equation:

$$2H_2O \rightleftharpoons H_3O^+ + OH^-$$

Although the hydrated hydrogen ion, the hydronium ion, is actually the form in which the hydrogen ion occurs, H^+ is used for the sake of simplicity, and the net equation for the ionization of water is:

$$H_2O \rightleftharpoons H^+ + OH^-$$

The ionization constant (K_i) for water can then be expressed as:

$$K_i = \frac{[H^+][OH^-]}{[H_2O]}$$

For water at 25°C, it is found experimentally that $[H^+] = [OH^-] = 1 \times 10^{-7}$ moles per liter. The concentration of OH^- must equal the concentration of H^+, since ionization of 1M of water produces 1M of H^+ and 1M of OH^-. Since the concentration of water, $[H_2O]$, is very large (55.55) compared with $[H^+]$ and $[OH^-]$, for all practical purposes, it is constant and the product, $K_i[H_2O]$, called K_w, is generally employed for the **ion product constant for water.**

$$K_i = \frac{[H^+][OH^-]}{[H_2O]}$$

$$K_i[H_2O] = K_w = [H^+][OH^-] = [1 \times 10^{-7}][1 \times 10^{-7}] = 1 \times 10^{-14}$$

The magnitude of K_w indicates the very small extent of the dissociation of water.

Even though K_w was determined for pure water, the product of the molar concentrations of hydrogen ions and hydroxyl ions, $[H^+][OH^-]$, in any aqueous solution (even in the presence of solutes), must remain constant at 1×10^{-14}. Any increase in the $[H^+]$ concentration must be accompanied by a corresponding decrease in $[OH^-]$ to keep K_w constant; conversely, any increase in the $[OH^-]$ concentration must be accompanied by a corresponding decrease in $[H^+]$ to keep K_w constant. For example, if 1×10^{-4}M of gaseous HCl is added to 1 liter of pure water, the hydroxide ion concentration decreases to 1×10^{-10}M.

$$HCl + H_2O \rightarrow H_3O^+ + Cl^-$$

$$[H^+] = 1 \times 10^{-4}$$

$$K_w = [1 \times 10^{-14}] = [1 \times 10^{-4}][OH^-]$$

$$\frac{1 \times 10^{-14}}{1 \times 10^{-4}} = [OH^-] = 1 \times 10^{-10}$$

Since the ion product constant of water must at all time equal 1×10^{-14}, quantitative calculations become easily possible for all strong acids and bases.

What is the $[H^+]$ in an aqueous solution which is 0.01M in NaOH? Since NaOH is a strong electrolyte, the $[OH^-]$ is 0.01M. The $[H^+]$ is obtained in the following way:

$$K_w = [H^+][OH^-] = 1 \times 10^{-14}$$

$$[H^+][0.01] = 1 \times 10^{-14}$$

$$[H^+] = \frac{1 \times 10^{-14}}{0.01} = 1 \times 10^{-12}M$$

Therefore, in this problem, increasing the base concentration results in a decrease in the concentration of the acid.

pH, pOH, pK

Although in many cases the molar concentration $[H^+]$ is frequently used, chemists often find it more convenient to express the $[H^+]$ and $[OH^-]$ in nonexponential numbers. An alternate expression involves the use of the negative logarithm of the quantity in question. The **pH** of an aqueous solution is defined as **the negative logarithm of $[H^+]$**. Thus,

$$pH = -\log [H^+] = \log 1/[H^+]$$
$$\text{and } [H^+] = 10^{-pH} = 1/10^{pH} = \text{antilog}[-pH]$$

The **pOH** of an aqueous solution is defined as **the negative logarithm of $[OH^-]$**. Thus,

$$pOH = -\log [OH^-] = \log 1/[OH^-]$$
$$\text{and } [OH^-] = 10^{-pOH} = 1/10^{pOH} = \text{antilog}[-pOH]$$

For example, if the $[H^+]$ is $1 \times 10^{-2}M$, the pH is:

$$pH = -\log [H^+] = -\log [1 \times 10^{-2}] = 2$$

$$[OH^-] = \frac{K_w}{[H^+]} = \frac{1 \times 10^{-14}}{1 \times 10^{-2}} = 1 \times 10^{-12}$$

$$pOH = -\log [1 \times 10^{-12}] = 12$$

The **sum** of pH and pOH **must always equal 14.** Note that in the above example pH(2) + pOH(12) = 14. At pH = 7, $[H^+] = 1 \times 10^{-7}$ and $[OH^-] = 1 \times 10^{-7}$. A solution in which the $[H^+] = [OH^-]$ is called a neutral solution. A solution in which $[H^+]$ is greater than $[OH^-]$ is called an acidic solution ($pH<7$), and a solution in which the $[OH^-]$ is greater than the $[H^+]$ is called a basic solution ($pH>7$). Remember that since pH is a logarithmic function, a pH change of one unit represents a tenfold change in hydrogen ion concentration. Thus, a solution of pH 1 is 100 times more acidic (or less basic) than a solution of pH 3. The values of pH and pOH at various H^+ and OH^- concentrations are shown in Table 1–4. In addition, the usual pH ranges of several common liquids are shown in Table 1–5.

Another defined quantity, called pK, is often used by chemists to again avoid exponential numbers when comparing equilibrium constants. The quantity **pK** is defined as **the negative logarithm of K_i.** Therefore, for acids, the quantity pK_a is used, and the quantity pK_b is used for bases.

$$pK_a = -\log K_a = \log 1/K_a$$
$$pK_b = -\log K_b = \log 1/K_b$$

For acetic acid (CH_3COOH), $K_a = 1.8 \times 10^{-5}$

$$pK_a = -\log 1.8 \times 10^{-5} = 4.75$$

Some typical acid and base dissociation constants and pK_a and pK_b values are shown in Table 1–6. Again note that $pK_a + pK_b$ must equal 14. Compounds with a high value for pK_a are weaker acids than those with a lower pK_a value. Therefore, the smallest pK_a and pK_b will be associated with the strongest acids and bases.

TABLE 1–4 pH and pOH of Aqueous Solutions at 25°C

$[H^+]$ mol/l	pH	↑	$[OH^-]$ mol/l	pOH
1	0		10^{-14}	14
10^{-1}	1		10^{-13}	13
10^{-2}	2		10^{-12}	12
10^{-3}	3	acidic	10^{-11}	11
10^{-4}	4		10^{-10}	10
10^{-5}	5		10^{-9}	9
10^{-6}	6		10^{-8}	8
10^{-7}	7	neutral	10^{-7}	7
10^{-8}	8		10^{-6}	6
10^{-9}	9		10^{-5}	5
10^{-10}	10		10^{-4}	4
10^{-11}	11	basic	10^{-3}	3
10^{-12}	12		10^{-2}	2
10^{-13}	13		10^{-1}	1
10^{-14}	14	↓	1	0

TABLE 1-5 pH RANGES OF COMMON LIQUIDS

LIQUID	pH RANGE
Human gastric juices	1.0–3.0
Lime juice	1.8–2.0
Lemon juice	2.2–2.4
Vinegar	2.4–3.4
Carbonated drinks	2.0–4.0
Orange juice	3.0–4.0
Tomato juice	4.0–4.4
Beer	4.0–5.0
Cow's milk	6.3–6.6
Human blood	7.3–7.5
Sea water	7.8–8.3
Household ammonia (1–5%)	10.5–11.5

BUFFER SOLUTIONS

As we have seen in our earlier discussion of pH and pOH, the addition of a strong acid or base to pure water causes a large change in the pH or pOH of the resulting solution. It is possible, however, to prepare aqueous solutions which maintain a nearly constant pH or pOH even when large quantities of acid or base are added. Solutions of this type are called **buffer solutions** or, commonly, **buffers.** Solutions which contain weak acids and the salt of the weak acid or weak bases and the salt of the weak base act as buffer solutions.

To illustrate the action of a buffer solution let us consider the acetic acid–sodium acetate solution and the related equilibrium:

$$CH_3COOH \rightleftharpoons H^+ + CH_3COO^-$$

TABLE 1-6 ACID AND BASE DISSOCIATION CONSTANTS AT 25°C

ACID	K_a	pK_a
HCl	∞	—
HF	6.7×10^{-4}	3.2
HCN	4×10^{-10}	9.4
CH_3CO_2H	1.8×10^{-5}	4.7
HNO_2	4.5×10^{-4}	3.3
H_2SO_4	$(K_{a1})\ \infty$	—
	$(K_{a2})\ 1.3 \times 10^{-2}$	1.9
H_2CO_3	$(K_{a1})\ 4.2 \times 10^{-7}$	6.4
	$(K_{a2})\ 4.7 \times 10^{-11}$	10.3
H_2S	$(K_{a1})\ 1 \times 10^{-7}$	7.0
	$(K_{a2})\ 1 \times 10^{-15}$	15.0
H_3PO_4	$(K_{a1})\ 7.1 \times 10^{-3}$	2.2
	$(K_{a2})\ 6.3 \times 10^{-8}$	7.2
	$(K_{a3})\ 4 \times 10^{-13}$	12.4
BASE	K_b	pK_b
NaOH	∞	—
NH_3	1.8×10^{-5}	4.7

Remember that acetic acid is only 1 to 2 per cent dissociated, so that the equilibrium lies to the left of the equation. The addition of a small amount of potassium hydroxide to this equilibrium system will result in the consumption of H^+, but the excess CH_3CO_2H present will dissociate to re-establish the equilibrium and replace the H^+ consumed. Correspondingly, the addition of a small amount of hydrochloric acid will react with acetate ion to induce a shift in the equilibrium to consume the added H^+ (from HCl) and some of the excess CH_3COO^- (added initially as $CH_3COO^-Na^+$), and again only a small change in pH will result.

The behavior of buffer solutions can be treated in a quantitative way as follows: Using the K_a (acid dissociation constant) for the weak acid, acetic acid, the following expression for H^+ can be derived.

$$[K_a] = \frac{[H^+][CH_3COO^-]}{[CH_3COOH]}$$

Solve for H^+:

$$[H^+] = \frac{[K_a][CH_3COOH]}{[CH_3COO^-]}$$

Take the $-\log$ of each side of the above expression:

$$-\log [H^+] = -\log K_a - \log \frac{[CH_3COOH]}{[CH_3COO^-]}$$

Since $-\log K_a$ has been defined as pK_a and $-\log [H^+]$ has been defined as pH, the above expression can be rewritten as:

$$pH = pK_a - \log \frac{[CH_3COOH]}{[CH_3COO^-]}$$

or for any weak acid-salt buffer solution:

$$pH = pK_a - \log \frac{[acid]}{[salt]}$$

or more generally:

$$pH = pK_a + \log \frac{[salt]}{[acid]}$$

The above equation is known as the **Henderson-Hasselbalch equation** and will be used many times in the discussion of biochemistry (p. 250).

For a buffer solution composed of a weak base-salt combination in which the equilibrium Base + $H_2O \rightleftharpoons [\text{Base H}]^+ + OH^-$ is involved, a similar expression can be derived:

$$pH = 14 - pK_b - \log \frac{[\text{Base H}]^+}{[\text{Base}]}$$

The use of these expressions is illustrated by considering the addition of acid and base to one liter of a solution which is 0.1M in CH_3CO_2H and 0.1M in $CH_3CO_2^-$. Initially the ratio $[CH_3CO_2H]/[CH_3CO_2^-]$ is 1, so that

$pH = pK_a - \log 1$ or $pH = pK_a$ (4.75). The addition of 0.01 mole of HCl shifts the equilibrium to make $[CH_3CO_2H] = 0.11M$ and $[CH_3CO_2^-] = 0.09M$. Now $pH = 4.75 - \log(0.11/0.09) = 4.75 - 0.09$ or 4.66. This is a decrease of only 0.09 pH unit, whereas addition of the same quantity of acid to a liter of water decreases pH from 7 to 2. The addition of 0.01 mole NaOH to one liter of this buffer shifts the equilibrium to make $[CH_3CO_2H] = 0.09M$ and $[CH_3CO_2^-] = 0.11M$. Now $pH = 4.75 - \log(0.09/0.11) = 4.75 + 0.08$ or 4.83. The same quantity of NaOH added to a liter of water increases pH from 7 to 12.

Buffers play an important role in many naturally occurring processes. All body fluids have definite pH values that must be maintained within fairly narrow ranges for proper physiological functions. The pH of the blood is normally between 7.35 and 7.45. If the pH of the blood falls below 7.0 or goes above 7.8, death occurs. Since many of the reactions that take place in our tissues form acid substances, the blood must have a mechanism to prevent such changes in pH. The equilibrium system involved in the buffer action of blood includes bicarbonates and carbonates, phosphates, and complex salts of proteins (p. 366).

TITRATION

Titration is the process of determining the concentration of a substance (such as an acid or base) in solution by the addition of a **standard solution** (one of known concentration) until the number of gram equivalents of the reactants are the same. Usually the standard solution is added, in known volumes, to a measured amount of the solution of

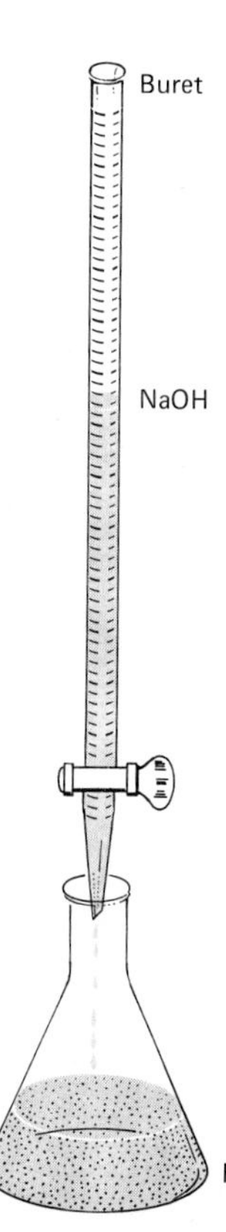

FIGURE 1-16 A laboratory titration involving the neutralization of hydrochloric acid by sodium hydroxide.

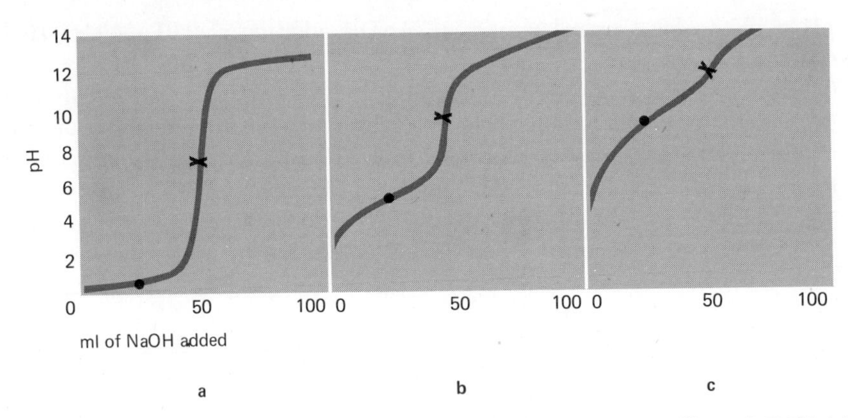

FIGURE 1–17 Titration curves of 50 ml of 1.0M HCl (*a*), CH_3CO_2H (*b*), and HCN (*c*) using 1.0M NaOH. In each case x marks the equivalence point of the titration and ● marks the half-equivalence point of the titration.

unknown concentration until enough reagent has been added to just react with the unknown present. This point (called the **equivalence point**) can be detected using indicators which change color upon addition of excess standard solution. The most common indicator used in titrations of acids and bases is **red phenolphthalein,** which is red in basic solution and colorless in acid solution. Figure 1–16 shows the apparatus commonly used in titration. The **buret,** a graduated tube, allows a regulated introduction of the standard solution into the flask containing the second reactant.

If a plot of the pH changes **versus** the volume of titrant added (equivalents of acid or base) is made, a **titration curve** is obtained. This is shown in Figure 1–17 for the titration of 50 ml of a 0.1M solution of a strong acid (HCl), a weak acid (CH_3COOH), and a very weak acid (HCN) using 1.0M NaOH as the titrant (standard solution). In the titration of HCl in which the Cl^- does not compete for the proton, the pH shows a very gradual change until the equivalence point is reached, then it climbs sharply. Such curves show a "sharp end point." In contrast to this behavior of strong acids, as the acid being titrated decreases in strength, the pH change at the equivalence point becomes less pronounced, again pointing out the buffering action of these types of solutions in resisting pH change. Similar titration curves for amino acids or proteins as buffers will be encountered in the biochemistry section of the text (p. 250).

IMPORTANT TERMS AND CONCEPTS

acid
base
buffers
covalent bonds
electrolyte
electronegativity
electronic configuration
hybrid orbitals
hydrogen bonds
ionic bonds
K_a
K_b
K_w
Lewis symbols
orbital
pH
pi-bond
pK
pOH
polar bond
resonance hybrid
sigma-bond
solute
solvent
tetrahedral molecule
titration
titration curve
valence electrons

with many reagents such as HCl, H_2SO_4, and PCl_5, whereas dimethyl ether is unreactive with these reagents.*

If the natures of the bonds and the organic groups in the isomers are quite similar, similar properties and reactivities can be expected. For example, C_4H_{10} can exist as the following isomers:

```
     H  H  H  H              H        H        H
     |  |  |  |              |        |        |
  H—C—C—C—C—H             H—C———————C———————C—H
     |  |  |  |              |        |        |
     H  H  H  H              H     H—C—H      H
                                      |
                                      H
```

n-Butane (b.p. 0°C)

2-Methylpropane (b.p. −10°C)

These two isomeric structures have similar boiling points and react with similar chemical reagents. As the number of carbon atoms in the molecule increases, the number of isomeric structures also increases, and it is the ability of carbon to form many isomeric structures that accounts in part for the large number of known organic compounds. Most inorganic compounds, with the exception of complex ions, do not form isomeric structures. Hence, for a particular combination of atoms in an inorganic molecule only one structure is possible.

The types of bonding in organic and inorganic compounds also differ and account for the large difference in some of the physical properties of organic and inorganic compounds. Whereas most inorganic compounds are composed of ions and held together by strong electrostatic forces, most organic compounds are composed of weak covalently bonded atoms and are relatively nonpolar materials. This difference in bonding is reflected in the physical properties such as boiling point, melting point, and solubility. Most inorganic compounds have high melting points and high boiling points (generally $> 1000°C$), whereas most organic compounds melt at temperatures less than 300°C and boil at temperatures less than 500°C. The high temperatures required to volatilize inorganic compounds indirectly measure the polarity of the bonds in the molecule.

Na^+Cl^-

Sodium chloride (m.p. 801°C) (b.p. 1413°C)

$CH_3C(=O)—NH_2$

Acetamide (m.p. 81°C) (b.p. 222°C)

Since most inorganic compounds are made up of ions held together electrostatically, it would be expected that inorganic compounds should be soluble in polar solvents and, as expected, most inorganic compounds are soluble in the polar solvent water. Water breaks the bond between the ions in the inorganic crystal and hydrates the individual ions. It is also found that these hydrated ions conduct an electric current and behave as good electrolytes. On the other hand, most organic compounds are insoluble in a polar solvent like water but are quite soluble in nonpolar solvents like ether, benzene, and hydrocarbons. Since dissolution of an organic compound into an organic solvent does not produce ions, most solutions of organic compounds do not conduct an electric current

*As will be obvious from later discussions of alcohols and ethers, the presence of the —OH group in the alcohol will account for its extensive reactivity compared to the ether which has no —OH groups.

TABLE 2–1 PROPERTIES OF ORGANIC AND INORGANIC COMPOUNDS

ORGANIC	INORGANIC
1. Low boiling points	1. High boiling points
2. Low melting points	2. High melting points
3. Low solubility in water	3. High solubility in water
4. High solubility in nonpolar solvents	4. Low solubility in nonpolar solvents
5. Flammable	5. Nonflammable
6. Covalent bonding	6. Ionic bonding
7. Solutions are nonconductors of electricity	7. Solutions are conductors of electricity
8. Exhibit isomerism	8. Isomerism is very limited

and are classified as nonelectrolytes.* The general properties of organic compounds relative to most inorganic compounds are summarized in Table 2–1.

THE ROLE OF CARBON IN ORGANIC CHEMISTRY

Since we have defined organic chemistry as "the chemistry of carbon compounds," the obvious question is—why define and separate a branch of chemistry for one element, such as carbon, and classify the chemistry of the other hundred or so elements as another branch of chemistry, namely, inorganic chemistry? The unique character and emphasis on the atom carbon can be summarized as follows:

1. Its position in the periodic table: Carbon is in the middle of the second period in the periodic table and has an atomic number of six. Consequently, it has six orbital electrons. Two of these orbital electrons make up the first ($1s^2$) shell of electrons, leaving four electrons in the outer valence shell available for bonding purposes. Carbon can attain a stable rare-gas configuration by losing four valence electrons to form C^{+4} (inert gas configuration of He), or can gain four valence electrons to form C^{-4} (inert gas configuration of Ne). Both processes (gain or loss of $4e^-$) are energetically very unfavorable for carbon, since carbon, being in the middle of the periodic table, is neither strongly electronegative nor strongly electropositive, and therefore has little tendency to form either cations (C^{+4}) or anions (C^{-4}). In fact, carbon forms bonds with other elements by sharing electrons (covalent bonds) and attains the inert gas configuration in this manner.

Carbon:

$$\text{Electronic configuration:}\quad \underbrace{1s^2}_{\text{Inner shell electrons}} \qquad \underbrace{2s^2\ 2p^2}_{\text{Valence bonding electrons}} \equiv \cdot\dot{\underset{\cdot}{C}}\cdot\,^{\dagger}$$

*Although an organic compound such as trimethyl amine, $(CH_3)_3\ddot{N}$, does not conduct an electric current, treatment of this amine with HCl produces a salt, $[(CH_3)_3\overset{+}{N}H]Cl^-$, which does conduct an electric current. Consequently, some organic compounds, which can be converted into ions by the appropriate acid or base reaction, can behave as conductors, but this is not the normal behavior of most organic compounds.

†In this type of representation, the symbol of the element (C) represents the nucleus and the inner shell (non-bonding electrons). The dots represent the valence electrons involved in covalent bond formation. This type of formula is called a Lewis formula (Chapter 1). It shows only the number of valence electrons involved in covalent bond formation and gives no indication of the spatial orientation of the atoms in the molecule.

Hydrogen:

$$\text{Electronic configuration:}\quad \underbrace{1s^1}_{\text{Bonding electron}} \equiv \text{H}\cdot$$

Methane:

$$CH_4 \equiv \begin{array}{c} \text{H} \\ \vdots \\ \text{H}\cdot\cdot\text{C}\cdot\cdot\text{H} \\ \vdots \\ \text{H} \end{array}$$

In a molecule like methane, each hydrogen shares an electron (forms a covalent bond) with carbon, and carbon in turn shares an electron with each hydrogen atom. In order to complete the bonding capacity of carbon, four hydrogen atoms are necessary. In addition, hydrogen, by sharing an electron, has attained the electronic configuration of He, and carbon has attained the electronic configuration of Ne.

2. Ability to bond with itself and to form multiple bonds with itself: Carbon, because of its small atomic radius and because of the strength of carbon–carbon bonds, has the striking property of being able to form bonds with itself. Although other atoms, such as silicon, are able to do this to some extent, carbon possesses this property more than any other element. This property accounts for the many organic compounds known in this branch of chemistry. For example, if two carbon atoms share an electron and form a covalent bond, the valence electrons can be represented as follows:

$$\cdot\dot{\underset{\cdot}{C}}\cdot + \cdot\dot{\underset{\cdot}{C}}\cdot \rightarrow \cdot\dot{\underset{\cdot}{C}}\cdot\cdot\dot{\underset{\cdot}{C}}\cdot \equiv \cdot\dot{\underset{\cdot}{C}}\text{—}\dot{\underset{\cdot}{C}}\cdot^{*}$$

If the other valences of the carbon atoms are filled by forming covalent bonds to hydrogen, the molecule ethane, C_2H_6, results:

$$\cdot\dot{\underset{\cdot}{C}}\text{—}\dot{\underset{\cdot}{C}}\cdot + 6\text{H}\cdot \rightarrow \text{H—}\begin{array}{c}\text{H}\\ | \\ \text{C} \\ | \\ \text{H}\end{array}\text{—}\begin{array}{c}\text{H}\\ | \\ \text{C} \\ | \\ \text{H}\end{array}\text{—H} \quad \text{or} \quad H_3CCH_3{}^{\dagger}$$

Ethane

Of course, if carbon were to share two electrons between each carbon atom, the following situation would result:

$$\cdot\dot{\underset{\cdot}{C}}\cdot + \cdot\dot{\underset{\cdot}{C}}\cdot \rightarrow \cdot\dot{C}::\dot{C}\cdot \equiv \cdot\dot{C}\text{=}\dot{C}\cdot$$

Now there are two covalent bonds between the two carbon atoms and only four valence electrons remain for additional bonding. If hydrogen atoms are used to complete the bonding capacity of carbon in this system, the molecule ethylene,

*A covalent bond formed by sharing an electron between two atoms can be represented by a dash for the sake of convenience.

†The structural formula in which all the covalent bonds are indicated by a dash is known as an *expanded structural formula*. For convenience, the dashes (covalent bonds), especially to hydrogen, are often omitted and the *condensed* structural formula is used.

C_2H_4, results. In ethylene, there is a double (or multiple) bond between the carbon

$$\cdot\dot{C}{=}\dot{C}\cdot + 4H\cdot \rightarrow \underset{H}{\overset{H}{}}\!\!>C{=}C<\!\!\underset{H}{\overset{H}{}} \quad \text{or} \quad H_2C{=}CH_2$$

Ethylene

atoms.* If this same sort of process is used to share three electrons between carbon, and then to use hydrogen atoms to complete any unused bonding capacity, the molecule acetylene, C_2H_2, results, as follows:

$$\cdot\dot{\underset{\cdot}{C}}\cdot + \cdot\dot{\underset{\cdot}{C}}\cdot \rightarrow \cdot C\vdots\vdots C\cdot \equiv \cdot C{\equiv}C\cdot$$

$$\downarrow 2H\cdot$$

$$H{-}C{\equiv}C{-}H \quad \text{or} \quad HC{\equiv}CH$$

Acetylene

Acetylene again contains a multiple bond between the carbon atoms, and in this case we have a triple bond between carbon atoms. Applying this same process further, the molecule C_2, C≡C, could be formed by sharing all four valence electrons between two carbon atoms. However, four bonds between two carbon atoms has not been observed, and only single-, double-, and triple-bonded carbon atoms have been found in organic compounds.

The process used in the preceding paragraphs could be repeated again and again using additional carbon and hydrogen atoms to give even longer carbon chains. In addition, carbon can share electrons, not only with itself and with hydrogen but with many other simple elements to form cyclic organic compounds as well as linear-chain compounds. Some examples of these various types of compounds are as follows:

$CH_3CH_2CH_3$	$H_2C{-}CH_2$ (ring with CH_2)	$CH_3CH_2CH_2CH_2CH_3$
Propane	Cyclopropane	Pentane
H_3CCl	CCl_4	$HCCl_3$
Methyl chloride	Carbon tetrachloride	Chloroform
six-membered ring of CH_2 groups	$H_3C{-}O{-}CH_3$	H_3CCH_2I
Cyclohexane	Dimethyl ether	Ethyl iodide

H_3CNH_2	$H_3C\overset{O}{\overset{\|}{C}}CH_3$	$H_2C{=}O$ (H, H on C)	$H_3CC{\equiv}N$
Methyl amine	Acetone	Formaldehyde	Acetonitrile

*As will become obvious later in this chapter and in the next chapter, the type of bonding, either single or multiple, between carbon atoms has a dramatic effect on the shape of the molecule and on the chemical reactivity of the organic compound.

THE SHAPES OF ORGANIC MOLECULES

Ionic compounds, such as those commonly found among inorganic molecules, are held together by electrostatic forces between positive and negative ions. Electrostatic forces of this type, such as in Na^+Cl^-, are exerted symmetrically in all directions, and the ions can be thought of as a point charge, or a sphere of unit charge on which the charge is distributed equally over the surface of the sphere.

In contrast to the nondirectional nature of electrostatic forces, covalent bonds are directional in nature and give a definite shape to the molecule which depends on the type of covalent bond. In the simple examples methane, ethane, ethylene, and acetylene, which were considered in the previous section, different shapes and bond angles are found in each case. In methane, the carbon atom is considered to be at the center of a regular tetrahedron and the four bonds to hydrogen are directed to the corners of the tetrahedron (see Fig. 2–1). The molecule can be pictured as follows:

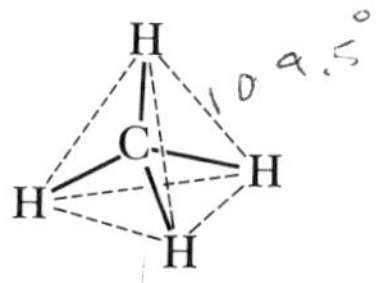

Methane

FIGURE 2–1

Other methods of defining chemical shapes and geometry, such as x-ray and electron-diffraction, have confirmed the regular tetrahedron shape of molecules such as CH_4 and CCl_4. The bond angles in molecules of this shape are 109.5°. When all the bonds to carbon are not identical, such as in chloroform, $CHCl_3$, the shape of the molecule is still tetrahedral. It is, however, no longer a regular tetrahedron, but a distorted tetrahedron with bond angles slightly different than 109.5°.

When carbon–carbon bonds are linked together in the formation of more complex molecules, such as ethane, propane, and so forth, the shape of the molecules is a series of tetrahedrons which share a common corner. The normal carbon–carbon single bond distance in molecules such as this is 1.54 Å.* Longer linear-chain molecules can be assembled by adding on additional tetrahedrons which share a common corner.

In a compound, such as ethylene, C_2H_4, the formation of the carbon–carbon double bond in the molecule imposes certain geometric requirements on the shape of the molecule. First, the introduction of the double bond limits rotation around the carbon–carbon bond. In compounds, such as ethane, which share a corner of a tetrahedron, there is free rotation† around the carbon–carbon single bond. However, introduction of the carbon–carbon double bond restricts any free rotation (360°), and for all practical purposes no rotation is allowed in this molecule unless the carbon–carbon double bond is broken. Secondly in a molecule which contains a carbon–carbon double bond the atoms attached to the carbon atoms with the double bond are also coplanar (all the carbon and hydrogen atoms in ethylene lie in the same plane) with bond angles of 120° and a carbon–carbon bond length of 1.34 Å. The bond angles and bond distances again may vary slightly depending upon what atoms are attached to carbon, but the gross overall features of the molecule will not change.

---

* 1 Å = 10^{-8} cm = 10^{-1} nm

† In actuality there is a small energy barrier to rotation, since the atoms do occupy space and must pass each other on rotation. However, in most compounds containing carbon–carbon single bonds this barrier is very small.

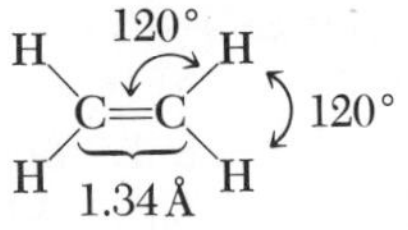

H 120° H
C=C 120°
H 1.34 Å H

Ethylene

In acetylene, C_2H_2, and other molecules containing a carbon–carbon triple bond, even greater deviations from the simple tetrahedral structures occur. X-ray and electron-diffraction methods have shown that compounds containing a —C≡C— linkage are linear molecules with a carbon–carbon bond length of 1.21 Å, as illustrated below for acetylene:

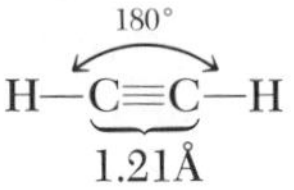

180°
H—C≡C—H
1.21Å

Acetylene

The formation of the multiple carbon–carbon bonds again allows no free rotation around the carbon–carbon bonds.

In cyclic organic compounds some deviation from the normal bond angles illustrated above may be expected, as the constraining of the carbon atoms into rings of certain sizes will force the atoms into unusual and strained shapes. For example, cyclopropane must be a planar molecule, since three points (the three carbon atoms) define a plane. The bond angles in cyclopropane must necessarily be equal, since all the atoms are identical, and have been shown to be 60°. Since the normal bond angle of a single

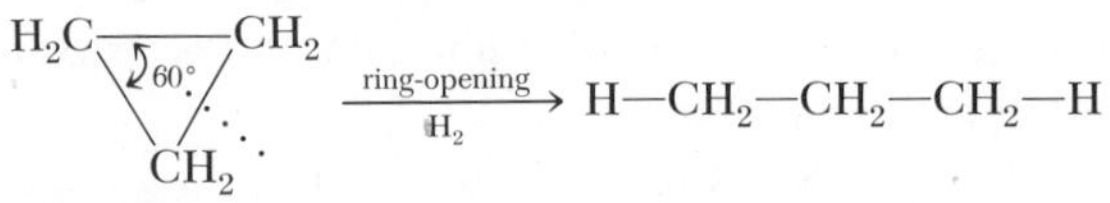

H_2C—CH_2 (60°) / CH_2 $\xrightarrow[H_2]{\text{ring-opening}}$ H—CH_2—CH_2—CH_2—H

Cyclopropane Propane

carbon-carbon bond is 109.5° (see Fig. 2–1), to constrain or compress these bond angles from 109.5° to 60° will introduce lots of strain into the molecule. Consequently, we might expect cyclopropane and any other highly strained compound to be particularly susceptible to ring-opening reactions, since after ring-opening the bond angles become approximately 109.5° again.

Experimentally, it is commonly found that the chemical susceptibility to ring-opening reactions does increase with increasing amount of ring strain, and cyclopropane does undergo ring-opening reactions with many chemical reagents. As the size of the ring increases, the bond angles increase, and the amount of strain decreases. Consequently, ring systems higher than cyclopropane are less prone to undergo ring-opening reactions. Ring systems higher than cyclopropane can be either planar or "puckered" (nonplanar) systems. In four- and five-membered rings there is some "puckering" of the ring, and

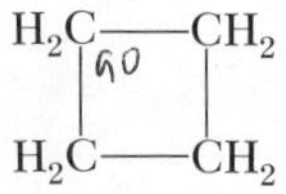

H_2C—CH_2 / H_2C—CH_2

Planar cyclobutane
∡ 90°

CH_2 / H_2C CH_2 / H_2C—CH_2

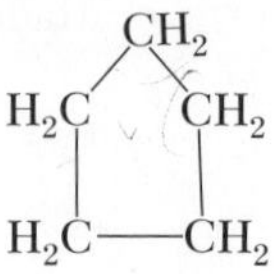

Planar cyclopentane
∡ 108°

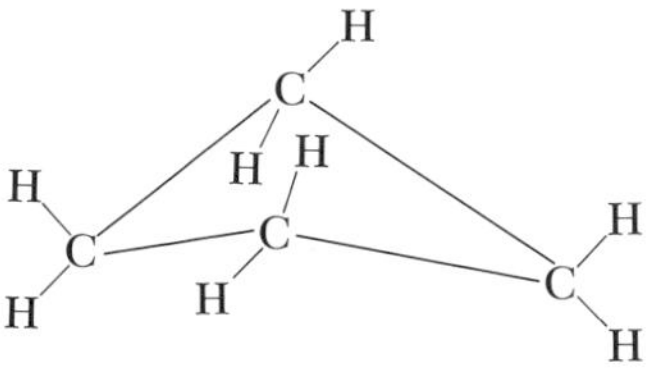

Nonplanar cyclobutane

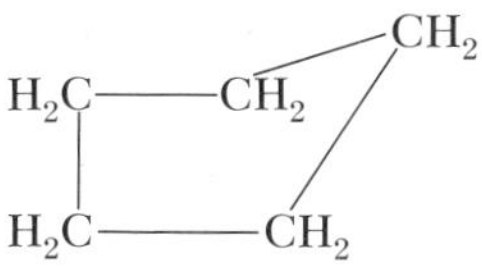

Nonplanar cyclopentane

these ring systems are not completely planar.* In higher ring systems, such as cyclohexane, bond angles of 120° would be expected if the molecule were a planar hexagon, and extensive strain and ring-opening properties would be anticipated for cyclohexane. However, cyclohexane has been found, both experimentally and by x-ray analysis, to be a nonstrained and nonplanar molecule, and, hence, is not susceptible to easy ring-opening reactions owing to the presence of bond angles of 109.5°. The cyclohexane molecule generally exists in two arrangements known as a "chair" and a "boat" form. The chair form in most cases is the preferred form, since it again minimizes steric repulsions of the hydrogen atoms in the molecule. The chair and boat forms are not isolable.

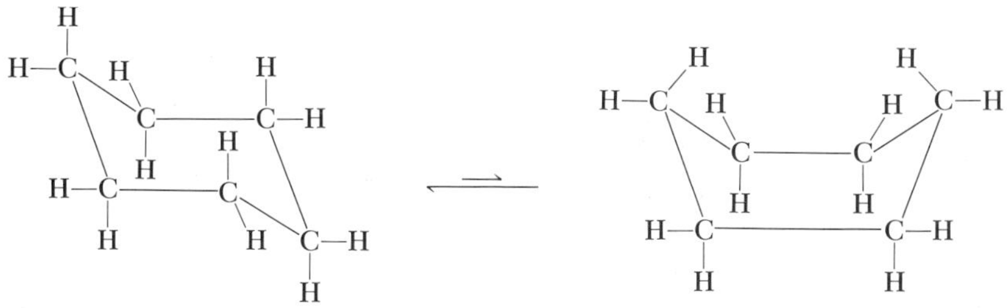

Chair form

Boat form

Many naturally occurring steroids (see Chapter 14), such as cholesterol, ergosterol, sex hormones, D vitamins, and sitosterols, contain the steroid nucleus (which is composed of fused cyclic rings) which can exist in the chair or boat form. Most steroids adapt the more stable chair form. Also, carbohydrates exist for the most part in a cyclic form (see Chapter 13), and the most stable form is the chair form, shown below for the α-anomer of D-glucose.

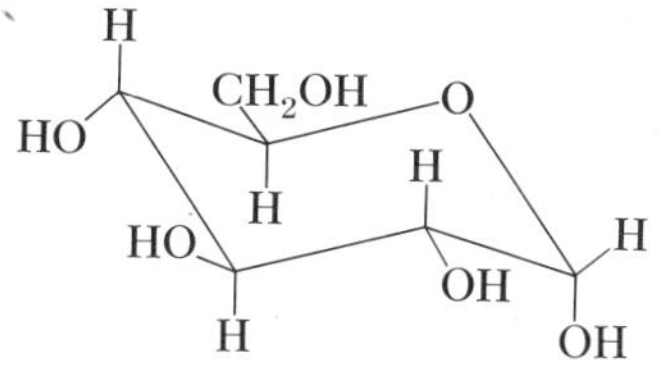

D-glucose

Larger ring systems are not as common as five- and six-membered rings, but they have been found in many natural products, and in recent years synthetic methods have been developed in the laboratory for preparing many larger ring systems. In these larger rings the molecule again exists preferentially in a "puckered" type form to minimize steric repulsions, and these systems are receiving much attention in present-day organic chemistry.

*In the planar compounds, there are unfavorable steric repulsions between the hydrogen atoms caused by compression of the hydrogen atoms trying to occupy the same space. In the "puckered" form, these steric repulsions are minimized, since the hydrogen atoms are staggered in space and this form is more stable (this can be seen more clearly by building a model of these compounds).

ORGANIC FUNCTIONAL GROUPS

Most of the compounds used as illustrations in this chapter thus far have contained only carbon and hydrogen atoms. Experimentally, compounds containing exclusively carbon and hydrogen make up only a small number of organic compounds. In addition, the chemical reactivity of compounds containing only carbon and hydrogen is less in most cases than compounds containing atoms such as oxygen, nitrogen, halogen, and sulfur, as well as carbon and hydrogen.

When the chemistry of organic compounds is considered in detail in the following chapters, it will become apparent that only certain bonds and organic groups participate in the chemical reaction, and that most of the carbon chain structure of the molecule remains unchanged in going from reactants to products. The following reaction of *t*-butyl alcohol is illustrative:

$$\underset{\substack{t\text{-Butyl}\\ \text{alcohol}}}{(CH_3)_3COH} + \underset{\substack{\text{Hydrogen}\\ \text{bromide}}}{HBr} \rightarrow \underset{\substack{t\text{-Butyl}\\ \text{bromide}}}{(CH_3)_3CBr} + \underset{\text{Water}}{HOH}$$

In this reaction the —OH group of the alcohol is lost, and its place on the carbon chain is taken by the bromine atom of the hydrogen bromide. In turn, the —OH lost by the alcohol combines with the hydrogen of the hydrogen bromide to produce water as the other product of this reaction.* The important point to consider here is that only a small portion of the organic molecule undergoes change. The bond between the carbon atom and the —OH group is broken, and a new bond between carbon and bromine is formed. Other than these simple changes, all the remaining bonds (all carbon and hydrogen in this case) in the organic compound remain unchanged. This is a characteristic feature of organic molecules, that only certain atoms or groups of atoms in an organic molecule determine the chemistry of the class of compounds containing that particular atom or group of atoms. *The atom or group of atoms that defines the structure of a particular class of organic compounds and determines its properties is called the* ***functional group.*** In the particular example above, the functional group in the alcohol is the —OH group, and the functional group in the product is the halogen atom, Br.

A large portion of organic chemistry is concerned with the transformation of one functional group into another. A basic understanding of organic chemistry is a mastery of the properties and reactivities of each type of functional group, and how one functional group can be transformed into another functional group of different properties and reactivity. In later chapters, using simple molecules, we shall learn to associate a particular set of properties with a particular functional group. When encountering a more complicated molecule, we may expect the properties of this molecule to roughly approximate the properties of the various functional groups contained in the molecule. For example, by understanding the simple properties of an alcohol like *t*-butyl alcohol, we can then extrapolate this knowledge to a more complicated alcohol, such as cholesterol, which contains an alcohol functional group and undergoes many of the same chemical reactions as *t*-butyl alcohol.†

---

*The student should be aware that this explanation is very simplified, and that the details in converting the alcohol to the bromide are more involved than merely exchanging groups or atoms. This will become apparent when we look at a mechanism for this type of reaction.

†It should be pointed out that the properties of a complicated molecule containing several different functional groups may be modified relative to a simple monofunctional compound, and a complete extrapolation of chemical behavior from simple to complicated compounds should not be expected. However, in most cases this simple extention from simple to complicated molecules using functional group chemistry yields surprisingly excellent results.

TABLE 2-2 SIMPLE FUNCTIONAL CLASSES OF ORGANIC COMPOUNDS

ILLUSTRATIVE EXAMPLE	NAME OF FUNCTIONAL CLASS	FUNCTIONAL GROUP
$CH_3CH_2CH_3$	Alkanes	—
$CH_3CH{=}CH_2$	Alkenes	$C{=}C$
$CH_3C{\equiv}CH$	Alkynes	$C{\equiv}C$
$CH_3CH_2CH_2OH$	Alcohols	$—OH$
$CH_3CH_2\overset{H}{C}{=}O$	Aldehydes	$—\overset{H}{C}{=}O$
$CH_3\overset{O}{\overset{\Vert}{C}}CH_3$	Ketones	$—\overset{O}{\overset{\Vert}{C}}—$
$CH_3CH_2\overset{O}{\overset{\Vert}{C}}OH$	Acids	$—\overset{O}{\overset{\Vert}{C}}OH$
$CH_3CH_2CH_2NH_2$	Amines	$—NH_2$
$CH_3OCH_2CH_3$	Ethers	$—C—O—C—$
$CH_3CH_2CH_2Br$	Halides	$—Br$

A summary of the typical functional groups to be taken up in later chapters is presented in the accompanying Table 2-2 with representative examples. Although it is not assumed that the student understands the chemistry of these functional groups, it is extremely helpful at this stage to learn the names of some of the simple functional classes, and especially to learn to associate a particular atom or group of atoms with a particular functional class.

ACIDS AND BASES IN ORGANIC CHEMISTRY

The terms acid and base can be defined in a number of ways. For our purposes we shall find it useful to review two different definitions of acids and bases, and to use the one which fits the problem at hand.

1. Brönsted-Lowry definition:

   An acid is a substance that gives up a proton, and a base is a substance that accepts a proton. For example:

$$HCl + \ddot{N}H_3 \rightleftharpoons NH_4^+ + Cl^-$$

Hydrogen chloride (acid) donates a proton to ammonia (base) which accepts the proton and is converted to the ammonium ion. The strength of an acid by this definition depends upon its tendency to donate a proton, and the strength of a base depends upon its tendency to accept a proton.

A similar type of acid-base chemistry can be used for many organic compounds. For example, in the reaction of acetic acid and sodium hydroxide to give sodium acetate, acetic acid gives up a proton to the hydroxide, and consequently fits the definition of an acid as a proton donor:

$$\underset{\text{Acetic acid}}{CH_3COOH} + NaOH \rightarrow \underset{\text{Sodium acetate}}{CH_3\overset{O}{\overset{\Vert}{C}}—O^-Na^+} + HOH$$

Similarly, if methyl amine reacts with hydrogen chloride, an acid-base reaction occurs with the amine accepting a proton (acting as a base) from the acid:

$$\underset{\text{Methyl amine}}{CH_3\ddot{N}H_2} + HCl \rightarrow CH_3\overset{+}{N}H_3\,Cl^-$$

Therefore, we can expect organic acids (compounds containing the —COOH group) and organic bases (compounds containing $—\ddot{N}H_2$ group) to behave similarly to inorganic acids and bases. Other organic molecules can act as proton donors (terminal acetylenes) and proton acceptors (alcohols, ethers, aldehydes, and ketones).

2. Lewis definition:

   An acid is a substance that can accept an electron pair to form a covalent bond, and a base is a substance that can donate an electron pair to form a covalent bond. For example, boron trifluoride and diethyl ether react to form an etherate:

$$\underset{\text{Boron trifluoride}}{BF_3} + \underset{\text{Diethyl ether}}{:\ddot{O}(C_2H_5)_2} \longrightarrow \underset{\text{Boron trifluoride etherate}}{F_3B:\ddot{O}(C_2H_5)_2}$$

In boron trifluoride, the boron has only six electrons and needs an additional pair to complete its octet. It accepts and shares an electron pair from the oxygen of the ether to form the etherate ($BF_3 \cdot Et_2O$). Hence, BF_3 is acting as an electron-pair acceptor, and the ether (oxygen) is acting as an electron-pair donor. Other Lewis acids, such as $AlCl_3$ and $SnCl_4$, and Lewis bases, such as alcohols ($—\ddot{O}H$) and amines ($—\ddot{N}H_2$), are important in catalyzing organic reactions and will reappear frequently throughout the text. A basic understanding of acids and bases according to the two definitions just presented facilitates understanding simple reaction mechanisms.

REACTION MECHANISMS AND REACTION INTERMEDIATES

To really understand an organic chemical reaction, it is important to know not only what happens (to know the reactant and products) but also how it happens (to know something about the intermediates formed, how they are formed, and how they recombine to form the products). The answer to this question—how does the reaction occur?—is called a reaction mechanism. It is a stepwise description which is advanced to account for the facts. In most cases it is difficult to prove a mechanism outright, since it is very difficult to detect or trap many of the transient intermediates or compounds that may make up some of the steps which lead to the final product. Nevertheless, by investigating many reactions using various chemical, spectroscopic, and stereochemical techniques, the organic chemist has learned that a few simple chemical intermediates can be used to account for many of the observed facts of organic chemistry. Also, by using these intermediates to explain chemical reactions, it becomes apparent that many unrelated reactions appear to proceed by the same or a similar mechanistic pathway, so that by having some knowledge of the more common mechanisms of organic chemistry we can relate and correlate many facts that appear on the surface to be totally unrelated. Obviously, if we have a knowledge of basic mechanisms, this will facilitate our understanding of organic chemistry; it is much better than trying to merely memorize all the facts.

Three fundamental types* of reaction intermediates can be used to explain many of the more common organic reactions. A simple dissociation of an electron pair can illustrate what these intermediates are, as follows:

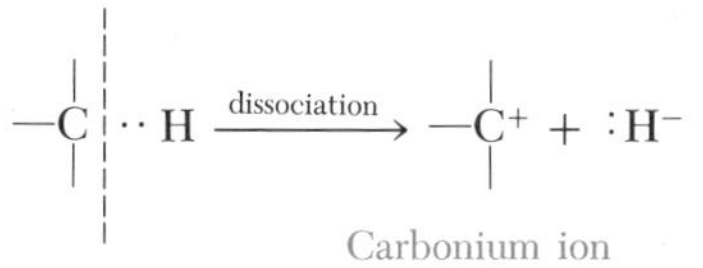

$$-\overset{|}{\underset{|}{C}}:H \xrightarrow{\text{dissociation}} -\overset{|}{\underset{|}{C}}^{+} + :H^{-}$$

Carbonium ion

[a carbonium ion is defined as a trivalent carbon bearing a positive charge and having a sextet of electrons]

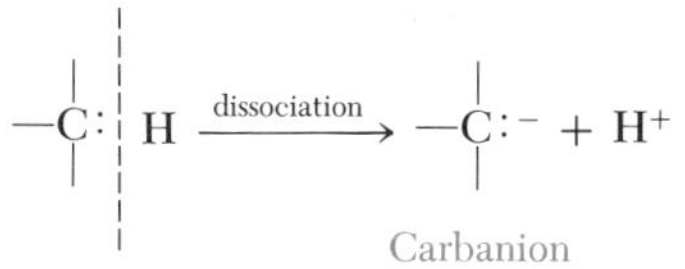

$$-\overset{|}{\underset{|}{C}}:H \xrightarrow{\text{dissociation}} -\overset{|}{\underset{|}{C}}:^{-} + H^{+}$$

Carbanion

[a carbanion is defined as a trivalent carbon bearing a negative charge and having an octet of electrons]

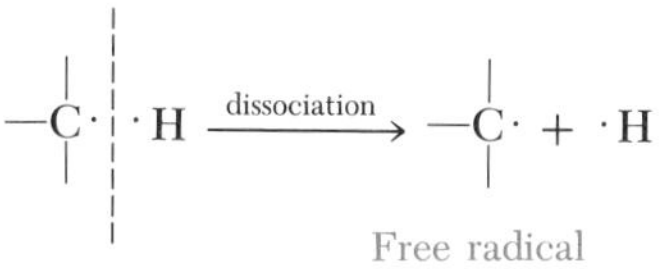

$$-\overset{|}{\underset{|}{C}}\cdot\cdot H \xrightarrow{\text{dissociation}} -\overset{|}{\underset{|}{C}}\cdot + \cdot H$$

Free radical

[a free radical is defined as a trivalent carbon atom having an odd electron and having zero charge]

This is not to imply that these kinds of intermediates are formed by a simple dissociation type process. In fact, they are not. They are usually formed via a chemical or acid-base reaction. For example, the previous conversion of an alcohol to a halide can now be more easily understood as illustrated below:

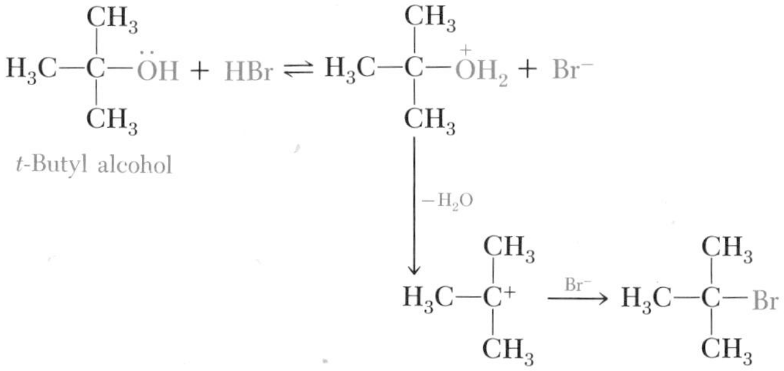

$$H_3C-\overset{CH_3}{\underset{CH_3}{C}}-\ddot{O}H + HBr \rightleftharpoons H_3C-\overset{CH_3}{\underset{CH_3}{C}}-\overset{+}{O}H_2 + Br^-$$

t-Butyl alcohol

$$\xrightarrow{-H_2O} H_3C-\overset{CH_3}{\underset{CH_3}{C^+}} \xrightarrow{Br^-} H_3C-\overset{CH_3}{\underset{CH_3}{C}}-Br$$

t-Butyl bromide

The oxygen of *t*-butyl alcohol reacts with hydrogen bromide in an acid-base reaction to produce the protonated alcohol, which loses a molecule of water to form the carbonium ion, which then captures a halide ion to form the final product. This is not the only possible reaction mechanism for this type of conversion, since the formation of the carbonium

*Other types of reaction intermediates are known than these three, but are generally less important for most common reactions. They will be introduced later, as needed, in the text.

ion will depend to some extent on the type of alcohol, and particularly on the stability of the carbonium ion formed. This will be discussed in more detail in Chapter 4.

Carbanions are generally formed by the acid-base reaction of an organic compound containing an acidic hydrogen with either an organic or an inorganic base. For example, the alkylation of diethyl malonate proceeds through a reaction mechanism involving a carbanion intermediate as shown below:

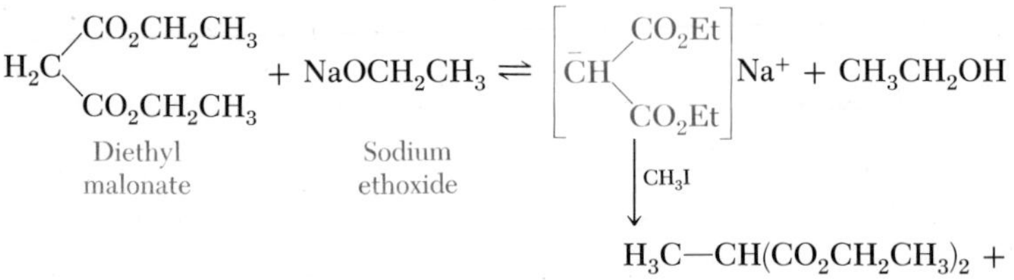

$$H_2C(CO_2CH_2CH_3)_2 + NaOCH_2CH_3 \rightleftharpoons \left[\bar{C}H(CO_2Et)_2\right]Na^+ + CH_3CH_2OH$$

Diethyl malonate   Sodium ethoxide

$$\xrightarrow{CH_3I} H_3C{-}CH(CO_2CH_2CH_3)_2 + NaI$$

The organic base, sodium ethoxide, removes an acidic hydrogen from diethyl malonate to form the carbanion, $\bar{C}H(CO_2CH_2CH_3)_2$, which in turn displaces an iodide ion from methyl iodide to give the final product.

Free radicals are generally formed either by a dissociation type process or by reaction with another free radical. Some reactions, such as the photochemical halogenation of alkanes, involve both processes, as illustrated below for the monochlorination of methane:

$$:\ddot{C}l:\ddot{C}l: \xrightarrow{h\nu} 2:\ddot{C}l\cdot$$

Chlorine   Chlorine atoms

$$:\ddot{C}l\cdot + CH_4 \rightarrow H:\ddot{C}l: + \cdot CH_3$$

$$:\ddot{C}l:\ddot{C}l: + \cdot CH_3 \rightarrow H_3C:\ddot{C}l: + \cdot\ddot{C}l:$$

In the first step, a chlorine molecule is irradiated ($h\nu$ is the symbol used for photochemical or radiant energy) by light of the appropriate wavelength, which dissociates* some of the chlorine molecules into chlorine atoms (free radicals). The chlorine atom then abstracts a hydrogen atom from the methane to give hydrogen chloride and a methyl free radical ($\cdot CH_3$). The methyl free radical then abstracts a chlorine atom from a chlorine molecule to give methyl chloride and regenerates a chlorine atom, which can go back and repeat this cycle.† The overall reaction then can be summarized in the following equation:

$$CH_4 + Cl_2 \xrightarrow{h\nu} CH_3Cl + HCl$$

Methane   Methyl chloride

The preceding examples illustrate that much more detail about a chemical reaction can be learned by investigating its reaction mechanism, rather than merely writing an equation which represents the reactants and products. Of course, it is not always easy to decide experimentally whether a reaction is proceeding through a carbonium ion, a carbanion, or a free-radical type of intermediate. In later chapters, the current reaction mechanisms known for the most common type of functional group reactions will be employed as an aid to understanding the fundamental concepts of an organic reaction.

*Only a small percentage of the chlorine molecules are actually dissociated by this irradiation process, as recombination of chlorine atoms can occur to regenerate the chlorine molecule.

†Most free-radical reactions proceed via a cyclic process. Consequently, in many of these types of reaction, only a small amount of an initiator (compound which dissociates easily to free radicals) is required to get the reaction going.

The criterion which is used to arrive at a particular mechanism will generally not be included, as this type of material is beyond the scope of this book. However, in most cases good theoretical and experimental reasoning is available for writing a particular mechanism for a reaction, and the student is referred to more advanced texts for more mechanistic detail.

Reaction mechanisms not only are important in organic chemistry but are of paramount importance in understanding biochemistry. For example, in the Lobry de Bruyn-von Eckenstein transformation of carbohydrates (see p. 215) and in the oxidative deamination of amino acids by amino acid oxidases (see p. 350), the key to understanding these reactions is a knowledge of the reaction intermediates involved.

IMPORTANT TERMS AND CONCEPTS

Brönsted acid
Brönsted base
carbanion
carbonium ion
covalent bond
double bond
free-radical
functional group
isomer
Lewis acid
Lewis base
reaction mechanism
single bond
structural formula
triple bond

QUESTIONS

1. Which of the following compounds are isomers?

(a) $CH_3CH_2CH_2OH$
(b) $CH_3CHClCH_3$
(c) $CH_3CH_2CH_3$
(d) $CH_3\overset{\overset{\displaystyle O}{\|}}{C}CH_2CH_3$
(e) $CH_3CH_2CH_2Cl$

2. Which of the following compounds are identical?

$$\text{(a)}\ \begin{array}{ccccccccc} & & H & & H & & H & & H & \\ & & | & & | & & | & & | & \\ H & — & C & — & C & — & C & — & C & — & H \\ & & | & & | & & | & & | & \\ & & H & & H & & H & & Cl & \end{array}$$

$$\text{(d)}\ \begin{array}{ccccccc} & & H & & H & & H & \\ & & | & & | & & | & \\ H & — & C & — & C & — & C & — & H \\ & & | & & | & & | & \\ & & H & & HCH & & Cl & \\ & & & & H & & & \end{array}$$

$$\text{(g)}\ \begin{array}{ccccccccc} & & H & & H & & H & & Cl & \\ & & | & & | & & | & & | & \\ H & — & C & — & C & — & C & — & C & — & H \\ & & | & & | & & | & & | & \\ & & H & & H & & H & & H & \end{array}$$

$$\text{(b)}\ \begin{array}{ccccccccc} & & H & & H & & Cl & & H & \\ & & | & & | & & | & & | & \\ H & — & C & — & C & — & C & — & C & — & H \\ & & | & & | & & | & & | & \\ & & H & & H & & H & & H & \end{array}$$

$$\text{(e)}\ \begin{array}{ccccccccc} & & H & & H & & H & & H & \\ & & | & & | & & | & & | & \\ Cl & — & C & — & C & — & C & — & C & — & H \\ & & | & & | & & | & & | & \\ & & H & & H & & H & & H & \end{array}$$

$$\text{(c)}\ \begin{array}{ccccccccc} & & H & & H & & H & & H & \\ & & | & & | & & | & & | & \\ H & — & C & — & C & — & C & — & C & — & H \\ & & | & & | & & | & & | & \\ & & H & & Cl & & H & & H & \end{array}$$

$$\text{(f)}\ \begin{array}{ccccccccccc} & & H & & H & & H & & H & & H & \\ & & | & & | & & | & & | & & | & \\ H & — & C & — & C & — & C & — & C & — & C & — & Cl \\ & & | & & | & & | & & | & & | & \\ & & H & & H & & H & & H & & H & \end{array}$$

3. Represent the structure of $CH_3CH_2CH_2CH_3$, using a Lewis formula.

4. Write expanded structural formulas for *all* the different isomers of:
(a) C_3H_8O
(b) $C_2H_3Br_3$
(c) $C_2H_4Br_2$
(d) C_3H_5Cl
(e) C_3H_8

5. Which of the following pairs of compounds are isomers?
   (a) $CH_3CH_2CH_2CH_3$ and $(CH_3)_2CHCH_3$
   (b) $CH_3CH{=}CH_2$ and $CH_3C{\equiv}CH$
   (c) 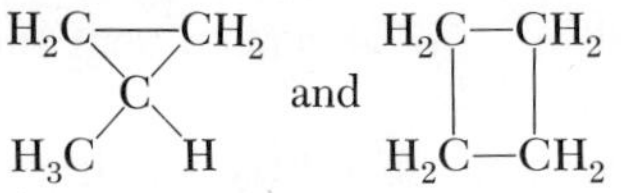 H_2C——CH_2 over $\overset{}{C}$ bearing H_3C and H (methylcyclopropane) and $H_2C{-}CH_2$ / $H_2C{-}CH_2$ (four-membered ring)
   (d) $CH_3CH_2CH_2\overset{\overset{H}{|}}{C}{=}O$ and $CH_3\overset{\overset{O}{\|}}{C}CH_2CH_3$
   (e) $CH_3CH_2CH_2\overset{\overset{O}{\|}}{C}OH$ and $CH_3CH_2\overset{\overset{O}{\|}}{C}OCH_3$

6. Classify each of the following compounds with respect to the shape of the molecule.
   (a) C_2Cl_4
   (b) CBr_4
   (c) C_2Cl_2
   (d) CH_2Cl_2
   (e) C_2F_6

7. In each of the following compounds, pick out the functional group and classify each compound into a functional class.
   (a) $CH_3CH_2\overset{\overset{O}{\|}}{C}CH_3$
   (b) $(CH_3)_2CHCH_2C(CH_3)_3$
   (c) $CH_3CH_2OCH_2CH_3$
   (d) $CH_3CH{=}CHCH_3$
   (e) $CH_3CH_2NH_2$
   (f) $CH_3CH_2CH_2\overset{\overset{OH}{|}}{C}HCH_3$
   (g) 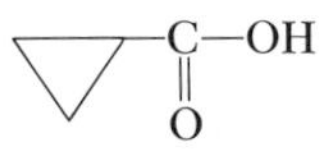 cyclopropyl$-\underset{\underset{O}{\|}}{C}{-}OH$
   (h) 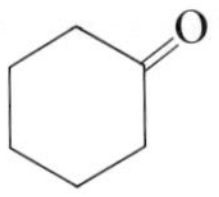 cyclohexanone ($=O$ on six-membered ring)
   (i) $CH_3CHBrCH_3$
   (j) $CH_3C{\equiv}CCH_3$

SUGGESTED READING

Caserio: Reaction Mechanisms in Organic Chemistry. Journal of Chemical Education, Vol. 42, pp. 570 and 627, 1965.

De LaMare and Vaughn: Detection and Reactions of Free Alkyl Radicals. Journal of Chemical Education, Vol. 34, p. 10, 1957.

Dence: Conformational Analysis or How Some Molecules Wiggle. Chemistry, Vol. 43, No. 6, p. 6, 1970.

Herron: Models Illustrating the Lewis Theory of Acids and Bases. Journal of Chemical Education, Vol. 30, p. 199, 1953.

Kurzer and Sanderson: Urea in the History of Organic Chemistry. Journal of Chemical Education, Vol. 33, p. 452, 1956.

Morris. Brönsted-Lowry Acid-Base Theory—A Brief Survey. Chemistry, Vol. 43, No. 3, p. 18, 1970.

Stewart: The Reactive Intermediates of Organic Chemistry. Journal of Chemical Education, Vol. 38, p. 308, 1961.

HYDROCARBONS

CHAPTER 3

The *objectives* of this chapter are to enable the student to:

1. Define the class of organic compounds known as hydrocarbons.
2. Describe and give examples of the class of hydrocarbons called alkanes.
3. Draw a structural formula of an alkane given the IUPAC name.
4. Write the IUPAC name of an alkane given the structural formula.
5. Recognize and name the most common alkyl groups.
6. Understand the concept of homologous series.
7. Appreciate the role of petroleum in modern society.
8. Understand the conversion of other functional groups to alkanes.
9. Understand the role of combustion as a source of useful energy.
10. Describe the general mechanism of a free-radical chain process.
11. Distinguish between substitution and addition types of reactions.
12. Explain the role of "cracking" in the petroleum industry.

The simplest organic compounds are hydrocarbons, which contain only the elements carbon and hydrogen. Replacement of a carbon-hydrogen bond by a functional group, such as those given in Chapter 2, gives rise to the various classes of organic compounds, and these functional classes may be thought of as derivatives of hydrocarbons.

Although hydrocarbons contain only two elements, these elements may be combined in several ways. For example, in the following compounds the carbon atoms may be linked together to form a linear chain or a ring. Also, the molecule may contain only carbon-

$CH_3CH_2CH_2CH_2CH_2CH_3$
n-Hexane
C_6H_{14}

Cyclohexane (ring of six CH_2 groups)
C_6H_{12}

Cyclohexene (ring of four CH_2 groups and CH=CH)
C_6H_{10}

$CH_3CH_2CH_2CH_2CH{=}CH_2$
1-Hexene
C_6H_{12}

$CH{\equiv}CH$
Acetylene
C_2H_2

carbon single bonds or may contain carbon-carbon multiple bonds. Even within a similar type of compound, such as *n*-hexane and cyclohexane which do not contain any multiple bonds, the ratio of carbon to hydrogen is not constant, and hence the molecular formula is not an indication of the type of hydrocarbon structure. For example, the molecular formula C_6H_{12} as shown previously could refer to either cyclohexane or 1-hexene. The chemistry of these two compounds is quite different even though they have the same ratio of carbon to hydrogen atoms.

Within the class of compounds known as hydrocarbons there are different degrees and types of chemical reactivity, and in order to classify the properties and chemical reactions of hydrocarbons, it is convenient to divide hydrocarbons into several subclasses. The basis for classification is the number of covalent bonds formed between the carbon atoms in the compounds. If only carbon-carbon single bonds are involved in the compound, the class is known as **alkanes,** or **saturated hydrocarbons.** The term "saturated" means that only one pair of electrons is shared covalently between any two bonded atoms in the molecule. Therefore, *n*-hexane and cyclohexane are alkanes by this definition. If the molecule contains multiple carbon-carbon bonds (more than one pair of electrons is shared covalently between any two bonded atoms), the compounds are classified on the basis of the number of multiple bonds between any two bonded atoms in the molecules. For example, cyclohexene and 1-hexene both contain a carbon-carbon double bond and would therefore be classed in the same category. However, acetylene contains a carbon-carbon triple bond and would not be classed with cyclohexene or 1-hexene. Compounds containing carbon-carbon double bonds are known as **alkenes,** and compounds containing carbon-carbon triple bonds are known as **alkynes.** Hence, cyclohexene and 1-hexene are alkenes and acetylene is an alkyne. These classifications of hydrocarbons are summarized in Table 3–1.

Although only simple examples of the classes of hydrocarbons are illustrated, it should be obvious that more complex hydrocarbons are possible either by introducing more carbon atoms or by introducing more than one alkene or alkyne linkage in the molecule. For example, 1,4-cyclohexadiene and 1,4-pentadiene contain two carbon-carbon double bonds per molecule, and diacetylene contains two carbon-carbon triple bonds per mole-

CH₂ / CH CH (double bonds) / CH CH / CH₂

1,4-Cyclohexadiene

$CH_2{=}CHCH_2CH{=}CH_2$

1,4-Pentadiene

$HC{\equiv}C{-}C{\equiv}CH$

Diacetylene

$CH_2{=}CH{-}C{\equiv}CH$

Vinyl acetylene

cule. These compounds, however, behave similar to simple alkenes and alkynes, except that instead of one mole of reagent, they require two moles of reactants in a chemical reaction. On the basis of our previous definition, they are also classified as alkenes and

TABLE 3–1 HYDROCARBONS

CLASS	DISTINGUISHING FEATURE	SIMPLEST EXAMPLE
Alkanes	$-\overset{\vert}{\underset{\vert}{C}}-\overset{\vert}{\underset{\vert}{C}}-$	CH_4, Methane
Alkenes	$>C{=}C<$	C_2H_4, Ethylene
Alkynes	$-C{\equiv}C-$	C_2H_2, Acetylene

alkynes, respectively. It is not the number of multiple centers in the molecule which determines its classification, but the *number of multiple bonds between any two bonded atoms*. In some special cases the molecule may contain two different types of multiple bonds, as illustrated in vinyl acetylene. This compound can behave as either an alkene or an alkyne, and its classification will depend on the type of reaction. Complex cases of this type are beyond the scope of the present treatment and will not be considered here, but the student should recognize that such compounds do exist and that in many cases the behavior of the compound can be predicted by reference to the chemistry of simple alkenes and alkynes.

ALKANES

The simplest hydrocarbon is methane:

$$\begin{array}{c}H\\ \vdots\\ H{:}C{:}H\\ \vdots\\ H\end{array} \quad \text{or} \quad \begin{array}{c}H\\ |\\ H-C-H\\ |\\ H\end{array} \quad \text{or} \quad CH_4$$

Methane

Higher members of this series can be developed by replacing one or more of the hydrogens by one or more carbon atoms and by completing the valence requirements of the added carbon atom with hydrogen atoms. Several examples of this stepwise development of a carbon chain are illustrated below:

$$\begin{array}{c}H\\ |\\ H-C-H\\ |\\ H\end{array} \xrightarrow[\text{of 1 carbon atom}]{\text{addition}} \begin{array}{c}H\ \ H\\ |\ \ \ |\\ H-C-C-H\\ |\ \ \ |\\ H\ \ H\end{array} \quad \text{or} \quad CH_3CH_3 \quad \text{or} \quad C_2H_6$$

Ethane

$$\begin{array}{c}H\\ |\\ H-C-H\\ |\\ H\end{array} \xrightarrow[\text{of 2 carbon atoms}]{\text{addition}} \begin{array}{c}H\ \ H\ \ H\\ |\ \ \ |\ \ \ |\\ H-C-C-C-H\\ |\ \ \ |\ \ \ |\\ H\ \ H\ \ H\end{array} \quad \text{or} \quad CH_3CH_2CH_3 \quad \text{or} \quad C_3H_8$$

Propane

$$\begin{array}{c}H\\ |\\ H-C-H\\ |\\ H\end{array} \xrightarrow[\text{of 3 carbon atoms}]{\text{addition}} \begin{array}{c}H\ \ H\ \ H\ \ H\\ |\ \ \ |\ \ \ |\ \ \ |\\ H-C-C-C-C-H\\ |\ \ \ |\ \ \ |\ \ \ |\\ H\ \ H\ \ H\ \ H\end{array} \quad \text{or} \quad CH_3CH_2CH_2CH_3 \quad \text{or} \quad C_4H_{10}$$

Normal butane

Another and easier way to view this process of building a carbon chain structure is that the propane carbon chain is developed from ethane by replacing a hydrogen of ethane by a $-CH_3$ group. Similarly, normal* butane is developed from propane by replacement of the terminal hydrogen with a $-CH_3$ group. Higher members of this series are built up in an analogous manner as shown above on the next page:

*__Normal__ refers to the fact that all of the carbon atoms in the chain are arranged in a linear manner. This is in contrast to other isomers having the same molecular formula, but in which the carbon atoms are arranged in other than a linear chain.

$$\begin{array}{ccccccccccc}
 & & H & & H & & H & & H & & \\
 & & | & & | & & | & & | & & \\
H & - & C & - & C & - & C & - & C & - & H \\
 & & | & & | & & | & & | & & \\
 & & H & & H & & H & & H & &
\end{array}
\xrightarrow[\text{—H by —CH}_3]{\text{replacement}}$$

Normal butane

$$\begin{array}{ccccccccccccc}
 & & H & & H & & H & & H & & H & & \\
 & & | & & | & & | & & | & & | & & \\
H & - & C & - & C & - & C & - & C & - & C & - & H \\
 & & | & & | & & | & & | & & | & & \\
 & & H & & H & & H & & H & & H & &
\end{array}
\quad \text{or} \quad CH_3CH_2CH_2CH_2CH_3 \quad \text{or} \quad C_5H_{12}$$

Normal pentane

$$\begin{array}{ccccccccccccc}
 & & H & & H & & H & & H & & H & & \\
 & & | & & | & & | & & | & & | & & \\
H & - & C & - & C & - & C & - & C & - & C & - & H \\
 & & | & & | & & | & & | & & | & & \\
 & & H & & H & & H & & H & & H & &
\end{array}
\xrightarrow[\text{—H by —CH}_3]{\text{replacement}}$$

Normal pentane

$$\begin{array}{ccccccccccccccc}
 & & H & & H & & H & & H & & H & & H & & \\
 & & | & & | & & | & & | & & | & & | & & \\
H & - & C & - & C & - & C & - & C & - & C & - & C & - & H \\
 & & | & & | & & | & & | & & | & & | & & \\
 & & H & & H & & H & & H & & H & & H & &
\end{array}
\quad \text{or} \quad CH_3CH_2CH_2CH_2CH_2CH_3 \quad \text{or} \quad C_6H_{14}$$

Normal hexane

In the compounds illustrated above, the molecular formulas conform to the general formula C_nH_{2n+2}, in which n is the number of carbon atoms in the molecule. This general formula fits all linear (straight chain) alkanes and can be used to predict the molecular formula of any linear alkane. For example, an alkane containing seven carbon atoms (heptane) would have a molecular formula C_7H_{16}; one with eight carbons (octane), C_8H_{18}; one with nine carbons (nonane), C_9H_{20}; one with ten carbons (decane), $C_{10}H_{22}$, and so on. Inspection of each of these structures shows that propane differs from butane by a —CH_2 unit; pentane differs from hexane by a —CH_2 unit; hexane differs from heptane by a —CH_2 unit, and so on. This —CH_2 unit is known as a methylene group. Each higher member in this series differs from the next lower or higher one by one more or one less methylene unit. A series of compounds of this type in which each member differs from the next higher or lower member by a constant increment is known as a **homologous series.**

A family of compounds in a homologous series exhibit characteristic features: (a) they all contain the same elements and can be represented by a single general formula; (b) each homolog differs from the one above and below it in the series by a —CH_2 unit; and (c) all the homologs show similar and closely related physical and chemical properties.

It is important to recognize that once we have established the physical and chemical properties of a few members of any homologous series, we can predict the properties of the other members of the series. Consequently, it is not necessary to investigate the properties of every single organic compound. Rather, the properties of each homologous class are studied with representative members of that series, and these properties are used to predict the behavior of the other members of this series.

MOLECULAR MODELS

One drawback to the representation of structure illustrated in the preceding section is that the structures are shown only in two dimensions. Sometimes it is advantageous to depict the three-dimensional features of the molecule. The chemist does this by the use of molecular models. Two types of molecular models are frequently used, and these are illustrated on the next page.

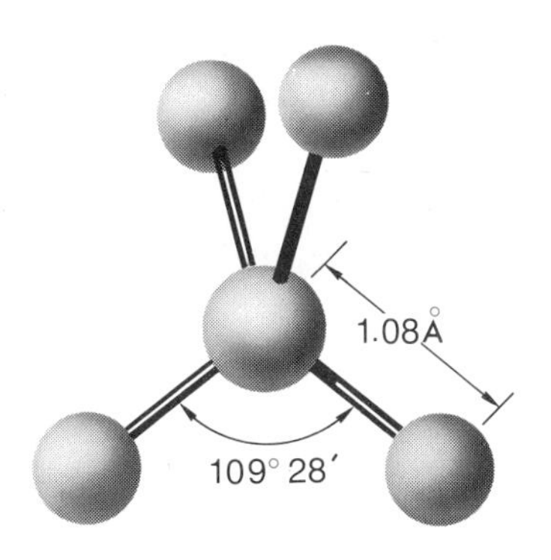

Ball and Stick Model Space-Filling Model

FIGURE 3-1 Methane.

The ball and stick model (Tinkertoy) shows clearly the bond angles in the molecule. However, the real bonded atoms are not spherical in shape as shown by this model and are not separated in space by rigid bonds. The space-filling model (called Van der Waals model) are exact scale models which show the relative size of each atom. However, the bond angles and bond distances are more difficult to see. Each type of model has its own utility, and the chemist uses the appropriate model depending upon the type of information desired.

The ball and stick models for some of the simple alkanes are illustrated below:

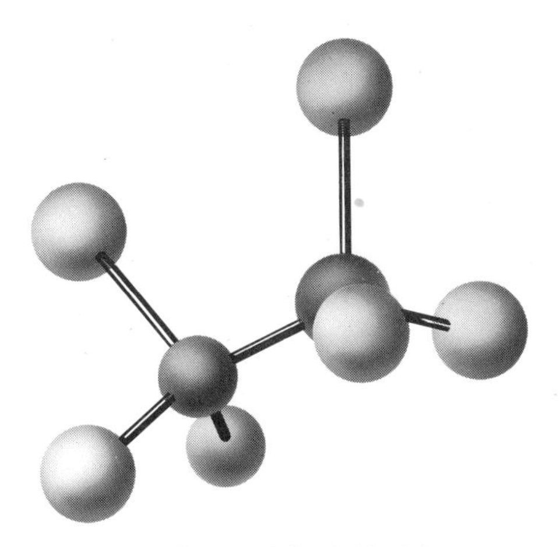

Ball and Stick Model Space-Filling Model

FIGURE 3-2 Ethane.

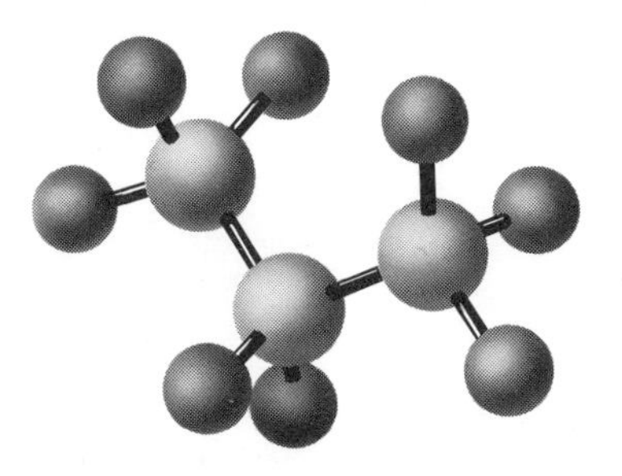

Ball and Stick Model Space-Filling Model

FIGURE 3-3 Propane.

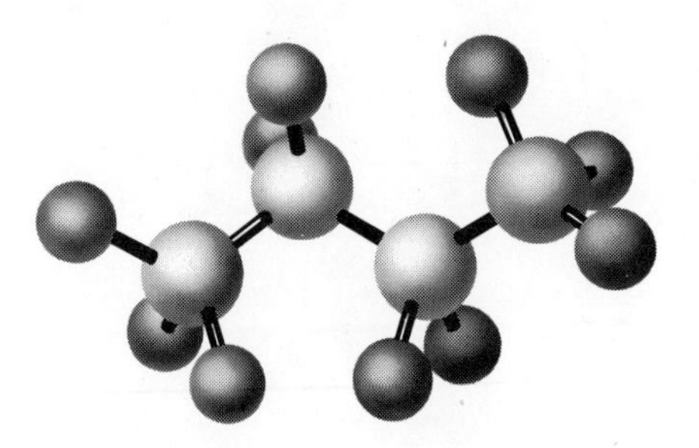

Ball and Stick Model

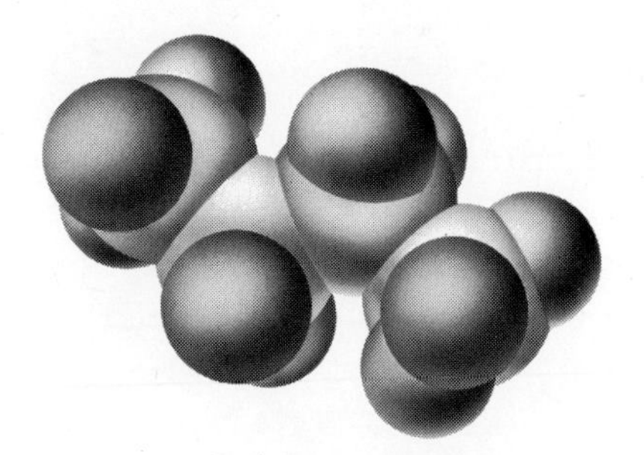

Space-Filling Model

FIGURE 3-4 Butane.

The cycloalkanes are best shown with the ball and stick type model and the first several members of this series, which were discussed in Chapter 2, are shown on page 49.

NOMENCLATURE

The nomenclature of the simple straight chain alkanes is straightforward. The first four members have common names (methane, ethane, propane, and butane), but the stem names of the higher members are derived from the number of carbon atoms in the chain, and the -ane ending is added to the stem name. Therefore, the names hexane, heptane, octane, and so forth are used for alkanes containing six, seven, and eight carbons in the chain.

In the preceding section normal butane was developed by replacing a terminal hydrogen in propane by a $-CH_3$ group. Since both ends of the propane molecule are equivalent, it doesn't matter which terminal hydrogen is replaced by the $-CH_3$ group.

$$\underset{\text{Propane}}{\begin{array}{c} \;\;H\;\;H\;\;H \\ \;\;|\;\;\;\;|\;\;\;\;| \\ H-C-C-C-H \\ \;\;|\;\;\;\;|\;\;\;\;| \\ \;\;H\;\;H\;\;H \end{array}} \xrightarrow[\text{group}]{+CH_3} \underset{\text{Normal butane}}{\begin{array}{c} \;\;H\;\;H\;\;H\;\;H \\ \;\;|\;\;\;\;|\;\;\;\;|\;\;\;\;| \\ H-C-C-C-C-H \\ \;\;|\;\;\;\;|\;\;\;\;|\;\;\;\;| \\ \;\;H\;\;H\;\;H\;\;H \end{array}} \xleftarrow{+CH_3} \underset{\text{Propane}}{\begin{array}{c} \;\;H\;\;H\;\;H \\ \;\;|\;\;\;\;|\;\;\;\;| \\ H-C-C-C-H \\ \;\;|\;\;\;\;|\;\;\;\;| \\ \;\;H\;\;H\;\;H \end{array}}$$

However, another possibility still exists for developing the carbon chain by this process. If the hydrogen of the $-CH_2-$ group in propane is replaced by a $-CH_3$ group, an isomer of normal butane is formed, namely, isobutane. The molecular structures of isobutane are shown in Figure 3-6.

$$\begin{array}{c} \;\;H\;\;H\;\;H \\ \;\;|\;\;\;\;|\;\;\;\;| \\ H-C-C-C-H \\ \;\;|\;\;\;\;|\;\;\;\;| \\ \;\;H\;\;H\;\;H \end{array} \xrightarrow[\text{group}]{+CH_3} \underset{\text{Isobutane}}{\begin{array}{c} H \\ | \\ H-C-H \\ \;\;H\;\;\;\;|\;\;\;\;H \\ \;\;|\;\;\;\;|\;\;\;\;| \\ H-C-C-C-H \\ \;\;|\;\;\;\;|\;\;\;\;| \\ \;\;H\;\;H\;\;H \end{array}} \text{ or } \begin{array}{c} CH_3 \\ | \\ CH_3CHCH_3 \end{array} \text{ or } C_4H_{10}$$

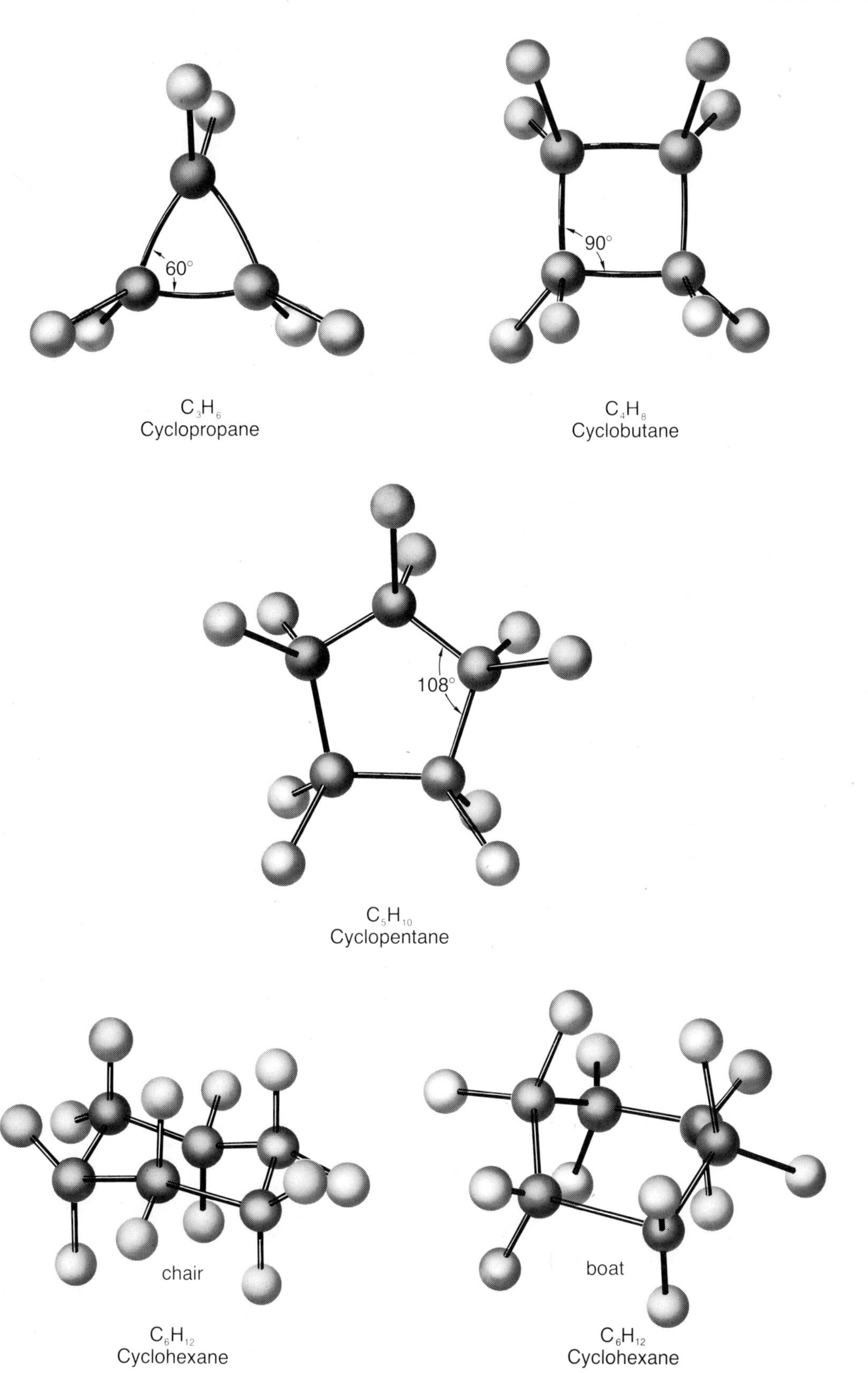

FIGURE 3-5 Cycloalkanes.

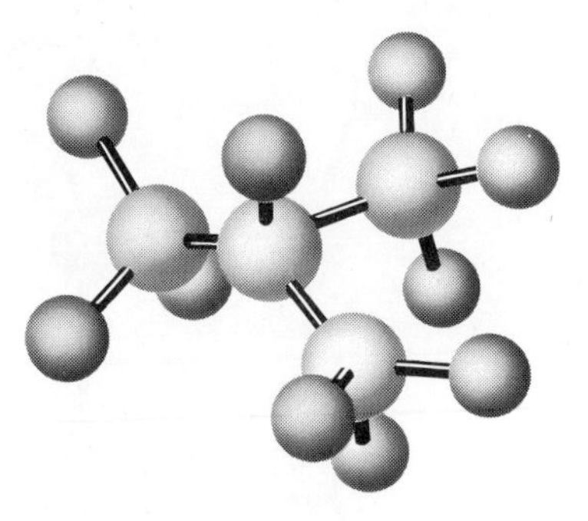

FIGURE 3-6 Isobutane.

In the case of propane, only two isomers of butane can be developed by replacing hydrogen by a —CH_3 group. With higher members of this series, many isomers are possible by replacement of the different hydrogens. For example, in the octane series, C_8H_{18}, eighteen structural isomers are possible. Of course, when the number of isomers of a particular carbon chain structure becomes high, the naming of these isomers becomes difficult, and an unambiguous nomenclature is necessary to avoid confusion. The naming system selected should be simple, but must provide a name that fits one and only one structure. The International Union of Pure and Applied Chemistry (IUPAC) has recommended a system which is used universally by organic chemists to name alkanes. The rules of this system are as follows:

1. The characteristic ending -ane is applied to the stem name to obtain the name of a linear saturated hydrocarbon.
2. For branched chain alkanes, the compound is named as a derivative of the hydrocarbon corresponding to the *longest continuous carbon chain* in the molecule.
3. Substituents (atoms or groups of atoms) are indicated by a suitable prefix and a number to indicate their position on the carbon chain.
4. Numbering of the longest continuous carbon chain (Rule 2) must be done in such a way that the numbers giving the position of the substituents are kept as low as possible.

For example, the following compound can be named in two different ways:

```
 1    2   3    4    5              5   4    3    2    1
CH3 CHCH2CH2CH3    or    CH3CHCH2CH2CH3
      |                        |
     CH3                      CH3
(A) 2-Methylpentane      (B) 4-Methylpentane
```

In both structures, (A) and (B), the longest continuous carbon chain contains five carbons; hence, the compound will be named as a derivative of pentane. The —CH_3 group (methyl group) is attached at position 2 in structure (A) or at position (4) in structure (B). Rule 4 demands that the lowest number be used in numbering substituents; hence, 2-methylpentane is the correct name for this compound—not 4-methylpentane. Similarly, the following compounds are correctly named as indicated following the IUPAC rules:

$$\begin{array}{l} \mathrm{CH_3\underset{\displaystyle |}{C}HCH_2CH_3} \\ \quad\mathrm{CH_2CH_3} \end{array}$$ 3-Methylpentane, not 2-ethylbutane

$$\begin{array}{l} \mathrm{CH_3CH_2\underset{\displaystyle |}{C}HCH_2CH_2CH_2CH_3} \\ \qquad\quad\mathrm{CH_3} \end{array}$$ 3-Methylheptane, not 5-methylheptane

When more than one substituent is present on the carbon chain, each substituent is given a number. For example, structure (C) is correctly named as 2,2-dimethylbutane—

$$\begin{array}{c} \mathrm{CH_3} \\ | \\ \mathrm{CH_3CCH_2CH_3} \\ | \\ \mathrm{CH_3} \end{array} \quad \text{or} \quad \mathrm{CH_3C(CH_3)_2CH_2CH_3}$$

(C)

2,2-dimethylbutane

not 2-dimethylbutane. Even though the substituents are identical, each must be given a number, and the number of identical substituents is also indicated by the prefix di-. The following examples further illustrate this nomenclature system:*

$$\begin{array}{l} \mathrm{CH_3CH_2\underset{\displaystyle |}{C}HCH_2CH_2\underset{\displaystyle |}{C}HCH_3} \\ \qquad\quad\mathrm{CH_3} \qquad\quad \mathrm{CH_3} \end{array}$$ 2,5-Dimethylheptane

$$\begin{array}{l} \mathrm{CH_3\underset{\displaystyle |}{C}HCH_2\underset{\displaystyle |}{C}HCH_2CH_3} \\ \quad\mathrm{CH_3} \quad \mathrm{CH_2CH_3} \end{array}$$ 4-Ethyl-2-methylhexane

$$\begin{array}{l} \qquad\quad\mathrm{CH_2CH_3} \\ \qquad\quad | \\ \mathrm{CH_3\underset{\displaystyle |}{C}HCH\underset{\displaystyle |}{C}HCH_2CH_3} \\ \quad\mathrm{CH_3} \quad \mathrm{CH_3} \end{array}$$ 2,4-Dimethyl-3-ethylhexane

In the preceding examples, the substituents (other than hydrogen atoms) were designated as methyl or ethyl. The names of these substituents were derived from the alkane containing the same number of carbon atoms by changing the **-ane** ending to **-yl.** These groups are derived from the parent alkane by removing one of the hydrogen atoms and are known as **alkyl** groups (from alk**ane** → alk**yl**). Therefore, methyl and ethyl groups can be formulated from methane and ethane as follows:

$$\begin{array}{c} \mathrm{H} \\ | \\ \mathrm{H{-}C{-}H} \\ | \\ \mathrm{H} \end{array} \rightarrow \begin{array}{c} \mathrm{H} \\ | \\ \mathrm{H{-}C{-}} \\ | \\ \mathrm{H} \end{array} \quad \text{or} \quad \mathrm{H_3C{-}}$$

Methane — Methyl group

$$\begin{array}{c} \mathrm{H\;\;H} \\ |\;\;\;\;| \\ \mathrm{H{-}C{-}C{-}H} \\ |\;\;\;\;| \\ \mathrm{H\;\;H} \end{array} \rightarrow \begin{array}{c} \mathrm{H\;\;H} \\ |\;\;\;\;| \\ \mathrm{H{-}C{-}C{-}} \\ |\;\;\;\;| \\ \mathrm{H\;\;H} \end{array} \quad \text{or} \quad \mathrm{CH_3CH_2{-}} \quad \text{or} \quad \mathrm{C_2H_5{-}}$$

Ethane — Ethyl group

* It is evident that by counting the number of carbon atoms indicated in the name given to the longest continuous chain and adding to this the number of carbons indicated in the substituent names, the total number of carbons in the compound is obtained. This is a good way to check that no carbons have been omitted in naming the compound.

In compounds containing more than two carbon atoms, the number of alkyl groups will depend on the number of different types of hydrogens in the molecule. From propane, for example, two different alkyl groups are possible:

$$\underset{\text{Propane}}{CH_3CH_2CH_3} \rightarrow \underset{\text{Normal propyl }(n\text{-propyl})}{CH_3CH_2CH_2-} \text{ or } nC_3H_7-$$

$$CH_3CH_2CH_3 \rightarrow \underset{\text{2-Propyl or isopropyl}}{CH_3\overset{|}{C}HCH_3} \text{ or } isoC_3H_7-$$

From higher homologues, even more possibilities can be formulated. In general, naming of alkyl groups by this method is only reasonable for groups containing a small number of carbons. This nomenclature is usually employed for compounds containing 1 to 4 carbon atoms. Table 3–2 summarizes some of the more important alkyl groups.

Use of alkyl groups in naming compounds provides another method of nomenclature which has much current usage for simple compounds, and the student should become acquainted with both the IUPAC and alkyl group systems, although it should be kept in mind that the IUPAC is the correct systematic method. The following examples illustrate both systems.

TABLE 3-2 ALKYL GROUPS

ALKYL GROUP	IUPAC NAME	COMMON NAME
CH_3-	Methyl	Methyl
CH_3CH_2-	Ethyl	Ethyl
$CH_3CH_2CH_2-$	Propyl	*n*-Propyl
$CH_3\underset{\vert}{C}HCH_3$	Methylethyl	Isopropyl°
$CH_3CH_2CH_2CH_2-$	Butyl	*n*-Butyl
$CH_3CH_2\underset{\vert}{C}HCH_3$	1-Methylpropyl	*s*-Butyl°°
$CH_3\underset{CH_3}{\underset{\vert}{C}}HCH_2-$	2-Methylpropyl	Isobutyl°
$CH_3\underset{CH_3}{\underset{\vert}{\overset{\vert}{C}}}CH_3$	Dimethylethyl	*t*-Butyl°°
$CH_3CH_2CH_2CH_2CH_2-$	Pentyl	*n*-Pentyl (Amyl)

° iso- refers to any structure having a terminal $CH_3\overset{\vert}{C}HCH_3$ grouping.

°° *s*- and *t*- refer to secondary and tertiary. In this system of nomenclature, the carbon having the unsatisfied valence is the focal point. Whether this carbon is referred to as primary, secondary, or tertiary depends upon whether it is attached to 1, 2, or 3 additional carbon atoms.

Compound	*Alkyl group name*	*IUPAC name*
$CH_3\underset{\underset{Br}{\vert}}{C}HCH_3$	isopropyl bromide	2-bromopropane
CH_3CH_2I	ethyl iodide	iodoethane
CH_3Cl	methyl chloride	chloromethane
$CH_3CH_2CH_2F$	*n*-propyl fluoride	1-fluoropropane
$CH_3-\overset{\overset{CH_3}{\vert}}{\underset{\underset{H}{\vert}}{C}}-CH_2Cl$	isobutyl chloride	1-chloro-2-methylpropane
$CH_3CH_2CH_2CH_2Br$	*n*-butyl bromide	1-bromobutane
$CH_3-\overset{\overset{CH_3}{\vert}}{\underset{\underset{CH_3}{\vert}}{C}}-Cl$	*t*-butyl chloride	2-chloro-2-methylpropane

Cycloalkanes are analogous to straight chain alkanes, except that the ends of the carbon chain are joined together in a ring. This process of linking the ends of the chain into a ring requires the use of one additional valence from each terminal carbon atom. Consequently, two less C—H bonds are formed, and the general formula for these compounds is C_nH_{2n}, where n equals the number of carbon atoms. The parent name of the cyclic hydrocarbon is derived by adding the prefix **cyclo-** to the linear alkane having the same number of carbon atoms as shown below:

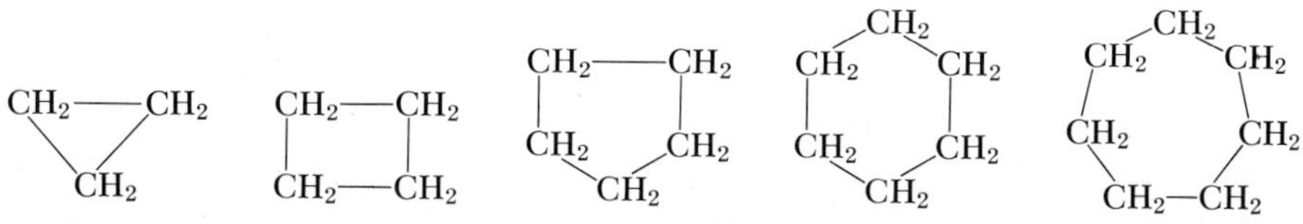

Cyclopropane	Cyclobutane	Cyclopentane	Cyclohexane	Cycloheptane
C_3H_6	C_4H_8	C_5H_{10}	C_6H_{12}	C_7H_{14}

(Each ring structure is drawn as CH_2 groups joined in a ring: 3, 4, 5, 6 and 7 CH_2 units respectively.)

For substituted cycloalkanes, the IUPAC rules again require that each substituent be given the lowest possible number, and that each substituent be named and indicated by the appropriate prefix. Some further examples are illustrated on page 54.*

SOURCES OF HYDROCARBONS

Natural gas and petroleum are the most important natural sources of hydrocarbons. Large deposits of these substances have been formed over the years by the gradual decomposition of marine life and other biological materials. These deposits usually accumulate under a dome-shaped layer of rock several thousand feet under the earth's surface (Fig. 3–7). When a hole is drilled through the rock layer, the pressure under the dome forces the gas or oil to the surface. After the pressure is released, pumps are required to bring the remaining oil to the surface.

*For convenience the ring structure is often drawn without detailing the carbon and hydrogen atoms of the ring. Each junction of the ring is meant to be a carbon atom, and if no other substituents are attached to the carbon atom it's implied that the remaining valences are satisfied by hydrogens.

Structure		Skeletal formula	Name
CH_2—CH_2, CH_2, $CHCH_3$, CH_2	or	CH_3	Methylcyclopentane
CH_2, CH_2, $C(CH_3)_2$, CH_2, CH_2, CH, CH_3	or	CH_3, CH_3, CH_3	1,1,3-Trimethylcyclohexane—not 1,5,5-trimethylcyclohexane
CH_2—CBr_2, CH_2—CBr_2	or	Br, Br, Br, Br	1,1,2,2-Tetrabromocyclobutane—not 1,1,4,4-tetrabromocyclobutane
CH_2, CH_2, CH_2, Cl, CH_2, C, CH_2, CH_3	or	Cl, CH_3	1-chloro-1-methylcyclohexane

Natural Gas

Natural gas is an excellent source of low molecular weight alkanes. Natural gas occurs in most parts of the United States, but most of it is produced in the southwest. In recent years a vast network of pipelines has been installed to carry natural gas from Texas to other parts of the United States.

A typical composition of natural gas is shown in Table 3–3. The propane and butane are removed by liquefaction before the gaseous fuel is introduced into the pipelines for distribution. The liquid propane and butane are stored under pressure in steel cylinders from which they are released as a gaseous fuel to be used in rural areas and in locations that are not supplied by natural gas mains.

Petroleum

The crude oil, or petroleum, obtained from oil wells is another rich source of hydrocarbons. The hydrocarbons in petroleum, in contrast to those of natural gas, are of higher molecular weight. Petroleum has been known for several centuries and has

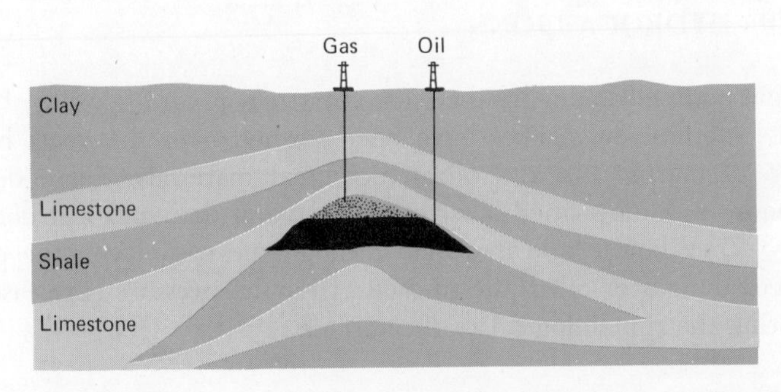

FIGURE 3-7 An illustration of an oil dome, or anticline, showing deposits of petroleum and natural gas.

TABLE 3-3 COMPOSITION OF NATURAL GAS

HYDROCARBON	TYPICAL COMPOSITION (per cent)
Methane	82
Ethane	10
Propane	4
Butane	2
Higher hydrocarbons	2

been used for many purposes, particularly as a fuel. It was not until recent years, however, that petroleum was separated into its hydrocarbon components. It is a very complex mixture of hydrocarbons, and its composition varies with the location of the oil field from which it is obtained. It contains mainly a mixture of alkanes, cycloalkanes, and aromatic hydrocarbons. In addition to the hydrocarbons, petroleum contains about 10 per cent by weight of sulfur, nitrogen, and oxygen compounds.

Petroleum is separated into its hydrocarbon fractions by the process of distillation. Since the forces of attraction between the individual hydrocarbon molecules are small, they can be converted into the gaseous state without decomposition at a fairly low temperature (boiling point). When the hydrocarbon vapors are cooled, they condense to a liquid form. This process is called distillation and can be used to separate hydrocarbons, because the forces of attraction between the molecules of one compound differ from those between other molecules and result in different distillation temperatures. Fractional distillation is commonly used in the petroleum industry. In this process, petroleum is separated by distillation into several fractions possessing different distilling temperatures. By the use of a fractionating column, more efficient continuous fractional distillation of petroleum can be achieved. A fractionating column consists of a tall column containing perforated plates or irregularly shaped glass or ceramic pieces designed to promote intimate contact between the distilling vapors and the refluxing liquid that condenses and runs back down the column. The effect of such a column is to concentrate the lower-boiling constituents in the vapors as they rise, and to enrich the reflux with the higher-boiling constituents. By proper construction and operation, the various petroleum fractions can be removed from different levels in the fractionating column (Fig. 3–8). In the petroleum industry, these distillation procedures are called the refining process. The distillation fractions from a typical petroleum are shown in Table 3–4.

Gases from Petroleum. The first products given off during the distillation of petroleum are the gaseous hydrocarbons containing from 1 to 5 carbon atoms. These hydrocarbons are both saturated and unsaturated, and are usually separated from each other by chemical methods. The unsaturated gases are used in the production of aviation gasoline, synthetic rubber, and other organic compounds. The saturated hydrocarbons, especially propane and butane, are liquefied and sold as **bottled gas.** Over two billion gallons of these liquefied gases from both petroleum and natural gas are used in the United States every year.

Petroleum Ether. The second fraction that is distilled from petroleum is called petroleum ether. This consists mainly of pentanes, hexanes, and heptanes, and is used extensively as fat solvents; paint, varnish, and enamel thinner; and dry-cleaning agents.

Gasolines. In the early days of the automobile industry simple distillation of petroleum gave more than enough gasoline to supply the demands. This type of gasoline is called **straight-run gasoline** and is composed essentially of alkanes plus minor amounts

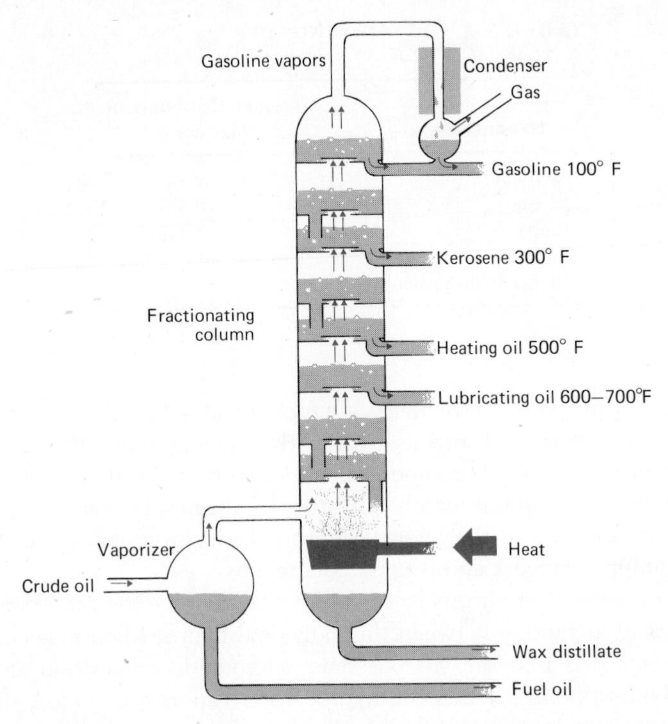

FIGURE 3-8 A diagram of a fractionating column, showing the various levels from which the petroleum fractions are removed.

of cycloalkanes and aromatic hydrocarbons. The composition of gasoline varied considerably and depended on the source of petroleum from which it was distilled.

The use of straight-run gasoline in an automobile engine causes the motor to knock, or to develop the familiar ping heard when a motor is accelerated too rapidly or when going up a steep grade. This knock results from uneven combustion of the air-gasoline mixture that is ignited by the spark plug in the cylinder of an internal combustion engine. It was found that the knocking characteristic of a gasoline depends on the structure of the constituent hydrocarbons. A predominance of straight chain hydrocarbons causes excessive knocking, whereas those with a high degree of branching prevent this knock. As a measure of the performance of a particular gasoline, two hydrocarbons were chosen as standards. These two hydrocarbons, ***n*-heptane** and trimethyl pentane, known as **isooctane,** are shown on page 57.

TABLE 3-4 DISTILLATION FRACTIONS FROM A TYPICAL PETROLEUM

NAME	COMPOSITION (per cent)	MOLECULAR SIZE	BOILING RANGE (°C)	USES
Gases	2	C_1–C_5	0	Fuel
Petroleum ethers	2	C_5–C_7	30–110	Solvents
Gasoline	32	C_6–C_{12}	30–200	Motor fuel
Kerosene	18	C_{12}–C_{15}	175–275	Diesel and jet fuel
Gas oil (fuel oil)	20	C_{15}—	250–400	Fuel
Lubricating oils and residue		C_{19}—	300—	Lubricants, paraffin wax, petrolatum, and asphalt

$CH_3CH_2CH_2CH_2CH_2CH_2CH_3$

n-Heptane

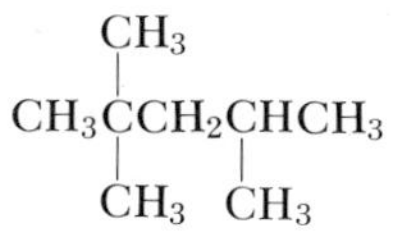

$$\begin{array}{c} CH_3 \\ CH_3\overset{|}{\underset{|}{C}}CH_2\underset{|}{C}HCH_3 \\ CH_3\ \ CH_3 \end{array}$$

Isooctane

The performance of a gasoline is expressed as its octane number, compared with an octane number of zero for *n*-heptane and 100 for isooctane. Mixtures of these two hydrocarbons are prepared that exactly match the knocking characteristics of a gasoline under test. The **octane number** of the gasoline is then equal to the percentage of isooctane in the mixture. A thorough study of the hydrocarbon components of gasoline indicates that the octane number increases with increasing content of branched chain hydrocarbons, unsaturated hydrocarbons, and aromatic hydrocarbons.

As the number of automobiles increased, the supply of straight-run gasoline was insufficient to meet the demands. In addition to the shortage of gasoline, the petroleum industry was faced with an excess of other fractions distilled from petroleum. It was found as early as 1912 that lower molecular weight hydrocarbons could be produced from the higher fractions by heating them to a temperature of from 400 to 500°C. At this temperature some of the bonds of the larger hydrocarbon molecules break to form lower boiling range fractions. If the temperature is increased beyond 500°C, small molecular weight gaseous hydrocarbons are produced. This process, called **thermo-cracking,** not only produces smaller hydrocarbons, but also yields unsaturated and aromatic hydrocarbons. More gasoline with a higher octane number was therefore produced by the cracking process. Research by petroleum engineers has resulted in the development of several cracking processes that employ petroleum fractions of lower molecular weight and higher molecular weight than those in gasoline. Catalysts have also been found that increase the yield of high octane gasoline from these petroleum fractions.

When cracking processes were first being developed, petroleum engineers were also attempting to increase the octane number of gasoline by other methods. The addition of fairly large proportions of benzene or ethyl alcohol (20 to 40 per cent) to a low octane gasoline reduced or completely prevented the knocking in a gasoline motor. Much smaller quantities of iodine or aniline were found to produce the same effect as benzene or ethyl alcohol. Further research revealed that metallic derivatives of the alkanes were the most efficient **antiknock agents,** with **lead tetraethyl** by far the most effective. One of the disadvantages of the use of tetraethyl lead is that its combustion product, lead oxide, is reduced to metallic lead which is deposited inside the cylinder head of the motor. When ethylene bromide is added with the tetraethyl lead, the compound lead bromide is formed, which is not readily reduced to metallic lead. A mixture containing approximately 65 per cent tetraethyl lead, 25 per cent ethylene dibromide, and 10 per cent ethylene dichloride plus a small amount of a dye is known commercially as **ethyl fluid.** From 1 to 3 milliliters of ethyl fluid is added per gallon to improve the antiknock properties of modern gasoline. The two common grades in use today are known as regular gasoline and ethyl gasoline, and in general the ethyl gasoline has a higher octane rating and contains more ethyl fluid than the regular gasoline. The toxicity of lead by-products from auto exhausts has caused much social concern in recent years. Several "lead-free" gasolines have appeared on the market, and government legislation has been proposed to ban lead additives from gasoline products. Undoubtedly, other additives will come under scrutiny, and the petroleum industry is attempting to develop other methods to increase octane rating without the use of harmful additives.

In recent years increased competition and increased research by the petroleum industry have resulted in the gradual improvement of quality and octane rating of gasolines. When fortified with ethyl fluid, the branched chain hydrocarbon trimethyl butane, known as **triptane,** exhibits in automobile engines a performance decidedly superior to that of isooctane. The development and use of additives such as TCP (tricresyl phosphate) and boron hydrides have resulted in extensive advertising campaigns.

Kerosene. In the early days of the petroleum industry the most important fraction from petroleum was kerosene. It was used for lighting purposes, for cooking, and for heating. With the advent of the electric light and the automobile, the demand for kerosene decreased and that for gasoline increased. **Kerosene** is composed of a mixture of saturated, unsaturated, and aromatic hydrocarbons containing from 12 to 15 carbon atoms. It is distilled from petroleum at a temperature of 175 to 275°C. Since the unsaturated hydrocarbons produce an inferior flame when kerosene is used for lighting purposes, they are usually removed by a refining process.

Gas Oil and Fuel Oil. The next higher boiling fraction after kerosene contains a mixture of hydrocarbons whose smallest members have 15 carbon atoms. This fraction contains gas oil, fuel oil, and diesel oil. The name **gas oil** is derived from the fact that this fraction was used originally to enrich water gas for use as a fuel. Large quantities of **fuel oil** are used in furnaces that burn oil, whereas **diesel oil** is used in diesel engines. This fraction may also be cracked to produce gasoline.

Lubricating Oils. Lubricating oils are produced from the fraction of petroleum that distills at the highest temperature, usually over 300°C. This fraction consists of hydrocarbons with 20 or more carbon atoms and can be separated into oils of different viscosity by fractional distillation. The **viscosity,** or consistency, of lubricating oils is directly related to the structure of their constituent hydrocarbons. For example, an increase in the length of the carbon chain results in increased viscosity and the higher boiling fractions have a higher viscosity than the lower boiling fractions. It is also well known that the viscosity of a lubricating oil increases as the temperature of the oil is decreased. Such processes as redistillation, refining, and dewaxing are used to prepare lubricating oils with different viscosities. One example of the importance of viscosity in oils is the switch from an oil of relatively high viscosity to one of low viscosity when preparing an automobile for winter driving. Recently, by the proper combination of oils and substances such as detergents, a superior multirange viscosity automobile oil has been produced. Such oils remain fluid at low temperature and possess a greater viscosity at high temperatures than regular motor oil.

The Residual Fraction. The residual material that is left after the removal of the distillable fractions of petroleum usually contains either asphalt or paraffin types of hydrocarbons. The nature of the residue has given rise to the terminology of **paraffin base motor oils, asphalt base motor oils,** and **mixed base oils. Paraffin wax** is prepared from the residue of paraffin base oil and consists of straight chain alkanes with 26 to 30 carbon atoms. This residue also yields petrolatum, which is commonly known as petroleum jelly or Vaseline. **Petrolatum** is a semisolid substance that is used as a pharmaceutical base for many salves and ointments. The asphalt type crude oil produces a residue containing pitch or asphalt that is used in roofing material, protective coatings, paving, and asphalt tiles for floors.

PHYSICAL PROPERTIES

The lower molecular weight alkanes, methane through butane, are gases at ordinary temperatures and pressures. The C_5 to C_{17} alkanes (m.p. $< 20°C$) are liquids at room

TABLE 3-5 ALKANES

NAME	M.P. °C	B.P. °C	MOL. FORMULA	SP. GRAVITY (as liquids)
Methane	−183	−162	CH_4	—
Ethane	−172	−89	C_2H_6	—
Propane	−187	−42	C_3H_8	—
n-Butane	−135	−0.5	C_4H_{10}	—
n-Pentane	−130	+36	C_5H_{12}	0.626
n-Hexane	−94	69	C_6H_{14}	0.659
n-Heptane	−90	98	C_7H_{16}	0.683
n-Octane	−57	126	C_8H_{18}	0.703
n-Nonane	−54	151	C_9H_{20}	0.718
n-Decane	−30	174	$C_{10}H_{22}$	0.729
n-Undecane	−26	196	$C_{11}H_{24}$	0.740
n-Dodecane	−10	216	$C_{12}H_{26}$	0.749
n-Tridecane	−6	235	$C_{13}H_{28}$	0.757
n-Tetradecane	+6	251	$C_{14}H_{30}$	0.764
n-Pentadecane	10	268	$C_{15}H_{32}$	0.769
n-Hexadecane	18	280	$C_{16}H_{34}$	0.775
n-Heptadecane	22	303	$C_{17}H_{36}$	0.777
n-Octadecane	28	308	$C_{18}H_{38}$	0.777
n-Nonadecane	32	330	$C_{19}H_{40}$	0.778
n-Eicosane	36	343	$C_{20}H_{42}$	0.778

temperature, and compounds containing more than C_{17} are solids at room temperature. The alkanes in the C_{26} and C_{36} range make up the substance known as paraffin wax, and these alkanes are sometimes referred to as paraffin hydrocarbons. Table 3–5 summarizes some of the physical properties of the more common alkanes.

The boiling points of the alkanes show a regular increase of approximately 20 to 30°C with the introduction of each new methylene group. Branching of the carbon chain lowers the boiling point of the alkane, and, with compounds that can exist as isomers, the straight-chain isomer is always the highest boiling isomer. For example, there are three isomers having the molecular formula, C_5H_{12}, as illustrated below:

		B.P. °C
n-Pentane	$CH_3CH_2CH_2CH_2CH_3$	36
Isopentane	$CH_3CH(CH_3)CH_2CH_3$	28
Neopentane	$CH_3C(CH_3)_2CH_3$	9.5

As the amount of branching increases, the boiling point decreases.

The saturated hydrocarbons are almost completely insoluble in water, but are soluble in organic solvents.

PREPARATION OF ALKANES

Many of the alkanes are actually obtained from petroleum and natural gas, either directly or by fractional distillation of petroleum products. In many cases, however, a particular compound or isomer may be required that is not available or easily obtained commercially from petroleum. Then the compound must be synthesized in the laboratory by the organic chemist. For the preparation of most organic compounds there are generally several possible routes whereby one functional group may be converted to another. The method selected by the organic chemist takes into consideration the yield and purity of the reaction product, the cost of the reagents, the difficulty in carrying out the reaction, and the ease of isolating the product. These factors will vary from compound to compound, and a method of synthesis for one type of compound may not be the best method for another. Consequently, some knowledge of these factors is necessary in order to select the best possible method to prepare a specific compound. The following are several of the more common methods to prepare alkanes.

Reduction of Alkenes (Hydrogenation)

The carbon-carbon double bond present in alkenes can add a mole of hydrogen (H_2) in the presence of a catalyst to give an alkane as the final product.

$$\rangle C{=}C\langle \; + H_2 \xrightarrow{\text{catalyst}} -\underset{H}{\overset{|}{C}}-\underset{H}{\overset{|}{C}}-$$

The catalysts most commonly used are palladium (Pd), platinum (Pt), and nickel (Ni). For most small-scale laboratory work the reaction is generally carried out in the **liquid phase** at temperatures below 100°C and pressures of 1 to 10 atmospheres of hydrogen. The reaction is a general one and can be employed in the preparation of both straight chain and cyclic alkanes. Some specific examples are illustrated below:

$$\underset{\text{Propene}}{CH_3CH{=}CH_2} + H_2 \xrightarrow{\text{Pt}} \underset{\text{Propane}}{CH_3CH_2CH_3}$$

$$\text{Cyclohexene} + H_2 \xrightarrow{\text{Pd}} \text{Cyclohexane}$$

$$\text{Methylcyclopentene } (CH_3) + H_2 \xrightarrow{\text{Pt}} \text{Methylcyclopentane } (CH_3)$$

The addition of hydrogen (reduction) to biologically important compounds generally occurs enzymatically. For example, the important enzyme lactic dehydrogenase catalyzes the reduction of pyruvic acid to lactic acid (see p. 322).

Reduction of Carbonyl Compounds

Reduction of aldehydes and ketones can lead to two different types of products, depending on the reagent used. Catalytic reduction with hydrogen gives an alcohol, whereas reduction with zinc amalgam in hydrochloric acid gives an alkane. This latter reaction is known as a **Clemmensen reduction.**[*] Similar conversion of a carbonyl group

---

[*] Many synthetic organic reactions have been named after the discoverer of the reaction. Although memorization of the reaction name is unimportant in understanding the chemistry involved, many of these reactions are commonly known by these names, and the student will find it useful to become acquainted with some of them.

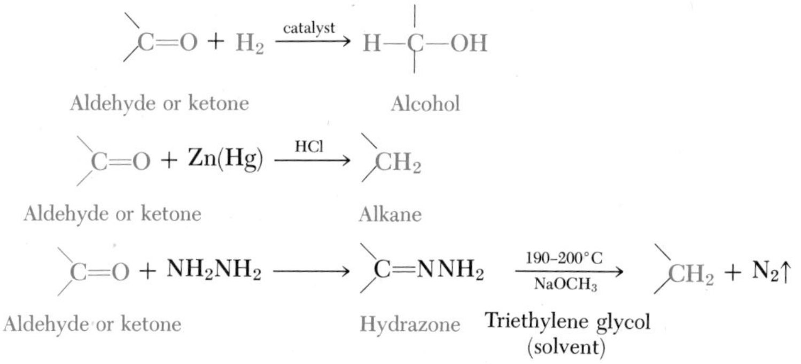

$$>C{=}O + H_2 \xrightarrow{\text{catalyst}} H{-}\overset{|}{\underset{|}{C}}{-}OH$$

Aldehyde or ketone — Alcohol

$$>C{=}O + Zn(Hg) \xrightarrow{HCl} >CH_2$$

Aldehyde or ketone — Alkane

$$>C{=}O + NH_2NH_2 \longrightarrow >C{=}NNH_2 \xrightarrow[NaOCH_3]{190\text{–}200^\circ C} >CH_2 + N_2\uparrow$$

Aldehyde or ketone — Hydrazone — Triethylene glycol (solvent)

to a methylene group under basic conditions can be accomplished by converting the carbonyl group to a hydrazone with hydrazine (NH_2NH_2) and heating the hydrazone with a base (sodium methoxide, $NaOCH_3$) at 200°C in triethylene glycol as a solvent. This type of reaction is known as a **Wolff-Kishner reduction.** The choice of using the Clemmensen or Wolff-Kishner method will depend on whether there are other groups in the molecule which are sensitive to either acid or base. Some examples of these reactions are illustrated below:

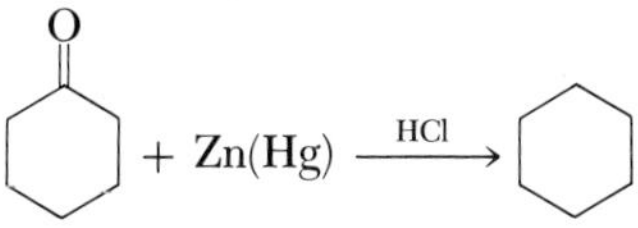

$$\text{Cyclohexanone} + Zn(Hg) \xrightarrow{HCl} \text{Cyclohexane}$$

Cyclohexanone — Cyclohexane

$$CH_3\overset{O}{\overset{\|}{C}}(CH_2)_5CH_3 + NH_2NH_2 \xrightarrow[200^\circ C]{NaOMe} CH_3(CH_2)_6CH_3$$

2-Octanone — Triethylene glycol — *n*-Octane

Electrolytic Decarboxylation of Carboxylic Acids. The decarboxylation of a solution of the acid salt may also be carried out electrolytically. This type of reaction is known as a **Kolbe reaction.** The reaction differs from other decarboxylation methods in that the main product obtained is not a hydrocarbon containing one less carbon atom than the salt, but a hydrocarbon containing two less carbon atoms than twice the number of carbons in the original salt, as illustrated for sodium acetate (the product is a dimer of the two alkyl groups in the salt):

$$2\,CH_3\overset{O}{\overset{\|}{C}}{-}O^-Na^+ + 2H_2O \xrightarrow[\text{current}]{\text{electric}} CH_3CH_3 + 2CO_2 + 2NaOH + H_2$$

Sodium acetate — Ethane

The mechanism of this reaction is believed to involve a loss of an electron by the carboxylate anion to give a free radical, which in turn loses carbon dioxide to give a methyl radical. Dimerization of two methyl radicals gives the product (ethane):

$$CH_3-\overset{\overset{O}{\|}}{C}-\ddot{\underset{..}{O}}:^- \xrightarrow[\text{anode}]{\text{at the}} 1e^- + CH_3-\overset{\overset{O}{\|}}{C}-\ddot{\underset{..}{O}}\cdot$$

Acetate anion — Free radical

$$H_3C\cdot\cdot\overset{\overset{O}{\|}}{C}-\ddot{O}: \longrightarrow O{=}C{=}O + \cdot CH_3$$

Methyl radical

$$2\cdot CH_3 \longrightarrow CH_3CH_3$$

Ethane

In biologic systems the coenzyme, thiamine pyrophosphate (TPP), functions as the catalyst in the oxidative decarboxylation of pyruvic acid to carbon dioxide and acetaldehyde (see p. 304). This type of biochemical decarboxylation prevents the accumulation of pyruvic acid in the blood and tissues.

CHEMICAL REACTIONS

The alkanes are the least reactive organic compounds. They are resistant to strong acids and alkalies and under ordinary conditions are resistant to oxidizing agents. For these reasons they are commonly employed as inert solvents in chemical reactions of the other functional groups. They do, however, react with halogens, nitric acid, and oxygen under special conditions.

Combustion

Hydrocarbons will react with oxygen when ignited in the presence of excess oxygen. The products of the combustion reaction are carbon dioxide and water. This reaction may be expressed as:

$$C_nH_{2n+2} + \frac{3n+1}{2}O_2 \rightarrow nCO_2 + (n+1)H_2O + \text{heat}$$

The heat given off in this reaction is called the heat of combustion. It is the utilization of this heat that accounts for much of the commercial use of hydrocarbons as heating fuels. Gasoline, which is composed of hydrocarbons containing 6 to 12 carbon atoms per molecule, burns similarly with oxygen in an automobile engine. The gases CO_2 and H_2O power the pistons of the engine, and the heat evolved is the heat carried away by the cooling system. Some specific examples are shown below:

$$CH_4 + 2O_2 \rightarrow CO_2 + 2H_2O + 213\ \text{kcal/mole}$$
$$C_2H_6 + 3\tfrac{1}{2}O_2 \rightarrow 2CO_2 + 3H_2O + 373\ \text{kcal/mole}$$

If combustion occurs in the absence of sufficient oxygen, carbon monoxide is formed. Carbon monoxide is the toxic gas found in the exhaust fumes of a car.

$$C_2H_6 + 2\tfrac{1}{2}O_2 \rightarrow 2CO + 3H_2O$$

The oxidation of organic compounds in biochemical systems occurs enzymatically to also give carbon dioxide, water, and cellular energy. Since the human cells could not adequately handle the large amounts of energy released in a combustion reaction, biologic oxidation is not completely analogous to combustion. It occurs in a stepwise sequence to release energy in small amounts useful to the cells. For example, the biologic oxidation of glucose (p. 325) releases large amounts of energy to be used in muscular work. Similarly, the enzymatic oxidation of fatty acids provides useful available energy for the human body (p. 336).

Vapor Phase Nitration

Other than combustion which degrades the alkane to smaller fragments, the most common reaction of alkanes is a substitution reaction in which a carbon-hydrogen bond is broken, and a new atom or group of atoms substitutes for the hydrogen. Under the proper reaction conditions, nitric acid will react with alkanes to replace a hydrogen of the alkane by a nitro (NO_2) group. These reactions require vigorous reaction conditions and are generally conducted in the vapor phase at temperatures of 400°C. Under these conditions the reaction is not entirely selective in only replacing a hydrogen by a nitro group. Some carbon-carbon bond breakage also occurs, and nitro derivatives containing less carbon atoms than the original alkane are also found. Some examples are given below:

$$\underset{\text{Ethane}}{CH_3CH_3} + \underset{\text{Nitric acid}}{HONO_2(HNO_3)} \xrightarrow{400^\circ C} \underset{\text{Nitroethane}}{CH_3CH_2NO_2}$$

$$\underset{\text{Propane}}{CH_3CH_2CH_3} + HONO_2 \xrightarrow{400^\circ C}$$

$$\underset{\text{1-Nitropropane}}{CH_3CH_2CH_2NO_2} + \underset{\text{2-Nitropropane}}{CH_3\underset{\displaystyle NO_2}{\underset{|}{C}}HCH_3} + \underset{\text{Nitroethane}}{CH_3CH_2NO_2} + \underset{\text{Nitromethane}}{CH_3NO_2}$$

Because of the carbon-carbon bond breakage reactions, the complexity of the product mixture increases considerably with the higher alkanes, and this process is only of commercial importance for the lower molecular weight alkanes. The simple nitroalkanes are useful as solvents for lacquers, as fuel additives, and as starting materials for the synthesis of other organic chemicals useful in drugs, insecticides, and explosives.

Halogenation

The halogens, fluorine, chlorine, and bromine, will react under the proper conditions with alkanes. Iodine does not react with alkanes. In contrast to chlorine and bromine which react at room temperature, fluorine reacts with alkanes almost explosively at room temperature. Fluorine undergoes a controlled reaction with alkanes only under very carefully controlled conditions, and because of its extreme reactivity it is generally not too useful for ordinary laboratory conditions.

Consequently, chlorine and bromine are the only practical halogens useful in the laboratory for reaction with alkanes. At higher temperatures or in the presence of sunlight, a chlorine or bromine atom can substitute for the hydrogen of an alkane. These reactions are known as **halogenation** reactions, or more specifically as **chlorination** or **bromination,** depending on whether a chlorine or a bromine atom substitutes for the hydrogen. The mechanism for this reaction is a free radical, as was outlined in Chapter 2 for the formation of methyl chloride from methane. Chlorine molecules absorb energy (either heat or light) and dissociate into chlorine atoms, which attack the alkane and remove a hydrogen atom and generate an alkyl free radical. The free radical then abstracts a chlorine atom from a chlorine molecule to form the halogenated product and a chlorine atom which can repeat the cycle to generate more halogenated alkane. The mechanism can be represented as follows:

$$:\ddot{Cl}-\ddot{Cl}: \xrightarrow[\text{light}]{\text{heat or}} 2:\ddot{Cl}\cdot$$

Chlorine molecule Chlorine atom

$$H:\underset{H}{\overset{H}{\ddot{C}}}:H + :\ddot{Cl}\cdot \longrightarrow H:\underset{H}{\overset{H}{\ddot{C}}}\cdot + H:\ddot{Cl}:$$

Methane Methyl radical

$$H:\underset{H}{\overset{H}{\ddot{C}}}\cdot + :\ddot{Cl}-\ddot{Cl}: \longrightarrow H:\underset{H}{\overset{H}{\ddot{C}}}:\ddot{Cl}: + :\ddot{Cl}\cdot$$

Methyl chloride

The overall reaction can be summarized as:

$$Cl_2 + CH_4 \xrightarrow[\text{sunlight}]{\text{heat or}} CH_3Cl + HCl$$

Only half of the halogen ends up in the organic product; the other half ends up as hydrogen chloride. Substitution reactions always give two products, whereas addition reactions (such as the hydrogenation of alkenes to give alkanes) give only one product.

Substitution reactions in many instances give rise to more than two products. In the case of the chlorination of methane to give methyl chloride, the product, methyl chloride, may react further (via the same type of free-radical mechanism) to give methylene chloride (CH_2Cl_2), which can react further to give chloroform ($CHCl_3$), which reacts further to give carbon tetrachloride (CCl_4). The reactions may be summarized as follows:

$$CH_3Cl + Cl_2 \rightarrow CH_2Cl_2 + HCl$$

Methyl chloride (chloromethane) Methylene chloride (dichloromethane)

$$CH_2Cl_2 + Cl_2 \rightarrow CHCl_3 + HCl$$

Chloroform (trichloromethane)

$$CHCl_3 + Cl_2 \rightarrow CCl_4 + HCl$$

Carbon tetrachloride (tetrachloromethane)

In actual practice, chlorination of methane gives all four products. The amounts of each depend on the amount of chlorine used and the reaction conditions.

Higher homologues increase the complexity of even the simple monohalogenation step. For example, in the chlorination of propane, two possible monochlorinated products are possible, depending upon which of the two different types of hydrogens chlorine substitutes for. In practice, both products are obtained. Further halogenation of these monochlorinated products produces an even more complex mixture. Because of this

$$Cl_2 + CH_3CH_2CH_3 \xrightarrow[\text{sunlight}]{\text{heat or}} CH_3CH_2CH_2Cl + CH_3CHClCH_3 + HCl$$

Propane *n*-Propyl chloride (1-chloropropane) Isopropyl chloride (2-chloropropane)

multiplicity of products, halogenation of alkanes is not a particularly useful laboratory reaction. However, in many commercial products where ultra pure products are not always required, the mixtures obtained are useful, and this type of reaction is commercially important for preparing halogenated alkanes used in such products as dry cleaning agents and fire extinguishers.

Pyrolysis (Cracking) of Alkanes

Although alkanes are generally some of the most stable organic compounds, they can be broken down (cracked into smaller fragments) by heating to high temperatures (400–700°C) in the absence of air (to avoid combustion). This is an important reaction in the petroleum industry for converting high molecular weight alkanes into fragments in the gasoline range, thereby increasing the amount of gasoline obtainable from crude petroleum. In the petroleum industry many catalysts have been developed which affect the "cracking" reaction at much lower temperatures than simple pyrolysis. A simple illustrative example of pyrolytic cracking is shown below:

$$CH_3CH_2CH_2CH_3 \xrightarrow{700°C} CH_2{=}CHCH_2CH_3 + CH_3CH{=}CHCH_3 + CH_3CH_3 + CH_2{=}CH_2 + CH_3CH{=}CH_2 + CH_4$$

Both carbon-carbon and carbon-hydrogen bonds are broken at these temperatures, and a complex mixture of products is obtained.

IMPORTANT TERMS AND CONCEPTS

addition reaction
alkane
Clemmensen reduction
combustion
cracking reaction
cycloalkane
halogenation
homologous series
hydrocarbon
hydrogenation
IUPAC rules
methylene group
petroleum
substitution reaction
Wolff-Kishner reduction

QUESTIONS

1. Draw out all the possible isomers for the compound having a molecular formula, C_6H_{14}. Give the correct IUPAC name to each of the compounds.

2. Which of the isomeric compounds in Question 1
   (a) contains a tertiary butyl group?
   (b) is isohexane?
   (c) contains an isopropyl group?
   (d) has the lowest boiling point?
   (e) has the highest boiling point?

3. Draw a correct structure for each of the following compounds:
   (a) 1,2-dimethylcyclohexane
   (b) 4-isopropyl-5-methyldecane
   (c) *n*-heptane
   (d) tertiary butyl bromide
   (e) 3-methyl-4-ethyl-5-isopropyloctane

4. Name the following compounds by the IUPAC system.
   (a) $(CH_3)_2CHC(CH_3)_2C(C_2H_5)_2$
   (b) $(CH_3)_2CHC(C_2H_5)_3$
   (c) $(CH_3)_3CCH_2CH(CH_3)CH_2CH(CH_3)_2$
   (d) cyclopropane ring with C_2H_5
   (e) cyclopentane ring with CH_3, CH_3 and C_2H_5

5. Explain what is wrong with *each* of the following names. Give the *correct* name for each compound.
   (a) 1,1-diethylbutane
   (b) 1,1,2-trimethylbutane
   (c) 4-isopropyl-5-methylhexane
   (d) 2-Dimethylbutane
   (e) 1,3,3-trimethylcyclobutane

6. An alkane was found to have a molecular weight of 58. On photochemical chlorination, two isomeric monochlorinated products were obtained. What is the structure of the original alkane? Name the substitution products by the IUPAC system. Write a reasonable mechanism to form the substitution products.

7. Write a chemical equation for each of the following chemical reactions. Indicate any necessary catalysts and reaction conditions. Name each of the organic products of the reaction.
   (a) complete combustion of propane
   (b) reduction of $CH_3\overset{O}{\overset{\|}{C}}CH_2CH(CH_3)_2$ by zinc amalgam and hydrochloric acid
   (c) electrolytic decarboxylation of sodium propionate ($CH_3CH_2CO_2Na$)

SUGGESTED READING

Conrad and Sabin: Motor Fuel Quality as Related to Refinery Processing and Antiknock Compounds. Journal of Chemical Education, Vol. 34, p. 262, 1957.

Evienx: The Geneva Congress on Organic Nomenclature, 1892. Journal of Chemical Education, Vol. 31, p. 326, 1954.

Ferguson: Ring Strain and Reactivity of Alicyclics. Journal of Chemical Education, Vol. 47, p. 46, 1970.

Hurd: The General Philosophy of Organic Nomenclature. Journal of Chemical Education, Vol. 38, p. 43, 1961.

Kimberlin: Chemistry in the Manufacture of Modern Gasoline. Journal of Chemical Education, Vol. 34, p. 569, 1957.

Larder: Historical Aspects of the Tetrahedron in Chemistry. Journal of Chemical Education, Vol. 44, p. 661, 1967.

Medeiros: Lead From Automobile Exhaust. Chemistry, Vol. 44, No. 10, p. 7, 1971.

Mills: Ubiquitous Hydrocarbons. Part I. Chemistry, Vol. 44, No. 2, p. 8, 1971. Part II. Chemistry, Vol. 44, No. 3, p. 12, 1971.

Nelson: The Origin of Petroleum. Journal of Chemical Education, Vol. 31, p. 399, 1954.

Nickerson: Antiknock Agents—Tetraethyl Lead—A Product of American Research. Journal of Chemical Education, Vol. 31, p. 560, 1954.

Orchin: Determining the Number of Isomers from a Structural Formula. Chemistry, Vol. 42, No. 5, p. 8, 1969.

Reti: Leonardo Da Vinci's Experiments on Combustion. Journal of Chemical Education, Vol. 29, p. 590, 1952.

Rossini: Hydrocarbons in Petroleum. Journal of Chemical Education, Vol. 37, p. 554, 1960.

Schmerling: The Mechanisms of the Reactions of Aliphatic Hydrocarbons. Journal of Chemical Education, Vol. 28, p. 562, 1951.

Shoemaker, d'Ouville, and Marschner: Recent Advances in Petroleum Refining. Journal of Chemical Education, Vol. 32, p. 30, 1955.

ALKENES AND ALKYNES

CHAPTER 4

The *objectives* of this chapter are to enable the student to:

1. Understand the concept of pi bonds in unsaturated hydrocarbons.
2. Define and explain geometrical isomerism.
3. Write structural formulas and IUPAC names of alkenes and alkynes.
4. Describe and understand the concept of elimination reactions.
5. Distinguish between primary, secondary, and tertiary alcohols.
6. Predict the direction of an elimination reaction.
7. Define the terms electrophilic reagent and nucleophilic reagent.
8. Describe and mechanistically understand Markownikoff's rule.
9. Describe the generalized mechanism of an electrophilic addition reaction.
10. Describe the concept of conjugation.
11. Understand conjugate addition reactions.
12. Describe the basic chemistry of alkynes.

Alkenes are distinguished from alkanes by the presence of the carbon-carbon double bond in the molecule. Other terms used to denote alkenes are **olefins** (from olefiant gas—the old name for ethylene) and **unsaturated hydrocarbons** (to denote the presence of a multiple bond in the molecules). The presence of the carbon-carbon double bond in the molecule makes these kinds of compounds highly reactive compared to alkanes.

In Chapter 2 the differences in bonding between alkanes and alkenes were briefly noted. In alkanes, which contain only carbon-carbon single bonds, essentially free rotation is allowed around the axis of the bond between the two carbon atoms. The orbitals used in the bonding of alkanes were sp^3 and gave a tetrahedral structure. However, in alkenes three sp^2 orbitals are used to form the carbon-hydrogen bonds and the carbon-carbon single bond. The use of sp^2 orbitals in bonding gives a planar arrangement of the carbon and hydrogen atoms involved* (see also Fig. 1–10.) The remaining carbon-carbon bond is formed by overlap of the $2p$ orbital remaining on each carbon atom as illustrated in Figure 4–1. Hence, the first carbon-carbon bond is formed by the overlap of two sp^2 orbitals (called a sigma [σ] bond) and the second carbon-carbon bond is formed by overlap of two $2p$ orbitals (called a pi [π] bond). The π-bond is much weaker than the σ-bond and is the first bond broken in the chemical reactions of most alkenes.

Resultant features of this multiple bonding in alkenes are: (1) The bond distance between the two carbon atoms of the multiple bond is 1.34 Å compared to 1.54 Å in an alkane. This is because two pairs of electrons pull the two nuclei closer together than

*The student is encouraged to use models to confirm this coplanar arrangement of the atoms and the restricted rotation in alkenes.

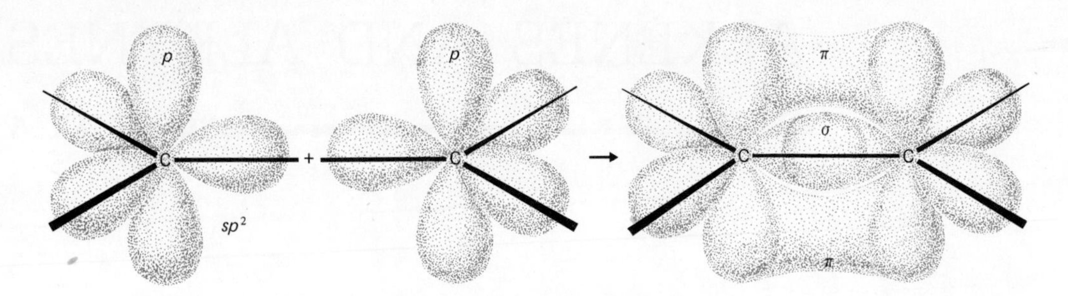

FIGURE 4-1 Bonding orbitals in a carbon-carbon double bond.

the one pair in alkanes; (2) Rotation about the carbon-carbon double bond is restricted. Although the atoms or groups of atoms attached to the double bond can vibrate or twist within small angles, no free rotation is allowed, since the π-bond would have to be broken to allow this rotation; (3) Because of restricted rotation around the multiple bond, a new type of isomerism is possible—namely, *geometrical isomerism* in which the two isomers differ only in the arrangement of the four atoms or groups of atoms that are attached to the multiple bond. This point will be discussed more fully later in this chapter.

NOMENCLATURE AND STEREOCHEMISTRY

Alkenes have the general formula C_nH_{2n}, and cycloalkenes have the general formula C_nH_{2n-2}. Two systems again are most commonly used to name these compounds. The correct nomenclature system is the IUPAC system. In this system the characteristic ending **-ene** is added to the stem name (formed by dropping the **-ane** ending from the hydrocarbon having the same number of carbon atoms); the carbon atoms in the *longest continuous chain containing the double bond* are numbered so that the carbon atoms of the double bond have the lowest possible numbers. Therefore the location of the double bond is indicated by the lowest possible number, and this number is placed in front of the name. Substituents are listed in front of the alkene name with the appropriate number to indicate their position. The other system of nomenclature is the use of common names for the lower members of this series. The common names are formed by dropping the **-ane** ending of the parent alkane and adding the ending **-ylene** to the remaining root stem. Some examples of these nomenclature systems are illustrated below with the common name in parentheses:

$H_2C{=}CH_2$ (drawn structural formula: two H on each carbon of C=C) or $H_2C{=}CH_2$

Ethene
(ethylene)

$H_3C{-}CH{=}CH_2$ (drawn structural formula) or $CH_3CH{=}CH_2$

Propene
(propylene)

$H_2C{=}CH{-}CH_2{-}CH_3$ (drawn structural formula) or $CH_2{=}CHCH_2CH_3$

1-Butene

$H_2C{=}C(CH_3){-}CH_3$ (drawn structural formula) or $CH_2{=}C(CH_3)CH_3$

2-Methylpropene
(isobutylene)

In olefins containing four carbon atoms or more, not only are isomers possible by varying the arrangement of the carbon atoms, as in 1-butene and isobutylene, but isomerism can also result from a shift in the position of the double bond without changing the carbon skeleton itself, as illustrated below for the pentenes:

$CH_2{=}CHCH_2CH_2CH_3$ 1-Pentene

$CH_3CH{=}CHCH_2CH_3$ 2-Pentene

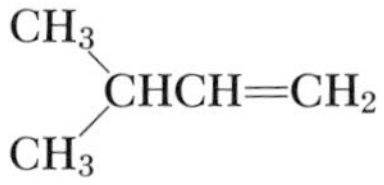

$(CH_3)_2CHCH{=}CH_2$ 3-Methyl-1-butene

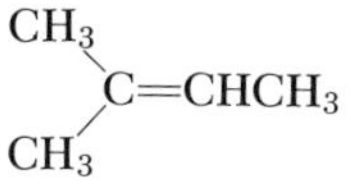

$(CH_3)_2C{=}CHCH_3$ 2-Methyl-2-butene

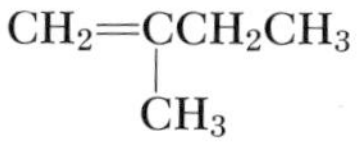

$CH_2{=}C(CH_3)CH_2CH_3$ 2-Methyl-1-butene

The only difference between 1-pentene and 2-pentene is the position of the double bond. Further examination of the pentene isomers has revealed that there are two different pentenes which have the structure $CH_3CH{=}CHCH_2CH_3$. The structure of these two pentenes has been shown to be:

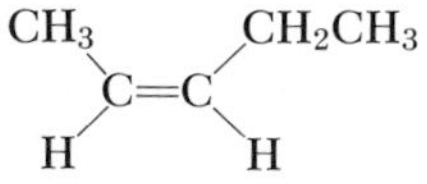

$CH_3(H)C{=}C(H)CH_2CH_3$ (CH₃ and CH₂CH₃ on the same side; H and H on the same side) and $CH_3(H)C{=}C(CH_2CH_3)H$ (CH₃ and CH₂CH₃ on opposite sides)

cis-2-Pentene (A) *trans*-2-Pentene (B)

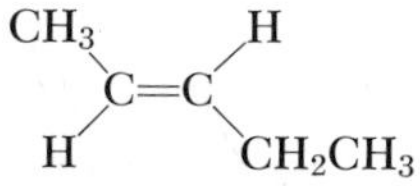

The only difference between these two isomers is that in one structure (A) the two hydrogens attached to the double bond are on the same side of the double bond (this is called a *cis* arrangement) and the two alkyl groups are also on the same (but opposite to the hydrogen) side of the double bond. In structure (B), the two hydrogens attached to the double bond are on opposite sides of the double bond (this is called a *trans*-arrangement), and the two alkyl groups are also *trans* across the double bond. Because of the restricted rotation around the double bond, these two compounds are fixed in respect to the arrangement of the atoms and alkyl groups around the double bond. Both the *cis*- and *trans*- compounds do exist, and both have been isolated and their structures unambiguously proved.

The necessary requirements for *geometrical isomerism* are: (1) that restricted rotation of some kind be present in the molecule. This may be either a double bond in olefins or the presence of a ring system which also prevents free rotation. Consequently, cyclic alkanes also exhibit geometric isomerism; (2) that neither carbon atom involved in the restricted rotation may hold identical groups. Some examples of geometric isomerism are shown on page 70.

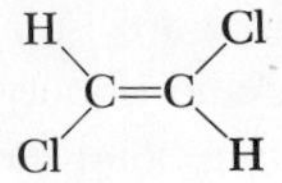

trans-1,2-Dichloroethene (H(Cl)C=C(H)Cl)

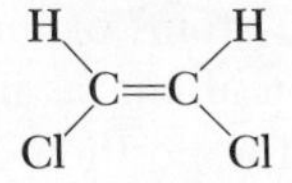

cis-1,2-Dichloroethene (H(Cl)C=C(Cl)H)

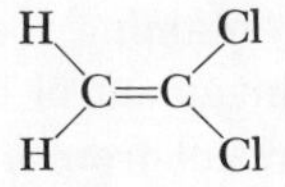

1,1-Dichloroethene (no geometric isomers) ($H_2C=CCl_2$)

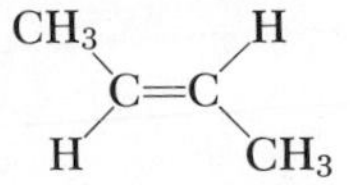

trans-2-Butene

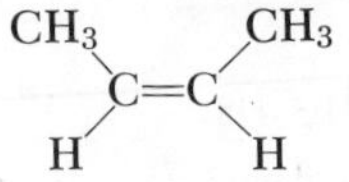

cis-2-Butene

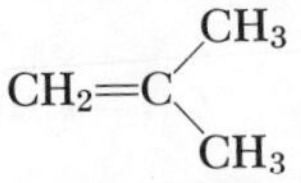

2-Methyl propene (no geometric isomers) ($CH_2=C(CH_3)_2$)

$CH_2=CHCH_2CH_3$

1-Butene (no geometric isomers)

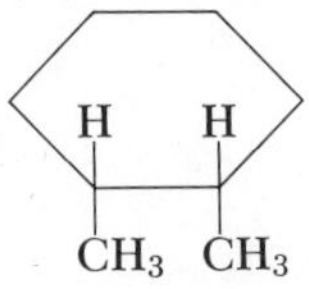

cis-1,2-Dimethylcyclohexane

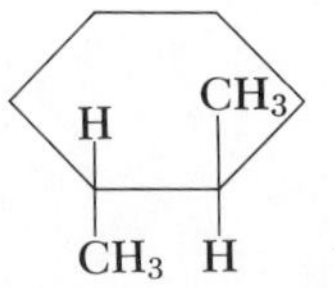

trans-1,2-Dimethylcyclohexane

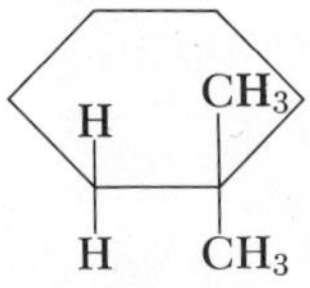

1,1-Dimethylcyclohexane (no geometric isomers)

Many naturally occurring biologic systems also exhibit geometrical isomerism. For example, a fatty acid, such as oleic acid [$CH_3(CH_2)_7CH=CH(CH_2)_7COOH$] is found in nature in the cis form. *Cis-* and *trans-* vitamin A (p. 239) are also important in the human visual cycle. Steroid systems (p. 234) are fused cyclic rings which also exhibit *cis-trans* isomerism with the *cis* configuration being the most common form.

In **cycloalkenes,** the possibility of *cis-trans* isomerism is limited by the constraints of the ring. Cycloalkenes of less than seven carbons in the ring exist only in the *cis*-arrangement. The corresponding *trans*-isomer in the smaller rings is too highly strained and is not capable of being isolated. Some common cycloalkenes are given below:

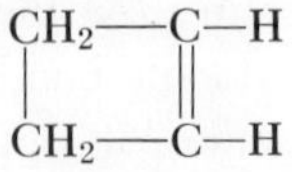

Cyclobutene

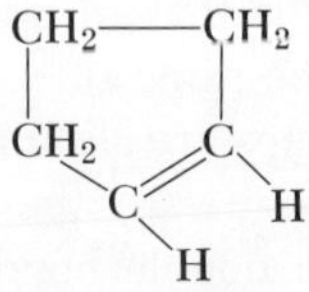

Cyclopentene

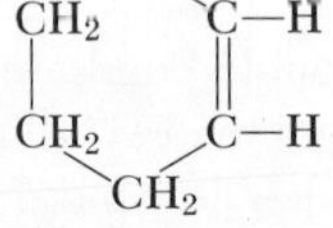

Cyclohexene

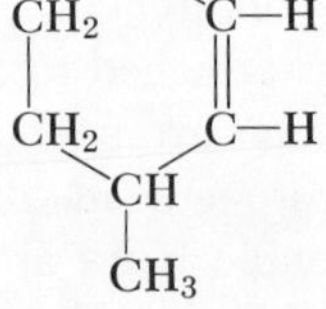

3-Methyl-1-cyclohexene

Cyclopropene, $H_2C<(CH=CH)$, is a known organic compound, but it is unstable at room temperature and undergoes rapid polymerization.

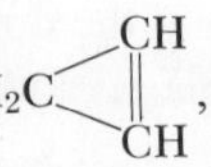

PHYSICAL PROPERTIES OF ALKENES

The alkenes are quite similar to the alkanes in physical properties. Alkenes containing 2 to 4 carbon atoms are gases; those containing 5 to 18 carbon atoms are liquids; and

TABLE 4-1 PHYSICAL PROPERTIES OF ALKENES AND DIENES

COMPOUND	STRUCTURE	M.P. °C	B.P. °C
Ethylene	$CH_2{=}CH_2$	−169	−104
Propylene	$CH_3CH{=}CH_2$	−185	−48
1-Butene	$CH_3CH_2CH{=}CH_2$	−185	−6
2-Butene (*cis*)	$CH_3CH{=}CHCH_3$	−139	+4
2-Butene (*trans*)	$CH_3CH{=}CHCH_3$	−106	+1
1-Pentene	$CH_3CH_2CH_2CH{=}CH_2$	−165	+30
1-Hexene	$CH_3(CH_2)_3CH{=}CH_2$	−140	+64
Cyclobutene	[ring structure]	—	+2
Cyclopentene	 [ring structure]	−135	+44
Cyclohexene	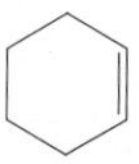 [ring structure]	−104	+83
Allene	$CH_2{=}C{=}CH_2$	−136	−35
1,3-Butadiene	$CH_2{=}CH{-}CH{=}CH_2$	−109	−4

those containing more than 18 carbons are solids. They are relatively insoluble in water but are soluble in concentrated sulfuric acid. The physical properties of some of the more common alkenes are summarized in Table 4-1.

PREPARATION OF ALKENES

The most general laboratory synthesis of alkenes involves a reaction known as an **elimination** reaction in which a molecule of water, hydrogen halide, or halogen is removed from adjacent carbon atoms in a saturated compound. The generalized reaction may be summarized as:

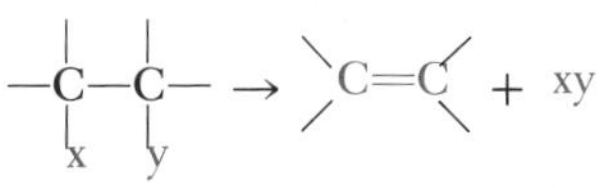

$$-\underset{x}{\overset{|}{\underset{|}{C}}}-\underset{y}{\overset{|}{\underset{|}{C}}}- \rightarrow \;>C{=}C<\; + \; xy$$

x = H or halogen
y = OH or halogen

In this process the group —x is removed in such a manner that it leaves its electron pair which then forms an additional bond (the double bond) between the carbon atoms involved.

Dehydration of Alcohols

When the molecule (xy) lost in an elimination reaction is water (HOH), the reaction is called a **dehydration reaction.** The generalized equation can then be represented as shown below:

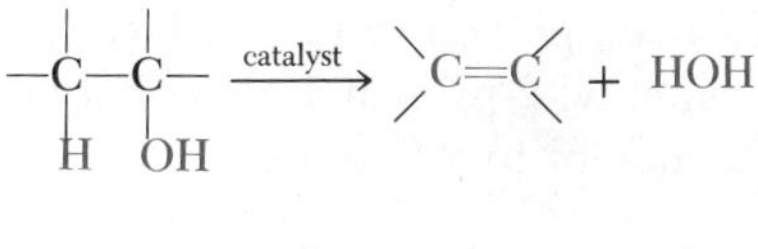

$$-\underset{H}{\overset{|}{\underset{|}{C}}}-\underset{OH}{\overset{|}{\underset{|}{C}}}- \xrightarrow{\text{catalyst}} \;>C{=}C<\; + \; HOH$$

The starting material which is dehydrated is an alcohol, and the alcohol can be either a straight chain or a cyclic alcohol. The catalysts employed for dehydration reactions are acidic, and all have a strong affinity for water. Acids such as concentrated sulfuric acid and phosphoric acid, or aluminum oxide and phosphorus pentoxide at high temperatures, are generally employed. For example, propene can be prepared from 1-propanol and cyclohexene can be prepared from cyclohexanol by this method. In some alcohols

$$\underset{\text{1-Propanol}}{CH_3CH_2CH_2OH} \xrightarrow{H_2SO_4} \underset{\text{Propene}}{CH_3CH{=}CH_2}$$

$$\underset{\text{Cyclohexanol}}{\begin{array}{ccccc} & & CH_2 & & \\ & CH_2 & & CH_2 & \\ & | & & | & H \\ & CH_2 & & C & \\ & & CH_2 & & OH \end{array}} \xrightarrow{H_2SO_4} \underset{\text{Cyclohexene}}{\begin{array}{ccccc} & & CH_2 & & \\ & CH_2 & & C{-}H & \\ & | & & \| & \\ & CH_2 & & C{-}H & \\ & & CH_2 & & \end{array}}$$

the elements of H_2O may be able to be lost in more than one way. For example, in 2-butanol either 1-butene or 2-butene may be formed by the elimination of water.

$$\begin{array}{c} H\;\;H\;\;H\;\;H \\ H{-}C{-}C{-}C{-}C{-}H \\ \boxed{H\;\;OH}\;H\;\;H \end{array} \xrightarrow{H_2SO_4} \underset{\text{1-Butene}}{\begin{array}{c} H \quad\; H\;\;H\;\;H \\ C{=}C{-}C{-}C{-}H \\ H \qquad\;\; H\;\;H \end{array}}$$

$$\begin{array}{c} H\;\;H\;\;H\;\;H \\ H{-}C{-}C{-}C{-}C{-}H \\ H\;\boxed{OH\;\;H}\;H \end{array} \xrightarrow{H_2SO_4} \underset{\text{2-Butene}}{\begin{array}{c} H\;\;H\;\;H\;\;H \\ H{-}C{-}C{=}C{-}C{-}H \\ H \qquad\quad H \end{array}}$$

In alcohols of this type, mixtures of both possible olefins are generally obtained. However, one olefin is usually formed as the major product, and in nearly all cases the predominant product is the *most highly substituted olefin*. In the previous example, 1-butene is a monosubstituted olefin (one hydrogen of ethylene has been substituted by an ethyl group), and 2-butene is a disubstituted olefin (two hydrogens of ethylene have been substituted by methyl groups). The ratio of 2-butene to 1-butene in this reaction is 4:1.

The ease of dehydration of alcohols depends upon the type of alcohol used. Alcohols are classified as primary, secondary, and tertiary by counting the number of alkyl groups bonded to the carbon atom bearing the —OH group, as follows:

$$\underset{\substack{\text{Primary alcohol}\\ \text{1-Butanol}}}{CH_3CH_2CH_2CH_2OH} \qquad \underset{\substack{\text{Secondary alcohol}\\ \text{2-Butanol}}}{\begin{array}{c} CH_3CH_2CHCH_3 \\ \quad\;\; | \\ \quad\;\; OH \end{array}} \qquad \underset{\substack{\text{Tertiary alcohol}\\ \text{2-Methyl-2-propanol}}}{\begin{array}{c} CH_3 \\ | \\ CH_3{-}C{-}OH \\ | \\ CH_3 \end{array}}$$

The ease of dehydration has been proved to be: tertiary > secondary > primary. Thus, it is easier to dehydrate 2-methyl-2-propanol to isobutylene than it is to dehydrate 1-butanol to 1-butene, or 2-butanol to 2-butene.

The following mechanism for the acid catalyzed dehydration of alcohols has been proposed:

$$CH_3CH_2\underset{\underset{\cdot\cdot}{:OH}}{\underset{|}{C}}HCH_3 + H_2SO_4 \rightleftharpoons CH_3CH_2\underset{\underset{\cdot\cdot}{^{+}OH_2}}{\underset{|}{C}}HCH_3 + HSO_4^-$$

Oxonium ion

$$CH_3CH_2\underset{\underset{\cdot\cdot}{^{+}OH_2}}{\underset{|}{C}}HCH_3 \rightleftharpoons CH_3CH_2\overset{+}{C}HCH_3 + H_2O$$

Carbonium ion

$$CH_3\underset{\underset{H}{|}}{C}H-\overset{+}{C}HCH_3 + HSO_4^- \rightarrow CH_3CH{=}CHCH_3 + H_2SO_4$$

$$CH_3CH_2\overset{+}{C}H-\underset{\underset{H}{|}}{C}H_2 + HSO_4^- \rightarrow CH_3CH_2CH{=}CH_2 + H_2SO_4$$

The first step involves protonation of the oxygen of the hydroxyl group to generate an oxonium ion, followed by loss of water to generate a carbonium ion. The carbonium ion can lose a proton (H^+) from either the adjacent $—CH_2$ group or $—CH_3$ group. The proton does not simply dissociate off but is lost to a base. In this case, the bisulfate ion can act as the base and abstract the proton. The electron pair left behind by the proton then forms the double bond of the olefin. In the process, the acid catalyst, H_2SO_4, is regenerated and can then repeat this cycle.

Dehydration (loss of H_2O) is not unique to the formation of a carbon-carbon double bond. Other functional groups, such as ether linkages, in the dehydration of sugars to form disaccharides, polysaccharides, glycogen, starch, and cellulose (see Chapter 13), or amide linkages, in the formation of protein molecules by the loss of water between amino acids (see p. 175), are important biochemical dehydrations that will be encountered later.

Dehydrohalogenation of Alkyl Halides

An elimination reaction, similar to the dehydration of alcohols, can occur with **alkyl halides** (compounds containing a halogen bonded to an alkyl group) in the presence of base to eliminate the elements of —HX (where X = halogen) and form an alkene. For example, if isopropyl chloride is treated with potassium hydroxide in alcohol, propene, potassium chloride, and water are formed as products. Similarly, cyclohexyl chloride yields cyclohexene as the organic product.

$$CH_3CHClCH_3 + KOH \xrightarrow{\text{alcohol}} CH_3CH{=}CH_2 + KCl + H_2O$$

Isopropyl chloride (2-chloropropane) → Propene

Cyclohexyl chloride (chlorocyclohexane) $+ KOH \xrightarrow{\text{alcohol}}$ Cyclohexene $+ KCl + H_2O$

The ease of removal of halogen in reactions of this type depends on the type of halogen eliminated and the structure of the alkyl halide. For example, tertiary butyl chloride undergoes elimination much easier than *n*-butyl chloride, and the ease of dehydrohalogenation is in the following order: tertiary alkyl halides > secondary > primary. Also, the ease of dehydrohalogenation for a particular type of carbon structure increases as the halogen is varied from fluorine through iodine. Consequently, it is easier to dehydrohalogenate isopropyl iodide than it is to dehydrohalogenate isopropyl chloride.

The generally accepted mechanism for dehydrohalogenation is as follows:

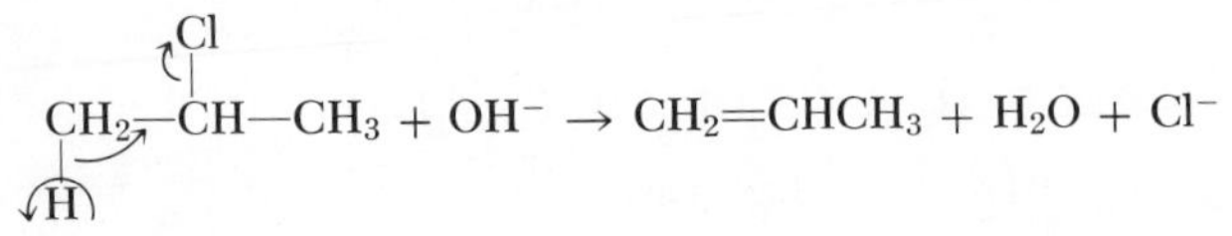

$$\underset{\underset{H}{|}}{CH_2}—\overset{\overset{Cl}{|}}{CH}—CH_3 + OH^- \rightarrow CH_2{=}CHCH_3 + H_2O + Cl^-$$

The base, OH^- in the example, abstracts a proton from the adjacent carbon atom to form water. The electron pair left behind by the proton being abstracted then forms the carbon-carbon double bond with the ejection of a chloride ion. The type of reaction is known as a β-elimination, since the proton is being abstracted from the carbon atom which is *beta* to the group being eliminated (Cl^-). In this example, it doesn't matter whether a proton is abstracted from either methyl group, since both groups are identical, and the same product would be formed in either case. However, in compounds which are not symmetrical, two different olefins can be produced. In 2-chlorobutane, both 1-butene and *cis*- and *trans*-2-butene can be formed as illustrated below:

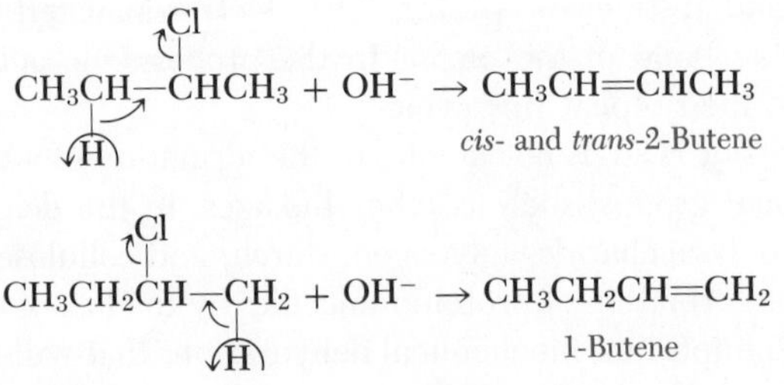

$$CH_3\underset{\underset{H}{|}}{CH}—\overset{\overset{Cl}{|}}{C}HCH_3 + OH^- \rightarrow CH_3CH{=}CHCH_3$$

cis- and *trans*-2-Butene

$$CH_3CH_2\overset{\overset{Cl}{|}}{CH}—\underset{\underset{H}{|}}{CH_2} + OH^- \rightarrow CH_3CH_2CH{=}CH_2$$

1-Butene

In compounds of this kind both products are generally formed, but the more highly substituted alkene (2-butene) is the predominant product. In this particular compound, 80% of the 2-butene and 20% of the 1-butene are produced.

REACTIONS OF ALKENES

In the previous chapter the reactions of alkanes were shown to be reactions of a substitution type in which the carbon-hydrogen bond is broken and a new bond is formed between carbon and a new atom or group of atoms. Two products are formed in reactions of this type; for example, methyl chloride and hydrogen chloride in the chlorination of methane form as follows:

$$\underset{\text{Methane}}{CH_4} + \underset{\text{Chlorine}}{Cl_2} \xrightarrow{h\nu} \underset{\text{Methyl chloride}}{CH_3Cl} + \underset{\text{Hydrogen chloride}}{HCl}$$

In contrast to this behavior of alkanes, *the most characteristic reaction of alkenes is addition to the double bond.* The π-bond of the olefin is broken and two new single bonds are formed in this process. *Only one product is formed in addition reactions of this type.* A generalized reaction scheme for an addition reaction can be represented by the following equation:

$$>C{=}C< + xy \rightarrow -\overset{|}{\underset{x}{\underset{|}{C}}}-\overset{|}{\underset{y}{\underset{|}{C}}}-$$

Addition reaction

The π-bond (the weaker of the two bonds) is broken and a new bond is formed between carbon and atom or group —x, and between carbon and atom or group —y, to yield the addition product, which now contains only single bonds between all atoms or groups of atoms. The molecule or compound xy can be one of many types of materials added to olefins. We shall only consider several simple types of compounds which have been added to olefins, but it should be kept in mind that this is a very general and widely used reaction by the organic chemist and one of the more important types of reactions in organic chemistry. In later chapters we shall again encounter addition reactions to other multiple bonds, such as $>C{=}O$, or $-C{\equiv}N$. The necessary requirement for an addition reaction is the presence of a multiple bond, and addition reactions are not necessarily limited to carbon-carbon multiple bonds.

Hydrogenation of Alkenes

In Chapter 3, the preparation of alkanes was shown to be possible by the catalytic hydrogenation of alkenes. This is a general reaction of alkenes and is important both

$$>C{=}C< + H_2 \xrightarrow{\text{catalyst}} -\overset{|}{\underset{H}{\underset{|}{C}}}-\overset{|}{\underset{H}{\underset{|}{C}}}-$$

in the laboratory and commercially in the preparation of many organic compounds. Dienes, which are discussed later in this chapter, also will undergo this reaction, but will add twice as much hydrogen. Hydrogenation of olefins can be carried out quantitatively, and has been employed also as a method to help elucidate the structure of more

$$CH_2{=}CH{-}CH{=}CH_2 + 2H_2 \xrightarrow{\text{catalyst}} CH_3CH_2CH_2CH_3$$

complex molecules. Quantitative hydrogenation of a compound containing carbon-carbon multiple bonds makes it possible to calculate how many carbon-carbon multiple centers are present in the molecule, and immediately the research chemist can tell if the compound is an alkene containing one, two, three, and so forth multiple bonds.

Catalytic hydrogenation is also of considerable commercial importance. Margarine and cooking shortenings are prepared by this type of reaction from vegetable oils. Vegetable oils are long chain unsaturated acids which contain one or more $>C{=}C<$ linkage. For example, oleic acid, $CH_3(CH_2)_7CH{=}CH(CH_2)_7COOH$, m.p. +16°C, linoleic acid, $CH_3(CH_2)_4CH{=}CH{-}CH_2CH{=}CH(CH_2)_7COOH$, m.p. −5°C, and linolenic acid, $CH_3CH_2CH{=}CHCH_2CH{=}CHCH_2CH{=}CH{-}(CH_2)_7COOH$, m.p. −11°C, are all converted to stearic acid, $CH_3(CH_2)_{16}COOH$, m.p. +71°C *via* the addition of one, two, or three moles of hydrogen, respectively. Note the increase in the melting point as the number of double bonds is decreased. Consequently, by controlling the extent of hydrogenation, a fat of the desired melting point can be obtained. Commercial fats such as "Crisco" and "Spry" contain approximately 20 to 25 per cent saturated fatty acids, 65 to 75 per cent oleic acid, and 5 to 10 per cent linoleic acid.

Addition of Halogen

Halogens behave similar to hydrogen and undergo 1,2-addition across the double bond. Generally, the halogen addition reaction is carried out in a solvent such as carbon tetrachloride (CCl_4) or glacial acetic acid (CH_3COOH). In practice, chlorine and bromine are the halogens generally employed. Fluorine can be added to olefins under special conditions but is too reactive for general laboratory use. Iodine adds reversibly to olefins, and most 1,2-vicinal iodides are unstable. Some characteristic halogen addition reactions are illustrated below.

$$\underset{\text{Propene}}{CH_2{=}CHCH_3} + Cl_2 \rightarrow \underset{\text{1,2-Dichloropropane}}{CH_2ClCHClCH_3}$$

$$\underset{\substack{\text{Isobutylene}\\ \text{(2-methylpropene)}}}{CH_3{-}\underset{\substack{|\\ CH_3}}{C}{=}CH_2} + Br_2 \rightarrow \underset{\text{1,2-Dibromo-2-methylpropane}}{CH_3{-}\overset{\substack{Br\\ |}}{\underset{\substack{|\\ CH_3}}{C}}{-}CH_2Br}$$

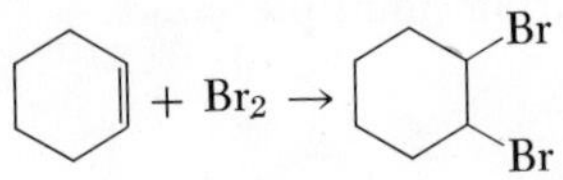

Cyclohexene $+ Br_2 \rightarrow$ 1,2-Dibromocyclohexane

The accepted mechanism for halogen addition is an ionic mechanism, as illustrated below for a bromine addition to propene:

$$:\ddot{Br}:\ddot{Br}: \rightleftharpoons :\ddot{Br}^+ + :\ddot{Br}:^-$$

$$:\ddot{Br}^+ + CH_2{=}CHCH_3 \rightarrow \underset{\text{I}}{BrCH_2\overset{+}{C}HCH_3} \quad or \quad \underset{\text{II}}{\overset{+}{C}H_2CHBrCH_3}$$

$$BrCH_2\overset{+}{C}HCH_3 \xrightarrow{Br^-} BrCH_2CHBrCH_3$$

or

$$BrCH_2\overset{+}{C}HCH_3 + Br_2 \rightarrow BrCH_2CHBrCH_3 + Br^+$$

The first step is proposed as the dissociation* of a bromine molecule into a bromonium ion (Br^+) and a bromide (Br^-) ion. The bromonium ion attacks the π-bond (the π-bond is a center of high electron density and will be attacked by electron-deficient species) to generate a carbonium ion, either I or II. Carbonium I is a secondary carbonium and carbonium II is a primary carbonium ion. The order of stability of carbonium ions is: tertiary > secondary > primary. Therefore, carbonium ion I is the intermediate preferentially formed. This carbonium ion (I) then either picks up bromide ion to produce the 1,2-addition product, or abstracts a bromide ion from another bromine molecule and regenerates a new bromonium ion which can repeat the cycle. A similar mechanism employing chloronium ions (Cl^+) can be written for the addition of chlorine to an olefin.

*The process can be viewed as merely an instantaneous unequal sharing of the electrons in a bromine molecule rather than a dissociation (which is probably an oversimplification), but for the sake of illustration the actual formation of Br^+ and Br^- will be used.

The addition of bromine to a $>C=C<$ bond is the basis for a simple chemical test for this functional group. Since a bromine solution in carbon tetrachloride is red, disappearance of this red color upon addition of an unknown organic compound is a strong indication that the unknown compound contains the $>C=C<$ linkage.

The addition of iodine to a $>C=C<$ linkage is also the basis of a test to determine the content of unsaturated fatty acids in lipids (p. 226). Since the amount of unsaturation of a fat or oil is proportional to the number of $>C=C<$ linkages in the fatty acid, the amount of iodine absorbed by a lipid can be employed to determine the degree of unsaturation. The degree of unsaturation is called the *iodine number*, and is defined as the grams of iodine absorbed by 100 grams of fat or oil. For example, the iodine number for oleic acid can be calculated as shown below.

$$\underset{\text{oleic acid}}{CH_3(CH_2)_7CH{=}CH(CH_2)_7COOH} + I_2 \rightarrow CH_3(CH_2)_7CHICHI(CH_2)_7COOH$$

$$\text{Iodine Number} = \frac{\text{Mol. Wt. } (I_2) \times \text{no. of C=C linkages} \times 100}{\text{Mol. Wt. (fatty acid)}}$$

$$= \frac{(254)\,(1)\,(100)}{282.5} = 90$$

For linoleic acid, the iodine number is 181. In general, a high iodine number indicates a high degree of unsaturation.

In practice, a known amount of the unsaturated fatty acid sample is treated with an *excess* of a standard iodine solution. After the addition reaction is over, the excess of iodine is determined by titration. The difference between the iodine added and the iodine remaining (determined by the titration) gives the number of moles of iodine consumed and consequently gives the number of double bonds in the olefinic acid.

Addition of Acids

Acids such as sulfuric acid and the hydrogen halides (HF, HCl, HBr, HI) can be added across a carbon-carbon double bond to give either alkyl hydrogen sulfates or alkyl halides, as illustrated below for ethylene:

$$\underset{\text{Ethylene}}{CH_2{=}CH_2} + \underset{\text{Sulfuric acid}}{H_2SO_4(HOSO_2OH)} \rightarrow \underset{\text{Ethyl hydrogen sulfate}}{CH_3CH_2OSO_2OH}$$

$$CH_2{=}CH_2 + HBr \rightarrow CH_3CH_2Br$$

A mechanism similar to the addition of halogens can be written in which the proton (H^+) of the acid adds to the double bond to give a carbonium ion which then picks up an anion (either OSO_2OH^-, F^-, Cl^-, Br^-, or I^-) to give the addition product. In ethylene, addition of the proton to either carbon gives the ethyl carbonium ion ($CH_3CH_2^+$), and

$$CH_2{=}CH_2 + HOSO_2OH \rightleftharpoons CH_3\overset{+}{C}H_2 + \overset{-}{O}SO_2OH \rightarrow CH_3CH_2OSO_2OH$$

only one product can result. However, in unsymmetrical olefins, the proton could add to give two different carbonium ions, and two possible products could result. In actuality, the proton will add to give the *more stable carbonium ion*, and the product resulting from the more stable carbonium ion will be the product produced. For example, in the addition of hydrogen iodide to propene, either isopropyl iodide or *n*-propyl iodide could be produced. The product actually formed is isopropyl iodide formed from the

$$CH_3CH{=}CH_2 + HI \rightleftharpoons CH_3\overset{+}{C}HCH_3 \quad \text{or} \quad CH_3CH_2\overset{+}{C}H_2$$

Propene — Secondary — Primary

$$CH_3\overset{+}{C}HCH_3 + I^- \rightarrow CH_3CHICH_3 \quad \text{Isopropyl iodide}$$

$$CH_3CH_2\overset{+}{C}H_2 + I^- \rightarrow CH_3CH_2CH_2I \quad n\text{-Propyl iodide}$$

more stable secondary carbonium ion. An empirical rule, called **Markownikoff's rule,** generalizes the addition of unsymmetrical reagents to unsymmetrical olefins as follows: *When an unsymmetrical reagent adds to an unsymmetrical olefin, the positive part of the unsymmetrical reagent becomes attached to the carbon atom of the double bond which bears the greatest number of hydrogen atoms.* This rule also predicts isopropyl iodide as the product in the above reaction.

Other unsymmetrical reagents, such as hypohalous acid (HOCl, HOBr, HOI), add similarly, as shown below for 1-butene:

$$CH_2{=}CHCH_2CH_3 + HOCl \rightarrow CH_2ClCHOHCH_2CH_3$$

1-Chloro-2-butanol

From the product formed, it is obvious that the hypohalous acids add to olefins as HO^-Cl^+ and not as H^+OCl^-. Compounds containing a vicinal halogen and hydroxyl groups are called **halohydrins;** hence, the addition product of HOCl to 1-butene is known as a chlorohydrin.

The mechanism of ionic addition to a carbon-carbon double bond has been shown to be similar for the addition of halogen, sulfuric acid, hydrogen halides, and hypohalous acids. However, the attacking positive species was different in several of these reactions and involved either a proton (H^+) or a chloronium (Cl^+), bromonium (Br^+), or iodonium ion (I^+). A general name to include all species of this type is the term **electrophilic reagent.** An electrophilic reagent is defined as an electron-seeking reagent. Therefore, it either is positively charged or is electron deficient. Relating back to acidity, an electrophilic reagent is a Bronsted or Lewis acid. Some common examples of electrophilic reagents are: acids (hydrogen halides, sulfuric acid), Ag^+, SO_2, and oxidizing agents (MnO_4^-, $Cr_2O_7^=$). The term used to define the opposite of an electrophilic reagent is **nucleophilic reagent.** A nucleophilic reagent can be defined as a nucleus-seeking (positive center seeking) reagent. It can also be defined as a Bronsted or Lewis base. Some common examples of nucleophilic reagents are $H_2O\!:$, $\ddot{N}H_3$, $\bar{O}H$, $\bar{C}N$, and reducing agents such as Zn and Na.

Many common organic mechanisms can be viewed as reactions between electrophilic and nucleophilic reagents.* For example, in the addition of hydrogen chloride to propene, the hydrogen chloride is an electrophilic reagent and the olefin is a nucleophilic reagent. Since the emphasis in the mechanistic interpretation of these addition reactions of acids is on the addition of the proton (the electrophilic reagent), *the characteristic reaction of olefins may be defined even more explicitly as electrophilic addition reactions.* These terms, electrophilic and nucleophilic, are useful in describing mechanistic features of chemical reactions and will be used as needed throughout the organic section of the book.

*The inorganic chemistry parallel of electrophilic and nucleophilic reactions would be an oxidation-reduction reaction.

DIENES

Dienes are organic compounds containing two double bonds. The characteristic molecular formula which describes dienes is C_nH_{2n-2}. The nomenclature of these compounds is similar to that of alkenes, except that the term **-diene** is added to the parent hydrocarbon prefix instead of **-ene** with simple alkenes. In addition, the position of the two double bonds must be specified using the lowest possible numbers.

Two double bonds may be incorporated into an organic compound in three possible ways, as illustrated for pentadiene:

$CH_2{=}CH{-}CH{=}CHCH_3$	$CH_2{=}C{=}CHCH_2CH_3$	$CH_2{=}CH{-}CH_2{-}CH{=}CH_2$
1,3-Pentadiene	1,2-Pentadiene	1,4-Pentadiene

In 1,4-pentadiene, the two double bonds behave independently. The second double bond has no influence on the first double bond, and the compound behaves like a simple alkene except that it requires twice as much reagent to saturate the molecule. Consequently, it will add two moles of hydrogen, two moles of hydrogen halide, two moles of halogen, and so forth.

$$\underset{\text{1,4-Pentadiene}}{CH_2{=}CH{-}CH_2{-}CH{=}CH_2} + 2H_2 \xrightarrow{\text{Pt}} \underset{n\text{-Pentane}}{CH_3CH_2CH_2CH_2CH_3}$$

$$\underset{\text{1,4-Pentadiene}}{CH_2{=}CH{-}CH_2CH{=}CH_2} + 2Br_2 \longrightarrow \underset{\text{1,2,4,5-Tetrabromopentane}}{BrCH_2CHBrCH_2CHBrCH_2Br}$$

$$\underset{\text{1,4-Pentadiene}}{CH_2{=}CHCH_2CH{=}CH_2} + 2HCl \longrightarrow \underset{\text{2,4-Dichloropentane}}{CH_3CHClCH_2CHClCH_3}$$

Compounds containing a 1,2-diene system are known as **allenes.** Most of their chemistry is similar to that of compounds containing only one double bond; but again they add two moles of reagent per allene molecule. They are generally more reactive than simple olefins and do undergo some reactions that are the result of the molecule containing the allenic double bond. However, most of these reactions are quite complex and beyond the scope of this treatment.

Compounds containing a 1,3-diene system are known as **conjugated dienes. Conjugated systems** are compounds containing *alternating double and single bonds*. Therefore, 1,3-butadiene; 1,3-pentadiene; 1,3,5-hexatriene; 1,3-cyclohexadiene; and cyclopentadiene

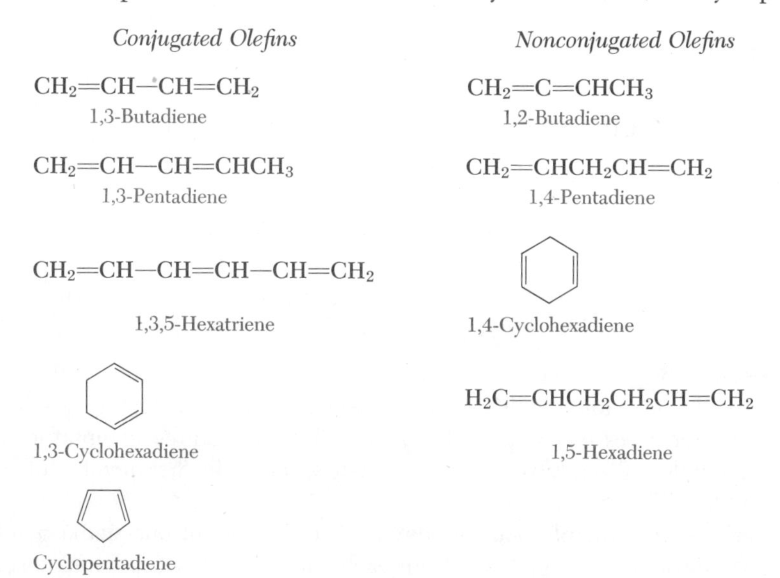

are all examples of a conjugated olefin, but 1,2-butadiene; 1,4-pentadiene; 1,4-cyclohexadiene; and 1,5-hexadiene are not conjugated systems.

Most diene systems can be prepared by methods similar to those employed for introducing a double bond into a simple olefin, except that the starting material must be polyfunctional (contain two or more functional groups). Some examples of the preparation of dienes are given below to illustrate these preparative methods:

$$\underset{\text{1,5-Dibromopentane}}{CH_2BrCH_2CH_2CH_2CH_2Br} + 2KOH \xrightarrow{\text{alcohol}} \underset{\text{1,4-Pentadiene}}{CH_2{=}CHCH_2CH{=}CH_2}$$

$$\underset{\text{1,4-Butanediol}}{HOCH_2CH_2CH_2CH_2OH} + Al_2O_3 \xrightarrow{\Delta} \underset{\text{1,3-Butadiene}}{CH_2{=}CH{-}CH{=}CH_2}$$

Conjugated dienes differ from other diene systems, not in the types of reagents that they react with, but in the mode of addition of these reagents. Whereas most dienes undergo 1,2-addition of electrophilic reagents similar to ethylene, conjugated dienes undergo 1,4-addition of electrophilic reagents. For example, if the amount of bromine is controlled, 1,3-butadiene will add only one mole of bromine to give a dibromide. Two dibromides are possible for this addition product. If 1,2-addition occurs, 3,4-dibromo-1-butene is the expected product. However, the product is actually found to be the

$$\underset{\text{1,3-Butadiene}}{CH_2{=}CH{-}CH{=}CH_2} + Br_2\text{ (1 mole)} \xrightarrow[\text{addition}]{\text{1,2-}} \underset{\text{3,4-Dibromo-1-butene}}{CH_2{=}CHCHBrCH_2Br}$$

$$\underset{\text{1,3-Butadiene}}{CH_2{=}CH{-}CH{=}CH_2} + Br_2\text{ (1 mole)} \xrightarrow[\text{addition}]{\text{1,4-}} \underset{\text{1,4-Dibromo-2-butene}}{CH_2BrCH{=}CHCH_2Br}$$

1,4-addition product, 1,4-dibromo-2-butene. In addition to the bromine adding in a 1,4-manner, the double bond is shifted to the 2,3- position in the carbon chain. This type of conjugative addition is common for conjugated dienes, and differentiates this kind of diene from other dienes. Some other examples of conjugative additions are given below:

$$\underset{\text{2,4-Hexadiene}}{CH_3CH{=}CH{-}CH{=}CHCH_3} + HCl \xrightarrow[\text{addition}]{\text{1,4-}} \underset{\text{2-Chloro-3-hexene}}{CH_3CH_2CH{=}CHCHClCH_3}$$

$$\underset{\text{1,3-Butadiene}}{CH_2{=}CH{-}CH{=}CH_2} + HBr \xrightarrow[\text{addition}]{\text{1,4-}} \underset{\text{1-Bromo-2-butene}}{CH_3CH{=}CHCH_2Br}$$

ALKYNES

Alkynes are organic compounds that contain a carbon-carbon triple bond. The general molecular formula which describes these compounds is C_nH_{2n-2}—the same as dienes, since both types of compounds contain the same degree of unsaturation. This class of compounds is also referred to as "acetylenes," after the first member of this series, acetylene itself, HC≡CH.

The carbon-carbon triple bond in alkynes is composed of one sigma bond formed by using an *sp*-orbital (see Fig. 1–12) from each carbon, and two π-bonds formed from

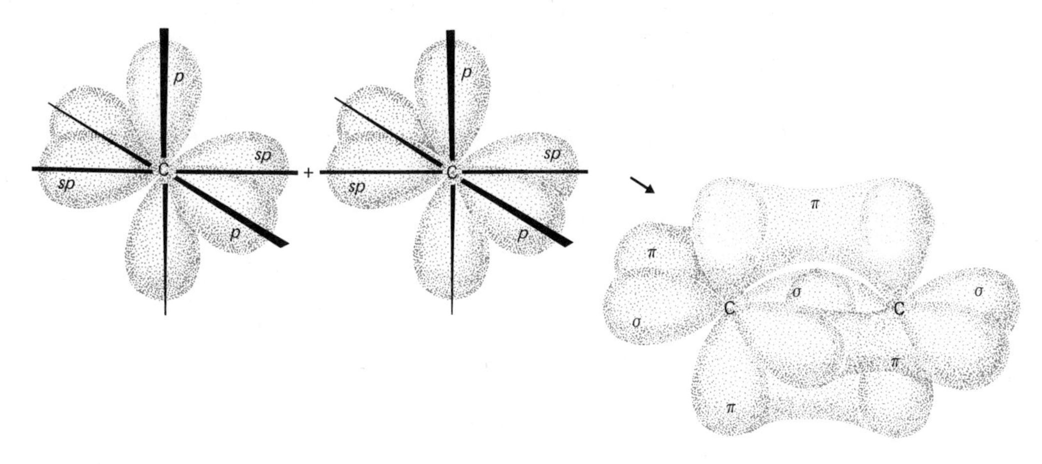

FIGURE 4-2 Schematic representation of bonding orbitals in a carbon–carbon triple bond.

the remaining *p*-orbitals on the respective carbon atoms. The two π-bonds are perpendicular to one another and enclose the carbon-carbon sigma bond in a cylinder of electron density (Fig. 4–2).

The alkynes are similar to alkanes and alkenes in physical properties. They are insoluble in water and soluble in organic solvents. Their boiling points are similar to those of the corresponding alkanes and alkenes, and the physical properties of some of the more common alkynes are listed in Table 4–2. Since the alkynes involve bonding using *sp*-orbitals, they are linear molecules. Consequently, *cis-trans* isomers are not possible for compounds such as 2-butyne ($CH_3C{\equiv}CCH_3$) as was found for the 2-butenes.

NOMENCLATURE

In the IUPAC nomenclature system the ending **-yne** is added to stem name, and the position of the triple bond is given the lowest number. Other substituents are named as before. Some of the lower homologues of acetylene are also named as derivatives of acetylene in which the —C≡C— group is named as acetylene, and the substituents attached to this group are denoted and attached to the acetylene name. Some typical examples are cited on page 82.

TABLE 4-2 PHYSICAL PROPERTIES OF ALKYNES

COMPOUND	STRUCTURE	M.P. °C	B.P. °C
Acetylene	$HC{\equiv}CH$	−82	−84 (subl.)
Propyne	$CH_3C{\equiv}CH$	−103	−23
1-Butyne	$CH_3CH_2C{\equiv}CH$	−126	+8
2-Butyne	$CH_3C{\equiv}CCH_3$	−32	+27
1-Pentyne	$CH_3CH_2CH_2C{\equiv}CH$	−106	+40
2-Pentyne	$CH_3C{\equiv}CCH_2CH_3$	−109	+56
1-Hexyne	$CH_3CH_2CH_2CH_2C{\equiv}CH$	−132	+71
2-Hexyne	$CH_3C{\equiv}CCH_2CH_2CH_3$	—	+84

Compound	*IUPAC Name*	*Common Name*
$HC{\equiv}CH$	Ethyne	Acetylene
$CH_3C{\equiv}CH$	Propyne	Methylacetylene
$CH_3CH_2C{\equiv}CH$	1-Butyne	Ethylacetylene
$CH_3C{\equiv}CCH_3$	2-Butyne	Dimethylacetylene
$CH_3CH(CH_3)C{\equiv}CH$	3-Methyl-1-butyne	Isopropylacetylene
$CH_3C(CH_3)_2{-}C{\equiv}CH$	3,3-Dimethyl-1-butyne	*t*-Butylacetylene

METHODS OF PREPARATION

Acetylene is prepared industrially by the hydrolysis of calcium carbide, which is prepared from lime and coke. This two-step process is outlined below:

$$\underset{\text{Coke}}{3C} + \underset{\text{Lime}}{CaO} \xrightarrow[\text{temps.}]{\text{high}} \underset{\text{Calcium carbide}}{CaC_2} + \underset{\text{Carbon monoxide}}{CO}$$

$$\underset{\text{Calcium carbide}}{CaC_2 + 2H_2O} \longrightarrow \underset{\text{Acetylene}}{HC{\equiv}CH} + \underset{\text{Calcium hydroxide}}{Ca(OH)_2}$$

Laboratory preparations of acetylenes are similar to those used for the preparation of alkenes and dienes.

Dehydrohalogenation

$$\underset{\text{2,3-Dichlorobutane}}{CH_3CHClCHClCH_3} + 2KOH \xrightarrow{\text{alcohol}} \underset{\text{2-Butyne}}{CH_3C{\equiv}CCH_3} + 2KCl + 2H_2O$$

The vicinal dihalides required in the previous reaction are conveniently prepared via the addition of halogens to alkenes. Thus, the synthesis of propyne may be envisioned as a two-step process starting from propene:

$$\underset{\text{Propene}}{CH_3CH{=}CH_2} + Br_2 \longrightarrow \underset{\text{1,2-Dibromopropane}}{CH_3CHBrCH_2Br}$$

$$\underset{\text{1,2-Dibromopropane}}{CH_3CHBrCH_2Br} + 2KOH \xrightarrow{\text{alcohol}} \underset{\text{Propyne}}{CH_3C{\equiv}CH}$$

Via Other Acetylenes

The carbon-hydrogen bonds in alkanes and alkenes are generally very stable to basic reagents. In contrast to this behavior, terminal acetylenes (those containing at least one hydrogen attached to the carbon-carbon triple bond) exhibit an acidic character. The hydrogen attached to the triple bond can be replaced when the terminal acetylene is treated with a strong base, such as metallic sodium. The acetylene is converted into an acetylide salt. Acetylides of this type react with alkyl halides to give a new acetylene as shown below:

$$\underset{\text{Acetylene}}{HC{\equiv}CH} + Na\ (1\ \text{mole}) \xrightarrow{\text{liq. } NH_3} \underset{\text{Sodium acetylide}}{HC{\equiv}C^-Na^+} + \tfrac{1}{2}H_2$$

$$\underset{\text{Propyne}}{CH_3C{\equiv}CH} + Na \xrightarrow{\text{liq. } NH_3} CH_3C{\equiv}C^-Na^+ + \tfrac{1}{2}H_2$$

$$CH_3C{\equiv}C^-Na^+ + \underset{\text{Ethyl bromide}}{CH_3CH_2Br} \longrightarrow \underset{\text{2-Pentyne}}{CH_3C{\equiv}CCH_2CH_3} + Na^+Br^-$$

The halide ion is displaced by the acetylide anion to produce the alkyne. For best results, the alkyl halide must be a primary alkyl halide. Secondary and tertiary halides undergo side reactions with the acetylide anion, and low yields of the alkyne are produced. This method is, however, quite versatile for preparing many homologues of acetylene.

REACTIONS OF ALKYNES

Since acetylenes contain two π-bonds, it might be expected that the characteristic reaction of acetylenes would also be an addition reaction. In fact, this prediction is realized experimentally, and alkynes undergo electrophilic addition reactions like olefins, except that two moles of reagent are required to saturate the two π-bonds. In many cases, addition reactions to acetylenes can be controlled to give an olefin derivative, which is formed via addition of one mole of reagent to a triple bond. Some of the more characteristic reactions of alkynes are summarized in the following sections.

Hydrogenation

Exhaustive catalytic (Pt, Pd, Ni) hydrogenation of alkynes gives alkanes as the final product. Partial hydrogenation gives an alkene. In compounds which can exhibit *cis-trans* isomerism, the *cis*-isomer is the predominant isomer formed on partial catalytic hydrogenation of an acetylene.

$$\underset{\text{Acetylene}}{HC{\equiv}CH} + 2H_2 \xrightarrow{Pd} \underset{\text{Ethane}}{CH_3CH_3}$$

$$\underset{\text{Propyne}}{CH_3C{\equiv}CH} + H_2 \xrightarrow{Ni} \underset{\text{Propene}}{CH_3CH{=}CH_2} \xrightarrow[Ni]{H_2} \underset{\text{Propane}}{CH_3CH_2CH_3}$$

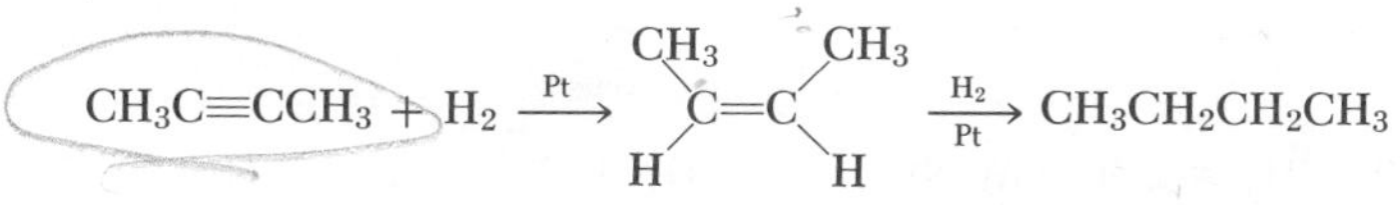

$$CH_3C{\equiv}CCH_3 + H_2 \xrightarrow{Pt} \underset{cis\text{-2-Butene}}{\overset{CH_3\quad CH_3}{\underset{H\qquad H}{C{=}C}}} \xrightarrow[Pt]{H_2} \underset{n\text{-Butane}}{CH_3CH_2CH_2CH_3}$$

Halogenation

Similarly to alkenes, chlorine and bromine add easily to a triple bond. Fluorine is generally too vigorous, and iodine generally does not form stable addition products. Again, the reaction may be carried out stepwise, as follows:

$$\underset{\text{Acetylene}}{HC{\equiv}CH} + Cl_2 \longrightarrow \underset{\text{1,2-Dichloroethylene}}{ClCH{=}CHCl} \xrightarrow{Cl_2} \underset{\text{1,1,2,2-Tetrachloroethane}}{CHCl_2CHCl_2}$$

$$\underset{\text{2-Pentyne}}{CH_3CH_2C{\equiv}CCH_3} + 2Cl_2 \longrightarrow \underset{\text{2,2,3,3-Tetrachloropentane}}{CH_3CH_2CCl_2CCl_2CH_3}$$

Addition of Hydrogen Halides

The addition of hydrogen halides (HF, HCl, HBr, and HI) leads first to vinyl* halides and then to 1,1-dihalides. These addition reactions follow Markownikoff's rule.

$$\underset{\text{Acetylene}}{HC{\equiv}CH} + HBr \longrightarrow \underset{\substack{\text{Vinyl bromide}\\ \text{(1-bromoethene)}}}{CH_2{=}CHBr} \xrightarrow{HBr} \underset{\text{1,1-Dibromoethane}}{CH_3CHBr_2}$$

$$\underset{\text{Propyne}}{CH_3C{\equiv}CH} + 2HCl \longrightarrow \underset{\text{2,2-Dichloropropane}}{CH_3CCl_2CH_3}$$

Addition of Water (Hydration)

Under suitable conditions water adds to acetylene to give the aldehyde, acetaldehyde. Until recently, this reaction was used as a commercial preparation of acetaldehyde, which is an important intermediate in the preparation of acetic acid (CH_3COOH) and other organic chemicals.

$$\underset{\text{Acetylene}}{HC{\equiv}CH} + H_2O \xrightarrow[HgSO_4]{H_2SO_4} \underset{\text{Acetaldehyde}}{CH_3CHO}$$

Formation of Metal Acetylides

As noted earlier in the formation of sodium acetylides, the acidic hydrogen of terminal acetylenes can be replaced by a metal ion to form an acetylide salt. With heavy metals, such as Ag^+ and Cu^+, these salts are easily formed and precipitate from solution. They are, however, explosive when dry and should be handled cautiously. Since only terminal acetylenes undergo salt formation, the formation of the metal acetylide can be used as a diagnostic test to distinguish a terminal acetylene from a nonterminal acetylene. For example, 1-butyne readily yields a Ag^+ or Cu^+ acetylide, whereas 2-butyne (which has no acidic hydrogen) does not give a precipitate of a metal acetylide under similar conditions.

$$\underset{\text{Acetylene}}{HC{\equiv}CH} + 2Ag(NH_3)_2^+ \rightarrow \underset{\text{White}}{AgC{\equiv}CAg\downarrow} + 2NH_4^+ + 2NH_3$$

$$\underset{\text{Acetylene}}{HC{\equiv}CH} + 2Cu(NH_3)_2^+ \rightarrow \underset{\text{Red}}{CuC{\equiv}CCu\downarrow} + 2NH_4^+ + 2NH_3$$

$$\underset{\text{1-Butyne}}{CH_3CH_2C{\equiv}CH} + Ag(NH_3)_2^+ \rightarrow \underset{\text{White}}{CH_3CH_2C{\equiv}CAg\downarrow} + NH_4^+ + NH_3$$

$$\underset{\text{2-Butyne}}{CH_3C{\equiv}CCH_3} + Ag(NH_3)_2^+ \rightarrow \text{No reaction}$$

* The $CH_2{=}CH{-}$ is known as the vinyl group. $CH_2{=}CHCl$ (vinyl chloride) is an important olefin in preparing the commercial polyvinyl chloride.

IMPORTANT TERMS AND CONCEPTS

alkyl halide
alkynes
cis isomer
conjugated diene
cycloalkene
dehydration
dehydrohalogenation
electrophilic addition reactions
electrophilic reagent
elimination reaction
geometrical isomerism
halohydrin
Markownikoff's rule
nucleophilic reagent
olefin
pi bond
σ bond
trans isomer
vinyl group

QUESTIONS

1. Draw a structural formula for each of the following compounds:
   (a) *trans*-2-hexene
   (b) *cis*-2,3-dichloro-2-butene
   (c) 1-methylcyclopentene
   (d) *trans*-1,2-dibromocyclohexane
   (e) 4-ethyl-1-octene
   (f) 3-hexyne
   (g) *cis*-diiodoethylene
   (h) 2-methyl-2-butene
   (i) 2-chloro-1,3-cyclohexadiene
   (j) 2-bromo-1,3-butadiene

2. Name each of the following compounds by the IUPAC system:
   (a) $CH_3CH_2C{\equiv}CH$
   (b) $CH_2{=}CHCBr{=}CHCH_3$
   (c) [cyclohexene ring with CH_3 substituent]
   (d) $(CH_3)_2C{=}CHCH_3$
   (e) $CH_3CH{=}CCl_2$
   (f) $CH_2{=}CF_2$
   (g) CH_3 and H on one carbon, Cl and CH_3 on the other: $(CH_3)(H)C{=}C(Cl)(CH_3)$
   (h) $CH_2{=}CHCH_2CH_2CH{=}CH_2$
   (i) $(CH_3)_2C{=}CHCH_2CH(CH_3)_2$
   (j) $CH_3CHClC{\equiv}CCH_3$

3. Draw out all the possible structural isomers for the compound having a molecular formula, C_5H_{10}. Name each isomer according to the IUPAC system.

4. Which of the isomeric compounds in Question 3
   (a) contain(s) no geometric isomers?
   (b) has(ve) no double bonds?
   (c) is (are) symmetrical?
   (d) has the highest boiling point?
   (e) contain(s) geometric isomers, but no double bonds?

5. Indicate what is wrong with each of the following names. Give the *correct* name for each of the following compounds:
   (a) 4-methyl-3-pentene
   (b) 6-methylcyclohexene
   (c) 2-isopropyl-1-propene
   (d) 4,6-dimethylcyclohexene
   (e) *cis*-4-heptene

6. Write equations for the reactions of 1-pentene with the following reagents:
   (a) Br_2
   (b) H_2SO_4
   (c) HI
   (d) HOCl
   (e) H_2/Ni

7. Give a chemical reaction that will distinguish between the following pairs. Draw the structures of the reaction products.
   (a) cyclopentane and $CH_3(CH_2)_2CH{=}CH_2$
   (b) $CH_3C{\equiv}CH$ and $CH_3CH{=}CH_2$
   (c) $CH_3(CH_2)_3C{\equiv}CH$ and $CH_3CH_2C{\equiv}CH_2CH_3$

8. Write equations for the reactions of 1-butyne with the following reagents:
   (a) excess Cl_2 (d) $Ag(NH_3)_2^+$
   (b) 1 mole Br_2 (e) excess H_2/Ni
   (c) excess HI

9. Using reactions discussed in this chapter and the preceding chapter, show how each of the following compounds could be prepared from isopropyl iodide:
   (a) propene (c) isopropyl chloride
   (b) propyne (d) 2,2-dibromopropane

10. Starting from propyne, show how each of the following compounds could be prepared:
   (a) propane (d) *cis*-2-butene
   (b) 2-butyne (e) isopropyl iodide
   (c) 2-chloropropene

11. A compound has a molecular formula, C_4H_6. When it is treated with excess hydrogen and a catalyst, a new compound, C_4H_{10} is formed. When it is treated with ammoniacal silver nitrate, a precipitate is formed. Give a possible structure for the original compound.

SUGGESTED READING

Cohen: The Shape of the 2p and Related Orbitals. Journal of Chemical Education, Vol. 38, p. 20, 1961.
Ihde: The Unraveling of Geometrical Isomerism and Tautomerism. Journal of Chemical Education, Vol. 36, p. 330, 1959.
Jones: The Markownikoff Rule. Journal of Chemical Education, Vol. 38, p. 297, 1961.
Traynham: The Bromonium Ion. Journal of Chemical Education, Vol. 40, p. 392, 1963.
Tucker: Catalytic Hydrogenation Using Raney Nickel. Journal of Chemical Education, Vol. 27, p. 489, 1958.

AROMATIC HYDROCARBONS

CHAPTER 5

The *objectives* of this chapter are to enable the student to:

1. Distinguish the class of hydrocarbons known as aromatics.
2. Understand the stability of aromatic hydrocarbons relative to other polyunsaturated hydrocarbons, such as alkenes and alkynes.
3. Describe the concept of resonance, resonance hybrids, and delocalization of electrons.
4. Write the structural formulas and names of aromatic hydrocarbons.
5. Give a definition of ortho, meta, and para isomers.
6. Understand the generalized concept of electrophilic aromatic substitution reactions.
7. Recognize an ortho-para directing group and a meta-directing group.
8. Predict the orientation in aromatic substitution reactions.
9. Distinguish between reactions which occur on the aromatic ring *versus* reactions which occur in the alkyl side chain.
10. Define a heterocyclic compound.
11. Recognize some of the basic five- and six-membered heterocyclic structures.

Early in the nineteenth century a class of organic compounds was isolated from aromatic substances such as oils of cloves, vanilla, wintergreen, cinnamon, bitter almonds, and benzoin. These compounds were pleasant smelling substances, and the term **aromatic** was given to this class of compounds to denote their aroma. In fact, many of these "aromatics" are still used in the perfumery and flavor extract industries because of their distinctive and pleasant odors.

The investigation of the chemistry of this class of compounds soon made it evident that the aromatic compounds were not related in an obvious manner to alkanes, alkenes, or alkynes and constituted a new class of hydrocarbons. Further chemical investigation also made it evident that all of the members of this class of hydrocarbons were structurally related to a cyclic hydrocarbon, **benzene,** which has the molecular formula C_6H_6. Benzene is not a pleasant smelling substance like many of its derivatives, and the original meaning of the term aromatic can only be loosely applied to benzene. However, the term has been carried over to include the chemistry of benzene and benzene derivatives; hence, this class of hydrocarbons is still called **aromatic hydrocarbons.**

Benzene, itself, was first isolated by Michael Faraday in 1825. Later, it was also shown to be a constituent of coal tar. Coal tar is a heavy black liquid obtained from the destructive distillation of coal at high temperatures. Coal tar has been the chief source of aromatic compounds and today still serves as the commercial source of many aromatics, although some aromatic hydrocarbons are now obtained from the petroleum industry.

The initial preparation of benzene by Faraday was carried out by pyrolyzing benzoic acid with lime. From the high ratio of carbon to hydrogen, it was expected that benzene would be highly unsaturated.

$$\underset{\text{Sodium benzoate}}{C_6H_5CO_2Na} + NaOH \xrightarrow[\Delta]{CaO} \underset{\text{Benzene}}{C_6H_6} + Na_2CO_3$$

Density, combustion analysis, and molecular weight studies established the molecular formula as C_6H_6. If a straight-chain structure is written for a compound of this molecular formula, several double bonds or triple bonds must be included in the carbon chain to attain the molecular formula C_6H_6. Surprisingly, however, it was found that benzene did not decolorize a solution of bromine in carbon tetrachloride (therefore, did not undergo addition of bromine), and it was not oxidized by potassium permanganate, a reagent which easily oxidizes alkenes and alkynes. Therefore, structures containing several double and/or triple bonds were ruled out for the structure of benzene.

Further investigation of benzene established the following facts: (1) catalytic hydrogenation of benzene indicates that benzene absorbs three moles of hydrogen and gives cyclohexane, C_6H_{12}, as the final product. The reaction follows:

$$\underset{\text{Benzene}}{C_6H_6} + 3H_2 \xrightarrow{Ni} C_6H_{12},$$

H₂C–CH₂ ring structure (six CH_2 groups in a ring)

Cyclohexane

Therefore, benzene must contain a cyclic six-membered ring of carbon atoms. Similarly, benzene adds three moles of chlorine in sunlight to give hexachlorocyclohexane, again consistent with a cyclic six-membered ring. These addition reactions also suggest that benzene contains an unsaturation which is equivalent to three double bonds.

$$\underset{\text{Benzene}}{C_6H_6} + 3Cl_2 \xrightarrow{\text{sunlight}} C_6H_6Cl_6,$$

(six HClC groups in a ring)

Hexachlorocyclohexane

(2) benzene can be chlorinated or brominated in the presence of a catalyst to give only one compound (C_6H_5X). Experimental evidence indicates that all six hydrogens in benzene are equivalent. If they were not equivalent, more than one isomer would be possible

$$C_6H_6 + X_2 \xrightarrow{Fe} C_6H_5X + HX$$

$(X = Cl, Br)$

from this kind of halogenation experiment. In addition, the halogenation reaction gives two products, indicating that this reaction is a **substitution** reaction and not an **addition** reaction.

To account for the above experimental facts, Friedrich Kekule in 1865 proposed the following structure for benzene. He suggested that benzene is composed of a hexagon of carbon atoms with alternating single and double bonds between the carbon atoms.

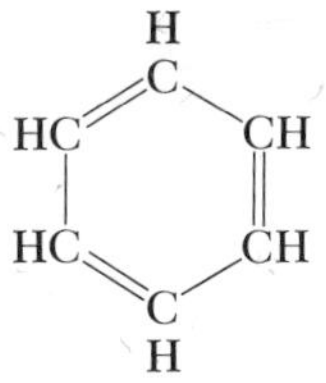

H
C
HC CH
HC CH
C
H

Kekule benzene

This structure could account quite nicely for the addition of three moles of hydrogen and chlorine to give cyclohexanes and for the formation of a monohalogenated derivative on halogenation with catalysis.

However, the **Kekule benzene structure** does not account for the stability of benzene to oxidizing agents like potassium permanganate. Also, this structure predicts two isomers for a 1,2-disubstituted benzene (I and II). In isomer I, the two carbon atoms holding

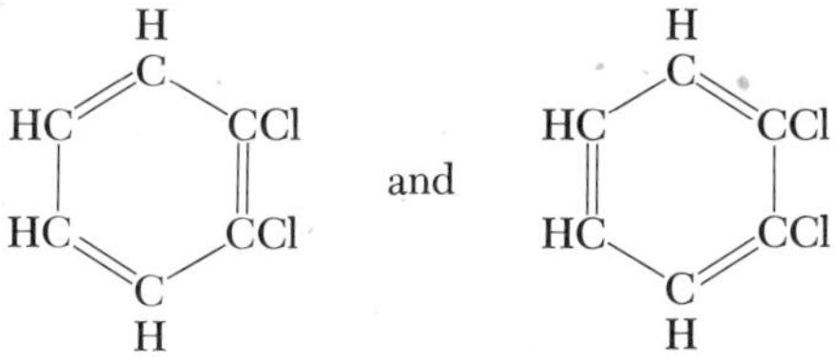

H
C
HC CCl
HC CCl
C
H
(I)

and

H
C
HC CCl
HC CCl
C
H
(II)

the chlorine atoms are bonded with a double bond, whereas in II they are bonded with a single bond. However, only one 1,2-dichlorobenzene has ever been isolated.

To circumvent these difficulties, Kekule further proposed that benzene is composed of a dynamic equilibrium between two equivalent structures (as shown) in which there

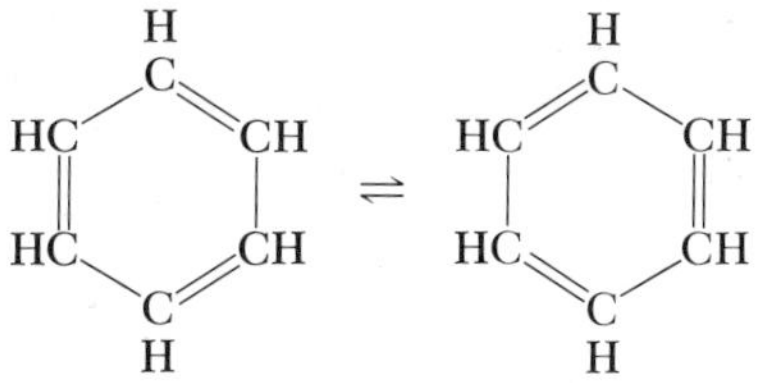

H
C
HC CH
HC CH
C
H

$\rightleftharpoons$

H
C
HC CH
HC CH
C
H

is a rapid oscillation of the double and single bonds. Kekule suggested that if the equilibrium were rapid enough, the two isomeric 1,2-disubstituted benzenes would be converted into each other at such a rate, so that only one isomer could be isolated. However, this rapid equilibrium still does not account for the stability of benzene to oxidizing agents and for its ease in undergoing substitution reactions rather than addition reactions.

The modern concept of the structure of benzene has evolved from this early chemical investigation and more recently from the use of x-ray analysis. X-ray diffraction has shown that benzene is a **planar hexagon** and that all the carbon-carbon bond distances are identical and equal to 1.39 Å. Consequently, the idea of alternating single and double bonds is unacceptable, since structures of this type cannot account for identical carbon-carbon bond lengths.

The modern structure of benzene is depicted as the formation of a six-membered ring of carbon atoms using the sp^2 orbitals of carbon. The resulting p orbitals (one on each carbon) then overlap to form a π-bond (similar to a π-bond in an alkene, except that the π-bond can be formed with one of two near neighbors). This π-bond is formed by the overlap of six p orbitals which results in a region of π-electron density above and below the hexagon structure, such that the hexagonal carbon skeleton is encased in a donut shape π-electron cloud. Thus, benzene can be depicted as shown in the following diagram:

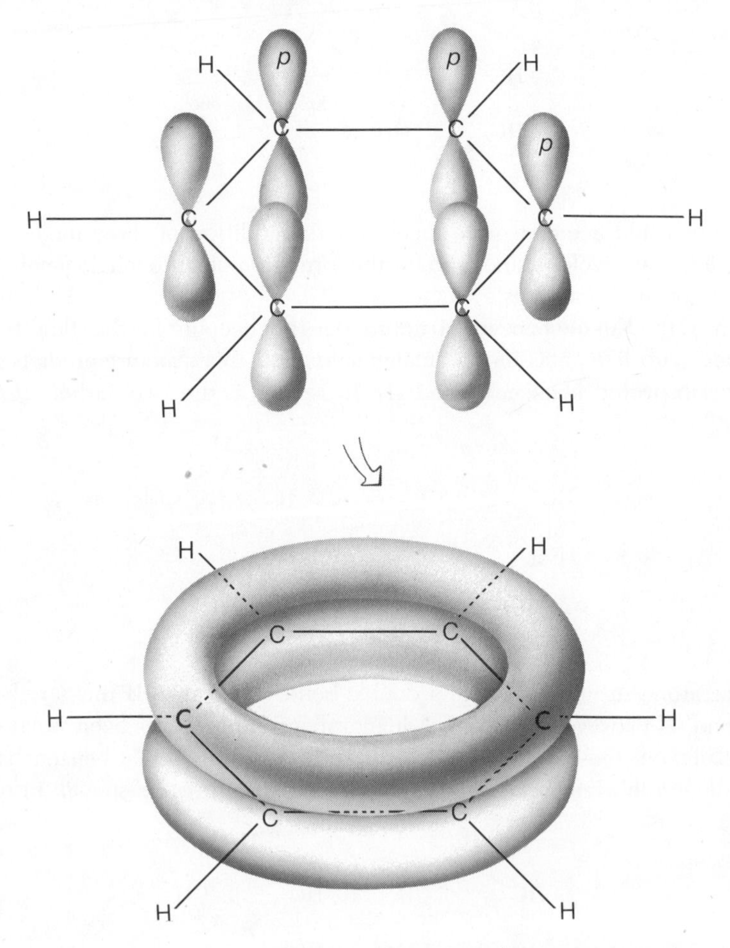

FIGURE 5-1 Representation of a π electron cloud in benzene.

Another representation commonly used is a double headed arrow between the two Kekule structures.

The double headed arrow indicates that these structures are resonance forms. Remember that resonance forms involve only a change in the distribution of electrons (see p. 13) and that the *real structure* is a composite of all the possible resonance forms. Thus, benzene is neither of the Kekule structures, nor is it simply an equal mixture of the two as proposed by Kekule, but it is a structure that lies between these two extremes and is commonly represented by the symbol ⌬. The dotted line indicates that the π-electron density is distributed evenly over the six carbon atoms of the ring.

The delocalization* of electrons stabilizes the molecule to such an extent that benzene does not react like a simple alkene.

It has been demonstrated that compounds like benzene, which have a continuous

*Electrons that occupy an orbital around more than two nuclei are called *delocalized electrons.*

Resonance hybrids

π-electron cloud encompassing all the carbon atoms of the ring system, exhibit unusual stability compared to analogous systems in which the electrons would be fixed in double bonds. In benzene this stability amounts to 36 kcal/mole compared to a cyclohexatriene having fixed double bonds. Consequently, any chemical reaction which occurs to disrupt

is 36 kcal /mole more stable than

this continuous π-electron cloud will be unfavored, since it will take 36 kcal/mole of energy more to carry out this reaction on benzene than on a cyclohexatriene type structure. Therefore, oxidation or addition reactions which would disrupt the continuous π-cloud are unfavorable energetically, and do not occur under ordinary conditions.

In addition to benzene and substituted benzenes, other aromatic hydrocarbons are known which have one or more **fused** or **condensed** rings. Again, the π-electron system is continuous over the entire carbon skeleton, and these compounds are more stable than would be expected for similar compounds containing fixed double bonds. Some of the more common condensed ring systems are illustrated below:

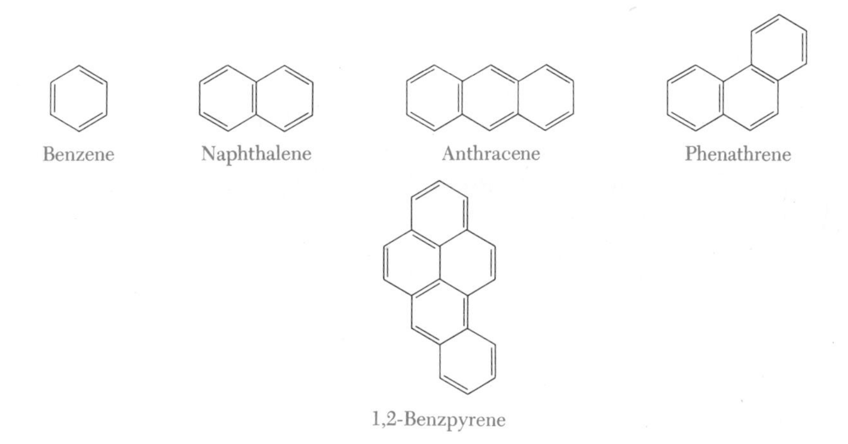

For convenience the bonds are shown as fixed, but it should be kept in mind that they are not fixed. Several systems containing condensed rings have been shown to be carcinogenic (will cause cancer) and should be handled with caution.

Many compounds of biochemical interest contain the benzene nucleus. For example, tetrahydrocannabinol (Chapter 2), lysergic acid diethylamide (Chapter 2), and mescaline, the essential amino acids, phenylalanine (see Chapter 15), tyrosine, and tryptophan, hormones such as estradiol (Chapter 14), and vitamins such as vitamin K (Chapter 14), all contain an aromatic ring as part of the complex molecule.

NOMENCLATURE

Many aromatic compounds are named by common names, or as derivatives of the parent hydrocarbon by naming the substituent attached to the ring followed by the name of the aromatic hydrocarbon.

CH_3 — Toluene (methylbenzene)

CH_2CH_3 — Ethylbenzene

NH_2 — Aniline (aminobenzene)

OH — Phenol (hydroxybenzene)

Cl — Chlorobenzene

NO_2 — Nitrobenzene

$CH{=}CH_2$ — Styrene (vinylbenzene)

$H_3C{-}\underset{}{\overset{H}{C}}{-}CH_3$ — Cumene (isopropylbenzene)

COOH — Benzoic acid

OCH_3 — Anisole (Methyl phenyl ether)

SO_3H — Benzene sulfonic acid

The C_6H_5 group is known as the **phenyl** group.

When two substituents are attached to the benzene ring, two systems of nomenclature are used. The position and number of each substituent can be indicated by the appropriate number and prefix. Alternatively, the relative positions of the two substituents can be indicated by the prefixes *ortho-* (*o-*), *meta-* (*m-*), and *para-* (*p-*) to indicate either a 1,2-; 1,3-; or 1, 4- position of the substituents relative to each other. For aromatic hydrocarbons containing more than two substituents, the numbering system is used. The manner in which the numbers are applied is not always consistent. Sometimes it is done alphabetically, and sometimes it is done by assigning the most important substituent the lowest number, and numbering the other substituents accordingly. Some examples of polysubstituted benzenes are shown below:

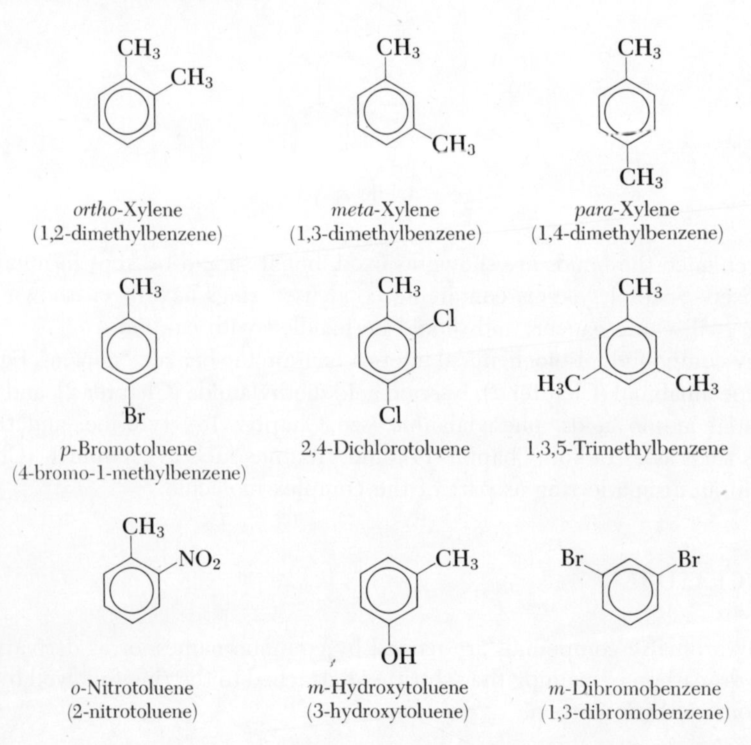

ortho-Xylene (1,2-dimethylbenzene)

meta-Xylene (1,3-dimethylbenzene)

para-Xylene (1,4-dimethylbenzene)

p-Bromotoluene (4-bromo-1-methylbenzene)

2,4-Dichlorotoluene

1,3,5-Trimethylbenzene

o-Nitrotoluene (2-nitrotoluene)

m-Hydroxytoluene (3-hydroxytoluene)

m-Dibromobenzene (1,3-dibromobenzene)

PHYSICAL PROPERTIES OF AROMATIC HYDROCARBONS

Benzene and its homologues are similar to other types of hydrocarbons with respect to their physical properties. They are insoluble in water but soluble in organic solvents. The boiling points of the aromatic hydrocarbons are slightly higher than those of the alkanes of similar carbon content. For example, *n*-hexane, C_6H_{14}, boils at 69°C, whereas benzene, C_6H_6, boils at 80°C. The planar structure and highly delocalized electron density in the aromatic hydrocarbon increase the forces acting between molecules and result in a higher boiling point. Also, the symmetrical structure of benzene permits better packing in the crystal, resulting in a higher melting point than the straight-chain alkane of similar carbon content. Benzene melts at +5.5°C, whereas *n*-hexane melts at −95°C. The physical properties of some of the common aromatic hydrocarbons are summarized in Table 5–1.

Aromatic hydrocarbons are quite flammable and should be handled with caution. Benzene is toxic when taken internally and must be used with proper precautions in any commercial process. Prolonged inhalation of its vapors results in a decreased production of red and white corpuscles in the blood which may prove fatal. Consequently, compounds of this class should only be handled under well-ventilated conditions. In addition, some of the more complex polynuclear aromatic hydrocarbons are carcinogenic and should be handled accordingly.

METHODS OF PREPARATION

Conversion of Alkanes to Aromatic Hydrocarbons

Although most of the benzene and benzene derivatives used commercially were obtained from coal tar, some of these compounds are now prepared by the conversion of aliphatic or cyclic petroleum products to aromatic hydrocarbons. Most of these conversions involve catalytic processes and/or high pressure processes, and they are not generally suitable for laboratory preparations. However, they are of commercial importance, because benzene and its simple derivatives are the basic starting materials for most of the commercial aromatic hydrocarbon industry. Two of these basic processes are illustrated below, and involve dehydrogenation or cyclization of aliphatic hydrocarbons:

$$\underset{\text{Methylcyclohexane}}{C_6H_{11}CH_3} \xrightarrow[\text{600°C, pressure}]{Al_2O_3 \cdot Mo_2O_3} \underset{\text{Toluene}}{C_6H_5CH_3} + 3H_2 \qquad \underset{n\text{-Heptane}}{CH_3(CH_2)_5CH_3} \xrightarrow[\text{600°C, pressure}]{Cr_2O_3 \cdot Al_2O_3} \underset{\text{Toluene}}{C_6H_5CH_3} + 4H_2$$

Friedel-Crafts Reaction

The French chemist Charles Friedel and the American chemist James Crafts discovered an alkylation reaction which could be used to prepare benzene derivatives. They found that when benzene is treated with an alkyl halide in the presence of a Lewis acid like aluminum chloride, an alkylated benzene is produced. For example, benzene and methyl chloride in the presence of aluminum chloride gives toluene. Ethylbenzene could be similarly prepared from benzene, ethyl bromide, and aluminum bromide.* The major limitation on this kind of reaction is that in many cases the aromatic hydrocarbon

*Although the aluminum halides are generally used, other acid catalysts such as H_2SO_4, BF_3, HF, and $SnCl_4$ have been used, depending on the type of starting material and its reactivity.

TABLE 5-1 Physical Properties of Aromatic Hydrocarbons

Compound	Structure	M.P. °C	B.P. °C
Benzene		+6	+80
Toluene	CH_3	−95	+111
Ethylbenzene	CH_2CH_3	−95	+136
Isopropylbenzene	$CH(CH_3)_2$	−96	+152
Naphthalene		+80	+218
Anthracene		+217	+355
Phenanthrene		+100	+340
o-Xylene	CH_3, CH_3	−25	+144
m-Xylene	CH_3, CH_3	−48	+139
p-Xylene	CH_3, CH_3	+13	+138

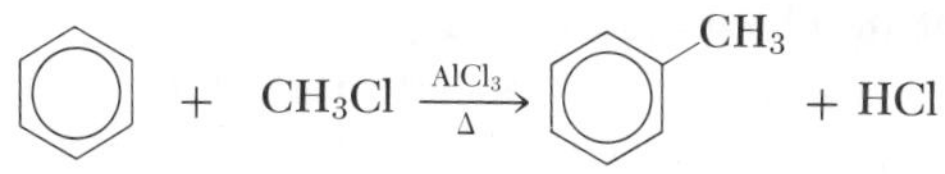

$$C_6H_6 + CH_3Cl \xrightarrow[\Delta]{AlCl_3} C_6H_5CH_3 + HCl$$

Benzene Methyl chloride Toluene

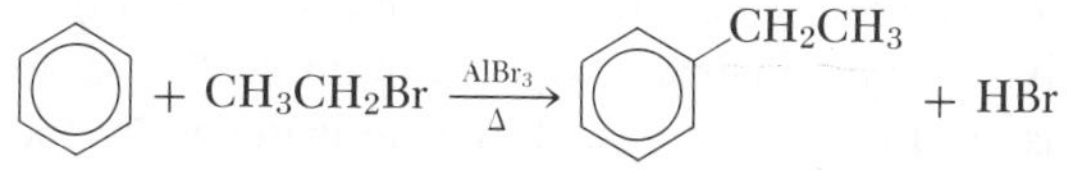

$$C_6H_6 + CH_3CH_2Br \xrightarrow[\Delta]{AlBr_3} C_6H_5CH_2CH_3 + HBr$$

Benzene Ethyl bromide Ethylbenzene

obtained contains an alkyl substituent that has a rearranged carbon skeleton. For example, the reaction of benzene, *n*-propyl chloride, and aluminum chloride would be expected to yield *n*-propylbenzene. However, cumene is the major product of this reaction.

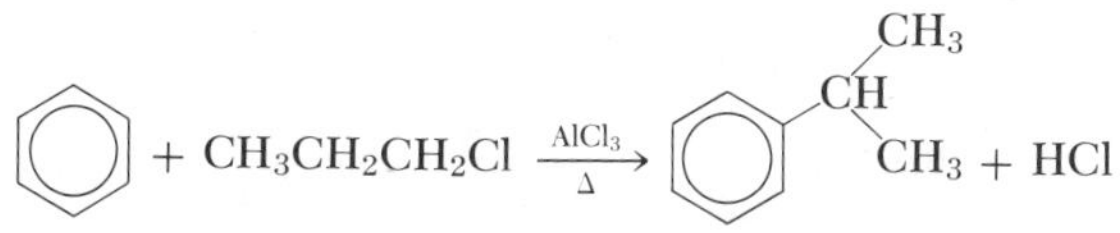

$$C_6H_6 + CH_3CH_2CH_2Cl \xrightarrow[\Delta]{AlCl_3} C_6H_5CH(CH_3)_2 + HCl$$

Benzene *n*-Propyl chloride Cumene (Isopropyl benzene)

Rearrangement of the carbon skeleton of the alkyl halide is usually observed with longer chain alkyl halides, and hence, this reaction is of limited use for the preparation of aromatic hydrocarbons containing a straight-chain alkyl substituent.

Although the Friedel-Crafts reaction historically involved alkyl halides as reactants, it has since been found that alkenes and alcohols also can be reacted with benzene in the presence of a suitable acid catalyst to give similar derivatives of aromatic hydrocarbons. Several examples are cited below:

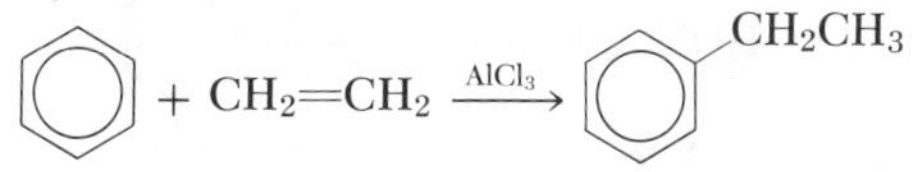

$$C_6H_6 + CH_2{=}CH_2 \xrightarrow{AlCl_3} C_6H_5CH_2CH_3$$

Benzene Ethylene Ethylbenzene

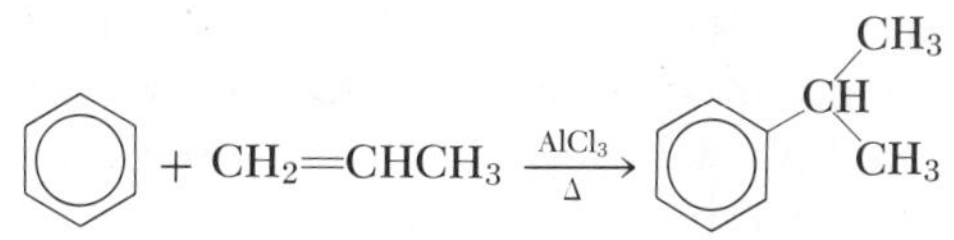

$$C_6H_6 + CH_2{=}CHCH_3 \xrightarrow[\Delta]{AlCl_3} C_6H_5CH(CH_3)_2$$

Benzene Propene Cumene

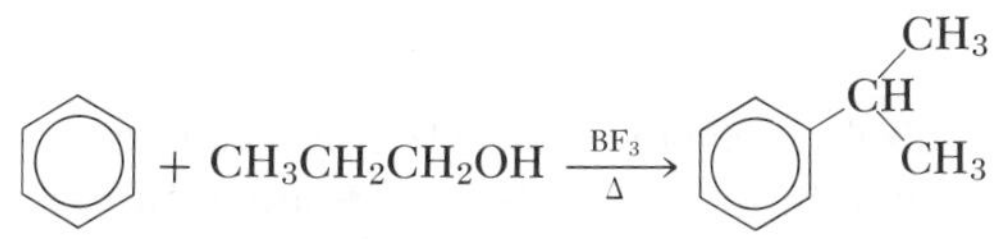

$$C_6H_6 + CH_3CH_2CH_2OH \xrightarrow[\Delta]{BF_3} C_6H_5CH(CH_3)_2$$

Benzene *n*-Propyl alcohol Cumene

Again, isomerization of the alkyl group occurs in the appropriate cases, and the same limitation exists for these reactants.

REACTIONS OF AROMATIC HYDROCARBONS

The most characteristic reactions of aromatic hydrocarbons are substitution reactions rather than addition reactions. Consequently, two products are formed in the reaction.

One is an organic product in which an atom or group of atoms has substituted for the hydrogen atom of the aromatic ring. The second product is usually an inorganic acid or water.

Halogenation

With chlorine or bromine, halogenation occurs readily in the presence of a Lewis acid catalyst. Usually, iron or an iron halide containing the same halide atom as the halogenating agent is used as the catalyst. Fluorine is generally too reactive to be used

$$C_6H_6 + Cl_2 \xrightarrow{FeCl_3} C_6H_5Cl + HCl$$

Benzene → Chlorobenzene

$$C_6H_6 + Br_2 \xrightarrow{FeBr_3} C_6H_5Br + HBr$$

Benzene → Bromobenzene

in substitution reactions of this type, and iodine is generally too unreactive to be successful by this method. The introduction of the first halogen (chlorine or bromine) into the aromatic ring makes the ring less susceptible to further attack and deactivates it, and the introduction of a second or third halogen atom becomes progressively more difficult.

Nitration

The introduction of a nitro group ($—NO_2$) into an aromatic ring can be readily carried out using a mixture of concentrated nitric and sulfuric acids. The introduction

$$C_6H_6 + HONO_2(HNO_3) \xrightarrow[50\text{–}60°C]{H_2SO_4} C_6H_5NO_2 + H_2O$$

Benzene, Nitric acid → Nitrobenzene

of a nitro group into the ring also deactivates the aromatic ring to further substitution, and more vigorous conditions must be used to introduce a second or third nitro group. Nitrobenzene has a harmful physiological effect on the red blood corpuscles and on the liver. Therefore, caution should be used in handling this material, and inhalation of its vapor should be avoided.

Sulfonation

The introduction of a sulfonic acid ($—SO_3H$) group can be accomplished by treating benzene with concentrated sulfuric acid at elevated temperatures, or with fuming sulfuric acid (sulfuric acid containing SO_3) at moderate temperatures.

$$C_6H_6 + HOSO_2OH + SO_3 \rightarrow C_6H_5SO_3H$$

Benzene, Fuming sulfuric acid → Benzenesulfonic acid

Sulfonic acids are strong acids and are usually water soluble. Their acidity is generally comparable to that of the mineral acids. In contrast to the other substitution reactions of aromatic hydrocarbons, sulfonation is a reversible reaction. If benzenesulfonic acid is treated with steam, benzene is regenerated. Sulfonic acid groups have been used as "blocking groups" in many syntheses to prevent a particular position on the ring from

$$C_6H_5SO_3H + H_2O\ (\text{steam}) \xrightarrow{\Delta} C_6H_6 + H_2SO_4$$

Benzenesulfonic acid

being substituted. At the end of the synthetic sequence the sulfonic acid group can be removed, and the "blocked" position becomes "free" again.

Although sulfonic acids themselves are important, derivatives of sulfonic acids have received extensive treatment, since many of them were found to exhibit physiological action. Sulfanilamide, the forerunner of the sulfa drugs, is a sulfonic acid derivative.

$$p\text{-}NH_2C_6H_4SO_2NH_2$$

Sulfanilamide

Sulfanilamide was found to be effective in the treatment of streptococcus infections, pneumonia, puerperal fever, gonorrhea, and gas gangrene. Unfortunately, sulfanilamide is only slightly soluble in water and tends to crystallize from aqueous solutions. When administered orally, the drug is absorbed and eventually carried to the kidney for excretion. When the dose is large, or under prolonged therapy, the kidneys are damaged by the accumulated sulfanilamide. Other toxic reactions, including methemoglobinemia, caused a search for derivatives that were less toxic. Sulfathiazole, sulfapyridine, sulfaguanidine, and sulfadiazine are among these derivatives.

Sulfathiazole: $p\text{-}NH_2C_6H_4SO_2NH\text{—}C_3H_2NS$ (thiazole ring: N—CH, S—CH)

Sulfapyridine: $p\text{-}NH_2C_6H_4SO_2NH\text{—}C_5H_4N$

Sulfaguanidine: $p\text{-}NH_2C_6H_4SO_2NH\text{—}C(=NH)NH_2$

Sulfadiazine: $p\text{-}NH_2C_6H_4SO_2NH\text{—}C_4H_3N_2$

Extensive studies of the therapeutic properties of each of the sulfa drugs were carried out following the initial therapeutic discovery of sulfanilamide in 1936. This resulted in better treatment and control of various infectious diseases. Sulfadiazine, for example, is less toxic than the other sulfa drugs, yet is one of the most effective in the treatment of pneumonia and staphylococcus infections. These studies also disclosed that some of the sulfa drugs are very poorly absorbed from the intestinal tract and can be used as intestinal antiseptics. However, in destroying organisms in the intestine, these drugs interfere with the synthesis of vitamin K and members of the vitamin B complex, such as p-aminobenzoic acid, biotin, and folic acid, and may produce vitamin deficiencies.

Sulfonic acids dissolve readily in water and ionize completely to the sulfonate, and the introduction of a sulfonic acid group

$$\underset{\text{Sulfonic acid}}{C_6H_5SO_3H} + H_2O \rightarrow \underset{\text{Sulfonate}}{C_6H_5SO_3^-} + H_3O^+$$

in the molecule is frequently used to increase the water solubility of many dyes, medicinals, and synthetic detergents.

Early synthetic detergents were prepared from alkylbenzenes by the following scheme:

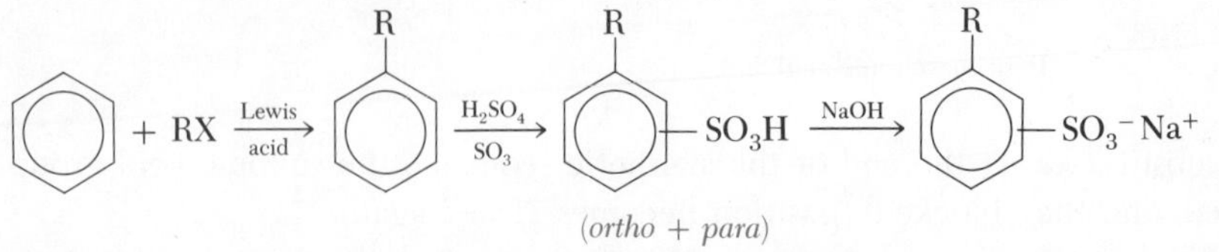

$$C_6H_6 + RX \xrightarrow[\text{acid}]{\text{Lewis}} C_6H_5R \xrightarrow[SO_3]{H_2SO_4} R{-}C_6H_4{-}SO_3H \xrightarrow{NaOH} R{-}C_6H_4{-}SO_3^-\,Na^+$$

(*ortho* + *para*)

The alkyl portion (R) was generally a highly branched C_{12}—C_{18} alkyl group. These synthetic detergents exhibited good detergent properties and were widely used until the late 1960's. By this time the increasing amount of "foam" which was appearing in the nation's rivers was recognized as due to the nonbiodegradability of these synthetic detergents. Bacteria which normally degrade hydrocarbons *via* oxidation could not degrade the highly branched side chains of the detergents. Eventually these types of detergents were banned by Federal law.

However, it was soon found that when the alkyl portion (R) of the molecule was a straight chain alkyl group that the synthetic detergent became biodegradable, and today a variety of these biodegradable type of synthetics are available on the market.

Friedel-Crafts Alkylation

Alkylation of an aromatic ring using an alkyl halide, alkene, or an alcohol in the presence of an acid catalyst was shown earlier in this chapter to give alkylated aromatic hydrocarbons. The major limitation was found to be rearrangement of the carbon skeleton of the alkyl group under the conditions of the reaction. In addition, the introduction of an alkyl group into an aromatic ring activates the ring to further substitution, and polysubstitution results. Usually, a mixture of alkylated aromatic hydrocarbons is obtained.

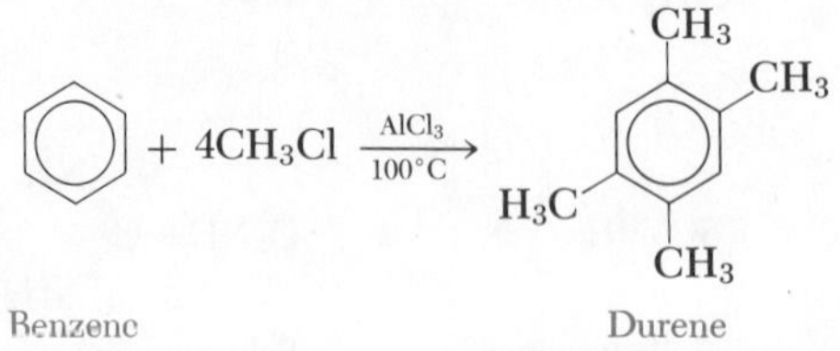

$$\underset{\text{Benzene}}{C_6H_6} + 4CH_3Cl \xrightarrow[100°C]{AlCl_3} \underset{\text{Durene}}{C_6H_2(CH_3)_4}$$

Friedel-Crafts Acylation

A reaction similar to alkylation of a benzene ring can be carried out using an acid chloride (RCOCl) and aluminum chloride to introduce the acyl group ($R\overset{O}{\overset{\|}{C}}$—) into an aromatic ring. The acyl group deactivates the ring toward further substitution, and no rearrangement of the acyl group occurs under the conditions of the reaction. Therefore, the limitations of the alkylation reaction can be avoided by using an acylation reaction.

Since the carbonyl group (—$\overset{O}{\overset{\|}{C}}$—) of the acyl group can be converted to a methylene (—CH_2—) group via a Clemmensen or Wolff-Kishner reduction (see Chapter 3), Friedel-Crafts acylation followed by reduction provides an easy two-step synthesis to aromatic hydrocarbons containing a straight-chain alkyl substituent, as shown on the next page for *n*-propyl benzene:

$$C_6H_6 + CH_3CH_2COCl \xrightarrow{AlCl_3} C_6H_5COCH_2CH_3 + HCl$$

Benzene — Phenyl ethyl ketone

$$C_6H_5COCH_2CH_3 + Zn(Hg) \xrightarrow{HCl} C_6H_5CH_2CH_2CH_3$$

Phenyl ethyl ketone — *n*-Propylbenzene

MECHANISM OF AROMATIC SUBSTITUTION REACTIONS

The generally accepted mechanism for aromatic substitution reactions is one involving electrophilic attack on the aromatic hydrocarbon. The π-electron system of the aromatic hydrocarbon is attacked by an electrophilic reagent to give a charged intermediate (A), which then loses a proton to a base to reform the aromatic π-system. Thus, the additional stability of the aromatic system gained from the continuous π-electron system is not lost. A generalized scheme which represents the main features of this mechanism is outlined below:

$$C_6H_6 + X^+ \rightarrow [C_6H_6X]^+ \xrightarrow{\text{base}} C_6H_5X + \text{base } H^+$$

Benzene — Electrophilic reagent — (A)

The electrophilic reagent X^+ can be any of the electrophiles formed by the following reactions:

Halogenation: $Y_2 + FeY_3 \rightleftharpoons Y^+[FeY_4]^-$

$Y = Cl, Br$

Nitration: $HONO_2 + H_2SO_4 \rightleftharpoons H_2O + NO_2^+[HSO_4]^-$

Nitric acid — Sulfuric acid

Sulfonation: $2H_2SO_4 \rightleftharpoons H_3O^+ + HSO_4^- + SO_3$

Friedel-Crafts:

(a) Alkylation: $RX + AlX_3 \rightleftharpoons R^+[AlX_4]^-$

$X =$ Halogen

(b) Acylation: $RCOX + AlX_3 \rightleftharpoons RCO^+[AlX_4]^-$

$X =$ Halogen

Thus, attack by a halonium ion (Cl^+, Br^+), a nitronium ion (NO_2^+), a proton (H^+), a carbonium ion (R^+), or an acyl cation (RCO^+) gives the charged intermediate (A). The proton is removed from (A) by the bases $[FeY_4]^-$, $[HSO_4]^-$, or $[AlX_4]^-$. The electron pair left behind on removal of the proton is used to reform the aromatic sextet of electrons.

The formation of the rearranged products in Friedel-Crafts alkylation reactions is due to the formation of carbonium ions as the electrophilic species in these reactions. The intermediate electrophile (R^+) will rearrange to a more stable carbonium ion (if possible) which in turn will behave as an electrophilic reagent. For example, in the reaction of *n*-propyl chloride, the initially formed *n*-propyl carbonium ion (A) rearranges (by migration of H^-) to the more stable secondary carbonium ion (B) which then attacks the benzene ring and leads to cumene as the final product. Hydride (H^-) shifts of this

$$CH_3CH_2CH_2Cl + AlCl_3 \rightarrow [CH_3CH_2\overset{+}{C}H_2]\ AlCl_4^-$$

n-propyl chloride

$$[CH_3\underset{H}{C}H-\overset{+}{C}H_2]AlCl_4^- \rightarrow [CH_3\overset{+}{C}HCH_3]\ AlCl_4^-$$

(A) (B)

$$[CH_3\overset{+}{C}HCH_3]\ AlCl_4^- + \text{benzene} \rightarrow C_6H_5CH(CH_3)_2 + HCl + AlCl_3$$

type are common in carbonium ion intermediates, and the rearrangement will always occur to give a more stable carbonium ion.

ORIENTATION IN AROMATIC SUBSTITUTION REACTIONS

The introduction of an atom or a group of atoms into an unsubstituted benzene ring presents no problem as to the position taken by the new atom or groups of atoms. Since all the carbon-hydrogen bonds in benzene are equivalent, substitution of a halogen atom, a nitro group, a sulfonic acid group, an alkyl group, or an acyl group can give only one product. However, if a second group is substituted on the ring, more than one isomer is possible, and the position of the new atom or group of atoms becomes important. For example, nitration of toluene could give any or all of the following products:

$$\text{Toluene } (C_6H_5CH_3) + HNO_3 \xrightarrow{H_2SO_4} o\text{-Nitrotoluene},\ m\text{-Nitrotoluene},\ p\text{-Nitrotoluene}$$

Toluene o-Nitrotoluene, m-Nitrotoluene, p-Nitrotoluene

From an extensive study of many electrophilic reactions of substituted benzene derivatives, a set of simple rules has been deduced which can be used to predict the expected product in reactions of this type.

Orientation Rules

The predominant products are predicted by the following orientation rules:

1. The position of the second substituent is determined by the group already present on the ring.
2. The atom or group of atoms already present on the ring may be divided into two classes.

Class A: Atoms or groups of atoms which orient the new group predominantly into the *ortho-* and *para-* positions. The members of this class are called *ortho-para* directors and include atoms or groups such as $—NH_2$, $—OH$, OCH_3, alkyl groups (CH_3, C_2H_5, and so forth), Cl, Br, and I. Except for the halogens, atoms and groups of this class activate the aromatic ring (relative to benzene) toward electrophilic substitution. Thus, toluene is nitrated more easily than benzene, whereas nitration of chlorobenzene will require more vigorous conditions than the nitration of benzene.

Class B: Atoms or groups of atoms which orient the new group predominantly into the *meta-* position. The members of this class are called *meta* directors and include groups such as NO_2, COOH, CHO, CN, CO_2CH_3, and SO_3H. Atoms and groups of this class all deactivate the aromatic ring (relative to benzene) toward electrophilic substitution.

Though these rules predict the major products, in most cases small amounts of the other isomers are also formed, but generally only to a minor extent.

Using these orientation rules, the nitration reaction of toluene would be expected to give mainly *ortho* and *para* nitrotoluenes, since the methyl group is an *ortho-para* director. Some other applications of these rules are outlined below:

$$C_6H_5NO_2 + HNO_3 \xrightarrow[\Delta]{H_2SO_4} m\text{-}C_6H_4(NO_2)_2$$

Nitrobenzene → *m*-Dinitrobenzene

$$C_6H_5CH_3 + H_2SO_4 \xrightarrow{SO_3} o\text{-}CH_3C_6H_4SO_3H + p\text{-}CH_3C_6H_4SO_3H$$

Toluene → *o*-Methylbenzene sulfonic acid + *p*-Methylbenzene sulfonic acid

$$C_6H_5CH_3 + CH_3COCl \xrightarrow{AlCl_3} o\text{-}CH_3C_6H_4C(=O)CH_3 + p\text{-}CH_3C_6H_4C(=O)CH_3$$

Toluene + Acetyl chloride → *o*-Methylacetophenone + *p*-Methylacetophenone

$$C_6H_5OCH_3 + Br_2 \xrightarrow{FeBr_3} o\text{-}BrC_6H_4OCH_3 + p\text{-}BrC_6H_4OCH_3$$

Anisole → *o*-Bromoanisole + *p*-Bromoanisole

Since the product which is formed in the electrophilic aromatic substitution reaction of a substituted benzene is determined by the incumbent substituent, the order in which the substituents are introduced becomes important. For example, bromination followed by nitration produces the *ortho* and *para* isomers. However, nitration followed by

Benzene $\xrightarrow[FeBr_3]{Br_2}$ Bromobenzene $\xrightarrow[H_2SO_4]{HNO_3}$ *o*-Nitrobromobenzene + *p*-Nitrobromobenzene

bromination gives mainly the *meta* isomer. Consequently, when designing an organic

Benzene $+ HNO_3 \xrightarrow{H_2SO_4}$ Nitrobenzene $\xrightarrow[FeBr_3]{Br_2}$ *m*-Nitrobromobenzene

synthesis which involves substituted benzene derivatives, the organic chemist must plan carefully the order in which the substituents are introduced, otherwise undesired isomeric products can be formed.

SIDE-CHAIN REACTIONS OF AROMATIC HYDROCARBONS

A substituted aromatic hydrocarbon can undergo two possible modes of chemical attack. It can undergo electrophilic aromatic substitution on the aromatic ring itself, or reaction can take place in the group (side-chain) attached to the aromatic ring. The mode of chemical reaction will be dependent on the type of side chain, the kind of chemical reagent, and the conditions used for carrying out the reaction. For example, toluene undergoes chlorination in the ring under ionic conditions; but under free-radical conditions (sunlight), chlorination occurs in the methyl group similarly to the free-radical substitution reaction of carbon-hydrogen bonds in alkanes (see Chapter 3).

Toluene $+ Cl_2 \xrightarrow{\text{sunlight}}$ Benzyl chloride* $+ HCl$

Toluene $\xrightarrow[FeCl_3]{Cl_2}$ *o*-Chlorotoluene + *p*-Chlorotoluene

In many cases, the reagent and conditions used will affect only the side-chain, and the aromatic ring will be unaffected by the chemical reaction. For example, aromatic hydrocarbons are resistant to oxidizing agents. Consequently, an alkyl group side-chain can be

*The $C_6H_5CH_2$ is commonly known as the benzyl group.

$$C_6H_5CH_3 + K_2Cr_2O_7 \xrightarrow{H_2SO_4} C_6H_5COOH$$

Toluene — Benzoic acid

$$CH_3C_6H_4CH_3 + K_2Cr_2O_7 \xrightarrow{H_2SO_4} HOOCC_6H_4COOH$$

p-Xylene — Terephthalic acid

$$C_6H_5CH_2CH_3 + KMnO_4 \xrightarrow{OH^-} C_6H_5CO_2^-K^+ \xrightarrow{H_2SO_4} C_6H_5COOH$$

Ethylbenzene — Potassium benzoate — Benzoic acid

oxidized to a carboxylic acid without oxidizing the aromatic ring. The alkyl group, regardless of its length, is degraded to the —COOH group. If more than one alkyl group is attached to the ring, a polyfunctional acid results on oxidation. Thus, *para*-xylene gives terephthalic acid, an important compound in the synthesis of Dacron.

Similarly, the aromatic ring is stable to chemical reduction. Thus, the acyl group can be reduced to a methylene group, as noted earlier in the Friedel-Crafts acylation reaction. Also, nitro groups can be reduced by tin and hydrochloric acid to give the amino group. Nitrobenzene gives aniline as a reduction product.

$$C_6H_5NO_2 + Sn + HCl \rightarrow C_6H_5NH_2$$

Nitrobenzene — Aniline

Derivatives of aromatic amines have important medicinal properties and are used as drugs. Nitration of an aromatic hydrocarbon followed by chemical reduction of the nitro group to the amino group provides a facile route to the intermediates used in many of these medicinal compounds.

HETEROCYCLIC COMPOUNDS

As has been stated previously, the two major types of organic compounds are the aliphatic and the cyclic. If the cyclic compounds are composed of rings of carbon atoms only, as in benzene and its derivatives, they are called **carbocyclic.** When atoms other than carbon are also included in the ring the compounds are termed **heterocyclic.** The most commonly occurring elements other than carbon in these ring structures are oxygen, nitrogen, and sulfur.

The basic ring structure of a heterocyclic compound is called the **heterocyclic nucleus.** In the majority of heterocyclic compounds the nucleus consists of five- or six-membered rings that contain either one or two elements other than carbon. Examples of the important heterocyclic nuclei and their derivatives will be considered in the following sections.

Five-Membered Rings

Several important heterocyclic compounds are derived from a heterocyclic ring made up of 1 oxygen and 4 carbon atoms. This ring is known as **furan,** and one of its most important derivatives is the α-aldehyde, **furfural.** For purposes of nomenclature, the rings are numbered counterclockwise starting with the element other than carbon. In the furan nucleus illustrated here, the carbon atoms adjacent to the oxygen are α, the next ones are called β carbons.

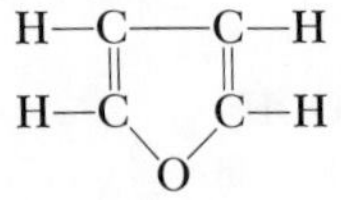

H—C——C—H
H—C   C—H
O

Furan

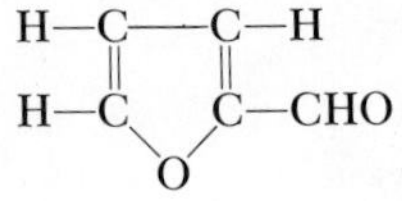

H—C——C—H
H—C   C—CHO
O

Furfural

β′ C4 —— 3C β
α′ C5   2C α
O

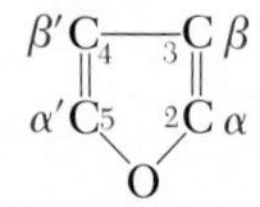

Furfural is produced commercially by the dehydrating action of acids on pentose sugars found in corn cobs, oat hulls, and straw. Furfural is a colorless liquid with a characteristic odor. It reacts with aniline to produce a red color and with phloroglucinol to produce a dark green precipitate. These color reactions may be used to test for the presence of furfural, or indirectly for pentose sugars.

Another important five-membered heterocyclic ring, containing nitrogen in place of oxygen, is **pyrrole.** It was originally obtained from animal matter but can be readily synthesized from the action of ammonia on a dicarboxylic hydroxy acid called glycaric acid.

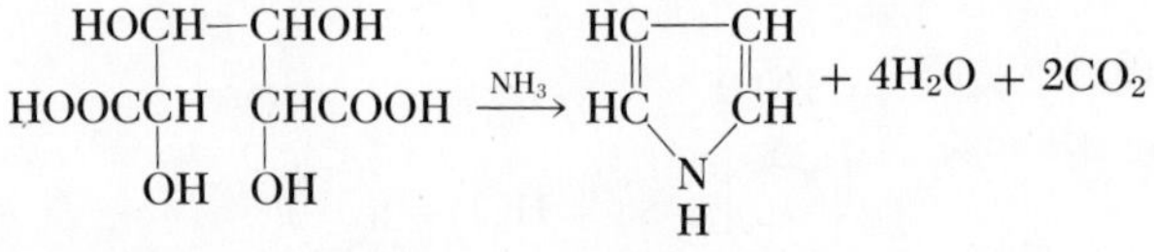

HOCH—CHOH
HOOCCH CHCOOH $\xrightarrow{NH_3}$ HC——CH / HC CH / N / H $+ 4H_2O + 2CO_2$
OH OH

Glycaric acid   Pyrrole

Pyrrole is a liquid that gradually forms dark-colored resins on exposure to the air. It is a constituent of many important naturally occurring substances such as hemoglobin, chlorophyll, amino acids, and alkaloid drugs.

The pyrrole nucleus is an integral constituent of the porphyrins, which are present in many natural pigments. A porphyrin closely related to hemoglobin and chlorophyll is protoporphyrin. The nitrogen atoms of the porphyrin combine with such metals as iron to form the heme of hemoglobin and with magnesium to form chlorophyll. The cytochromes, which are enzymes involved in biological oxidations and reductions, are also related to the porphyrins and contain groups similar to heme.

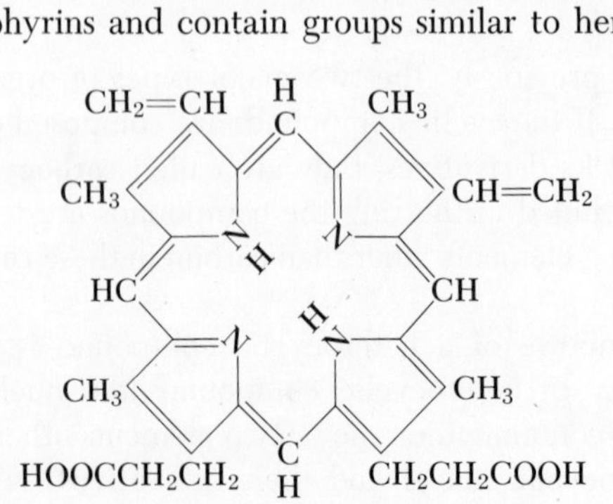

CH_2=CH   H C CH_3
CH_3 CH=CH_2
N H   N
HC   CH
N   H N
CH_3 CH_3
$HOOCCH_2CH_2$ C H CH_2CH_2COOH

Protoporphyrin

If pyrrole is condensed with a benzene ring, another type of heterocyclic nucleus called **indole** is produced. Indole and 3-methyl indole, which is called **skatole,** are formed during the putrefaction of proteins in the large intestine. They are responsible for the characteristic odor of feces. One of the most important derivatives of the indole nucleus is the amino acid **tryptophan.** Tryptophan is present in most proteins and is an essential constituent of the diet of growing animals.

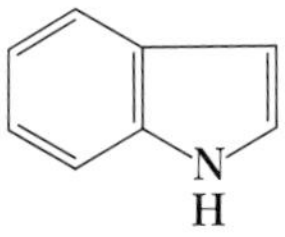

CH_3 $CH_2CHCOOH$ / NH_2

N H   N H   N H

Indole 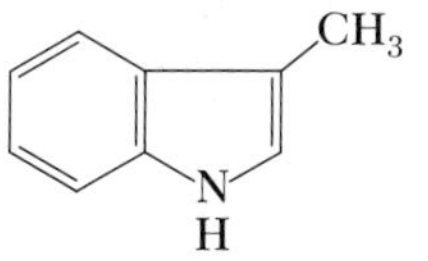   Skatole 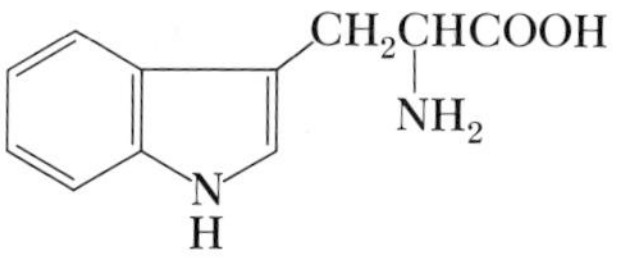   Tryptophan

The other example of a five-membered heterocyclic ring, in which sulfur takes the place of oxygen or nitrogen, is **thiophene.** This compound occurs as an impurity in the benzene obtained from coal tar. It is produced commercially by a reaction of butane and sulfur at high temperatures.

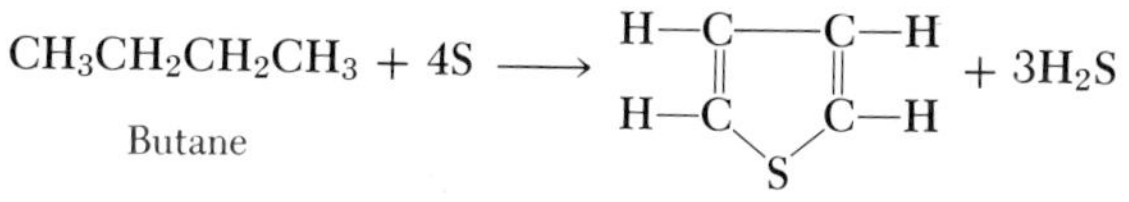

$$CH_3CH_2CH_2CH_3 + 4S \longrightarrow \begin{matrix} H—C——C—H \\ \| \quad\quad \| \\ H—C \quad C—H \\ S \end{matrix} + 3H_2S$$

Butane   Thiophene

Five-Membered Rings with Two Heterocyclic Atoms. There are several examples of heterocyclic rings containing two elements other than carbon. Oxazole, imidazole, and thiazole are similar to furan, pyrrole, and thiophene, except that they have a nitrogen in another position in each ring. Pyrazole is similar to pyrrole, except that an extra nitrogen is in the 2 position.

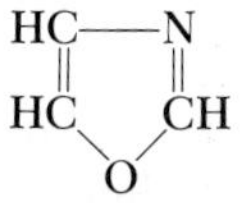

$$\begin{matrix} HC——N \\ \| \quad \| \\ HC \quad CH \\ O \end{matrix} \qquad \begin{matrix} HC——N \\ \| \quad \| \\ HC \quad CH \\ N \\ H \end{matrix}$$

Oxazole 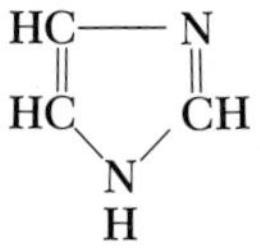   Imidazole

$$\begin{matrix} N——CH \\ \| \quad \| \\ HC \quad CH \\ S \end{matrix} \qquad \begin{matrix} HC——CH \\ \| \quad \| \\ HC \quad N \\ N \\ H \end{matrix}$$

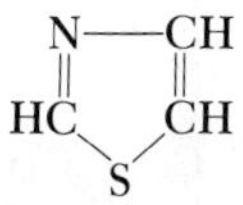

Thiazole 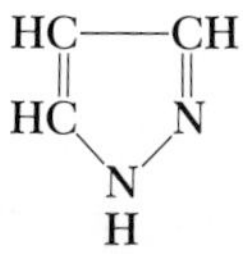   Pyrazole

The imidazole ring is found in important compounds such as the purines, which will be considered later, and the amino acid histidine, which is essential to animal nutrition. The thiazole ring is present in the vitamin thiamine and in the penicillins, which are valuable antibiotic agents used in medicine.

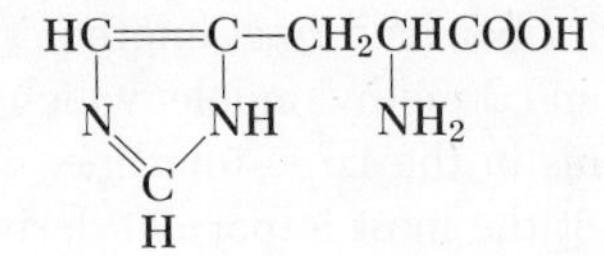

HC═C—CH$_2$CHCOOH
N NH NH$_2$
C
H

Histidine

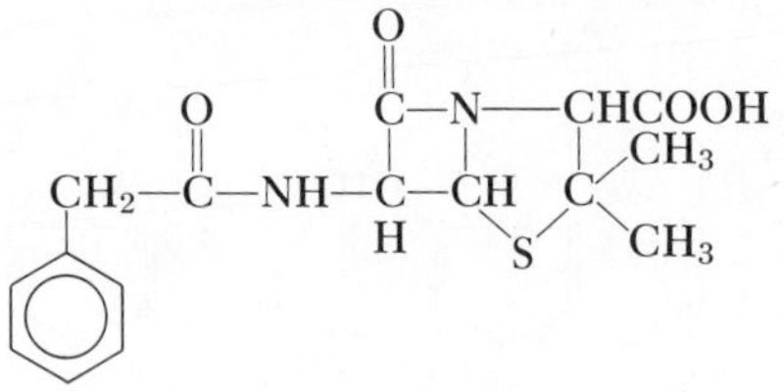

O O
C—N—CHCOOH
CH$_2$—C—NH—C—CH C CH$_3$
H S CH$_3$

Penicillin G

Six-Membered Rings

The most common six-membered rings containing one element other than carbon are pyran and pyridine.

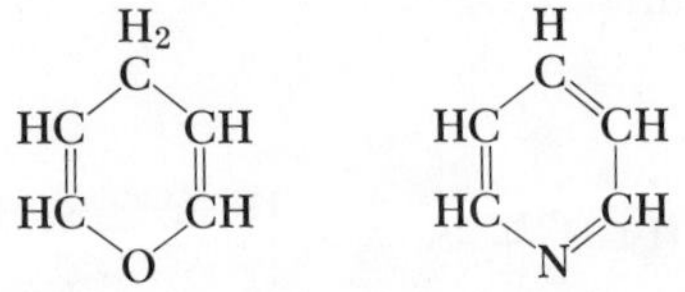

H$_2$ H
C C
HC CH HC CH
HC CH HC CH
O N

Pyran Pyridine

The pyran ring is present in anthocyanin, which is responsible for the color of flowers, and in rotenone, a plant material that is used as an insecticide. The benzopyran ring is found in many plants and is an integral part of the α-tocopherol molecule. Wheat germ oil is especially rich in α-tocopherol, or vitamin E, which is necessary for normal growth and reproduction in animals.

Pyridine is a common heterocyclic compound obtained from coal tar. Pyridine is a liquid with a characteristically disagreeable odor. It behaves as a weak base and is a good solvent for both organic and inorganic compounds. It is used to manufacture pharmaceuticals such as sulfa drugs, antihistamines, and steroids. Pyridine serves as a denaturant for ethyl alcohol, as a rubber accelerator, and in the preparation of a waterproofing agent for textiles. The methyl pyridines are known as picolines and may be oxidized to the corresponding picolinic acids. The acid obtained from the oxidation of β-picoline, or 3-methyl pyridine, is known as **nicotinic acid.**

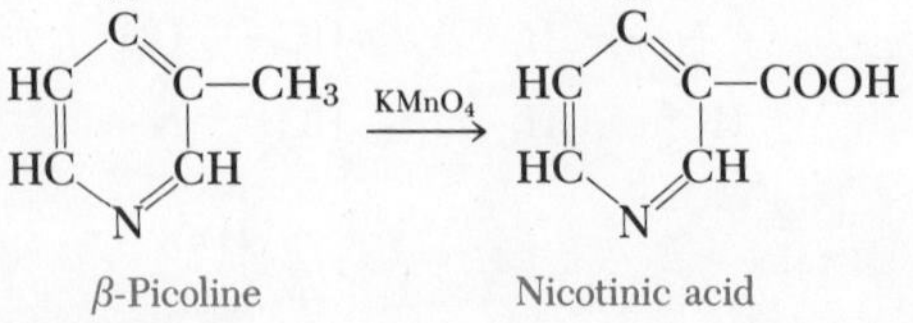

H H
C C
HC C—CH$_3$ $\xrightarrow{KMnO_4}$ HC C—COOH
HC CH HC CH
N N

β-Picoline Nicotinic acid

Nicotinic acid and its amide are members of the vitamin B complex. Consumption of a diet lacking in these compounds results in a deficiency disease called pellagra. Nicotinamide is used by the body for the manufacture of coenzymes I and II, which are essential for the proper functioning of certain dehydrogenating enzymes. Isonicotinic acid hydrazide, known as **isoniazid,** is an effective tuberculostatic drug.

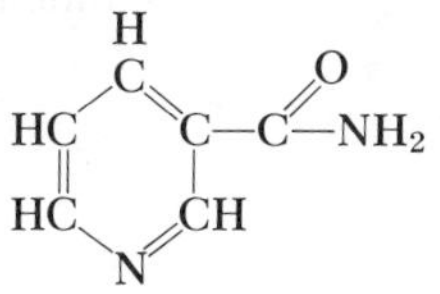

Nicotinamide

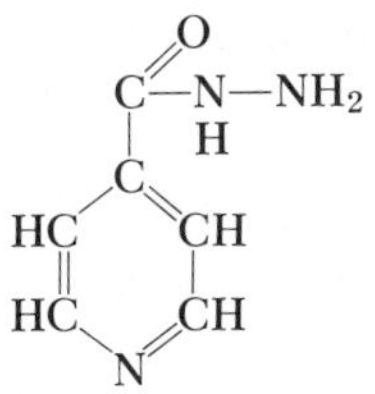

Isonicotinic acid hydrazide (isoniazid)

Six-Membered Rings with Two Hetero Atoms. There are three six-membered rings containing two nitrogen atoms. Pyridazine contains the nitrogens in the 1,2 position, pyrimidine in the 1,3, and pyrazine in the 1,4 positions of the ring. Pyrimidine is by far the most important of these heterocyclic rings.

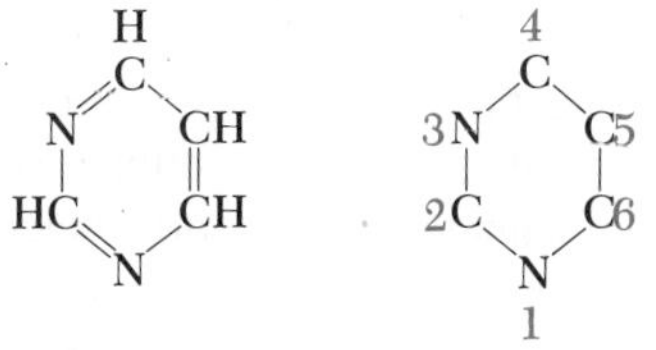

Pyrimidine

Pyrimidine nucleus

Pyrimidine derivatives are found in the nucleoproteins, which are essential constituents of all living cells. Three important pyrimidines in nucleoproteins are cytosine (2-oxy,4-amino pyrimidine), uracil (2,4-dioxy pyrimidine), and thymine (2,4-dioxy,5-methyl pyrimidine). Other important derivatives of pyrimidine are **thiamine,** or vitamin B_1, which contains a pyrimidine nucleus joined to a thiazole nucleus, and the **barbiturates.**

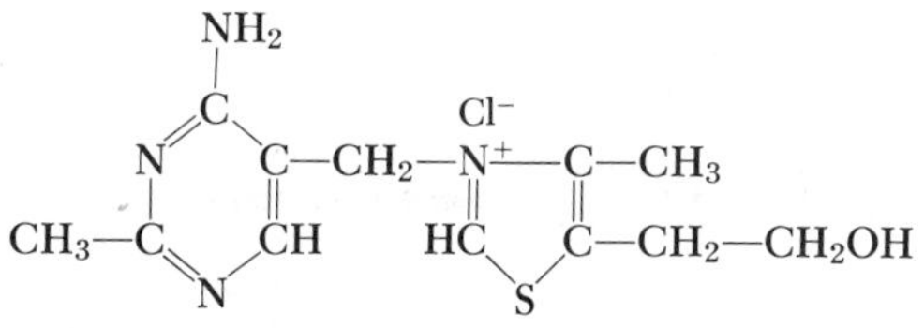

Thiamine

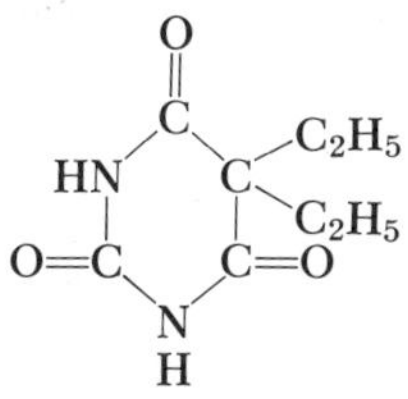

Barbital

Thiamine is one of the most important members of the vitamin B complex; its absence from the diet produces the deficiency disease known as beriberi or polyneuritis. **Barbital,** originally called Veronal, has been used as a hypnotic for about fifty years. Several derivatives of barbital have been prepared to increase the speed of action and decrease the unpleasant after-effects of the drug. **Phenyl barbital, Amytal,** and **Seconal** are examples of these barbital derivatives. In recent years the barbiturates have been used in such large quantities by the public that an attempt is being made to control their distribution.

Another class of compounds closely related to the pyrimidines are the **purines.** The purine's nucleus is composed of a pyrimidine ring fused to an imidazole ring and is usually represented as follows:

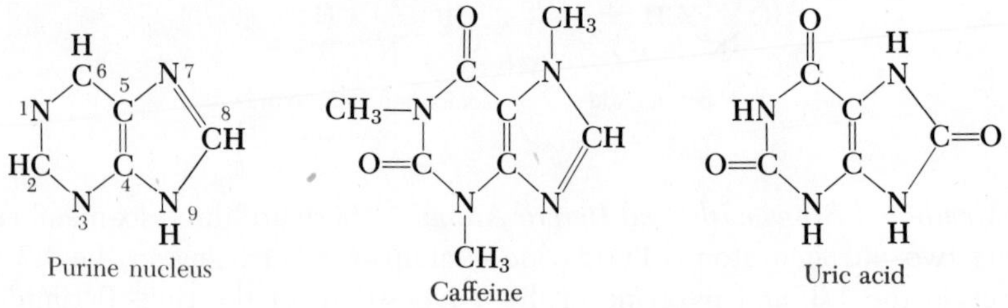

Purine nucleus

Caffeine

Uric acid

Both purines and pyrimidines that contain an oxygen atom attached to the ring at the 2, 6, or 8 position may exist in either the keto (C═O) or enol (C—OH) form. This phenomenon is known as **keto-enol tautomerism.** Although these compounds are represented as keto derivatives, they may shift to the enol form, especially when they are combined with other molecules in nucleic acids and nucleoproteins.

The purines present in nucleoproteins are adenine (6-amino purine) and guanine (2-amino, 6-oxy purine). The end-product of metabolism or oxidation of purines in the body is **uric acid,** which is 2,6,8-trioxypurine. Uric acid is present in the blood and is excreted in the urine. Under abnormal conditions uric acid may form insoluble deposits, or stones, in the kidney or bladder, or may crystallize in the joints and cause the painful condition known as gout. The stimulants **theophylline** in tea and **caffeine** in tea and coffee are methylated purines. Theophylline is used in certain heart conditions and in bronchial asthma. It is also a constituent of Dramamine, which is used for the prevention of sea sickness or air sickness. Caffeine is a very active stimulant, which is consumed in large quantities in coffee, cocoa, and cola based drinks.

IMPORTANT TERMS AND CONCEPTS

acylation
alkylation
aromatic hydrocarbon
barbiturate
benzyl group
delocalized electrons
electrophilic aromatic substitution
Friedel-Crafts reaction
furan
heterocyclic compound
meta-directing groups
nitration
ortho-para directing groups
phenyl group
pyran
pyridine
pyrrole
resonance hybrid
sulfa drug
sulfonation

QUESTIONS

1. Draw a structural formula for each of the following compounds:
   (a) toluene
   (b) naphthalene
   (c) *p*-nitrotoluene
   (d) *m*-bromotoluene
   (e) benzyl bromide
   (f) *m*-diethylbenzene
   (g) *o*-dinitrobenzene
   (h) 2,4-dinitrotoluene
   (i) 1,3,5-trimethylbenzene
   (j) 1,4-dichloro-2,5-dibromobenzene

2. Draw out all the possible structural isomers for aromatic compounds with the following molecular formulas and name each of the isomers:
   (a) C_8H_{10}
   (b) C_8H_9Cl
   (c) $C_7H_6Br_2$
   (d) $C_6H_3Cl_2Br$
   (e) C_7H_6ClBr

3. Write equations for each of the following reactions, and name each of the organic products obtained:
   (a) benzene + isopropyl chloride ($AlCl_3$ catalyst)
   (b) toluene + Br_2 (Fe catalyst)
   (c) toluene + H_2SO_4 + SO_3
   (d) nitrobenzene + concentrated nitric acid + sulfuric acid (heat)
   (e) chlorobenzene + bromine (Fe catalyst)
   (f) benzenesulfonic acid + Br_2 (Fe catalyst)
   (g) toluene + nitric acid + sulfuric acid
   (h) ethylbenzene + H_2SO_4 + SO_3
   (i) toluene + acetyl chloride ($AlCl_3$ catalyst)
   (j) toluene + bromine (sunlight)

4. Show how the following conversions may be carried out in the laboratory. More than one step may be necessary. Give any necessary catalysts.
   (a) benzene to *m*-chloronitrobenzene
   (b) benzene to *n*-butylbenzene
   (c) benzene to benzoic acid
   (d) benzene to *m*-nitroethylbenzene
   (e) benzene to *m*-bromoacetophenone

5. When $(CH_3)_2CHCH_2Cl$ is reacted with benzene and aluminum chloride, the isolated hydrocarbon is *t*-butylbenzene. Outline a mechanism to explain the formation of this product.

6. How many tetrachlorobenzenes are there? Name them.

SUGGESTED READING

Carter: The History of Barbituric Acid. Journal of Chemical Education, Vol. 28, p. 524, 1951.
Duewell: Aromatic Substitution. Journal of Chemical Education, Vol. 43, p. 138, 1966.
Fedor: Major Aromatics in Transition, 1961–1965. Chemical and Engineering News, Part I, p. 116, March 20, 1961. Part II, p. 130, March 27, 1961.
Ferguson: The Orientation and Mechanism of Electrophilic Aromatic Substitution. Journal of Chemical Education, Vol. 32, p. 42, 1955.
Gero: Kekule's Theory of Aromaticity. Journal of Chemical Education, Vol. 31, p. 201, 1954.
Hartough: The Chemical Nature of Thiophene and Its Derivatives. Journal of Chemical Education, Vol. 27, p. 500, 1950.
Marsi and Wilen: Friedel-Crafts Alkylation. Journal of Chemical Education, Vol. 40, p. 214, 1963.
Noller: A Physical Picture of Covalent Bonding and Resonance in Organic Chemistry. Journal of Chemical Education, Vol. 27, p. 504, 1950.
Varshni: Directive Influence of Substituents in the Benzene Ring. Journal of Chemical Education, Vol. 30, p. 342, 1953.
Waack: The Stability of the Aromatic Sextet. Journal of Chemical Education, Vol. 39, p. 469, 1962.
Willemant: "Charles Friedel." Journal of Chemical Education, Vol. 26, p. 2, 1949.

ALCOHOLS

CHAPTER 6

The *objectives* of this chapter are to enable the student to:

1. Recognize an alcohol and a phenol.
2. Write the structure and names of alcohols and phenols.
3. Explain the anomalous physical properties of alcohols.
4. Explain the enhanced acidity of aromatic alcohols (phenols).
5. Describe the conversion of other functional groups into alcohols or phenols.
6. Distinguish the different modes of hydration of olefins used in alcohol formation.
7. Describe the basic functional group reactions of alcohols and phenols.
8. Recognize some of the more important alcohols and phenols and their application in modern society.

The basic elements of the compounds contained in the preceding chapter were carbon and hydrogen. The addition of a third element, oxygen, greatly extends the number of types of possible organic compounds. Oxygen has six electrons in its valence shell and must therefore share two electrons to attain the stable inert gas configuration. The manner in which oxygen shares electrons with carbon and hydrogen determines the type of organic compound. For example, when oxygen shares one electron with carbon and one electron with hydrogen, the type of organic compound called an **alcohol** results. If each valence electron is shared with a separate carbon atom, a class of compounds called **ethers** is produced. If each electron is shared with the same carbon atom to form a carbon-oxygen double bond, an aldehyde or ketone results, depending upon the other groups or atoms bonded to carbon. The various types of covalent bonding possible for oxygen are illustrated below:

$$\begin{array}{ccccccc} & H & & H & & & \\ & | & & | & & & \\ H- & C & - & C & -O-H \\ & | & & | & & & \\ & H & & H & & & \end{array}$$
Ethyl alcohol

$$\begin{array}{ccccc} & H & & H & \\ & | & & | & \\ H- & C & -O- & C & -H \\ & | & & | & \\ & H & & H & \end{array}$$
Dimethyl ether

$$\begin{array}{cccc} & H & & H \\ & | & & | \\ H- & C & - & C{=}O \\ & | & & \\ & H & & \end{array}$$
Acetaldehyde

$$\begin{array}{ccccccc} & H & & O & & H & \\ & | & & \| & & | & \\ H- & C & - & C & - & C & -H \\ & | & & & & | & \\ & H & & & & H & \end{array}$$
Acetone

Ethers, aldehydes, and ketones will be discussed in detail in subsequent chapters.

Of all the classes of organic compounds, alcohols are probably the best known. For centuries it has been recognized that alcoholic beverages contain ethyl, or grain, alcohol. Also, most temporary automobile antifreezes contain methyl, or wood alcohol. Most permanent type antifreezes contain ethylene glycol (CH_2OHCH_2OH), an alcohol containing two hydroxyl groups.

Alcohols are also very important biologically, since the alcohol group occurs in a variety of compounds associated with biologic systems. For example, most sugars contain several hydroxyl (—OH) groups (see Chapter 13), and starch, cellulose, and glycogen contain thousands of hydroxyl groups. Cholesterol, hormones, and other related steroids also contain the alcohol functional group (see Chapter 14), as well as fat soluble vitamins such as vitamin A and vitamin D (Chapter 14). In fact, the reaction of the alcohol functional group in vitamin A (p. 238) is important in the human visual process. Not only alcohols themselves, but also derivatives of alcohols are important biologically. For example, fats (p. 225) which are simple derivatives of the important trihydroxy alcohol, glycerol, $CH_2OHCHOHCH_2OH$, and phosphatides (p. 232), which are complex derivatives of glycerols, are also important in biologic processes.

Alcohols may be considered derivatives of hydrocarbons in which a hydrogen of the hydrocarbon has been replaced (substituted) by a hydroxyl group. Another way of considering alcohols is to view them as the organic analogs of water, in which one of the hydrogen atoms of water has been replaced by an alkyl or aryl group. If the latter way of viewing alcohols is accepted, the chemical and physical behavior of alcohols might be anticipated to be similar in some respects to water.

H—OH	R—OH	Ar—OH
Water	Aliphatic alcohol	Aromatic* alcohol

Before studying the chemistry of alcohols, it is convenient to subdivide the aliphatic alcohols into three classes, primary, secondary, and tertiary. Each class is dependent upon the hydroxyl group being attached to a carbon that is covalently bonded to one, two, or three other carbon atoms. Many reactions of alcohols can be more easily discussed employing this type of subclassification, and it is only for this reason that the alcohols are classified in the aforesaid manner.

$$\underset{\text{Primary alcohol}}{R-\overset{\overset{H}{|}}{\underset{\underset{H}{|}}{C}}-OH} \qquad \underset{\text{Secondary alcohol}}{R-\overset{\overset{H}{|}}{\underset{\underset{R}{|}}{C}}-OH} \qquad \underset{\text{Tertiary alcohol}}{R-\overset{\overset{R}{|}}{\underset{\underset{R}{|}}{C}}-OH}$$

NOMENCLATURE

Alcohols are named in the IUPAC system by replacing the **-e** ending of the corresponding alkane with the characteristic **-ol** ending of the alcohols. Other substituents are named and their positions on the carbon chain indicated by the appropriate number and prefix. Since the hydroxyl group may appear at more than one position on the carbon chain, its position must also be indicated by the number of the carbon atom to which it is attached. Common names are used for the lower members of this series and are formulated by naming the alkyl group attached to the hydroxyl group, followed by the term alcohol. Still another system names alcohols as derivatives of carbinol (CH_3OH).

* The symbol Ar— is used to denote an aromatic ring. In this case, if Ar— = C_6H_5, the aromatic alcohol would be phenol. C_6H_5OH.

In this system, the alkyl or aryl groups which replace the carbon-hydrogen bonds in the methyl group are named, followed by the term carbinol. Some typical examples illustrating these various nomenclature and classification systems are illustrated below:

Compound	*Name*	*Classification*
CH_3OH	Methanol Methyl alcohol Carbinol	Primary
CH_3CH_2OH	Ethanol Ethyl alcohol Methyl carbinol	Primary
$CH_3CH_2CH_2OH$	1-Propanol *n*-Propyl alcohol Ethyl carbinol	Primary
$CH_3CHOHCH_3$	2-Propanol Isopropyl alcohol Dimethyl carbinol	Secondary
$CH_3CH_2CH_2CH_2OH$	1-Butanol *n*-Butyl alcohol *n*-Propyl carbinol	Primary
$(CH_3)_3COH$	2-Methyl-2-propanol *t*-Butyl alcohol Trimethyl carbinol	Tertiary

The aromatic alcohols are named as derivatives of the parent compound phenol (or carbolic acid). Some typical examples are represented below:

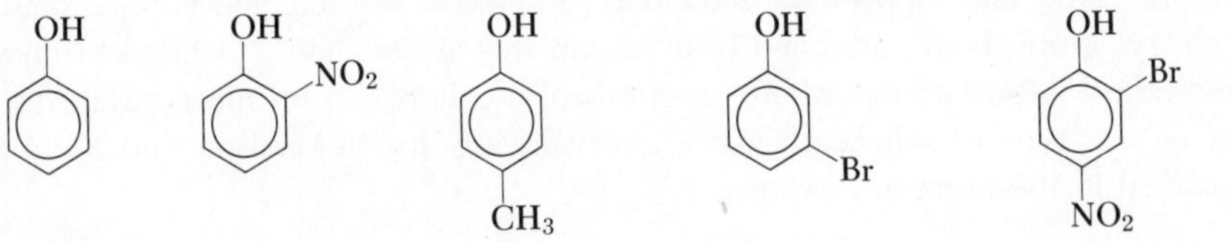

OH NO2 OH CH3 OH Br OH Br NO2 OH

Phenol — *o*-Nitrophenol — *p*-Methylphenol (*p*-cresol) — *m*-Bromophenol — 2-Bromo-4-nitrophenol

The methyl derivatives are also known by the common name, **cresols.**

Phenols also play an important biologic role. One of the important amino acids, tyrosine, is a phenol derivative (see p. 248). Also, the amino acid phenylalanine is biologically oxidized to a phenol derivative, dopa. Dopa is currently being used in the

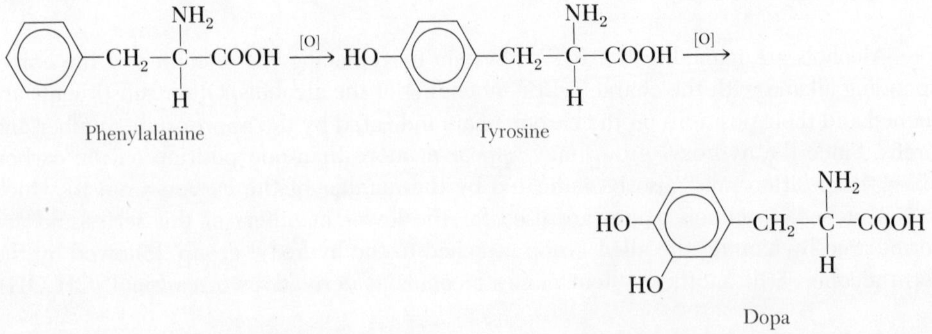

$$C_6H_5-CH_2-\underset{H}{\overset{NH_2}{C}}-COOH \xrightarrow{[O]} HO-C_6H_4-CH_2-\underset{H}{\overset{NH_2}{C}}-COOH \xrightarrow{[O]}$$

Phenylalanine — Tyrosine

$$(HO)_2C_6H_3-CH_2-\underset{H}{\overset{NH_2}{C}}-COOH$$

Dopa

control of Parkinson's disease and has shown great success in restoring normal life to persons affected with this disease.

PHYSICAL PROPERTIES OF ALCOHOLS

The introduction of a hydroxyl group into a hydrocarbon has a pronounced effect on the physical properties of the compound. In contrast to hydrocarbons, which are insoluble in water, the short-chain alcohols (methanol through the butanols) are soluble in water. As the number of carbons in the alcohol increases, the solubility in water decreases, and the physical properties approach those of the saturated hydrocarbons.

The boiling points of alcohols are also abnormally high compared to the saturated hydrocarbons of comparable molecular weight. For example, ethanol, CH_3CH_2OH (mol. wt. 46), has a boiling point of +78°C, whereas propane, $CH_3CH_2CH_3$ (mol. wt. 44), boils at −42°C. The large increase in boiling point is due to the presence of the hydroxyl group. That the increase is not due merely to the presence of the oxygen atom can be proved by comparing ethanol to its isomer, dimethyl ether, CH_3OCH_3 (mol. wt. 46), which boils at −24°C. Therefore, the change in boiling point must be due to the presence of the —OH group. The effect of this group has been attributed to the presence of hydrogen bonds formed between alcohol molecules in the liquid state, similar to the type of hydrogen bonding encountered in water.

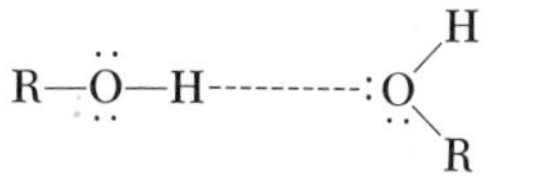

R—Ö—H----------:O(H)(R)   H(H)Ö:----------H—Ö—H

Hydrogen-bonded alcohol 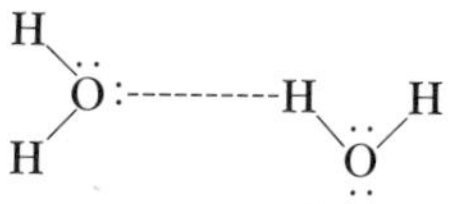   Hydrogen-bonded water

This type of association can also be used to explain the solubility of the short-chain alcohols in water. When these alcohols are dissolved in water, hydrogen bonding occurs between the hydroxyl group of the alcohol and the hydroxyl group of the water.

The aromatic alcohol, **phenol,** is slightly soluble in water, but very soluble in alcohol, ether, and other organic solvents. In contrast to aliphatic alcohols, which are weaker acids (Ka $\sim 10^{-16}$) than water (Ka $= 10^{-14}$), phenols are much stronger acids than water. Because phenol is $\sim 10^4$ stronger acid than water, it is soluble in aqueous sodium hydroxide and forms a sodium salt of the phenol. Aliphatic alcohols, however, do not form stable salts with aqueous sodium hydroxide, since water is $\sim 10^2$ stronger acid than the alcohol and displaces the following equilibrium to the left.

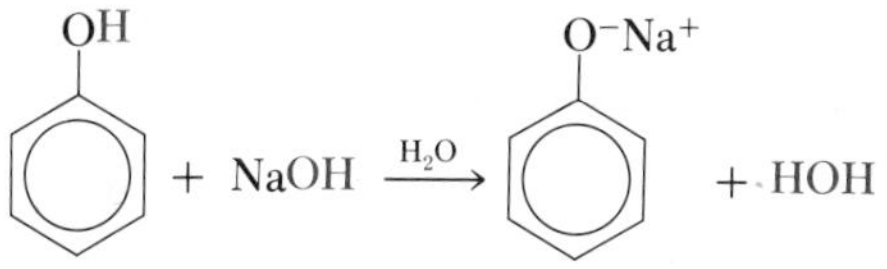

$$C_6H_5OH + NaOH \xrightarrow{H_2O} C_6H_5O^-Na^+ + HOH$$

Phenol   Sodium hydroxide   Sodium phenoxide

$$CH_3CH_2OH + NaOH \rightleftharpoons CH_3CH_2O^-Na^+ + HOH$$

Ethanol   Sodium ethoxide

The increased acidity of phenol (relative to aliphatic alcohols) is mainly due to resonance stabilization of the resulting anion that is formed. The negative charge on oxygen can be delocalized (spread out) over the aromatic ring. No such stabilization of

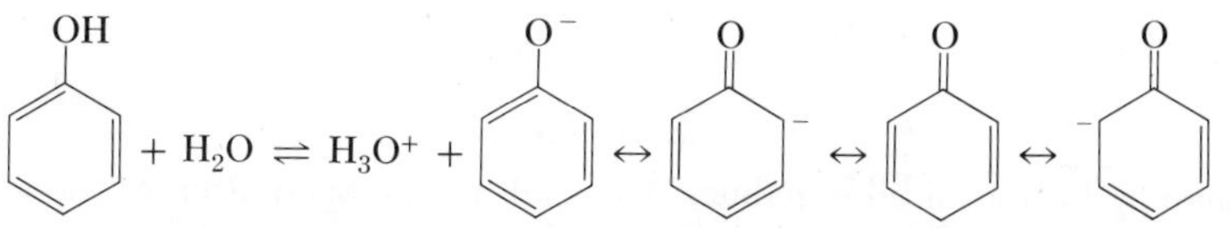

$$C_6H_5OH + H_2O \rightleftharpoons H_3O^+ + C_6H_5O^- \leftrightarrow \leftrightarrow \leftrightarrow$$

the alkoxide ion from an aliphatic alcohol is possible, hence the formation of the anion (loss of a proton) is not as favored as with phenols. The introduction of nitro groups

TABLE 6–1 PHYSICAL PROPERTIES OF ALCOHOLS AND PHENOLS

COMPOUND	STRUCTURE	M.P. °C	B.P. °C
Methyl alcohol	CH_3OH	−97	+65
Ethyl alcohol	CH_3CH_2OH	−114	+78
n-Propyl alcohol	$CH_3CH_2CH_2OH$	−126	+97
Isopropyl alcohol	$(CH_3)_2CHOH$	−89	+82
n-Butyl alcohol	$CH_3CH_2CH_2CH_2OH$	−90	+118
Isobutyl alcohol	$(CH_3)_2CHCH_2OH$	−108	+108
t-Butyl alcohol	$(CH_3)_3COH$	+25	+83
Cyclohexanol	OH	+24	+162
Phenol	OH	+41	+182
o-Cresol	CH_3, OH	+31	+191
m-Cresol	CH_3, OH	+12	+202
p-Cresol	CH_3, OH	+35	+202
Benzyl alcohol	CH_2OH	−15	+205

increases the acidity of phenols, and 2,4,6-trinitrophenol (picric acid) is a fairly strong acid (Table 6–1).

OH | OH, CH_3 | OH, NO_2 | OH, O_2N, NO_2, NO_2

Phenol $Ka \cong 10^{-10}$ | *p*-Cresol $Ka \cong 10^{-10}$ | *p*-Nitrophenol $Ka \cong 10^{-7}$ | Picric acid $Ka = 10^{-1}$

METHODS OF PREPARATION OF ALCOHOLS

Hydrolysis of Alkyl Halides

Alkaline hydrolysis of alkyl halides, in which the halogen atom of the alkyl halide is displaced by a hydroxyl group, is a useful method for preparing alcohols. A generalized representation of this reaction is shown on the following page:

$$R{-}X + OH^- \xrightarrow{H_2O} R{-}OH + X^-$$

X = Cl, Br, I
R = Alkyl group

Aqueous sodium or potassium hydroxide, or aqueous silver oxide* (Ag_2O), is generally used as the hydrolyzing agent. The reactivity of the halides in this reaction is RI > RBr > RCl. Aromatic halides, in which the halogen is directly attached to the ring, are inert to this type of displacement reaction. However, halides such as benzyl halides (C_6H_5-CH_2Cl) do undergo this reaction. A competing side reaction in this method of preparation is the elimination of hydrogen halide from the alkyl halide to give an olefin. The extent of the elimination reaction depends on the type of base used, the nature of the halide,

$$R'CH_2CH_2X + OH^- \xrightarrow{H_2O} R'CH{=}CH_2 + H_2O + X^-$$

X = Cl, Br, I
R′ = Alkyl group

the type of alkyl group, and the temperature of the reaction. Elimination increases with temperature; increases as the size of the halide being displaced increases; increases as the strength of the base increases; and increases in the order primary < secondary < tertiary for the type of alkyl group. Therefore, the best yields of the alcohols are obtained from primary alkyl halides using silver oxide at moderate temperatures. Some typical examples are outlined below:

$$CH_3CH_2CH_2I + Ag_2O \xrightarrow{H_2O} CH_3CH_2CH_2OH + AgI\downarrow$$

n-Propyl iodide → *n*-Propyl alcohol

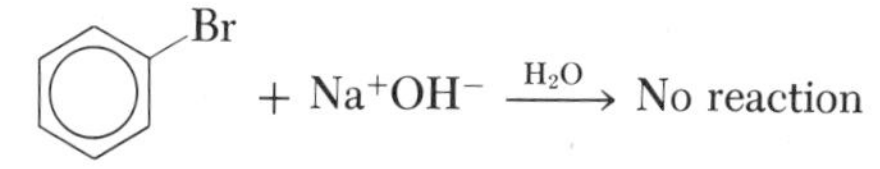

$$C_6H_5Br + Na^+OH^- \xrightarrow{H_2O} \text{No reaction}$$

Bromobenzene

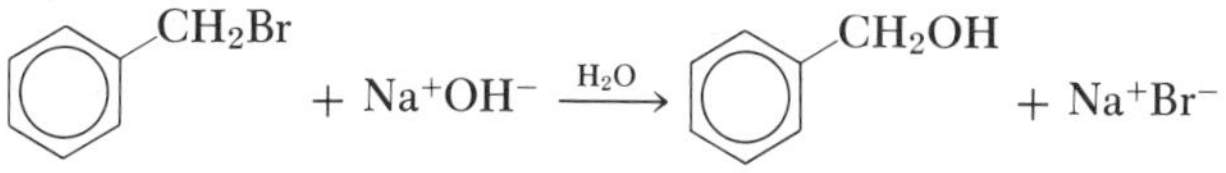

$$C_6H_5CH_2Br + Na^+OH^- \xrightarrow{H_2O} C_6H_5CH_2OH + Na^+Br^-$$

Benzyl bromide → Benzyl alcohol

Hydration of Olefins

Alkenes undergo hydration in the presence of strong acids such as sulfuric acid to give alcohols as the final product. The hydration reaction follows Markownikoff's rule; therefore, secondary and tertiary alcohols are formed except in the case of ethylene. Some typical examples are shown below:

$$CH_2{=}CH_2 + H_2O \xrightarrow{H_2SO_4} CH_3CH_2OH$$

Ethylene → Ethyl alcohol

$$CH_3CH{=}CH_2 + H_2O \xrightarrow{H_2SO_4} CH_3CHOHCH_3$$

Propylene → Isopropyl alcohol

$$(CH_3)_2C{=}CH_2 + H_2O \xrightarrow{H_2SO_4} (CH_3)_3COH$$

Isobutylene → *t*-Butyl alcohol

*The aqueous solution of silver oxide may be represented as AgOH (silver hydroxide). The silver oxide method promotes less elimination than the aqueous alkali method.

The mechanism of the reaction is similar to the mechanism observed in the electrophilic addition reactions of olefins. The alkyl hydrogen sulfate initially formed undergoes

$$CH_3CH{=}CH_2 + H_2SO_4 \rightleftharpoons CH_3\overset{+}{C}HCH_3 + HSO_4^-$$

$$CH_3\overset{+}{C}HCH_3 + HSO_4^- \rightarrow CH_3\underset{\displaystyle OSO_2OH}{\underset{|}{C}}HCH_3$$

$$CH_3\underset{\displaystyle OSO_2OH}{\underset{|}{C}}HCH_3 + H_2O \rightarrow CH_3\underset{\displaystyle OH}{\underset{|}{C}}HCH_3 + H_2SO_4$$

displacement by water to give the alcohol. Since the first step involves the addition of the electrophile H^+, these reactions would be expected to follow Markownikoff's rule.

Isopropyl and *t*-butyl alcohols are prepared commercially from petroleum fractions in this manner *via* the appropriate secondary and tertiary carbonium ions.

Electrophilic addition of water also plays an important role biologically in the metabolism of carbohydrates (Chapter 20). For example, as part of the Krebs cycle (p. 324), fumaric acid is converted to malic acid and *cis*-aconitic acid is converted to isocitric acid. Both of these conversions involve the enzymatic hydration of a carbon-carbon double bond.

$$\begin{matrix} H & & COOH \\ & C{=}C & \\ HOOC & & H \end{matrix} \xrightarrow[\text{Enzyme}]{H_2O} HOOC-\underset{\displaystyle OH}{\underset{|}{\overset{\displaystyle H}{\overset{|}{C}}}}-\underset{\displaystyle H}{\underset{|}{\overset{\displaystyle H}{\overset{|}{C}}}}-COOH$$

Fumaric acid → Malic acid

$$\begin{matrix} HOOC & & COOH \\ & C{=}C & \\ HOOCCH_2 & & H \end{matrix} \xrightarrow[\text{Enzyme}]{H_2O} HOOCCH_2\underset{\displaystyle COOH}{\underset{|}{C}}H-\overset{\displaystyle OH}{\overset{|}{C}}HCOOH$$

cis-aconitic acid → Isocitric acid

Until 1955 the preparation of primary alcohols such as 1-butanol *via* the hydration of alkenes was an unfulfilled goal of the organic chemist. The anti-Markownikoff addition of water to olefins was successfully developed by Professor Herbert C. Brown at Purdue University by the alkaline hydrogen peroxide oxidation of organoboranes. The organoboranes could be easily formed by the reaction of diborane* with olefins.

$$\underset{\text{1-Butene}}{CH_3CH_2CH{=}CH_2} + \underset{\text{Diborane}}{B_2H_6} \rightarrow \underset{\text{Tri-}n\text{-butylborane}}{2\,(CH_3CH_2CH_2CH_2)_3B}$$

$$2(CH_3CH_2CH_2CH_2)_3B \xrightarrow[OH^-]{H_2O_2} \underset{\text{1-Butanol}}{6CH_3CH_2CH_2CH_2OH} + \underset{\text{Boric acid}}{H_3BO_3}$$

As outlined in this reaction sequence, the boron always adds to the least substituted carbon of the olefinic bond. Consequently, with terminal olefins, primary alcohols always result. This type of anti-Markownikoff hydration is commonly called *hydroboration*. Some other typical examples of the hydroboration technique of hydration are shown on the following page:

*Diborane can be conveniently generated by the reaction of sodium borohydride and boron trifluoride: $3NaBH_4 + BF_3 \rightarrow 2B_2H_6 + 3NaF$.

$$CH_3CH_2\underset{\displaystyle CH_3}{\underset{|}{C}}{=}CH_2 + B_2H_6 \rightarrow \left[CH_3CH_2\underset{\displaystyle CH_3}{\underset{|}{C}}HCH_2\right]_3 B \xrightarrow[OH^-]{H_2O_2} CH_3CH_2\underset{\displaystyle CH_3}{\underset{|}{C}}HCH_2OH$$

2-Methyl-1-butene 2-Methyl-1-butanol

$$(CH_3)_2C{=}CH_2 \xrightarrow{B_2H_6} [(CH_3)_2\ CHCH_2]_3B \xrightarrow[OH^-]{H_2O_2} (CH_3)_2CHCH_2OH$$

Isobutylene 2-Methyl-1-propanol

When cyclic olefins are hydroborated, the stereochemistry of the addition of the B—H bond can be determined. Detailed mechanistic studies have shown the B—H bond to add in a *cis* manner. Subsequent oxidation by H_2O_2/OH^- merely replaces the boron by the —OH group.

$$\text{Methylcyclohexene} + B_2H_6 \rightarrow [\text{2-methylcyclohexyl-B}] \xrightarrow[OH^-]{H_2O_2} \text{trans-2-Methylcyclohexanol}$$

Methylcyclohexene *trans*-2-Methylcyclohexanol

Reduction of Carbonyl Compounds

Compounds containing a carbonyl group ($-\overset{O}{\overset{\|}{C}}-$), such as aldehydes, ketones, esters, and carboxylic acids, are reduced by lithium aluminum hydride* to give alcohols after hydrolysis of the initially formed complex. Aldehydes, esters, and carboxylic acids give primary alcohols, and ketones give secondary alcohols. Tertiary alcohols cannot be prepared by this method.

The initial reaction between the carbonyl group and $LiAlH_4$ gives an aluminum complex, which liberates the alcohol on acid hydrolysis as outlined below:

$$4\ CH_3\overset{H}{\overset{|}{C}}{=}O + LiAlH_4 \xrightarrow{ether} \left[CH_3\underset{H}{\underset{|}{\overset{H}{\overset{|}{C}}}}{-}O\right]_4 AlLi \xrightarrow[H_2O]{H^+} 4CH_3CH_2OH$$

Acetaldehyde Complex Ethyl alcohol

Some typical examples of the versatility of this reducing agent are given below:

$$4\ CH_3\overset{O}{\overset{\|}{C}}CH_3 + LiAlH_4 \xrightarrow{ether} \left[(CH_3)_2CH{-}O\right]_4 AlLi \xrightarrow[H_2O]{H^+} 4\ CH_3CHOHCH_3$$

Acetone Isopropyl alcohol

$$4\ CH_3COOH + 5\ LiAlH_4 \xrightarrow{ether} [CH_3CH_2{-}O]_4AlLi \xrightarrow[H_2O]{H^+} 4\ CH_3CH_2OH$$

Acetic acid Ethyl alcohol

$$2\ CH_3CH_2CO_2CH_3 + LiAlH_4 \xrightarrow{ether} [CH_3CH_2CH_2O]_2Al[OCH_3]_2Li \xrightarrow{H^+/H_2O} 2\ CH_3CH_2CH_2OH$$

Methyl propionate *n*-Propyl alcohol

* Lithium aluminum hydride ($LiAlH_4$) is a versatile reducing agent developed and employed by the organic chemist in the past twenty years. It is sensitive to moisture, however, and gives off hydrogen in contact with water. It should therefore be handled with extreme caution.

Lithium aluminum hydride is especially useful in preparing polyfunctional compounds containing an alcohol group. It is a selective reducing agent and does not reduce all functional groups. For example, it does not reduce carbon-carbon double bonds and does not reduce aromatic rings. Therefore, the following selective synthesis of unsaturated or aromatic containing alcohols can be easily carried out in the laboratory. Because of its selectivity and its powerful reducing action, lithium aluminum hydride is one of the most common reagents used in the modern organic laboratory.

$$CH_3CH{=}CHCH_2\overset{\overset{O}{\|}}{C}CH_3 \xrightarrow[(2)\ H^+/H_2O]{(1)\ LiAlH_4/ether} CH_3CH{=}CHCH_2CHOHCH_3$$

$$C_6H_5\overset{\overset{O}{\|}}{C}{-}CH_3 \xrightarrow[(2)\ H^+/H_2O]{(1)\ LiAlH_4/ether} C_6H_5\overset{\overset{OH}{|}}{C}HCH_3$$

METHODS OF PREPARATION OF PHENOLS

The methods of preparation of aliphatic alcohols outlined previously are generally unsatisfactory for preparing phenols in the laboratory. The hydrolysis reaction, however, has been used commercially to prepare phenol. The reaction requires high temperatures and pressure. The initially formed salt is subsequently converted to phenol.

$$\underset{\text{Chlorobenzene}}{C_6H_5Cl} + 2NaOH \xrightarrow[\text{pressure}]{\text{high temp.}} \underset{\text{Sodium phenoxide}}{C_6H_5O^-Na^+} + Na^+Cl^- + H_2O$$

$$\underset{\text{Sodium phenoxide}}{C_6H_5O^-Na^+} + CO_2 \xrightarrow{H_2O} \underset{\text{Phenol}}{C_6H_5OH} + Na_2CO_3$$

A similar hydrolysis process, called the Raschig process, was developed in Germany. Chlorobenzene is converted to phenol by treatment with steam at high temperatures in this process.

$$\underset{\text{Chlorobenzene}}{C_6H_5Cl} + H_2O \xrightarrow{400^\circ C} \underset{\text{Phenol}}{C_6H_5OH} + HCl$$

Fusion Method

A more classical method for preparing phenol on both a laboratory and a commercial scale involves the fusion of aromatic sulfonic acids with alkali. Since the sulfonic acids can be obtained directly from the aromatic hydrocarbon via sulfonation, this method provides a convenient synthesis directly from the aromatic hydrocarbon, as outlined on the following page for phenol:

$$C_6H_6 + H_2SO_4 \xrightarrow{SO_3} C_6H_5SO_3H \xrightarrow{Na^+OH^-} C_6H_5SO_3^-Na^+$$

Benzene — Benzene sulfonic acid — Sodium benzene sulfonate

$$C_6H_5SO_3^-Na^+ + NaOH \xrightarrow[300°C]{fusion} C_6H_5O^-Na^+ + Na_2SO_3 + H_2O$$

Sodium benzene sulfonate — Sodium phenoxide

$$C_6H_5O^-Na^+ + H^+Cl^- \longrightarrow C_6H_5OH + Na^+Cl^-$$

Sodium phenoxide — Phenol

The sodium salt of the phenol is converted to the free phenol by treatment with acid.

REACTIONS OF ALCOHOLS AND PHENOLS

Alcohols and phenols can undergo two types of reactions which involve the hydroxyl group. A reaction can occur to cleave the oxygen-hydrogen bond, or a reaction can occur to cleave the carbon-oxygen bond resulting in loss of the hydroxyl group. Reactions which involve oxygen-hydrogen bond cleavage are considered first.

Salt Formation

Because of their increased acidity, phenols will form salts with aqueous alkalis.

$$C_6H_5OH + Na^+OH^- \rightarrow C_6H_5O^-Na^+ + H_2O$$

Phenol — Sodium phenoxide

Aliphatic alcohols will not form salts with aqueous alkali. As shown earlier, the free phenol can be regenerated by treatment of the salt with acids such as hydrogen chloride. Although aliphatic alcohols do not form salts with aqueous alkalis, both alcohols and phenols will form salts with active metals such as sodium, potassium, magnesium, and so forth. These reactions are similar to the reaction of water with active metals to give an alkali metal hydroxide and hydrogen.

$$CH_3CH_2OH + Na \rightarrow CH_3CH_2O^-Na^+ + \tfrac{1}{2}H_2\uparrow$$

Ethyl alcohol — Sodium ethoxide

$$C_6H_5OH + K \rightarrow C_6H_5O^-K^+ + \tfrac{1}{2}H_2\uparrow$$

Phenol — Potassium phenoxide

$$H_2O + Na \rightarrow HO^-Na^+ + \tfrac{1}{2}H_2\uparrow$$

It is interesting to note that the salts of alcohols (called alkoxides) are strong bases when used in a nonaqueous solvent. Just as HO^- is a strong inorganic base, alkoxide ions (RO^-, where R is an alkyl group) are strong organic bases. In aqueous solution, the alkoxides are hydrolyzed back to the alcohol; therefore, they must be used in a nonaqueous medium.

$$CH_3CH_2O^-Na^+ + H_2O \rightarrow CH_3CH_2OH + Na^+OH^-$$

Sodium ethoxide — Ethyl alcohol

Ester Formation

In the presence of an acid catalyst, both alcohols and phenols will react with a carboxylic acid to form an ester. Experiments carried out using an alcohol containing oxygen-18 instead of the normal oxygen-16 have shown that the oxygen of the alcohol ends up as one of the oxygen atoms in the ester. Therefore, only the oxygen-hydrogen bond of the alcohol is broken, and the hydroxyl group used in forming the by-product water comes from the carboxylic acid and not the alcohol. This reaction is outlined schematically below:

$$R{-}O^{18}{-}H + HO{-}\overset{\overset{\displaystyle O}{\|}}{C}{-}R' \xrightarrow{H^+} R{-}O^{18}{-}\overset{\overset{\displaystyle O}{\|}}{C}{-}R' + HOH$$

Alcohol — Carboxylic acid — Ester

Some typical examples of ester formation are illustrated below:

$$CH_3CH_2OH + CH_3CO_2H \xrightarrow{H^+} CH_3\overset{\overset{\displaystyle O}{\|}}{C}OCH_2CH_3 + H_2O$$

Ethyl alcohol — Acetic acid — Ethyl acetate

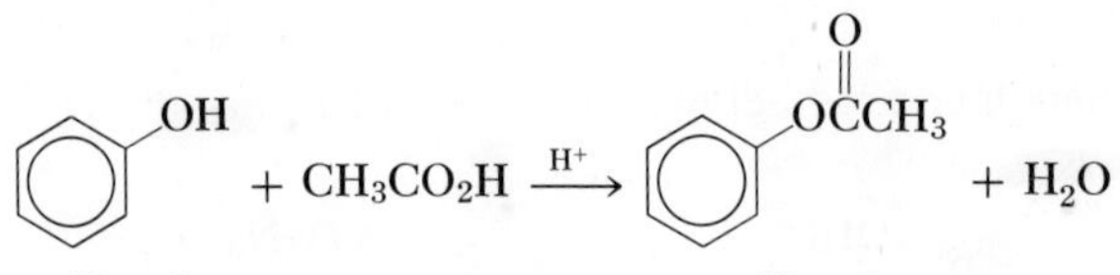

$$C_6H_5{-}OH + CH_3CO_2H \xrightarrow{H^+} C_6H_5{-}O\overset{\overset{\displaystyle O}{\|}}{C}CH_3 + H_2O$$

Phenol — Phenyl acetate

$$CH_3OH + C_6H_5{-}\overset{\overset{\displaystyle O}{\|}}{C}{-}OH \xrightarrow{H^+} C_6H_5{-}\overset{\overset{\displaystyle O}{\|}}{C}OCH_3 + H_2O$$

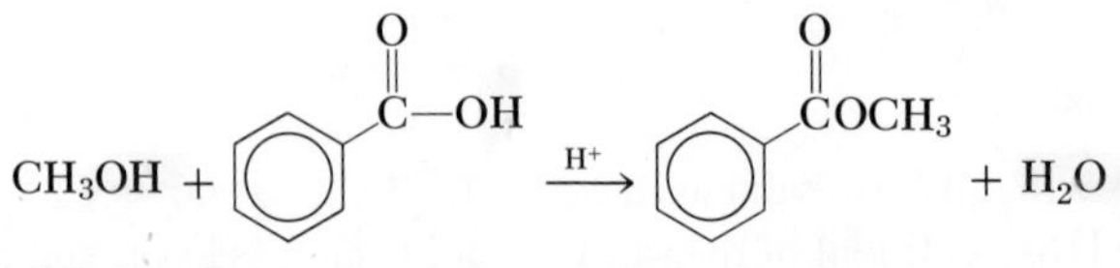

Methyl alcohol — Benzoic acid — Methyl benzoate

Inorganic acids also form esters with alcohols. For example, glyceryl trinitrate (nitroglycerine), which is the explosive ingredient in dynamite, is produced by the esterification of glycerol with nitric acid.

$$\begin{matrix} CH_2OH \\ | \\ CHOH \\ | \\ CH_2OH \end{matrix} + 3\,HONO_2 \rightarrow \begin{matrix} CH_2ONO_2 \\ | \\ CHONO_2 \\ | \\ CH_2ONO_2 \end{matrix} + 3H_2O$$

Glycerol — Nitric acid — Glyceryl trinitrate

Similarly, phosphate esters, which are of extreme importance in biochemistry, can also be produced by the esterification of alcohols with phosphoric acid.* In biochemical

*Either 1, 2, or 3 of the acidic hydrogens of phosphoric acid can be esterified.

$$HO-\overset{\overset{O}{\|}}{\underset{\underset{OH}{|}}{P}}-OH + R-OH \overset{H^+}{\rightleftharpoons} HO-\overset{\overset{O}{\|}}{\underset{\underset{OH}{|}}{P}}-OR + H_2O$$

Phosphoric acid — Alcohol — Phosphate ester

systems, a more complex phosphoric acid derivative will be used to form complex phosphate esters.

REPLACEMENT OF THE HYDROXYL GROUP

Dehydration

Alcohols undergo loss of water (dehydration) in the presence of acid catalysts under the proper conditions. These reactions have been considered earlier under the preparation of olefins.

$$\underset{n\text{-Propyl alcohol}}{CH_3CH_2CH_2OH} + Al_2O_3 \xrightarrow{\Delta} \underset{\text{Propylene}}{CH_3CH{=}CH_2}$$

Conversion to Alkyl Halides

The hydroxyl group of alcohols can be replaced with a halogen atom by several types of reagents. With hydrogen chloride, the reaction is generally carried out using zinc chloride ($ZnCl_2$) as a catalyst. The order of reactivity is tertiary $>$ secondary $>$ primary alcohols. With tertiary alcohols the reaction is extremely rapid even at room temperature, whereas secondary alcohols require several minutes to react at room temperature, and primary alcohols do not give any appreciable reaction at room temperature. This difference in reactivity is the basis of the **Lucas test** used to distinguish primary, secondary, and tertiary alcohols.

$$ROH + HCl \xrightarrow{ZnCl_2} RCl + H_2O$$

The order of reactivity of the hydrogen halides in this type of reaction is HI $>$ HBr $>$ HCl. The mechanism of this reaction is dependent on the type of alcohol undergoing displacement of the hydroxyl group. With primary and secondary halides the mechanism involves protonation of the hydroxyl group and subsequent displacement of water by a halide ion, as outlined below:

$$R\ddot{O}H + HX \rightarrow R\overset{+}{\ddot{O}}H_2 + X^-$$

X = halogen
R = primary or secondary

$$X^- \longrightarrow R-\overset{+}{\ddot{O}}H_2 \rightarrow X-R + H_2O$$

It is much easier to displace water than hydroxide ion (HO^-), hence the function of the proton catalyst is to facilitate the displacement reaction. With most tertiary halides, the protonated alcohol undergoes ionization to form a carbonium ion, which then picks up a halide ion to form the alkyl halide. In this case ionization gives the much more stable tertiary carbonium ion.

$$R\!-\!\overset{R'}{\underset{R''}{\overset{|}{\underset{|}{C}}}}\!-\!\ddot{\underset{..}{O}}H + HX \rightarrow R\!-\!\overset{R'}{\underset{R''}{\overset{|}{\underset{|}{C}}}}\!-\!\overset{+}{\underset{..}{O}}H_2 \xrightarrow{-H_2O} R\!-\!\overset{R'}{\underset{R''}{\overset{|}{\underset{|}{C^+}}}} \xrightarrow{X^-} R\!-\!\overset{R'}{\underset{R''}{\overset{|}{\underset{|}{C}}}}\!-\!X$$

X = halogen $+ X^-$
R,R′,R″ = Alkyl groups

With secondary or primary alcohols, ionization would generate the less stable secondary or primary carbonium ions. The energy required for this type of ionization process is higher for generating the secondary and primary carbonium ions than it is for the displacement of water from the protonated alcohol; consequently, the reaction proceeds via the lowest energy pathway, which is displacement of water by the halide anion.

Other types of halogenating agents also cause displacement of the hydroxyl group from an alcohol. The most common reagents used are phosphorus trichloride (PCl_3), phosphorus pentachloride (PCl_5), and thionyl chloride ($SOCl_2$). Thionyl chloride is the most practical reagent, since the byproducts of the reaction (SO_2 and HCl) are gases, thus facilitating the isolation and purification of the alkyl halide. Some typical examples are cited below:

$$3\,CH_3CH_2OH + PCl_3 \rightarrow 3\,CH_3CH_2Cl + H_3PO_3$$
Ethyl alcohol — Ethyl chloride — Phosphorus acid

$$CH_3CHOHCH_3 + PCl_5 \rightarrow CH_3CHClCH_3 + POCl_3 + HCl\uparrow$$
Isopropyl alcohol — Isopropyl chloride — Phosphorus oxychloride

$$CH_3CH_2CH_2OH + SOCl_2 \rightarrow CH_3CH_2CH_2Cl + SO_2\uparrow + HCl\uparrow$$
n-Propyl alcohol — *n*-Propyl chloride

Phenols generally do not undergo displacement of the hydroxyl group attached to the aromatic ring using these types of reagents.

OTHER REACTIONS OF ALCOHOLS AND PHENOLS

In addition to the reactions involving cleavage of the oxygen-hydrogen bond of the hydroxyl group and displacement of the hydroxyl group by carbon-oxygen bond cleavage, alcohols and phenols can undergo other kinds of reactions which involve the carbon-hydrogen bonds of the alcohol or the phenol. In phenols most of these reactions involve electrophilic aromatic substitution reactions similar to those observed earlier for benzene. The hydroxyl group attached to the aromatic ring activates the aromatic ring toward electrophilic attack, and phenol undergoes substitution by electrophilic reagents much more easily than either benzene or toluene. The hydroxyl group is an *ortho-para* director. Some typical reactions of phenol involving substitution on the aromatic ring are illustrated below:

Halogenation

$$C_6H_5OH + 3Br_2 \rightarrow C_6H_2Br_3OH + 3HBr$$

Phenol — 2,4,6-Tribromophenol

The hydroxyl group so activates the ring that substitution by halogen occurs at all the possible *ortho* and *para* positions. Special precautions must be used to obtain the monohalogenated product.

Nitration

$$\text{Phenol (C}_6\text{H}_5\text{OH)} + HNO_3 \rightarrow \textit{o}\text{-Nitrophenol} + \textit{p}\text{-Nitrophenol}$$

Phenol　　*o*-Nitrophenol　*p*-Nitrophenol

Nitric acid itself can be used to nitrate the activated aromatic ring.

Sulfonation

Sulfonation of phenol with sulfuric acid yields mainly the *ortho*-substituted product at low temperatures and mainly the *para*-substituted product at higher temperatures.

$$\text{Phenol} + H_2SO_4 \xrightarrow{15^\circ C} \textit{o}\text{-Hydroxybenzene sulfonic acid}$$

$$\text{Phenol} \xrightarrow[H_2SO_4]{100^\circ C} \textit{p}\text{-Hydroxybenzene sulfonic acid}$$

Phenol　　*o*-Hydroxybenzene sulfonic acid

p-Hydroxybenzene sulfonic acid

Oxidation Reactions

Oxidizing agents such as alkaline potassium permanganate or potassium dichromate and sulfuric acid oxidize primary and secondary alcohols to carboxylic acids and ketones, respectively. This reaction will be considered in more detail in Chapters 9 and 10. Under the normal conditions of this reaction, tertiary alcohols are not oxidized. Some typical examples are given below:

$$CH_3CH_2OH + K_2Cr_2O_7 \xrightarrow{H_2SO_4} CH_3CHO \rightarrow CH_3\overset{O}{\overset{\|}{C}}OH$$

Ethyl alcohol　　Acetaldehyde　Acetic acid

$$CH_3CHOHCH_3 + KMnO_4 \xrightarrow{NaOH} CH_3\overset{O}{\overset{\|}{C}}CH_3$$

Isopropyl alcohol　　Acetone

IMPORTANT ALCOHOLS AND PHENOLS

Many important laboratory, industrial, and biologically related compounds contain more than one functional group per molecule. Some typical examples containing a hydroxyl group are shown on the next page.

$HOCH_2CH_2OH$ — Ethylene glycol

$HOCH_2CHOHCH_2OH$ — Glycerol

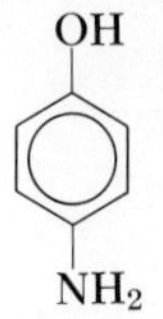

p-Aminophenol (benzene ring with OH and NH_2 in para positions)

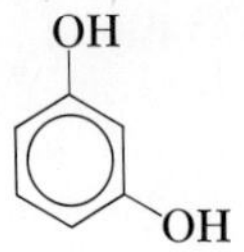

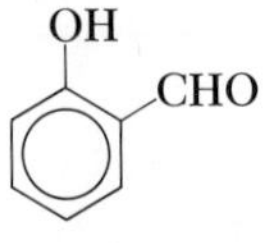

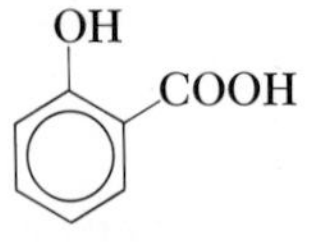

Resorcinol (benzene ring with two OH groups in meta positions)

Salicylaldehyde (benzene ring with OH and CHO in ortho positions)

Salicylic acid (benzene ring with OH and COOH in ortho positions)

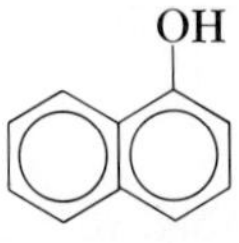

α-Naphthol

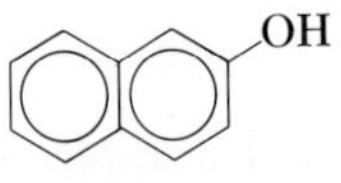

β-Naphthol

Some of the more important alcohols and their uses are discussed in the following pages of this chapter.

Important Alcohols

Methyl Alcohol. Methyl alcohol is commonly called wood alcohol because it was once exclusively produced by the destructive distillation of wood. A synthetic process for the production of methyl alcohol from carbon monoxide and hydrogen was developed in 1923 and has largely supplanted the wood distillation method. In this process, carbon monoxide and hydrogen react under a high pressure and a temperature of 350°C in the presence of zinc and chromium oxide catalysts.

$$\underset{\text{Carbon monoxide}}{CO} + 2H_2 \xrightarrow[Cr_2O_3]{ZnO} \underset{\text{Methyl alcohol}}{CH_3OH}$$

Methyl alcohol is a colorless, volatile liquid with a characteristic odor. It is used as a denaturant for ethyl alcohol, as an antifreeze for automobile radiators, and as the raw material for the synthesis of other organic compounds. About 40 per cent of all the methyl alcohol that is made in this country is used in the preparation of formaldehyde, which is a starting material for many plastics. When taken internally, methyl alcohol is poisonous, small doses producing blindness by degeneration of the optic nerve, whereas large doses are fatal. The taste, odor, and poisonous properties of wood alcohol make it a desirable denaturing agent to be added to ethyl alcohol to prevent its use in beverages. During the prohibition era in the United States, many persons were blinded and others died after drinking ethyl alcohol denatured in this fashion. An individual who was blinded temporarily after drinking a small amount of methyl alcohol was said to be "blind drunk."

Ethyl Alcohol. Ethyl alcohol is commonly known as alcohol, or as grain alcohol since it may be made by fermentation of various grains. It is prepared commercially by two major methods. One involves the fermentation of the sugars and starch of common grains, potatoes, or black-strap molasses. The yeast used in fermentation contains enzymes that catalyze the transformation of more complex sugars into simple sugars, and then into alcohol and carbon dioxide, as shown in the following reactions:

$$\underset{\text{Complex sugar}}{C_{12}H_{22}O_{11}} + H_2O \xrightarrow{\text{enzymes}} \underset{\text{Simple sugar}}{2\ C_6H_{12}O_6} \xrightarrow{\text{enzymes}} \underset{\text{Ethyl alcohol}}{4\ C_2H_5OH} + \underset{\text{Carbon dioxide}}{4\ CO_2}$$

 Enzymes and fermentation will be studied more completely in the section on biochemistry. The other method is a synthetic method that makes use of the reaction of the unsaturated hydrocarbon ethylene with sulfuric acid, followed by a hydrolysis reaction.

$$\underset{\text{Ethylene}}{H_2C{=}CH_2} + H_2SO_4 \rightarrow \underset{\text{Ethyl sulfate}}{CH_3CH_2OSO_3H}$$

$$\underset{\text{Ethyl sulfate}}{CH_3CH_2OSO_3H} + H_2O \rightarrow \underset{\text{Ethyl alcohol}}{CH_3CH_2OH} + H_2SO_4$$

The ethylene used in this process is produced by the cracking of petroleum hydrocarbons.

Ethyl alcohol is a colorless, volatile liquid with a characteristic pleasant odor. Industrial ethyl alcohol contains approximately 95 per cent alcohol and 5 per cent water. It is difficult to remove all the water from alcohol, since in simple distillation processes a constant boiling mixture of 95 per cent alcohol and 5 per cent water is formed. Methods for removing this water have been developed, since the solvent properties of pure, absolute alcohol are considerably different from those of industrial alcohol. For example, when benzene is mixed with 95 per cent alcohol and the mixture is distilled, the final fraction consists of absolute ethyl alcohol. A large proportion of the industrial ethyl alcohol produced each year is used as an antifreeze in automobile radiators. It is an excellent solvent for many substances and is used in the preparation of medicines, flavoring extracts, and perfumes. Alcohol is widely used in the hospital as an antiseptic, a vehicle for medications, and as a rubbing compound to cleanse the skin and lower a patient's temperature.

The concentration of alcohol in beverages is usually expressed as per cent or "proof." The relationship between proof and per cent alcohol concentration may be shown in the following examples. The common 100 proof whiskey is 50 per cent alcohol, whereas the standard laboratory 95 per cent alcohol is 190 proof. Beer and wine contain from 3 to 20 per cent, whereas whiskey, rum, vodka, and gin contain from 35 to 45 per cent alcohol.

When ethyl alcohol is taken internally, it is rapidly absorbed and oxidized. It may therefore be used as a readily available source of energy and is often employed to overcome shock or collapse. If large quantities are taken, it causes a depression of the higher nerve centers, mental confusion, lack of muscular coordination, lowering of normal inhibitions, and eventually stupor.

Ethylene Glycol. All the alcohols so far considered have been monohydroxy alcohols. Since it is possible to replace a hydrogen atom on more than one carbon atom in a hydrocarbon with a hydroxyl group, it is possible to have polyhydroxy (polyhydric) alcohols. The simplest polyhydric alcohol would be one formed by replacing a hydrogen on each of the two carbons of ethane by a hydroxyl group. This compound is called **ethylene glycol.** It is prepared by oxidizing ethylene to ethylene oxide, and subsequently hydrolyzing the oxide to ethylene glycol, as shown in the following equations:

$$\underset{\text{Ethylene}}{H_2C{=}CH_2} + O_2 \xrightarrow[\text{catalyst}]{\text{Ag}} \underset{\text{Ethylene oxide}}{\begin{array}{c} \;\;H\quad\;\; H \\ \;\;|\quad\;\;\; | \\ H{-}C{-}{-}C{-}H \\ \diagdown\,\diagup \\ O \end{array}}$$

$$\underset{\text{Ethylene oxide}}{\begin{array}{c} \;\;H\quad\;\; H \\ \;\;|\quad\;\;\; | \\ H{-}C{-}{-}C{-}H \\ \diagdown\,\diagup \\ O \end{array}} + H_2O \rightarrow \underset{\text{Ethylene glycol}}{\begin{array}{c} \;\;H\;\; H \\ \;\;|\;\;\;\, | \\ H{-}C{-}C{-}H \\ \;\;|\;\;\;\, | \\ \;\;O\;\;\; O \\ \;\;H\;\;\; H \end{array}}$$

Ethylene glycol is water soluble and has a very high boiling point compared to those of methyl and ethyl alcohols. These properties make it an excellent permanent, or nonvolatile, type of antifreeze for automobile radiators. Antifreeze preparations such as **Prestone** and **Zerex** consist of ethylene glycol plus a small amount of a dye. A more recent development is the so called extended life antifreeze, which is left in a car's radiator the year round and may be effective for more than two years. The nonvolatile properties of these products, such as **Dowgard** and **Telar,** are also based on their content of ethylene glycol.

Glycerol. The most important trihydric alcohol is glycerol, which is sometimes called glycerin. It is an essential constituent of fat (an ester of glycerol and fatty acids) and may be prepared by the hydrolysis of fat as represented in the following equation:

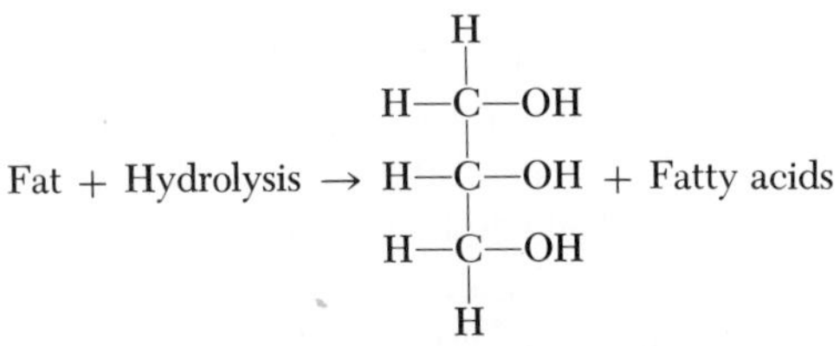

$$\text{Fat} + \text{Hydrolysis} \rightarrow \begin{array}{c} \text{H} \\ | \\ \text{H—C—OH} \\ | \\ \text{H—C—OH} \\ | \\ \text{H—C—OH} \\ | \\ \text{H} \end{array} + \text{Fatty acids}$$

Glycerol

Glycerol is obtained commercially as a by-product of the manufacture of soap, and from a synthetic process that uses propylene from the catalytic cracking of petroleum as a starting material. Glycerol is a syrupy, sweet-tasting substance that is soluble in all proportions of water and alcohol. It is nontoxic and is often used for the preparation of liquid medications. Since it has the ability to take up moisture from the air, it tends to keep the skin soft and moist when applied in the form of cosmetics and lotions.

Phenol. Phenol has strong antiseptic properties, and as a class of compounds phenols are active germicides. However, because of its extreme toxicity, phenol itself is rarely used as an antiseptic. It is caustic, causes blistering of the skin, and is a violent poison when taken internally. Dilute solutions of the cresols (Lysol), however, are used in hospitals as disinfectants.

Phenol is used as a standard for comparison of other germicides and the efficiency of other antiseptics is measured in arbitrary units called the *phenol coefficient.* For example, a 1 per cent solution of a germicide that is as effective as a 5 per cent solution of phenol in destroying an organism is assigned a phenol coefficient of 5.

Other important commercial phenol derivatives are hexyl resorcinol, used in mouthwashes, and hexachlorophene, used in soaps, deodorants, and toothpastes.

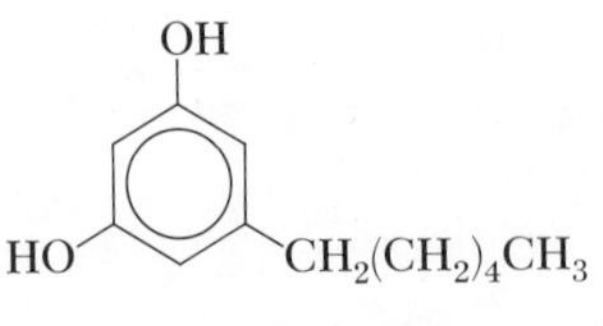

Hexyl resorcinol

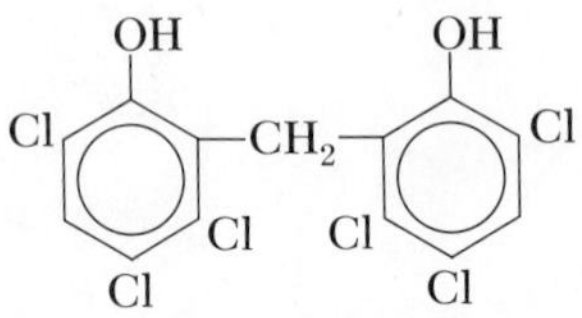

Hexachlorophene

The phenolic group also occurs in plants. The two most well known plant products which contain a phenolic ring are marijuana (Chapter 2) and morphine.

IMPORTANT TERMS AND CONCEPTS

alcohol
cresol
ester
ethylene glycol
glycerol
grain alcohol
hydration of olefins
hydroboration
hydrogen bonding
Lucas test
phenol
phosphate esters
primary alcohol
salicylic acid
secondary alcohol
tertiary alcohol
wood alcohol

QUESTIONS

1. Draw out all the possible structural isomers of an alcohol which has a molecular formula $C_5H_{12}O$. Name each of these isomers according to the IUPAC system.

2. Which of the structural isomers in Question 1
   (a) are primary alcohols?
   (b) are secondary alcohols?
   (c) are tertiary alcohols?
   (d) on dehydration will give 2-pentene?

3. Give each of the following compounds an appropriate name:
   (a) $(CH_3)_2CHCH_2CH_2CH_2OH$
   (b) $(CH_3CH_2)_3COH$
   (c) 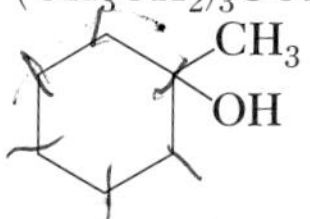 CH_3 OH
   (d) $CH_3CH_2CH(CH_3)CH(OH)CH(C_2H_5)CH_2CH_3$
   (e) 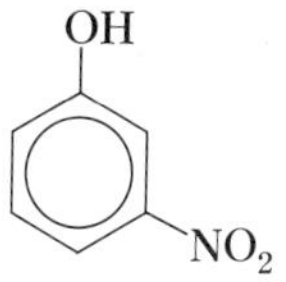 OH NO_2
   (f) 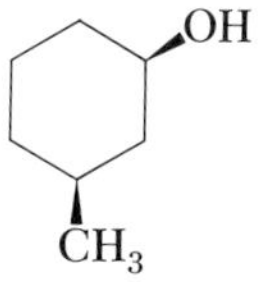 OH CH_3
   (g) $CH_3CHOHCHOHCH_2OH$
   (h) $(CH_3)_2CHCHBrCH_2CH_2OH$
   (i) 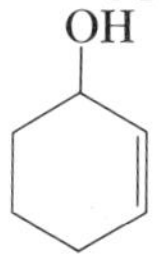 OH
   (j) 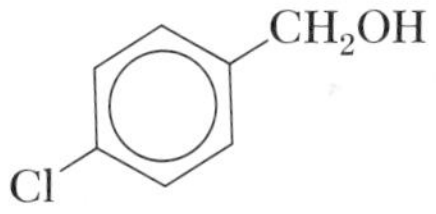 CH_2OH Cl

4. What is wrong with each of the following names? Give the correct name for each compound.
   (a) 1,1-dimethyl-1-butanol
   (b) 4-methyl-4-pentanol
   (c) 5-nitrophenol
   (d) 2-ethyl-2-propanol
   (e) 4-pentene-2-ol

5. Write structural formulas for each of the following compounds.
   (a) isobutyl alcohol
   (b) isopropyl alcohol
   (c) *t*-butyl alcohol
   (d) 3-pentanol
   (e) 3,3-dimethylcyclohexanol
   (f) 2,4,4-trimethyl-2-heptanol
   (g) *meta*-bromophenol
   (h) 1-bromo-2-methyl-3-hexanol
   (i) *cis*-3-ethylcyclopentanol
   (j) *trans*-2-pentene-1-ol

6. Which of the following alcohols are easily oxidized?
   (a) $CH_3CH_2CH_2OH$
   (b) $CH_3CHOHCH_2CH_3$
   (c) $(CH_3)_3COH$
   (d) $(CH_3)_3CCH_2OH$
   What is the functional class of compounds formed when *each* of the above alcohols is oxidized?

7. Write equations for the preparations of the following compounds. Give any necessary catalysts. More than one step may be necessary.
   (a) isopropyl alcohol from *n*-propyl bromide
   (b) 2-butanol from 2-butyne
   (c) *m*-bromophenol from benzene
   (d) 1-phenylethanol from benzene
   (e) tertiary butyl alcohol from isopropyl bromide

8. Write equations for the reaction of isopropyl alcohol with the following reagents:
   (a) hydrogen iodide
   (b) potassium
   (e) acetic acid
   (d) Al_2O_3/heat
   (e) $K_2Cr_2O_7/H_2SO_4$

9. Write equations for the reaction of phenol with the following reagents:
   (a) Cl_2
   (b) potassium
   (c) nitric acid
   (d) propionic acid (CH_3CH_2COOH)
   (e) sodium hydroxide

10. Arrange the following compounds in order of decreasing acidity:

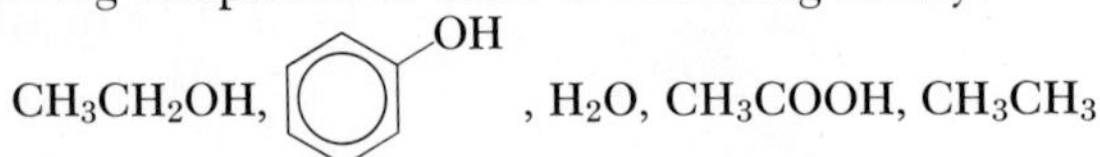

11. Arrange the following compounds in order of decreasing solubility in water:

$$CH_3CH_2OH,\ CH_3(CH_2)_6CH_2OH,\ CH_3(CH_2)_6CH_3,\ HOCH_2CH_2OH$$

12. A compound with molecular formula $C_4H_{10}O$ reacts with sodium to liberate hydrogen. It reacts with $ZnCl_2$/HCl to give an oily suspension. When passed over Al_2O_3 at high temperatures, it is converted to a new compound, C_4H_8, which, when treated with H_2SO_4/H_2O, gives the original compound, $C_4H_{10}O$. What are the structures of these compounds?

SUGGESTED READING

Ferguson: Hydrogen Bonding and Physical Properties of Substances. Journal of Chemical Education, Vol. 33, p. 267, 1956.

Lesser: Glycerin—Man's Most Versatile Chemical Servant. Journal of Chemical Education, Vol. 26, p. 327, 1949.

Snell: Soap and Glycerol. Journal of Chemical Education, Vol. 19, p. 172, 1942.

Weaver: Glycerol. Journal of Chemical Education, Vol. 29, p. 524, 1952.

Webb: Hydrogen Bond, "Special Agent." Chemistry, Vol. 41, No. 6, p. 16, 1968.

ETHERS

CHAPTER 7

The *objectives* of this chapter are to enable the student to:

1. Recognize an ether or epoxide.
2. Write the structural formulas and names of the common ethers and epoxides.
3. Describe the preparation of simple ethers and epoxides.
4. Recognize the difference in reactivity between ethers and epoxides.
5. Describe the basic chemical reactions of ethers.
6. Explain the ring opening reactions of epoxides.
7. Recognize some of the important ethers and epoxides and their current uses.

Ethers are closely related to the alcohols and may be considered a derivative of an alcohol in which the hydrogen of the hydroxyl group has been replaced by an alkyl or aryl group. Consequently, ethers may be considered to be organic derivatives of water (HOH) in which both hydrogens have been replaced by alkyl or aryl groups. In addition, cyclic ethers are possible in which the oxygen atom is part of the cyclic structure. The three-membered ring structures containing an oxygen atom as part of the ring are also called **epoxides.** As noted earlier in the chapter on alcohols, ethers are isomeric with the alcohols containing the same number of carbon atoms.

Ethers are named either by common names or by naming the two alkyl or aryl groups linked to the oxygen atom, followed by the word "ether." If one of the alkyl or aryl groups has no simple name, the compound is named as an **alkoxy** derivative.* Some typical examples of ethers are

CH_3OCH_3
Dimethyl ether

$CH_3CH_2OCH_2CH_3$
Diethyl ether

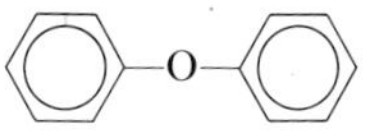

C_6H_5—O—C_6H_5
Diphenyl ether

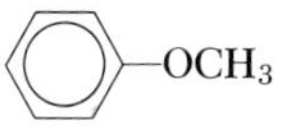

C_6H_5—OCH_3
Methyl phenyl ether (anisole)

$CH_3OCH_2CH_3$
Methyl ethyl ether

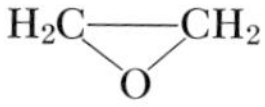

H_2C—CH_2 (bridged by O)
Ethylene oxide

Tetrahydrofuran

Furan

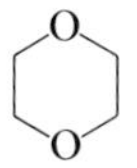

Dioxane

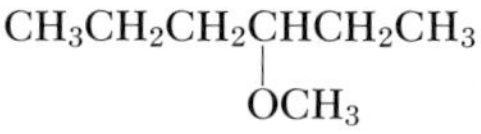

$CH_3CH_2CH_2CHCH_2CH_3$ with OCH_3 on C-3
3-Methoxyhexane

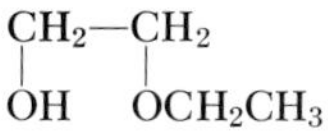

CH_2—CH_2 with OH and OCH_2CH_3
2-Ethoxyethanol (ethyl cellosolve)

$CH_2{=}CHOCH{=}CH_2$
Divinyl ether

*The —OR group is known as an alkoxy group. The name of this type of group is formed from the hydrocarbon name of the —R group by dropping the **-ane** and adding **-oxy.**

TABLE 7-1 PHYSICAL PROPERTIES OF ETHERS

COMPOUND	STRUCTURE	M.P. °C	B.P. °C
Dimethyl ether	CH_3OCH_3	−140	−25
Diethyl ether	$CH_3CH_2OCH_2CH_3$	−116 (−123)	+35
Methyl ethyl ether	$CH_3OCH_2CH_3$	—	+8
Di-*n*-propyl ether	$CH_3CH_2CH_2OCH_2CH_2CH_3$	−122	+91
Diisopropyl ether	$(CH_3)_2CHOCH(CH_3)_2$	−60	+68
Tetrahydrofuran	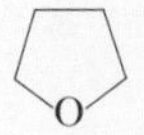 O	−108	+66
Anisole	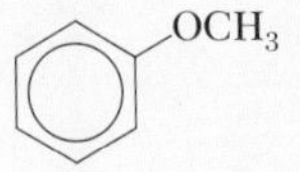 OCH_3	−37	+154
Diphenyl ether	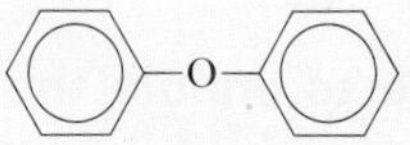 O	+27	+259

Since ethers are essentially hydrocarbons with a single oxygen atom, their physical properties would be expected to parallel those of hydrocarbons. They are colorless, insoluble in water, soluble in acids (whereas alkanes are not), soluble in organic solvents, and in general have densities and boiling points similar to hydrocarbons of corresponding molecular weight. It is interesting to note that dimethyl ether, which is isomeric with ethanol, is a gas at room temperature, whereas ethanol is a liquid at room temperature. Ethers, which have no hydroxyl group, cannot form hydrogen bonds as can alcohols; consequently, the boiling points are not abnormally high like the alcohols. The physical properties of some of the more common ethers are summarized in Table 7-1.

METHODS OF PREPARATION OF ETHERS

Dehydration of Alcohols

Alcohols can lose a molecule of water in two different ways. **Intramolecular*** loss of water gives an alkene, whereas **intermolecular*** loss of water gives an ether, as shown below:

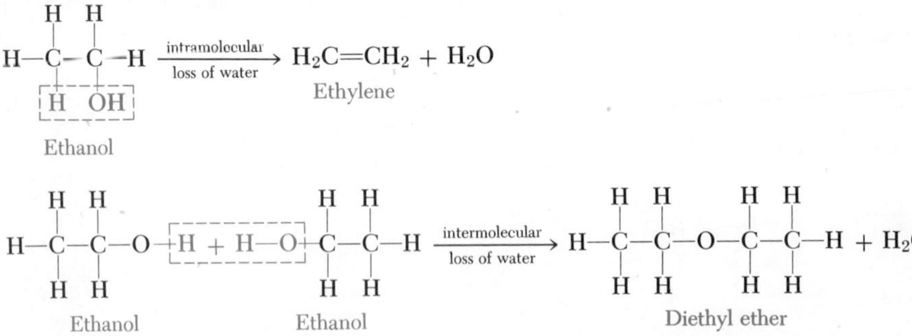

$$\underset{\text{Ethanol}}{H_2C(H)-CH_2(OH)} \xrightarrow[\text{loss of water}]{\text{intramolecular}} \underset{\text{Ethylene}}{H_2C{=}CH_2} + H_2O$$

$$\underset{\text{Ethanol}}{CH_3CH_2-O-H} + \underset{\text{Ethanol}}{H-O-CH_2CH_3} \xrightarrow[\text{loss of water}]{\text{intermolecular}} \underset{\text{Diethyl ether}}{CH_3CH_2-O-CH_2CH_3} + H_2O$$

The conditions under which the reaction is carried out determine the course of the dehydration reaction. When sulfuric acid is the dehydrating agent and an initial mixture of the alcohol and sulfuric acid is maintained at 140°C (while additional alcohol is added

*An intramolecular loss of water is a loss of the elements of H_2O from within one molecule. An intermolecular loss of the H_2O is the elimination of the elements of H_2O between two different molecules.

at the rate at which ether distills from the mixture), the reaction then gives mostly ether as the product. However, the olefin is the principal product when the alcohol and excess sulfuric acid mixture is maintained at 180°C.

$$RCH_2CH_2OH \xrightarrow[180°C]{H_2SO_4} RCH{=}CH_2$$

$$RCH_2CH_2OH \xrightarrow[140°C]{H_2SO_4} RCH_2CH_2OCH_2CH_2R$$

A carbonium ion mechanism can be used to explain these results. The alcohol undergoes protonation to form the oxonium ion. Displacement of water from the protonated alcohol gives the ether, whereas abstraction of a proton from the β-carbon followed by loss of water gives the olefin. In the absence of excess alcohol, this latter process is favored, and the alkene becomes the major product.

$$RCH_2CH_2\ddot{O}H + HOSO_2OH \rightleftharpoons RCH_2CH_2\overset{+}{O}H_2 + HSO_4^-$$

$$RCH_2CH_2\text{—}\overset{+}{O}H_2 + RCH_2CH_2\ddot{O}H \rightarrow RCH_2CH_2OCH_2CH_2R + H_2O + H^+$$

or

$$RCH(H)\text{—}CH_2\text{—}\overset{+}{O}H_2 + HSO_4^- \rightarrow RCH{=}CH_2 + H_2SO_4 + H_2O$$

In the preparation of ethers by this method, only symmetrical ethers are obtained. In addition, aryl ethers do not usually undergo this reaction, and it is limited to aliphatic ethers.

Williamson Synthesis of Ethers

A more general preparation of both symmetrical and unsymmetrical ethers is a reaction discovered by the British chemist Alexander Williamson. The reaction involves a displacement of halide ion from an alkyl halide by an alkoxide ion (obtained from an alcohol or phenol). The generalized scheme of this reaction, and several specific examples, are illustrated below:

$$\underset{\text{Metal alkoxide}}{RO^-M^+} + \underset{\text{Alkyl halide}}{R'X} \rightarrow \underset{\text{Ether}}{ROR'} + \underset{\text{Salt}}{M^+X^-}$$

M^+ = Na, K (usually)
R = alkyl or aryl

$$\underset{\text{Sodium methoxide}}{CH_3O^-Na^+} + \underset{\text{Ethyl iodide}}{CH_3CH_2I} \rightarrow \underset{\text{Methyl ethyl ether}}{CH_3OCH_2CH_3} + Na^+I^-$$

$$\underset{\text{Sodium phenoxide}}{C_6H_5\text{—}O^-Na^+} + \underset{\text{Methyl iodide}}{CH_3I} \rightarrow \underset{\text{Anisole}}{C_6H_5\text{—}OCH_3} + Na^+I^-$$

$$\underset{\text{Sodium methoxide}}{CH_3O^-Na^+} + \underset{\text{Bromobenzene}}{C_6H_5\text{—}Br} \rightarrow \text{No reaction}$$

Compared to the ease of displacement of halide ion from an alkyl halide by alkoxide, it is difficult to displace halide ion from an aromatic ring by alkoxide ion. Therefore, in preparing mixed aromatic-aliphatic ethers, such as anisole, the reaction must be carried out using sodium phenoxide displacement on methyl iodide. Reversal of the types of reagents does not give anisole.

The cyclic three-member ethers (epoxides) can also be prepared by an intramolecular type of Williamson synthesis. Addition of hypohalous acids to alkenes gives halohydrins, which undergo intramolecular cyclization in the presence of base to give epoxides, as outlined below:

$$\underset{\text{Ethylene}}{CH_2{=}CH_2} + \underset{\text{Hypochlorous acid}}{HOCl} \rightarrow \underset{\text{Ethylene chlorohydrin}}{CH_2(OH){-}CH_2(Cl)}$$

$$CH_2(OH){-}CH_2(Cl) + KOH \rightarrow \left[CH_2(O^-){-}CH_2(Cl) \right] \rightarrow \underset{\text{Ethylene oxide}}{H_2C{-}CH_2 \text{ (bridged by O)}} + K^+Cl^- + H_2O$$

Commercially, ethylene oxide is prepared from ethylene and oxygen in the presence of a silver gauze catalyst.

$$\underset{\text{Ethylene}}{CH_2{=}CH_2} + \tfrac{1}{2}O_2 \xrightarrow[\Delta,\ \text{pressure}]{\text{Ag catalyst}} \underset{\text{Ethylene oxide}}{H_2C{-}CH_2 \text{ (bridged by O)}}$$

REACTIONS OF ETHERS AND EPOXIDES

Except for the saturated hydrocarbons, ethers are the most unreactive of any of the simple functional groups. They are stable to dilute acids and bases and are also resistant to many oxidizing and reducing agents. They are similar to alkanes in their lack of chemical reactivity. This lack of chemical reactivity, however, makes ethers quite suitable as solvents for many chemical reactions. Diethyl ether and tetrahydrofuran are two of the most common solvents used today in the organic chemical laboratory. They are, however, quite flammable materials and caution must be exercised when using them. In addition, ethers form nonvolatile peroxides on standing. These peroxides are explosive in the dry state; consequently, ether solutions should never be evaporated or distilled to dryness unless these peroxides have been removed. Under more vigorous reaction conditions, ethers do undergo cleavage reactions with concentrated mineral acids, and the aromatic ethers do undergo ring substitution reactions.

Reaction with Acids

Ethers dissolve in cold concentrated sulfuric acid to form oxonium salts. This is a reversible reaction, and the ether can be regenerated on neutralization of the acid. With the hydrogen halides, ethers can be cleaved at high temperatures.

$$R-\ddot{\underset{..}{O}}-R' + H_2SO_4 \rightleftharpoons \left[R-\underset{..}{\overset{\overset{\Large H}{|}}{O}}-R'\right]^+ HSO_4^-$$

Hydrogen iodide and hydrogen bromide are usually used for this purpose. With one equivalent of the hydrogen halide, an alcohol and an alkyl halide are produced. With excess hydrogen halide, two moles of alkyl halide are produced if both groups attached to the oxygen atom of the ether are aliphatic. If one of the groups is aromatic, a mole of the corresponding phenol is formed, since phenols are not converted to aryl halides by hydrogen halides. Some typical cleavage reactions are illustrated in the following examples:

$$\underset{\text{Diethyl ether}}{CH_3CH_2OCH_2CH_3} + HI\ (1\ \text{mole}) \xrightarrow{\Delta} \underset{\text{Ethyl iodide}}{CH_3CH_2I} + \underset{\text{Ethyl alcohol}}{CH_3CH_2OH}$$

$$CH_3CH_2OCH_2CH_3 + HI\ (\text{excess}) \xrightarrow{\Delta} 2CH_3CH_2I$$

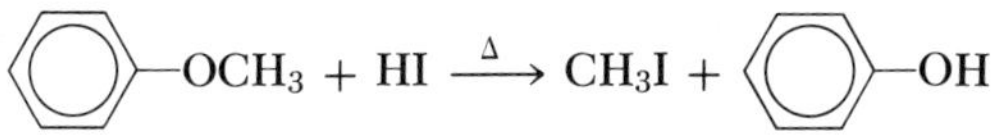

$$\underset{\text{Anisole}}{C_6H_5-OCH_3} + HI \xrightarrow{\Delta} \underset{\text{Methyl iodide}}{CH_3I} + \underset{\text{Phenol}}{C_6H_5-OH}$$

Aromatic Substitution Reactions of Aromatic Ethers

If one of the groups attached to the ether oxygen atom is an aromatic group, the usual halogenation, nitration, sulfonation, alkylation, and acylation reactions can be carried out on the aromatic ring without affecting any cleavage of the ether linkage. The alkoxy group of such a mixed ether is an *ortho-para* directing group and also activates the ring toward electrophilic substitution. Therefore, mixed aromatic-aliphatic ethers undergo ring substitution reactions under very mild conditions compared to benzene and alkyl substituted benzenes. A typical bromination of phenyl ethyl ether (phenetole) is illustrated below, but it should be kept in mind that similar reactions occur for nitration, sulfonation, and Friedel-Crafts alkylation and acylation:

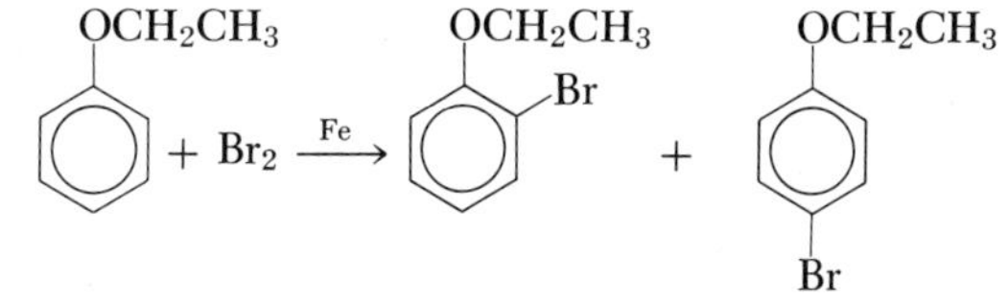

$$\underset{\text{Phenetole}}{C_6H_5OCH_2CH_3} + Br_2 \xrightarrow{Fe} \underset{o\text{-Bromophenetole}}{o\text{-}BrC_6H_4OCH_2CH_3} + \underset{p\text{-Bromophenetole}}{p\text{-}BrC_6H_4OCH_2CH_3}$$

Reactions of Epoxides

In contrast to the chemical inertness of simple ethers, epoxides contain a strained three-membered ring and are generally much more chemically reactive than ordinary ethers. As might be expected, the vast majority of reactions of epoxides are ring-opening reactions to relieve the strain in the three-membered ring. Both acid-catalyzed and base-catalyzed ring-opening reactions are known, and several such reactions are illustrated for ethylene oxide:

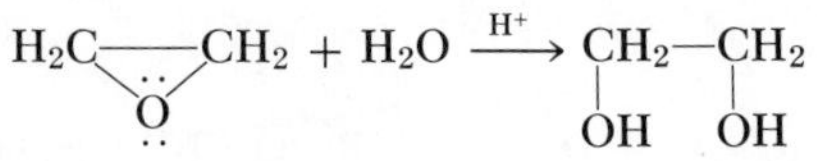

$$H_2C\underset{O}{—}CH_2 + H_2O \xrightarrow{H^+} \underset{OH}{CH_2}—\underset{OH}{CH_2}$$

Ethylene oxide — Ethylene glycol

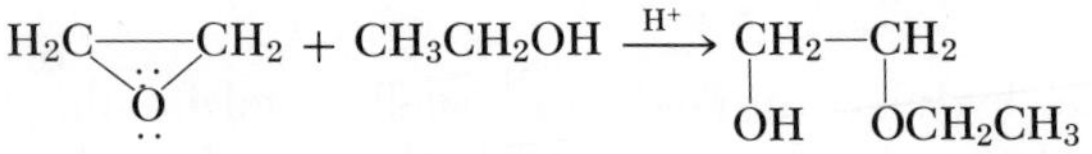

$$H_2C\underset{O}{—}CH_2 + CH_3CH_2OH \xrightarrow{H^+} \underset{OH}{CH_2}—\underset{OCH_2CH_3}{CH_2}$$

Ethylene oxide — Ethanol — 2-Ethoxyethanol (cellosolves)

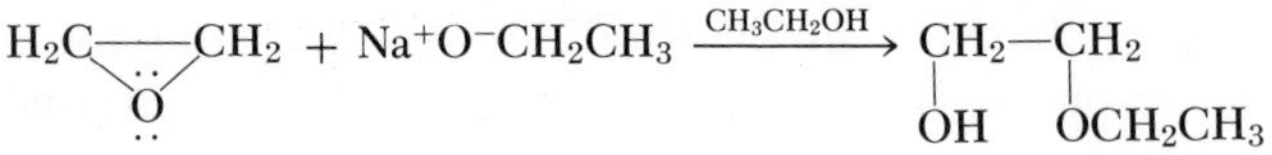

$$H_2C\underset{O}{—}CH_2 + Na^+O^-CH_2CH_3 \xrightarrow{CH_3CH_2OH} \underset{OH}{CH_2}—\underset{OCH_2CH_3}{CH_2}$$

Ethylene oxide — Sodium ethoxide — 2-Ethoxyethanol

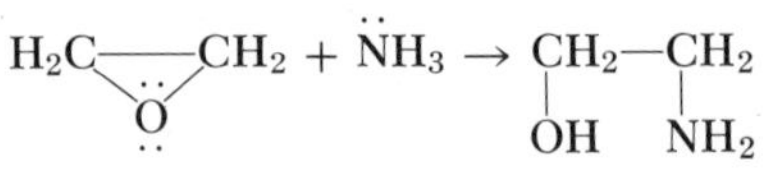

$$H_2C\underset{O}{—}CH_2 + \ddot{N}H_3 \rightarrow \underset{OH}{CH_2}—\underset{NH_2}{CH_2}$$

Ethylene oxide — Ethanolamine

The mechanism for these catalyzed ring-opening reactions can be rationalized as follows: The first step in the acid-catalyzed reactions involves simple protonation of oxygen to give the oxonium ion (I). Attack by the nucleophile (H_2O)

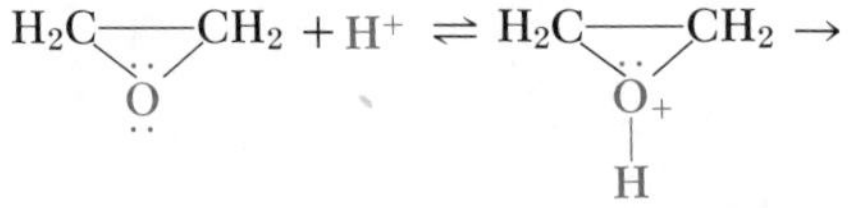

$$H_2C\underset{O}{—}CH_2 + H^+ \rightleftharpoons H_2C\underset{\underset{H}{O^+}}{—}CH_2 \rightarrow$$

(I)

$$H_2\ddot{O}: + H_2C\underset{\underset{H}{O^+}}{—}CH_2 \rightarrow \underset{^+\ddot{O}H_2}{CH_2}—\underset{OH}{CH_2} \xrightarrow{H_2O} \underset{OH}{CH_2}—\underset{OH}{CH_2} + H_3O^+$$

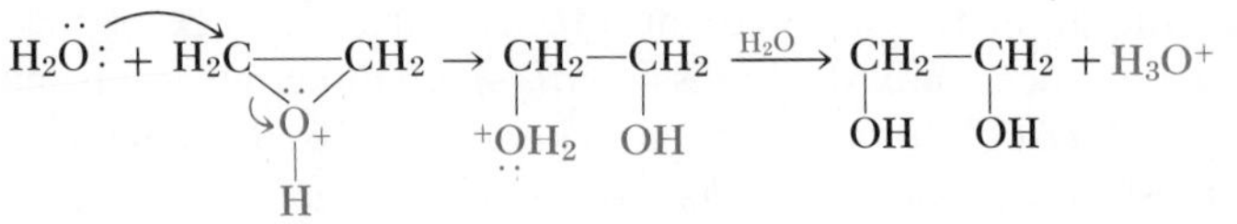

(II)

on carbon causes carbon-oxygen bond cleavage to give (II), which, on loss of a proton to another water molecule, gives the product, ethylene glycol.

In the base-catalyzed reaction, the first step involves attack by the nucleophile ($CH_3CH_2O^-$) on the ether carbon atom with simultaneous carbon-oxygen bond cleavage to give an anion (III), which abstracts a proton from the solvent (CH_3CH_2OH) to give the final product.

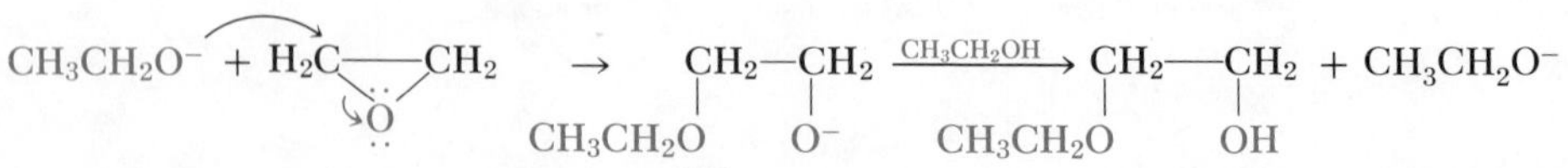

$$CH_3CH_2O^- + H_2C\underset{O}{—}CH_2 \rightarrow \underset{CH_3CH_2O}{CH_2}—\underset{O^-}{CH_2} \xrightarrow{CH_3CH_2OH} \underset{CH_3CH_2O}{CH_2}—\underset{OH}{CH_2} + CH_3CH_2O^-$$

(III)

Epoxides, particularly ethylene oxide, undergo a wide variety of ring-opening reactions similar to the ones described earlier, and ethylene oxide is one of the most important building blocks of the chemical industry.

SOME IMPORTANT ETHERS AND THEIR USES

Diethyl ether, which is often called ethyl ether or simply ether, is extensively used as a general anesthetic. It is easy to administer and causes excellent relaxation of the muscles. Blood pressure, pulse rate, and rate of respiration as a rule are only slightly affected. The main disadvantages are its irritating effect on the respiratory passages and its aftereffect of nausea. More recently, methyl propyl ether has been used as a general anesthetic. It has been claimed that this substance, called Neothyl, is less irritating and more potent than ethyl ether.

Diethyl ether is also an excellent solvent for fats and is often used in the laboratory for the extraction of fat from foods and animal tissue. In general, ethers are good solvents for fats, oils, gums, resins, and most functional derivatives of hydrocarbons. Ethylene oxide is considered an internal ether. It differs from most ethers in being completely soluble in water. It is used as a fumigating agent for seeds and grains and as the starting material in the preparation of the antifreeze ethylene glycol, the cellosolve solvents ($ROCH_2CH_2OH$, where R is an alkyl group) used in varnishes and lacquers, dioxane, and many other useful solvents and fibers.

Ethers occur also in many biologically active molecules. Tetrahydrocannabinol, mescaline, and morphine all contain the ether linkage, and this functionality is also present in DNA and RNA nuclei acids as a furan ring oxygen in the sugar portion of the molecule.

IMPORTANT TERMS AND CONCEPTS

anisole
cellosolve
diethyl ether
epoxide
ethylene oxide
Williamson reaction

QUESTIONS

1. Draw a structural formula for each of the following compounds:
   (a) methyl isopropyl ether
   (b) propylene oxide
   (c) *p*-nitroanisole
   (d) ethyl ether
   (e) 2-bromo-4-ethoxyhexane

2. Name each of the following compounds:
   (a) [benzene ring with OCH_3 and CH_3 substituents]
   (b) $(CH_3)_2CH{-}O{-}CH(CH_3)_2$
   (c) $CH_3CH_2O{-}C_6H_5$ [benzene ring]
   (d) $HOCH_2CHOHCH_2OH$
   (e) $CH_3OCH_2CH_2OCH_3$

3. Write equations for the reactions of anisole with the following reagents:
   (a) cold conc. H_2SO_4
   (b) excess HI/heat
   (c) $KMnO_4$
   (d) sodium
   (e) HNO_3/H_2SO_4

4. Show by equations how the following conversions may be carried out. More than one step may be necessary.
   (a) methyl *n*-propyl ether from *n*-propyl iodide
   (b) phenyl ethyl ether from phenol
   (c) cyclohexene oxide from cyclohexene
   (d) methyl isopropyl ether from propene
   (e) ethylene glycol from ethyl bromide

5. Write equations for the reactions of ethylene oxide with the following reagents:
   (a) *n*-propyl alcohol/H^+
   (b) sodium methoxide/CH_3OH
   (c) methylamine (CH_3NH_2)
   (d) ethylene glycol/H^+
   (e) ethanolamine

SUGGESTED READING

Beecher: Anesthesia. Scientific American, Vol. 196, No. 1, p. 70, 1957.
Krantz: Volatile Anesthetics and Analgesics. Journal of Chemical Education, Vol. 37, p. 169, 1960.

HALOGEN DERIVATIVES OF HYDROCARBONS

CHAPTER 8

The *objectives* of this chapter are to enable the student to:

1. Recognize an alkyl or aryl halide.
2. Write the structures and names of the halogen derivatives of hydrocarbons.
3. Describe the basic methods employed for the introduction of halogen into organic compounds.
4. Explain the concept of nucleophilic displacement reactions.
5. Define an organometallic compound.
6. Recognize Grignard reagents.
7. Recognize some simple fluorocarbon compounds and their current applications.
8. Recognize some important fungicides and insecticides.

In previous chapters compounds have been discussed that contain a carbon-halogen bond. These compounds resulted from the reaction of alcohols with halogenating agents and from the addition of hydrogen halides to olefins. In addition, they were found to be the basic starting materials in some of the preparations of olefins, alcohols, and ethers. The utility of this class of compounds should now be evident, and the purpose of this chapter is to correlate, extend, and amplify somewhat, the properties, reactions, and uses of the halogen derivatives of hydrocarbons.

Halogen derivatives of hydrocarbons can be simply defined as *compounds in which a hydrogen of a hydrocarbon has been replaced by a halogen*, in which the halogen can be fluorine, chlorine, bromine, or iodine. The carbon atom bonded to the halogen atom may be a carbon that is part of an alkyl, vinyl, acetylenic, aromatic, or heterocyclic system. Some examples of the various types of halogen derivatives are shown below:

CH_3CH_2Br	$CH_2{=}CHCl$	$HC{\equiv}CCl$	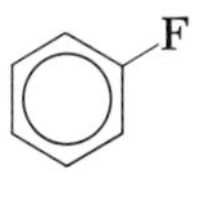 (benzene ring with F)	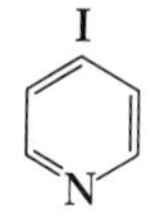 (pyridine ring with I at 4-position, N)
Ethyl bromide	Vinyl chloride	Chloroacetylene	Fluorobenzene	4-Iodopyridine

The most widely used types of halogen derivatives are the ones containing the halogen bonded either to an alkyl group (called **alkyl halides**) or to an aromatic ring (called **aryl halides**), and most of the emphasis in this chapter will be on these two types

TABLE 8–1 PHYSICAL PROPERTIES OF HALIDES

COMPOUND	STRUCTURE	M.P. °C	B.P. °C
Methyl fluoride	CH_3F	−142	−78
Methyl chloride	CH_3Cl	−98	−24
Methyl bromide	CH_3Br	−94	+4
Methyl iodide	CH_3I	−66	+42
Methylene chloride	CH_2Cl_2	−95	+40
Chloroform	$CHCl_3$	−64	+62
Carbon tetrachloride	CCl_4	−23	+77
Vinyl chloride	$CH_2{=}CHCl$	−154	−13
Tetrachloroethylene	$CCl_2{=}CCl_2$	−22	+121

of halides. Although many reactions of alkyl and aryl halides may appear quite similar, in the most important reaction (displacement of the halogen atom by another atom or group of atoms) of halides these two types of halides behave quite differently. Aryl halides are generally quite resistant to displacement reactions under ordinary conditions, whereas most alkyl halides undergo displacement reactions very easily. This contrasting behavior will be discussed in more detail later in this chapter.

PROPERTIES OF HALIDES

Within any series of alkyl or aryl halides, the boiling points increase with increasing molecular weight; consequently, the boiling points increase in the order: fluorides < chlorides < bromides < iodides, as illustrated below for the ethyl halides and phenyl halides. The physical properties of some of the more common halides are summarized in Table 8–1.

CH_3CH_2F	CH_3CH_2Cl	CH_3CH_2Br	CH_3CH_2I
B.P. −38°C	B.P. 12°C	B.P. 38°C	B.P. 72°C
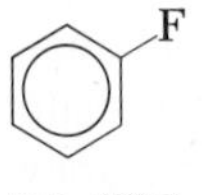 C_6H_5F	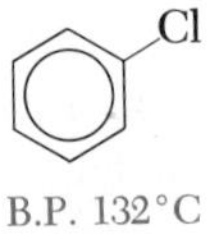 C_6H_5Cl	 C_6H_5Br	 C_6H_5I
B.P. 85°C	B.P. 132°C	B.P. 156°C	B.P. 187°C

The organic halides are insoluble in water. The monofluoro and monochloro derivatives are less dense than water, and the monobromo and monoiodo compounds are more dense than water. Similarly to alkanes, the halide derivatives are insoluble in cold concentrated sulfuric acid, and extraction by this reagent can be used to remove alcohol, alkene, and ether impurities from organic halides.

METHODS OF PREPARATION OF ORGANIC HALIDES

Several of the preparations of halides have been discussed in earlier chapters and will be noted here to correlate this material, but these previous methods will not be discussed in detail again, and the student is urged to go back and review these preparations in the previous chapters.

Direct Halogenation

Direct halogenation of alkanes was shown earlier to lead via a free-radical process to a mixture of halogenated hydrocarbons, and, except for a few special cases, is not generally applicable to preparing alkyl halides. Indirect methods are used to prepare the commercially important halides **carbon tetrachloride** and **chloroform.** Carbon tetra-

$$CH_4 + Cl_2 \xrightarrow{h\nu} CH_3Cl + CH_2Cl_2 + CHCl_3 + CCl_4$$

chloride is prepared industrially by the chlorination of carbon disulfide using antimony pentachloride as a catalyst. Carbon tetrachloride is a colorless liquid, insoluble in water, soluble in organic solvents, and more dense than water. It is extensively used in the

$$\underset{\text{Carbon disulfide}}{CS_2} + 3Cl_2 \xrightarrow{SbCl_5} \underset{\text{Carbon tetrachloride}}{CCl_4} + \underset{\text{Sulfur monochloride}}{S_2Cl_2}$$

laboratory as an extraction solvent and is used commercially as a solvent for oils and greases. Because of its high solvent power, it has also been used as a dry-cleaning agent and as a household cleaning agent. In contrast to the hydrocarbons, alcohols, and ethers, carbon tetrachloride is nonflammable and has found wide use as a fire-extinguishing agent. However, the oxidation of carbon tetrachloride at the temperatures of a fire gives phosgene ($COCl_2$), a toxic gas, and fire extinguishers containing carbon tetrachloride should only be used where adequate ventilation is available.

Chloroform ($CHCl_3$) is obtained commercially by the reduction of carbon tetrachloride using iron and steam as the reducing agent. Chloroform is a sweet-smelling

$$\underset{\text{Carbon tetrachloride}}{CCl_4} + H_2O \xrightarrow{Fe} \underset{\text{Chloroform}}{CHCl_3} + HCl$$

volatile liquid (b.p. 62°C) that once was widely used as an anesthetic. Toxic effects sometimes result from its use, however, and it has been replaced by other anesthetics. Chloroform undergoes photochemical (sunlight) oxidation to give phosgene unless stabilized, and commercially available chloroform contains $\frac{3}{4}$% ethanol as a stabilizer to prevent air oxidation.

$$\underset{\text{Chloroform}}{CHCl_3} + O_2 \xrightarrow{\text{sunlight}} \underset{\text{Phosgene}}{COCl_2}$$

Although direct halogenation of alkanes is unsuitable for preparing alkyl halides, it is the most widely used method for preparing aryl halides (see reactions of aro-

$$\underset{\text{Benzene}}{C_6H_6} + Br_2 \xrightarrow{Fe} \underset{\text{Bromobenzene}}{C_6H_5Br} + HBr$$

matic hydrocarbons) containing chlorine or bromine. Substituted aryl halides can also be prepared satisfactorily by this method.

$$\underset{\text{R = alkyl group or alkoxy group}}{R\text{-}C_6H_5} + Br_2 \xrightarrow{Fe} o\text{-}Br\text{-}C_6H_4\text{-}R + p\text{-}Br\text{-}C_6H_4\text{-}R$$

Addition of Hydrogen Halides to Olefins

The addition of hydrogen halides (HF, HCl, HBr, and HI) to olefins has been shown previously (see reactions of alkenes) to give alkyl halides. The addition reaction follows Markownikoff's rule.

$$CH_3CH{=}CH_2 + HCl \rightarrow CH_3CHClCH_3$$

Propene — Isopropyl chloride

Replacement of Other Functional Groups

Hydroxyl Group of Alcohols

The hydroxyl group of alcohols can be replaced by a halide atom using either the hydrogen halides, phosphorus halides (PCl_3 and PCl_5), or thionyl chloride (see reactions of alcohols). Fluorides are generally not prepared by this method, but chlorides, bromides, and iodides are attainable. Phenols do not usually undergo this type of reaction.

$$ROH + HX \rightarrow RX + H_2O$$
$$X = Cl, Br, I$$

Halogen Exchange Reactions

Chlorides and bromides, particularly, are starting materials in preparing many alkyl iodides and alkyl fluorides by exchanging either the chlorine or bromine atom for an iodine or fluorine atom. The most common type of exchange medium for introducing iodine is sodium iodide in acetone. The reverse of this reaction is actually the favored pathway (i.e., displacement of iodide by chloride or bromide). However, sodium chloride and sodium bromide are insoluble in acetone (whereas sodium iodide is soluble) and precipitate from solution as formed; therefore, the reverse reaction is not possible in this case. This type of displacement reaction proceeds best when the alkyl group is primary.

$$CH_3CH_2CH_2CH_2Br + NaI \xrightleftharpoons{\text{acetone}} CH_3CH_2CH_2CH_2I + Na^+Br^- \downarrow$$

n-Butyl bromide — *n*-Butyl iodide

Alkyl fluorides can also be prepared via the corresponding halides by treatment with metal fluorides, such as silver fluoride (AgF), potassium fluoride (KF), mercuric fluoride (HgF_2), antimony fluorides (SbF_3 and SbF_5), and cobalt trifluoride (CoF_3). The type of fluorinating agent used depends upon the ease of the exchange reaction. The introduction of fluorine into an aromatic ring by this type of exchange reaction requires drastic

$$CHBr_3 + HgF_2 \longrightarrow CHF_3$$

Bromoform — Fluoroform

$$ClCH_2CH_2SCH_2CH_2Cl + 4AgF \xrightarrow{50°C.} FCH_2CH_2SCH_2CH_2F$$

B,B'-Dichloroethyl sulfide — *B,B'*Difluoroethyl sulfide

conditions and is not widely used. In general, special methods are required to introduce fluorine into an aromatic ring.

REACTIONS OF HALIDES

The most common type of reaction of alkyl halides involves displacement of the halogen atom. This type of reaction may be rationalized as the attack of a negative ion (nucleophile) on an alkyl halide to bring about displacement of a halide ion. Because of the great variety of nucleophiles that will cause displacement of a halide ion from an alkyl halide, numerous types of functional derivatives can be prepared by this kind of reaction. Consequently, alkyl halides are one of the basic starting materials for many of the general classes of hydrocarbon derivatives. Some of the more important of such displacement reactions are summarized in the following equations. The general reaction may be summarized as follows:

$$N^- + R—X \rightarrow N—R + X^-$$

N = anion (nucleophile)
X = halogen

Nucleophilic Displacement of Halide Ion

Alcohol formation:	$R—X + \bar{O}H \rightarrow R—OH + X^-$
Ether formation:	$R'—X + \bar{O}R \rightarrow R'—O—R + X^-$
Amine formation:	$R—X + \bar{N}H_2 \rightarrow R—NH_2 + X^-$
Ester formation:	$R'—X + RCO_2^- \rightarrow R'—O—\overset{O}{\overset{\parallel}{C}}—R + X^-$
Acetylene formation:	$R'—X + RC{\equiv}C^- \rightarrow R'—C{\equiv}CR + X^-$
Nitrile formation:	$R—X + \bar{C}N \rightarrow R—CN + X^-$
Mercaptan formation:	$R—X + \bar{S}H \rightarrow R—SH + X^-$

In this type of displacement reaction, the reactivity is dependent on the type of nucleophile and the kind of alkyl group. For any given anion, the tendency toward displacement decreases in the order: primary > secondary > tertiary. Aromatic halides, unless activated by the presence of one or more nitro groups located *ortho* or *para* to the halogen atom, do not generally undergo this type of displacement reaction under normal conditions. Benzyl halides, of course, behave like alkyl halides and undergo this type of displacement reaction easily. Vinyl halides are likewise inert to this kind of displacement reaction under normal conditions.

Let us consider a specific example of this type of reaction, namely, the conversion of methyl bromide to methyl alcohol by hydroxide.

$$CH_3Br + {}^-OH \rightarrow CH_3OH + Br^-$$

The mechanism of this reaction has been shown to involve simultaneous formation of the $—\overset{|}{\underset{|}{C}}—OH$ bond and cleavage of the $—\overset{|}{\underset{|}{C}}—Br$ bond as shown below in *B*. The hydroxide attacks the carbon bearing the bromine from the backside (*A*) and the bromine

$${}^-OH + H_3C—Br \;(A) \rightarrow \left[HO\text{---}CH_3\text{---}Br\right] (B) \rightarrow HO—CH_3 \;(C) + Br^-$$

leaves from the front side (*C*). In this situation (*B*), the two electronegative Br and OH groups assume positions as far apart as possible. Consequently, their mutual repulsions are minimized, and this situation is the lowest energy (most favorable) mechanistic pathway. After departure of the bromide ion, the final product (*C*) has been inverted; i.e., the molecule has been essentially turned inside out and the hydroxyl group is on the opposite face of the molecule (relative to the other groups attached to carbon). For most molecules inversion at carbon by this type of reaction does not cause any change in the molecule; however, if the carbon atom is an asymmetrical[*] carbon, the resultant product will have the opposite configuration. This point will be discussed in more detail in Chapter 13, since the configuration of the molecule is closely related to its physiological activity.

Since attack by the nucleophile involves backside attack on carbon, the other groups attached to this carbon play a significant role in the ease of this reaction. As the size of these groups attached to carbon increases, the ease of approach of the nucleophile decreases, since it encounters increased resistance to approach at the backside of the carbon being attacked. Consequently, the ease of attack is primary > secondary > tertiary, since this order is also the order of increasing size of these groups.

Although alkyl halides are generally not encountered in living systems, other types of groups, such as phosphate, pyrophosphate, and water can function in similar types of displacement reactions, and these types of reactions will frequently be observed when you study the chemistry of carbohydrates and amino acids in later sections of the text. For example, the nucleoside thymidine is formed by the displacement of phosphate by thymine.

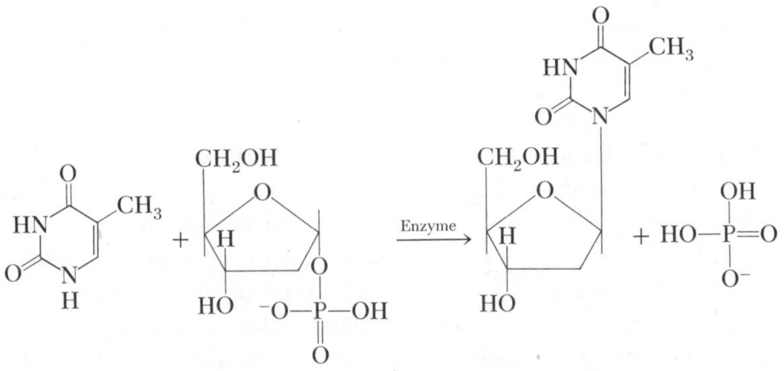

Enzyme

Thymine

Thymidine

Note that the entering thymine group takes its position on the oppositive face of the ring from which the phosphate has left.

Formation of Organometallic Reagents

Compounds in which a metal is bonded to carbon are called organometallic compounds. Depending upon the metal involved, these compounds may be mainly ionic or mainly covalent. In recent years, the preparation of organometallic compounds and the study of their chemistry has been one of the most fascinating and productive areas of chemistry. Compounds containing most of the known metals have been prepared and studied. Several of these organometallic compounds are well known to most people. For example, tetraethyl lead, $(C_2H_5)_4Pb$, is an important antiknock ingredient added to most gasolines. Mercurochrome and Merthiolate are two mercury derivatives used as antiseptics.

Three important biologic compounds which contain metals as part of the molecule are chlorophyll, which is important in photosynthesis and contains magnesium (see p. 327), heme, which is important as an oxygen carrier in the blood and contains iron (see p. 368), and vitamin B_{12}, which contains cobalt (see p. 309).

[*]An asymmetric carbon is a carbon atom which has four different groups attached to it.

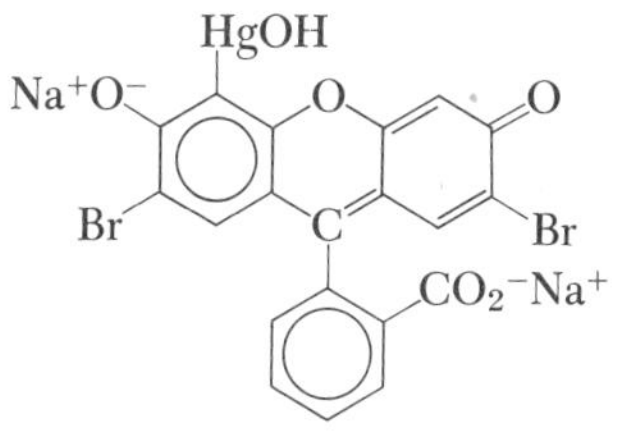

Mercurochrome

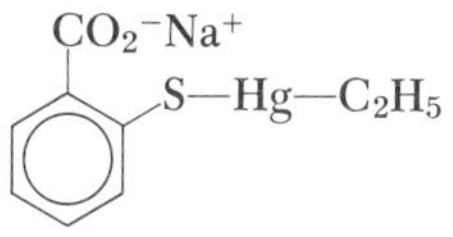

Merthiolate

Of the numerous and diversified kinds of organometallic compounds, the magnesium and lithium compounds have been the most useful to the organic chemist in the laboratory. These reagents are usually prepared by an exchange reaction in an inert solvent such as ether or tetrahydrofuran. Generally, they are not isolated, but are generated *in situ* and used as intermediates for preparing other compounds.

The preparation and use of organomagnesium compounds were developed by the French chemist, Victor Grignard, and this type of compound is known as a **Grignard reagent.** Both alkyl and aryl halides form a Grignard reagent when reacted with metallic magnesium in anhydrous ether, as shown below:

$$CH_3CH_2I + Mg \xrightarrow{\text{ether}} [CH_3CH_2]^- \overset{+}{Mg}I$$

Ethyl iodide — Ethyl magnesium iodide

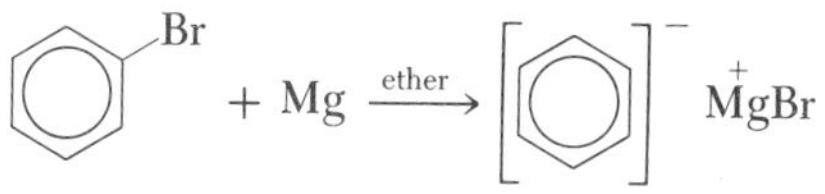

$$C_6H_5Br + Mg \xrightarrow{\text{ether}} [C_6H_5]^- \overset{+}{Mg}Br$$

Bromobenzene — Phenyl magnesium bromide

These types of reagents are sensitive to hydrolysis and are hydrolyzed by water or acids to hydrocarbons.

$$CH_3CH_2MgI \xrightarrow{H_2O} CH_3CH_3 + Mg(OH)I$$

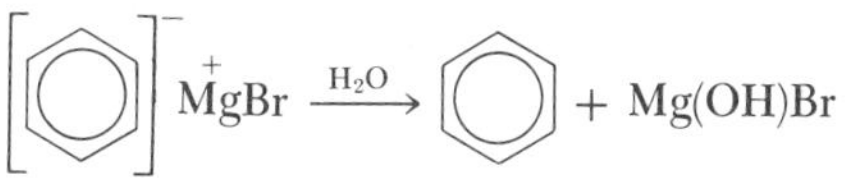

$$[C_6H_5]^- \overset{+}{Mg}Br \xrightarrow{H_2O} C_6H_6 + Mg(OH)Br$$

Organolithium compounds are prepared and behave similarly to the magnesium compounds.

$$R{-}X + 2Li \xrightarrow{\text{ether}} R^- \, Li^+ + Li^+X^-$$

R = alkyl or aryl

Both Grignard and lithium reagents are highly reactive compounds and undergo reaction with many functional groups (particularly carbonyl-containing functional groups) to give useful products. Applications of these organometallic reagents will be introduced in subsequent chapters.

Electrophilic Substitution Reactions of Aryl Halides

Although aromatic halides undergo nucleophilic substitution reactions only with extreme difficulty, they undergo the normal type of electrophilic substitution reactions without too much difficulty. The halogen atoms are *ortho-para* directing, and because of the inductive effect of the halogen compared to hydrogen, aryl halides are deactivating as compared to benzene. Therefore, bromobenzene gives *ortho-* and *para*-dibromobenzene on bromination, but the rate of bromination of bromobenzene is slower than the rate of bromination of benzene. Nitration, sulfonation, acylation, and alkylation proceed similarly.

$$C_6H_5Br + Br_2 \xrightarrow{FeBr_3} o\text{-}C_6H_4Br_2 + p\text{-}C_6H_4Br_2$$

Bromobenzene — *o*-Dibromobenzene — *p*-Dibromobenzene

FLUORINE COMPOUNDS

Because of the great reactivity of fluorine, it is not normally useful as a general purpose fluorinating agent except under special conditions. Consequently, organic compounds containing fluorine are usually prepared by special methods. The most important fluorine compounds are those in which all the hydrogen atoms have either been replaced by fluorine (this class is known as fluorocarbons) or by a combination of fluorine and other halogens. Substances containing many fluorine atoms are generally inert to oxidizing and reducing agents and to most acids. Consequently, they find extensive use in refrigeration, aerosols, lubricants, electrical insulators, and in the plastics industry.

The most widely used fluorine compounds are the methane and ethane derivatives sold under the trade names Freon, Genetron, and Ucon. Some typical examples are shown below:

CF_2Cl_2	$CFCl_3$	$CFCl_2CF_2Cl$	$CF_2—CF_2$ / $CF_2—CF_2$ (ring)
Freon-12 Dichlorodifluoromethane	Freon-11 Trichlorofluoromethane	Freon-113 1,1,2-Trifluoro-trichloroethane	Freon-C318 Octafluorocyclobutane

Freon-12 is volatile (b.p. $-28°C$) and nontoxic, and is used as a refrigerant in the majority of household and commercial refrigeration units and air conditioners. Freon-C318 (octafluorocyclobutane) is odorless, tasteless, nontoxic, and extremely stable to hydrolysis, and has application in the aerosol industry. Trifluorobromomethane, CF_3Br, is a nontoxic fire extinguishing agent and has replaced many of the carbon tetrachloride extinguishers which produce the toxic gas phosgene when used to extinguish fires.

Probably the best known of the fluorine containing compounds is the polymer known as **Teflon,** which is made by polymerization of tetrafluoroethylene, $CF_2{=}CF_2$. Teflon contains the repeating unit $(—CF_2—CF_2—)_n$. This fluorocarbon polymer is inert to most chemical reagents, has excellent electrical insulating properties, maintains its lubricating properties over a wide range (-50 to $+300°C$), and has "non-sticking" properties for most materials. This latter characteristic has been exploited in making Teflon coated frying pans, cookie sheets, pie tins, and so forth, which can be used for cooking and baking without grease, as foodstuffs do not adhere to Teflon.

Another important property of fluorocarbons is their nonwettability. Materials and fabrics coated with a fluorocarbon become water and oil repellant. Fabrics treated with fluorochemicals become stain resistant, and this particular property has been put to practical use in the manufacture of "Scotch Guard" coated fabrics.

Biologically, fluorine has been shown to enhance the physiological activity of the steroidal molecule. For example, in prednisolone, introduction of a fluorine atom at the 9-position increases the potency of this steroid by a factor of 100 to 250 (see p. 384).

Similarly, the fluorophosphates such as diisopropylfluorophosphate are extremely active nerve gases. They react rapidly with the enzyme cholinesterase and thereby prevent

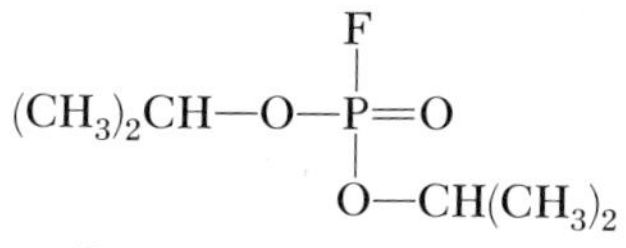

$$(CH_3)_2CH{-}O{-}P(=O)(F){-}O{-}CH(CH_3)_2$$

Diisopropylfluorophosphate

the transmission of nervc impulses which control many of the human vital functions.

MISCELLANEOUS HALOGEN COMPOUNDS

Halogen compounds play an important role in the chemical industry as synthetic starting materials. In addition, many halogen containing compounds find application as fumigants (CH_3Br) for controlling various insects and rodents, and as fungicides and pesticides. Some of the more common halogen-containing pesticides are shown below:

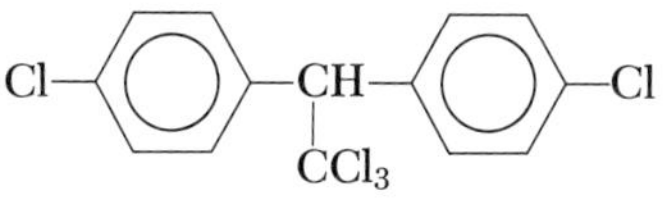

DDT

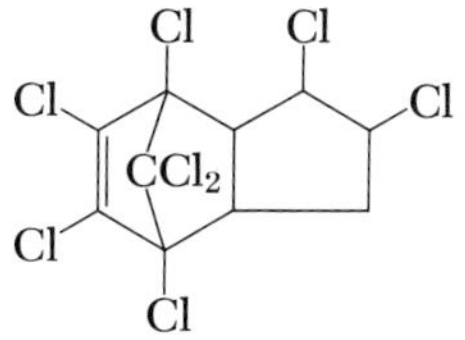

Chlordan

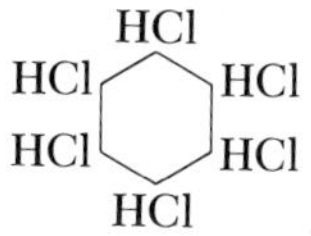

Lindane

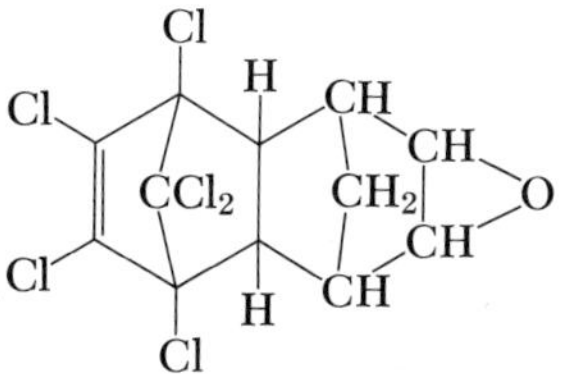

Dieldrin

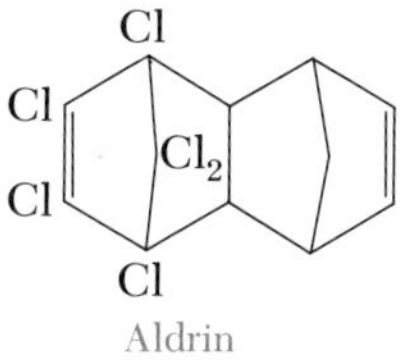

Aldrin

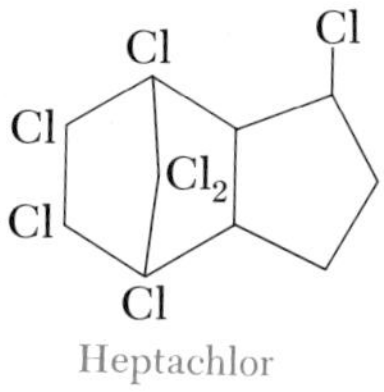

Heptachlor

In recent years the use of these chlorinated hydrocarbons has aroused public interest and criticism. Although successful control of malaria and the boll weevil have been the virtues of DDT and related compounds, more recent data has shown that these substances are accumulated in high concentration in fish and fowl. This increase in DDT concentration in the animal food chain has threatened the survival of several species of birds and fishes by effecting their reproductive cycle. Consequently, much controversy has arisen in recent years over the continued use of these chlorinated hydrocarbons, and renewed interest in biodegradable insecticides, chemosterilants, and insect attractants has arisen in the search for other means to control insects.

The most successful method thus far has been to utilize chemical compounds by which insects communicate with one another. Chemical compounds of this type are called **pheromones** referring to compounds which are hormone-like compounds secreted by an individual or organism which evoke a specific physiological reaction in another individual or organism of the same species. There are different classes of pheromones, divided according to their biological function. The most useful class thus far is that of **sex pheromones**, commonly called **sex attractants.** The basic idea in using this class of pheromones is to isolate the chemical substance with which one sex attracts the other, and then to bait a trap with this pheromone. Until recently, it was almost impossible to isolate and identify these pheromone compounds from insects, since any particular insect contains only extremely small amounts of the pheromone. However, sophisticated instrumentation and new techniques for handling small amounts of chemical compounds have allowed the chemist and biologist in recent years to isolate and identify many pheromone-like compounds.

In their practical use pheromones are then utilized to bait an insect trap. The insect is then attracted to the trap by the pheromone compound. In addition to the pheromone, the trap also contains an insecticide to kill the insect or a chemosterilant which, in the final analysis, would remove the species from the reproduction cycle.

The sensitivity of the sex pheromones is almost unbelievable. In the moth, as little as 30 molecules of the female pheromone will attract a male moth.

Some of the more common pheromones which have been identified in recent years are shown below.

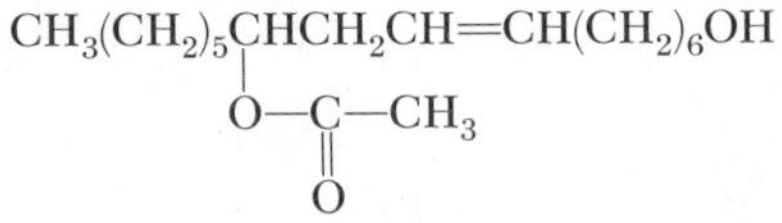

$$CH_3(CH_2)_5\underset{\displaystyle O-\underset{\displaystyle \overset{\|}{O}}{C}-CH_3}{\overset{|}{C}H}CH_2CH{=}CH(CH_2)_6OH$$

Gypsy moth

$$CH_3(CH_2)_2CH{=}CHCH{=}CH(CH_2)_9OH$$

Silkworm moth

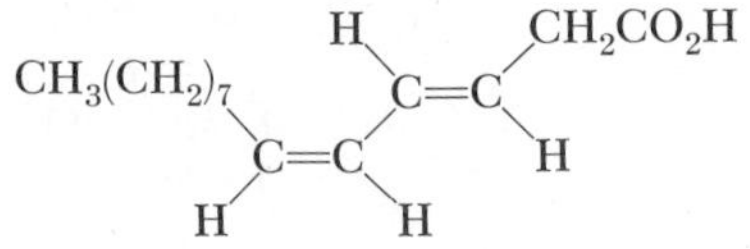

$$CH_3(CH_2)_7CH{=}CH{-}CH{=}CHCH_2CO_2H$$

(first C=C: H, H shown cis; second C=C: H and CH_2CO_2H shown trans)

Black carpet beetle

$$\begin{array}{l} \quad\quad\quad\ \ CH_3 \\ H_2C-C\cdots\cdots CH_2CH_2OH \\ \ \ |\quad\ \ | \\ H_2C-C\cdots\cdots \underset{\displaystyle CH_3}{\underset{|}{C}}{=}CH_2 \\ \quad\quad\ \ H \end{array}$$

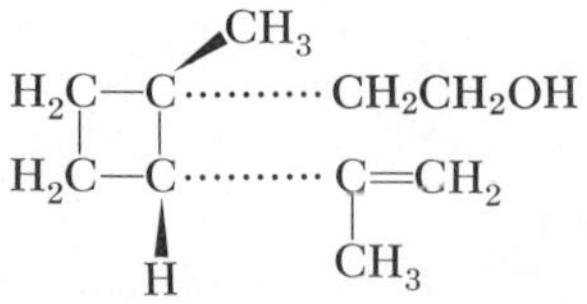

Boll weevil

Much research effort is currently being expended in this area of chemistry, since pheromones could help to eradicate harmful insects and, at the same time, prevent damaging ecological problems. Hopefully, a particular pheromone can be found to eradicate only harmful insect species without endangering useful insects. Most current insecticides are not discriminating and kill both harmful and useful insects. In addition, the pheromones will not accumulate in plants, birds, and fish as DDT has been reported to do and will not damage the ecology by their use.

IMPORTANT TERMS AND CONCEPTS

aldrin
alkyl Halide
aryl Halide
Chlordan
chloroform
DDT
dieldrin
freon
Grignard reagent
lindane
nucleophilic displacement reaction
organometallic
pheromones
phosgene
Teflon

QUESTIONS

1. Draw a structural formula for each of the following compounds:
   (a) tertiary butyl bromide
   (b) *p*-iodotoluene
   (c) cyclohexyl bromide
   (d) 1,1,2-trichlorotrifluoroethane
   (e) carbon tetrabromide

2. Name each of the following compounds:
   (a) $CHCl_3$
   (b) $CHCl{=}CCl_2$
   (c) (benzene ring)–CH_2Cl
   (d) (benzene ring)–Cl
   (e) CH_3CHICH_3

3. Indicate what is wrong with each of the following names. Give the correct name for each compound.
   (a) 2,4,4-trichloropentane
   (b) 2-ethyl-4-chloropentane
   (c) 1,4,5-tribromobenzene
   (d) *trans*-3,4-dichloro-3-pentene
   (e) 5-chlorotoluene

4. Write equations for the reactions of *n*-propyl bromide with the following reagents:
   (a) sodium ethoxide
   (b) potassium cyanide
   (c) Mg/ether
   (d) silver oxide/H_2O
   (e) sodium acetylide

5. Write equations for the reactions of chlorobenzene with the following reagents:
   (a) HNO_3/H_2SO_4
   (b) Mg/ether
   (c) sodium methoxide
   (d) sodium cyanide
   (e) $CH_3COCl(AlCl_3)$

6. Write equations for the preparation of the following compounds. Give any necessary catalysts. More than one step may be necessary.
   (a) ethane from ethyl iodide
   (b) *n*-propyl fluoride from *n*-propyl alcohol
   (c) cyclopentyl bromide from cyclopentanol
   (d) benzyl cyanide from benzyl alcohol
   (e) 2,2-dibromopropane from propene
   (f) 2-chloropropene from propene
   (g) *p*-bromonitrobenzene from benzene
   (h) 1-bromo-1-phenylethane from β-phenylethanol
   (i) ethyl mercaptan (CH_3CH_2SH) from ethylene
   (j) vinyl chloride from ethanol

SUGGESTED READING

C. & E. News Special Report: Fluorocarbons. Chemical and Engineering News, July 18, p. 92, 1960.

Finger: Recent Advances In Fluorine Chemistry. Journal of Chemical Education, Vol. 28, p. 49, 1951.

Jacobson and Beroza: Insect Attractants. Scientific American, Vol. 211, No. 8, p. 20, 1964.

Joiner: Pesticide Residue Control Under The Food, Drug, and Cosmetic Act. Journal of Chemical Education, Vol. 38, p. 370, 1961.

Keller: The DDT Story. Chemistry, Vol. 43, p. 8, 1970.

Lykken: Chemical Control of Pests. Chemistry, Vol. 44, p. 18, 1971.

McBee and Roberts: Organic Fluorine Chemicals. Journal of Chemical Education, Vol. 32, p. 13, 1955.

Preparing Fluorocarbons—A New Method. Chemistry, Vol. 43, p. 30, 1970.

Rheinboldt: Fifty Years of the Grignard Reaction. Journal of Chemical Education, Vol. 27, p. 476, 1950.

Wilson: Pheromones. Scientific American, Vol. 208, No. 5, p. 100, 1963.

ALDEHYDES AND KETONES

CHAPTER 9

The *objectives* of this chapter are to enable the student to:

1. Recognize an aldehyde and a ketone.
2. Write the structures and names of the more common aldehydes and ketones.
3. Describe the conversion of other functional groups into the carbonyl functional group.
4. Describe the reactions of carbonyl compounds.
5. Explain mechanistically the nucleophilic addition reactions of aldehydes and ketones.
6. Recognize an acetal and a hemiacetal.
7. Explain the aldol condensation reactions of aldehydes.
8. Explain the haloform reaction and its diagnostic utility.
9. Recognize the important aldehydes and ketones and their current applications.

Compounds containing the carbonyl group ($-\overset{\displaystyle O}{\overset{\|}{C}}-$) are known as **aldehydes** or **ketones.** If one of the atoms linked to the carbonyl group is a hydrogen atom, the compound is an aldehyde ($-\overset{\displaystyle O}{\overset{\|}{C}}-H$). The other atom or group attached to the carbonyl group of an aldehyde may be hydrogen, alkyl, or aryl. In the case of ketones, both of the groups attached to the carbonyl group are either alkyl or aryl. Cyclic ketones also exist in which the carbonyl group is part of the ring. Cyclic aldehydes are not possible. In the next chapter other combinations of atoms or groups attached to the carbonyl group will be introduced.

NOMENCLATURE

In the IUPAC system, the characteristic ending for aldehydes is **-al,** and the characteristic ending for ketones is **-one.** These endings are added to the stem name of the hydrocarbon having the same number of carbon atoms. As usual, the compound is named as a derivative of the longest continuous chain of the carbon atoms, including the carbonyl functional group. In the case of aldehydes, the **—CHO** group must always appear at the end of the chain and is always indicated by the number **1** (the lowest number), although this number does not appear in the name. In the case of ketones, however, the carbonyl group may appear at various positions in a carbon chain, and its position must be designated by the lowest possible number. All other substituents are indicated by the appropriate number and prefix to indicate their positions on the carbon chain.

Common names are also used to name the aldehydes and ketones. Aldehydes are generally named as derivatives of the corresponding acid *to which they can be oxidized.* The **-ic** ending of the acid is dropped and replaced by the term **aldehyde.** Ketones, with the exception of acetone, are named according to the alkyl or aryl groups attached to the carbonyl function, followed by the word **ketone.** Some typical examples are illustrated below (the common name is in parentheses):

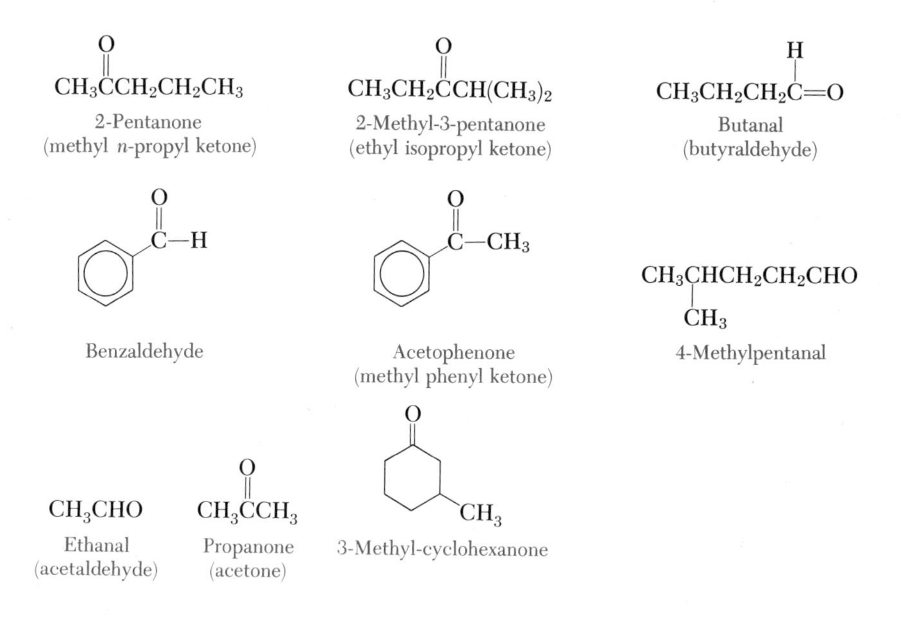

PHYSICAL PROPERTIES

Formaldehyde is a gas at room temperature (20°C). All other simple aliphatic and aromatic aldehydes are colorless liquids. Most of the lower molecular weight aldehydes have a sharp odor, but the odor becomes more fragrant as the molecular weight increases. The aromatic aldehydes have been used as flavoring agents and perfumes. Benzaldehyde, for example, is a constituent of the seeds of bitter almonds and was once called "oil of bitter almond." It is a colorless liquid with a pleasant almond-like odor.

All of the more common lower molecular weight ketones are liquids. Acetone, the simplest ketone, is a moderately low boiling liquid (b.p. 56°C). With few exceptions all other ketones are either liquids or solids (Table 9–1).

The lower molecular weight aldehydes and ketones are soluble in water. As the length of the carbon chain increases, the water solubility decreases. Limited solubility occurs around 5 to 6 carbon atoms. As expected, aldehydes and ketones dissolve in the normal organic solvents.

The carbonyl group is a polar group (oxygen being more electronegative than carbon), and the direction of electron density is toward the oxygen atom as indicated on page 150. The carbon atom is the positive end of the dipole, and the oxygen atom is the negative end.

TABLE 9-1 PHYSICAL PROPERTIES OF ALDEHYDES AND KETONES

COMPOUND	STRUCTURE	M.P. °C	B.P. °C
Formaldehyde	HCHO	−92	−21
Acetaldehyde	CH_3CHO	−123	+21
Propionaldehyde	CH_3CH_2CHO	−81	+49
Chloral	CCl_3CHO	−58	+98
Acetone	CH_3COCH_3	−95	+56
Methyl ethyl ketone	$CH_3COCH_2CH_3$	−86	+80
Cyclohexanone	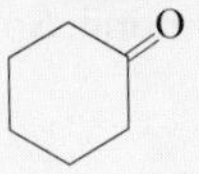 (cyclohexane ring with =O)	−45	+157
Acetophenone	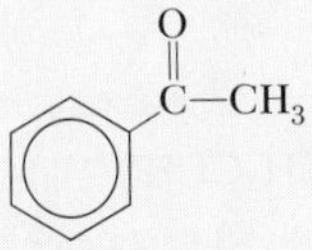 (benzene ring)–C(=O)–CH_3	+20	+202
Benzophenone	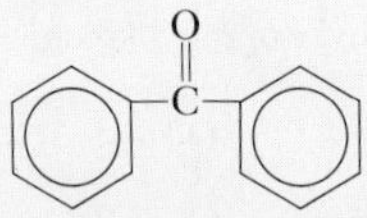 (benzene ring)–C(=O)–(benzene ring)	+48	+306
Acrolein	$CH_2{=}CHCHO$	−88	+53
Crotonaldehyde	$CH_3CH{=}CHCHO$	−77	+104
Benzaldehyde	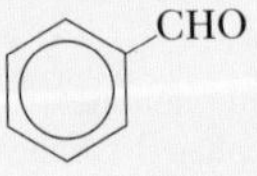 (benzene ring)–CHO	−26	+179
Salicylaldehyde	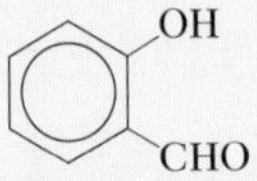 (benzene ring) with OH and CHO	−7	+197

$$-\overset{\delta+}{C}{=}\overset{\delta-}{O}$$

Carbonyl group

Because of the polarity of the carbonyl group, the boiling points of aldehydes and ketones are higher than either the hydrocarbon or the ether of corresponding molecular weight. However, aldehydes and ketones cannot form hydrogen bonds, and therefore their boiling points are lower than the alcohols and acids of comparable molecular weight. This relationship is illustrated below for a series of compounds of similar molecular weights.

$CH_3CH_2CH_2CHO$	$CH_3\overset{O}{\overset{\Vert}{C}}CH_2CH_3$	$CH_3(CH_2)_3CH_3$	$C_2H_5OC_2H_5$
B.P. 76°C	B.P. 80°C	B.P. 36°C	B.P. 34°C

$CH_3(CH_2)_2CH_2OH$	CH_3CH_2COOH
B.P. 118°C	B.P. 141°C

METHODS OF PREPARATION

Oxidation

Under the proper conditions, primary alcohols can be oxidized to give aldehydes, and secondary alcohols can be oxidized to give ketones.

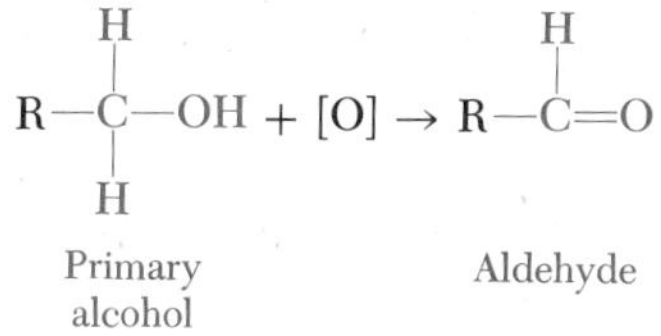

$$R-\underset{H}{\overset{H}{C}}-OH + [O] \rightarrow R-\overset{H}{C}=O$$

Primary alcohol — Aldehyde

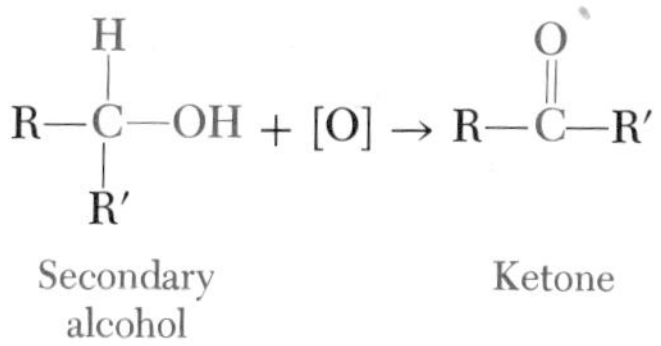

$$R-\underset{R'}{\overset{H}{C}}-OH + [O] \rightarrow R-\overset{O}{\overset{\|}{C}}-R'$$

Secondary alcohol — Ketone

These reactions can be viewed as a "dehydrogenation" process, since the elements of H_2 are lost in going from reactant to product. In the case of aldehydes, the process is somewhat complicated by the fact that aldehydes are more easily oxidized than alcohols. Consequently, as the aldehyde is formed, it undergoes further oxidation to give a carboxylic acid. Therefore, the aldehyde must be removed from the oxidation zone to prevent this additional oxidation to the acid. Two methods are employed to accomplish this purpose. For low boiling aldehydes, the oxidation is generally carried out using either

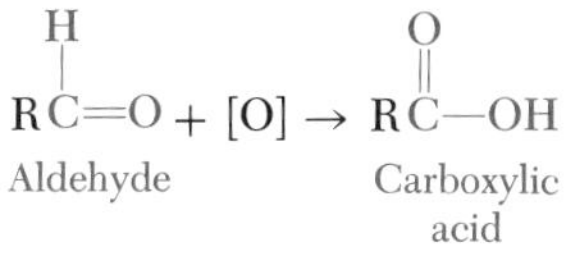

$$R\overset{H}{C}=O + [O] \rightarrow R\overset{O}{\overset{\|}{C}}-OH$$

Aldehyde — Carboxylic acid

potassium permanganate or potassium dichromate solutions. Since aldehydes boil lower than the corresponding alcohols, the aldehyde can be distilled out of the reaction mixture as it is formed, thus preventing further oxidation. The preparation of acetaldehyde by this method is illustrated in the following equation:

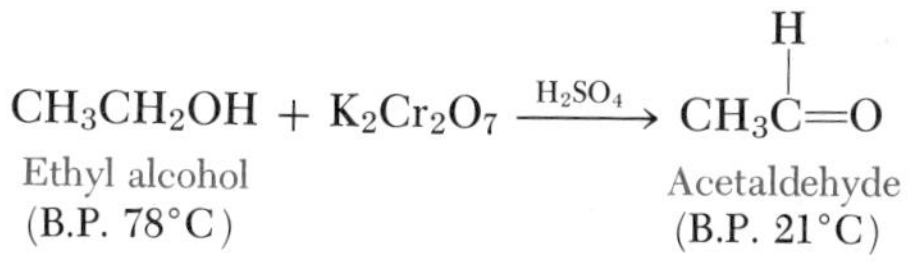

$$CH_3CH_2OH + K_2Cr_2O_7 \xrightarrow{H_2SO_4} CH_3\overset{H}{C}=O$$

Ethyl alcohol (B.P. 78°C) — Acetaldehyde (B.P. 21°C)

For higher boiling aldehydes, which can not be conveniently distilled from the reaction mixture, the oxidation is carried out by passing the alcohol vapor over metallic copper at 200 to 300°C. The contact time of the aldehyde in the hot oxidation zone is very short, and additional oxidation to the acid is minimized.

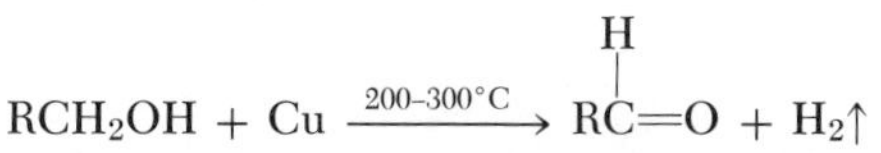

$$RCH_2OH + Cu \xrightarrow{200\text{–}300°C} R\overset{H}{C}=O + H_2\uparrow$$

Formaldehyde is prepared industrially by a variation of this type of oxidation reaction.

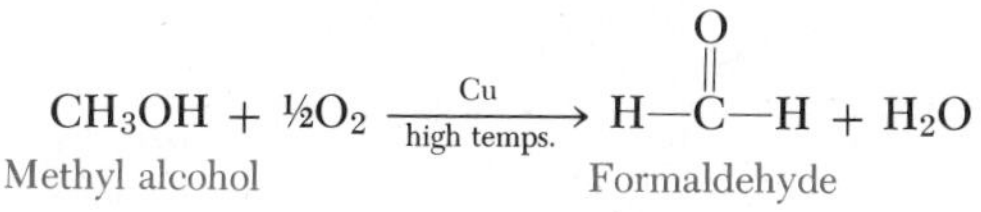

$$CH_3OH + \tfrac{1}{2}O_2 \xrightarrow[\text{high temps.}]{Cu} H-\overset{O}{\overset{\|}{C}}-H + H_2O$$

Methyl alcohol — Formaldehyde

Oxygen is used in this process to convert the hydrogen into water. The formaldehyde-water mixture obtained is passed into water, and the solution formed (40% formaldehyde) is sold under the trade name "Formalin."

The problem of further oxidation in the preparation of ketones from secondary alcohols is nonexistent under the normal conditions of the oxidation reactions. Ketones are quite resistant to further oxidation, except under extreme conditions, and the oxidation of the secondary alcohol stops very cleanly at the ketone stage. Some examples to illustrate the scope and versatility of this kind of reaction are shown below:

$$\underset{\text{Isopropyl alcohol}}{CH_3CHOHCH_3} + K_2Cr_2O_7 \xrightarrow{H_2SO_4} \underset{\text{Acetone}}{CH_3\overset{\overset{O}{\|}}{C}CH_3}$$

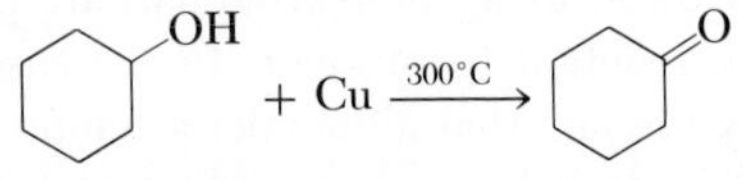

$$\underset{\text{Cyclohexanol}}{C_6H_{11}OH} + Cu \xrightarrow{300°C} \underset{\text{Cyclohexanone}}{C_6H_{10}{=}O}$$

$$\underset{\text{1-Phenylethanol}}{C_6H_5CHOHCH_3} + KMnO_4 \xrightarrow{OH^-} \underset{\text{Acetophenone}}{C_6H_5\overset{\overset{O}{\|}}{C}{-}CH_3}$$

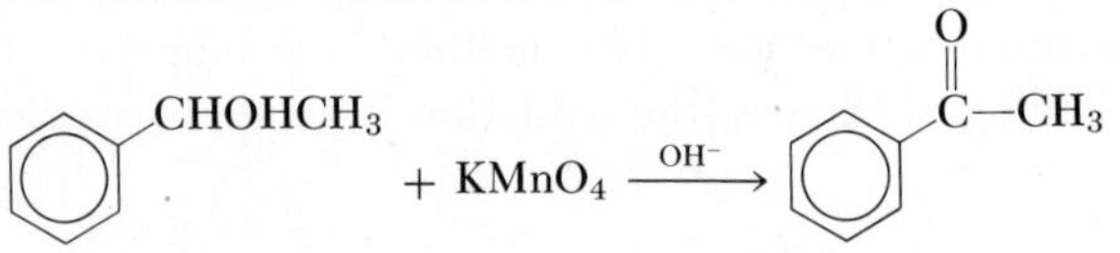

In the case of compounds containing an alkyl group, the alcohol is obtained by hydration (see reactions of alkenes) of the appropriate olefin. Indirectly then, the ketone is prepared via a two-step synthesis from the olefin, as shown below for the preparation of 2-butanone from 1-butene:

$$\underset{\text{1-Butene}}{CH_3CH_2CH{=}CH_2} + H_2SO_4 \xrightarrow{H_2O} \underset{\text{2-Butanol}}{CH_3CH_2\overset{\overset{OH}{|}}{C}HCH_3} \xrightarrow[H_2SO_4]{K_2Cr_2O_7} \underset{\text{2-Butanone}}{CH_3CH_2\overset{\overset{O}{\|}}{C}CH_3}$$

Friedel-Crafts Acylation

Ketones containing an aromatic ring can be conveniently prepared by the Friedel-Crafts acylation of an aromatic hydrocarbon (see reactions of aromatic hydrocarbons). Two typical examples are illustrated below:

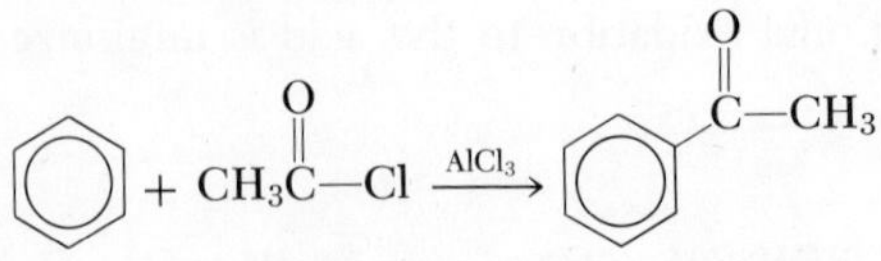

$$\underset{\text{Benzene}}{C_6H_6} + \underset{\text{Acetyl chloride}}{CH_3\overset{\overset{O}{\|}}{C}{-}Cl} \xrightarrow{AlCl_3} \underset{\text{Acetophenone}}{C_6H_5\overset{\overset{O}{\|}}{C}{-}CH_3}$$

$$\underset{\text{Benzene}}{C_6H_6} + \underset{\text{Benzoyl chloride}}{C_6H_5\overset{\overset{O}{\|}}{C}{-}Cl} \xrightarrow{AlCl_3} \underset{\text{Benzophenone}}{C_6H_5\overset{\overset{O}{\|}}{C}C_6H_5}$$

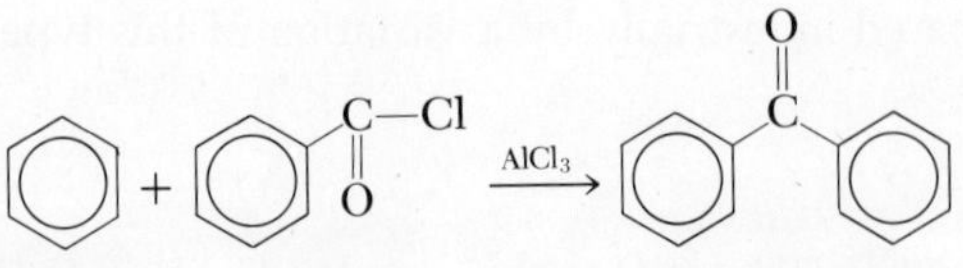

SPECIAL METHODS

In addition to the two general types of reactions discussed previously, special methods are used to prepare a particular aldehyde or ketone, particularly on an industrial scale. A few of these industrial methods will be briefly outlined.

Hydrolysis of Dihalides

Hydrolysis of a —CX_2— (X = halogen) gives a carbonyl group (—C(=O)—). This type of reaction has been particularly useful in the preparation of benzaldehyde and some benzaldehyde derivatives from toluene and toluene derivatives. Free-radical halogenation is used to prepare the dihalide. The preparation of benzaldehyde from toluene by this method is shown below:

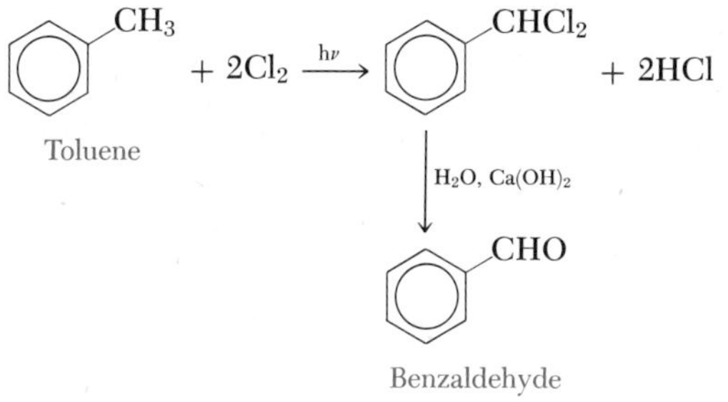

$$C_6H_5CH_3 + 2Cl_2 \xrightarrow{h\nu} C_6H_5CHCl_2 + 2HCl$$

Toluene

$$C_6H_5CHCl_2 \xrightarrow{H_2O,\ Ca(OH)_2} C_6H_5CHO$$

Benzaldehyde

Oxo Process

Aldehydes can be directly obtained from olefins by a reaction called the **Oxo process.** In this reaction, the olefin is treated with carbon monoxide and hydrogen at high temperatures and pressures in the presence of a catalyst such as dicobalt octacarbonyl $[Co(CO)_4]_2$. The terminal aldehyde is generally the predominant product. This reaction is currently receiving much attention industrially as a cheap method for preparing aldehydes on a commercial scale. Two typical examples are illustrated below:

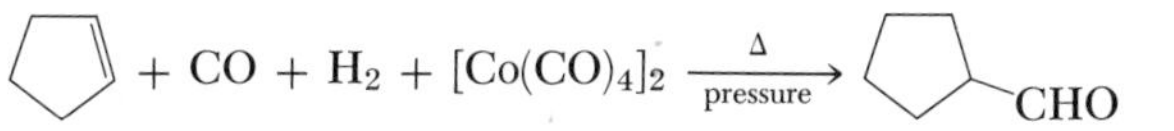

$$\text{cyclopentene} + CO + H_2 + [Co(CO)_4]_2 \xrightarrow[\text{pressure}]{\Delta} \text{cyclopentyl-}CHO$$

$$CH_3(CH_2)_3CH{=}CH_2 + CO + H_2 + [Co(CO)_4]_2 \xrightarrow[\text{pressure}]{\Delta} CH_3(CH_2)_4CH_2CHO + CH_3(CH_2)_3CH(CH_3)CHO$$

Major product — Minor product

Acetaldehyde

Acetaldehyde was originally produced by the hydration of acetylene in the presence of mercuric sulfate and sulfuric acid. The unsaturated vinyl alcohol intermediate is not stable and undergoes a shift of a hydrogen atom to give the aldehyde.

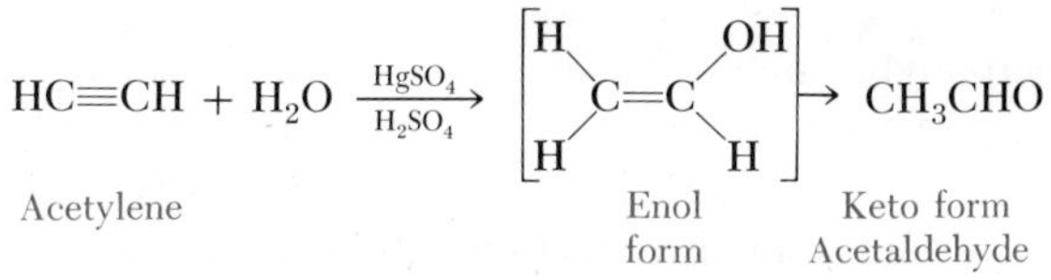

$$HC{\equiv}CH + H_2O \xrightarrow[H_2SO_4]{HgSO_4} \left[\begin{matrix} H & & OH \\ & C{=}C & \\ H & & H \end{matrix} \right] \rightarrow CH_3CHO$$

Acetylene — Enol form — Keto form Acetaldehyde

More recently, the direct oxidation of ethylene to acetaldehyde using a palladium chloride catalyst has become important (called the Wacker process) and has supplanted the older classical method for the preparation of acetaldehyde.

$$H_2C{=}CH_2 + \tfrac{1}{2}O_2 \xrightarrow[H_2O]{PdCl_2} CH_3CHO$$

Ethylene Acetaldehyde

REACTIONS OF CARBONYL COMPOUNDS

Oxidation

As noted in the preparation of aldehydes, oxidation of aldehydes occurs very readily to give a carboxylic acid. Even mild oxidizing agents bring about the oxidation of aldehydes. Two such mild oxidizing reagents are **Tollen's reagent,** an ammoniacal solution of silver nitrate, and **Fehling's** or **Benedict's solution,** alkaline solutions of copper sulfate.* Oxidation of an aldehyde with Tollen's reagent produces metallic silver (usually as a silver mirror), whereas Fehling's or Benedict's solution gives a red precipitate of cuprous oxide. These simple visual tests make it easy to distinguish aldehydes from ketones, since ketones are not oxidized by these reagents. Hence, no silver mirror or red precipitate is obtained on treatment of a ketone with either Tollen's reagent or Fehling's and Benedict's solutions. The overall reaction of these reagents with aldehydes is summarized below:

$$RCHO + 2Ag(NH_3)_2^+ + 2OH^- \longrightarrow RCO_2^-NH_4^+ + 2Ag\downarrow + 3NH_3 + H_2O$$

Aldehyde; Tollen's reagent (colorless); Acid salt; Silver metal

$$RCHO + 2Cu^{++}\text{ (complex)} \xrightarrow{NaOH} RCO_2^-Na^+ + Cu_2O\downarrow$$

Aldehyde; Fehling's or Benedict's solution (blue); Acid salt; Red precipitate

This reaction is the basis of the Fehling and Benedict tests for determining the presence of sugar in urine.

The metabolic oxidation of sugars, which are the most important biologic aldehydes, allows the body to convert the sugar into energy. An important step in the Embden-Meyerhof pathway (p. 322) oxidation of glucose is the oxidation of glyceraldehyde-3-phosphate.

$$\begin{array}{l} CHO \\ | \\ CHOH \\ | \\ CH_2{-}O{-}Ⓟ \end{array} \xrightarrow[\text{dehydrogenase}]{\text{phosphatorise}} \begin{array}{l} O \\ \| \\ C{-}O{-}Ⓟ \\ | \\ CHOH \\ | \\ CH_2O{-}Ⓟ \end{array}$$

1,3-Di-PO_4 glyceric acid

Reduction

Reduction of carbonyl compounds yields either hydrocarbons or alcohols, depending upon the reducing agent used. With either a Clemmensen reagent or a Wolff-Kishner reagent (see preparation of alkanes), reduction occurs to give a hydrocarbon group. However, catalytic hydrogenation or reduction with lithium aluminum hydride produces

*Fehling's solution also contains sodium potassium tartrate, and Benedict's solution contains sodium citrate. These salts form tartrate and citrate complexes with the cupric ion and help keep the cupric ion in solution.

alcohols. Aldehydes give primary alcohols, and ketones give secondary alcohols. Some typical examples are illustrated below:

$$4\ CH_3\overset{\overset{O}{\|}}{C}CH_3 + LiAlH_4 \xrightarrow{\text{ether}} [(CH_3)_2CH{-}O\,]_4AlLi \xrightarrow[H_2O]{H^+} CH_3CHOHCH_3$$

Acetone — Isopropyl alcohol

$$CH_3CH_2CHO + H_2 \xrightarrow{Pt} CH_3CH_2CH_2OH$$

Propionaldehyde — *n*-Propyl alcohol

In the utilization of glycogen (p. 319) by the body and the conversion of glucose into energy (p. 322), reduction of pyruvic acid to lactic acid is one of the important steps involved.

$$CH_3\overset{\overset{O}{\|}}{C}COOH \xrightarrow[\text{dehydrogenase}]{\text{lactic}} CH_3CHOHCOOH$$

Pyruvic acid — Lactic acid

Addition Reactions

The most characteristic type of reaction of aldehydes and ketones is an addition reaction across the carbon-oxygen double bond. A similar type of reaction is characteristic of compounds containing carbon-carbon multiple bonds (see reactions of alkenes and alkynes). Addition reactions to carbon-carbon double bonds were found to be electrophilic addition reactions, in which the electrophile was added to the carbon atom in the first step of the reaction sequence. With carbonyl groups, however, the polarization of the bond dipole is such that the carbon end of the dipole is the positive end. Therefore, it might be expected that the carbon atom of the carbonyl group would be attacked

$$\gt \overset{\delta+}{C}{=}\overset{\delta-}{O}$$

by a nucleophilic type of reagent, and that the addition reactions of aldehydes and ketones would be nucleophilic addition reactions rather than the electrophilic addition reactions characteristic of alkenes.

The investigation of the addition reactions of carbonyl compounds has confirmed this type of prediction, and aldehydes and ketones undergo a great variety of nucleophilic addition reactions. The generalized mechanism for these nucleophilic additions is outlined below:

$$N^{-}: + \gt C{=}\ddot{O}: \rightarrow N{-}\overset{|}{\underset{|}{C}}{-}\ddot{\underset{..}{O}}:^{-} \quad [1]$$

Nucleophile — (I)

$$N{-}\overset{|}{\underset{|}{C}}{-}\ddot{\underset{..}{O}}:^{-} + HN \rightarrow N{-}\overset{|}{\underset{|}{C}}{-}\ddot{\underset{..}{O}}H + N^{-}: \quad [2]$$

(II)

Overall reaction:

$$HN + \rangle C{=}O \rightarrow N{-}\overset{|}{\underset{|}{C}}{-}OH$$

The first step involves attack by the nucleophile, $N\overline{:}$, on the positive end of the carbonyl dipole to give the anion, (I). This anion (I) then abstracts a proton from the nucelophilic agent, HN, to give the final addition product, (II), and the nucleophile, $N\overline{:}$, which can then repeat the cycle. The overall reaction (the combination of equations [1] and [2]) is the addition of the nucleophilic reagent, HN, across the carbonyl group. In some cases, the initially formed addition product, (II), can undergo a dehydration reaction (loss of H_2O), so that the actual product isolated may not always appear exactly like (II). Some typical addition reactions of aldehydes and ketones are outlined below.

Addition of Hydrogen Cyanide

The addition of hydrogen cyanide gives an α-hydroxy nitrile,* called a cyanohydrin. These types of compounds are valuable in the preparation of hydroxy acids, amino acids, and sugars.

$$\rangle C{=}O + H^+CN^- \rightarrow \rangle C\langle^{OH}_{CN}$$

Cyanohydrin

$$CH_3\overset{O}{\overset{\|}{C}}{-}H + HCN \rightarrow CH_3\overset{OH}{\overset{|}{\underset{\underset{H}{|}}{C}}}{-}CN \xrightarrow[H^+]{H_2O} CH_3CHOHCOOH$$

Acetaldehyde — Hydrogen cyanide — Cyanohydrin — Lactic acid

When ammonium cyanide (NH_4CN) is used, the resultant hydrolysis product is an α-amino acid. This procedure, called the Strecker reaction, is useful in the synthesis of

$$CH_3\overset{O}{\overset{\|}{C}}{-}H + NH_4CN + H_2O \rightarrow CH_3\underset{\underset{NH_2}{|}}{C}HCOOH$$

Acetaldehyde — Alanine

the simpler α-amino acids. This reaction may be viewed at first as conversion of the aldehyde to a cyanohydrin followed by displacement of the hydroxyl group by the amino group and hydrolysis of the nitrile group to the carboxylic acid group. These steps are summarized on the next page for the formation of alanine from acetaldehyde.

* Organic compounds containing the $—C{\equiv}N$ grouping are called either cyanides (relating them to the inorganic analogues) or nitriles.

$$CH_3\overset{\overset{O}{\|}}{C}{-}H + NH_4CN \rightarrow CH_3\underset{\underset{H}{|}}{\overset{\overset{OH}{|}}{C}}{-}CN + NH_3$$

Acetaldehyde

$$CH_3\underset{\underset{H}{|}}{\overset{\overset{OH}{|}}{C}}CN + NH_3 \rightarrow CH_3\underset{\underset{H}{|}}{\overset{\overset{NH_2}{|}}{C}}{-}CN + H_2O$$

$$CH_3\underset{\underset{H}{|}}{\overset{\overset{NH_2}{|}}{C}}{-}CN + 2H_2O \rightarrow CH_3\underset{\underset{H}{|}}{\overset{\overset{NH_2}{|}}{C}}{-}COOH + NH_3$$

Alanine

Addition of Organometallic Reagents

Organometallic reagents, particularly Grignard and lithium reagents, also add across the carbonyl group. Hydrolysis of the initially formed complex gives alcohols as the final product. Aldehydes, except for formaldehyde, give secondary alcohols, and ketones give tertiary alcohols. This type of reaction provides one of the best laboratory methods for preparing secondary and tertiary alcohols. The generalized scheme of this kind of reaction is outlined below with several specific examples to illustrate the scope and versatility of this reaction:

$$RMgX + \;\rangle C{=}O \rightarrow \left[R{-}\underset{|}{\overset{|}{C}}{-}OMgX\right] \xrightarrow[H_2O]{H^+} R{-}\underset{|}{\overset{|}{C}}{-}OH + Mg(OH)X$$

(not isolated)

$$\underset{\text{Acetaldehyde}}{CH_3\overset{\overset{H}{|}}{C}{=}O} + CH_3CH_2MgBr \rightarrow \left[CH_3\underset{\underset{CH_2CH_3}{|}}{\overset{\overset{H}{|}}{C}}{-}OMgBr\right] \xrightarrow[H_2O]{H^+} \underset{\text{2-Butanol}}{CH_3\overset{\overset{OH}{|}}{C}HCH_2CH_3}$$

$$\underset{\text{Acetophenone}}{C_6H_5{-}\overset{\overset{O}{\|}}{C}{-}CH_3} + CH_3MgI \rightarrow \left[C_6H_5{-}\underset{\underset{CH_3}{|}}{\overset{\overset{OMgI}{|}}{C}}{-}CH_3\right] \xrightarrow[H_2O]{H^+} \underset{\text{2-Phenyl-2-propanol}}{C_6H_5{-}\underset{\underset{CH_3}{|}}{\overset{\overset{OH}{|}}{C}}{-}CH_3}$$

Addition of Water

Most carbonyl compounds will form hydrates in water, but the reaction is a reversible one, and the equilibrium lies far to the left. Attempted isolation of the hydrates gives only the starting carbonyl compound back. Aldehydes and ketones which have groups

$$\rangle C{=}O + HOH \rightleftharpoons \;\rangle C\langle^{OH}_{OH}$$

attached to the carbonyl group that are highly electron attracting do form stable hydrates. The number of stable hydrates of this type, however, is not very large.

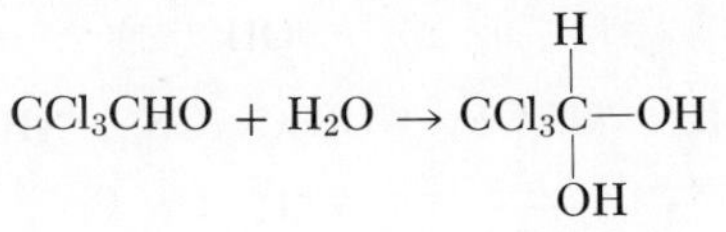

$$CCl_3CHO + H_2O \rightarrow CCl_3C(H)(OH)-OH$$

Chloral — Chloral hydrate (stable)

Addition of Alcohols

Compounds called **hemiacetals** and **acetals** are formed by reacting aldehydes with alcohols in the presence of anhydrous HCl. One molecule of the alcohol adds to the carbonyl group to form a **hemiacetal,** which, in turn, can form an **acetal** by reaction with another molecule of alcohol.

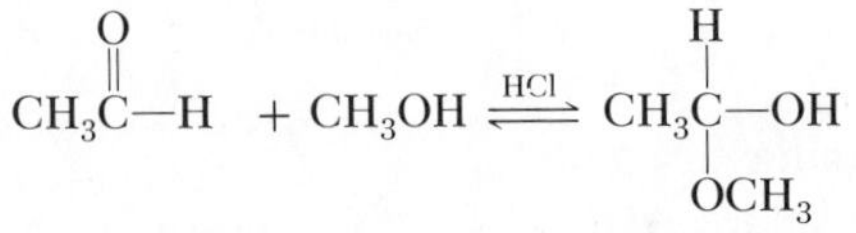

$$CH_3\overset{O}{\overset{\|}{C}}-H + CH_3OH \underset{}{\overset{HCl}{\rightleftharpoons}} CH_3C(H)(OCH_3)-OH$$

Acetaldehyde — Hemiacetal

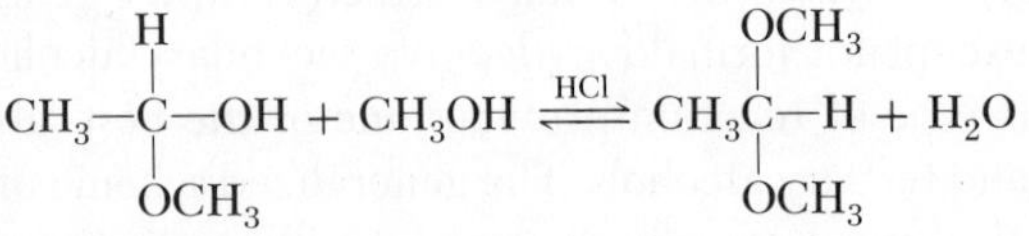

$$CH_3-C(H)(OCH_3)-OH + CH_3OH \overset{HCl}{\rightleftharpoons} CH_3C(OCH_3)(OCH_3)-H + H_2O$$

Acetal

Acetals behave like ethers in their reactivity. They are stable to base but react with dilute aqueous acid to regenerate the alcohol and aldehyde from which they were prepared. Consequently, acetal formation is frequently used to protect an aldehyde group in a chemical reaction in which the aldehyde function might be oxidized.

The acetal structure occurs in a number of important biologic compounds. Some sugars, such as glucose, ribose, and fructose, exist in nature as cyclic hemiacetals (see also p. 212).

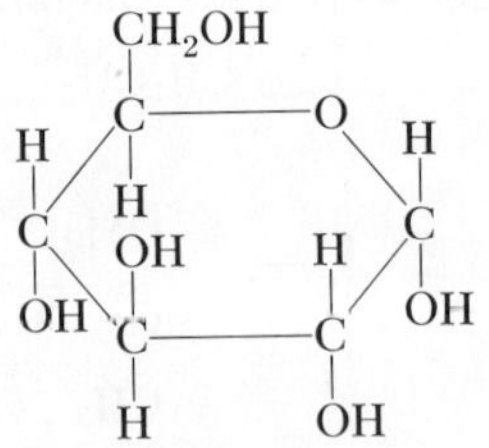

[Structure: six-membered ring with CH_2OH, H, OH substituents]

α-D-Glucopyranose

Other carbohydrates, like starch, glycogen, and cellulose, also contain cyclic acetal linkages (see Chapter 13). Nucleosides, nucleotides, and the nucleic acids DNA and RNA also contain the acetal linkage (see Chapter 16). These important biologic examples of the acetal linkage will be discussed in more detail in the biochemistry section of the text.

Addition of Ammonia and Ammonia Derivatives

Except for the reaction of ammonia and formaldehyde, most aldehydes and ketones do not give stable addition products with ammonia. However, ammonia derivatives, in which a hydrogen of an —NH bond has been replaced by another group, do undergo addition to aldehydes and ketones, followed by loss of water, as shown on the following page:

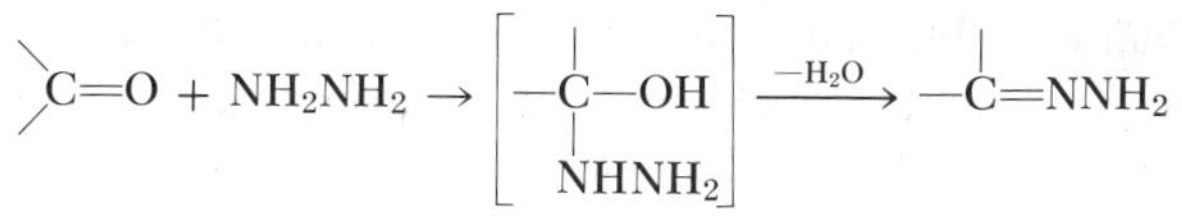

$$>C=O + NH_2NH_2 \rightarrow \left[-\overset{|}{\underset{NHNH_2}{C}}-OH \right] \xrightarrow{-H_2O} -\overset{|}{C}=NNH_2$$

Aldehyde or ketone   Hydrazine   Addition product   A hydrazone

Other hydrazine derivatives behave similarly.

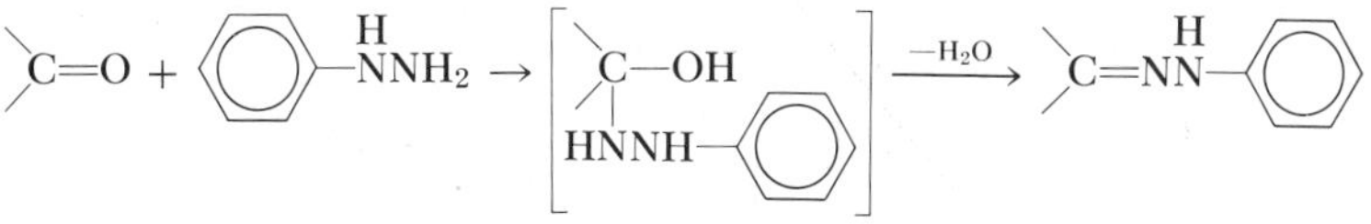

$$>C=O + C_6H_5-\overset{H}{N}NH_2 \rightarrow \left[>\underset{HNNH-C_6H_5}{C}-OH \right] \xrightarrow{-H_2O} >C=N\overset{H}{N}-C_6H_5$$

Phenyl hydrazine   A phenyl hydrazone

Hydroxylamine, NH_2OH, in which the hydrogen of an —NH bond of ammonia has been replaced by —OH, behaves similarly to the hydrazines to give a class of compounds called **oximes.** Most **hydrazones** and **oximes** are solid, crystalline compounds of a characteristic melting point. The melting

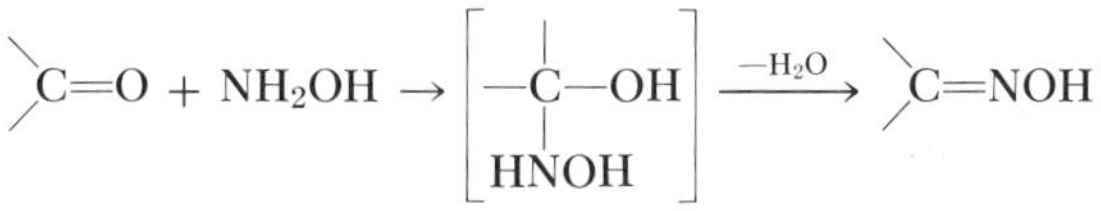

$$>C=O + NH_2OH \rightarrow \left[-\overset{|}{\underset{HNOH}{C}}-OH \right] \xrightarrow{-H_2O} >C=NOH$$

Hydroxylamine   Addition product   An oxime

points of these types of carbonyl derivatives are helpful in attempting to elucidate the structure of an aldehyde or ketone material.

Condensation Reactions

Another class of reactions that carbonyl compounds readily undergo is **condensation reactions.** The term that describes this class of reactions means precisely what it implies—two molecules of the aldehyde or ketone "condense" together. The actual isolated product may be the condensed product or a product derived from the condensed product by the elimination of water. Most of the condensation reactions of carbonyl compounds are base catalyzed.

Aldol Condensation

Aldehydes and some ketones which have an α-hydrogen atom undergo a self-condensation reaction in the presence of dilute sodium hydroxide. The name for this type of condensation is **aldol condensation,** after the common name "aldol" for the product obtained from acetaldehyde in the reaction shown below.

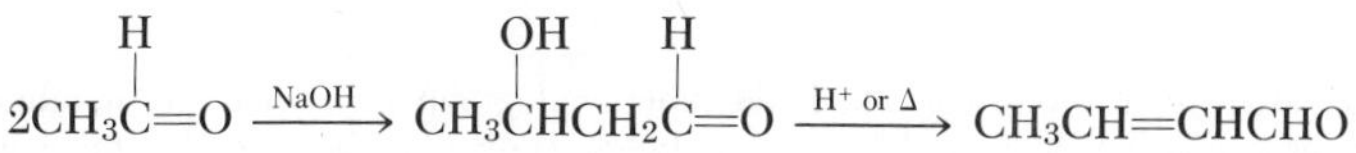

$$2CH_3\overset{H}{\overset{|}{C}}=O \xrightarrow{NaOH} CH_3\overset{OH}{\overset{|}{C}}HCH_2\overset{H}{\overset{|}{C}}=O \xrightarrow{H^+ \text{ or } \Delta} CH_3CH=CHCHO$$

Acetaldehyde   Aldol   Crotonaldehyde

The β-hydroxy aldehyde obtained readily loses a molecule of water on heating or with acid to give an α, β unsaturated* aldehyde. The generalized mechanism for this type of condensation reaction, and related condensation reactions, is outlined below:

$$\underset{(\alpha\text{-hydrogen})}{RCH_2\overset{H}{\overset{|}{C}}{=}O} + Na^+OH^- \rightleftharpoons \underset{(I)}{\left[R\bar{C}H\overset{H}{\overset{|}{C}}{=}O\right]} Na^+ + H_2O$$

$$R{-}\underset{H}{\overset{CHO}{C^-}} + \underset{H}{\overset{O}{\overset{\|}{C}}}{-}CH_2R \rightarrow \underset{(II)}{\left[R{-}\underset{H}{\overset{OHC}{C}}{-}\underset{H}{\overset{O^-}{C}}{-}CH_2R\right]} \xrightarrow{H_2O} \underset{(III)}{R{-}\underset{H}{\overset{OHC}{C}}{-}\underset{H}{\overset{OH}{C}}{-}CH_2R} + OH^-$$

The hydrogen atoms on the α-carbon of the aldehyde have increased acidity because of the electron-attracting ability of the carbonyl group. Also, the resultant carbanion can be resonance stabilized as shown below:

$$R{-}CH_2{-}\overset{O}{\overset{\|}{C}}{-}H + OH^- \rightleftharpoons R{-}\bar{C}H{-}\overset{O}{\overset{\|}{C}}{-}H \leftrightarrow R{-}CH{=}\overset{O^-}{\overset{|}{C}}{-}H$$

Consequently, the hydroxide ion removes one of the α-hydrogens to give a carbanion, (I). This carbanion, (I), is a nucleophile and attacks the carbonyl carbon atom of a second molecule of the aldehyde to give (II), which in turn abstracts a proton from water to give the β-hydroxy aldehyde, (III), and regenerates the catalyst, OH^-. The overall reaction is the nucleophilic addition of one aldehyde molecule to a second aldehyde molecule, and condensation reactions of this type are merely a more complex case of the nucleophilic addition reactions discussed earlier in this chapter.

Since the α-hydrogen atoms are the acidic hydrogens which are removed by the base, a β-hydroxy aldehyde always results when an aldehyde is used as the reactant. For higher homologues, such as propanal and so forth, similar condensation results, and a similar mechanism can be used to account for the products.

$$\underset{\text{Propanal}}{2CH_3CH_2CHO} + Na^+OH^- \rightarrow \underset{\text{2-Methyl-3-hydroxypentanal}}{CH_3CH_2\overset{OH}{\overset{|}{C}H}\underset{CH_3}{\underset{|}{C}H}CHO} \xrightarrow{\Delta} \underset{\text{2-Methylpent-2-enal}}{CH_3CH_2CH{=}\underset{CH_3}{\underset{|}{C}}CHO}$$

If a mixture of aldehydes, both of which contain α-hydrogen, are subjected to the aldol reaction, a mixture of four products results. However, if one aldehyde contains *no* α-hydrogen, only two products are possible, and if a large excess of the aldehyde with no α-hydrogens is used, then a high yield of one product is formed. This type of aldol reaction is called *a crossed aldol condensation* and is illustrated on the following page:

* The carbon atom adjacent to the aldehyde functional group is labeled the α (alpha) carbon. The remaining carbon atoms on the chain are labeled β (beta), γ (gamma), δ (delta), and so forth, as one progresses along the chain away from the aldehyde group.

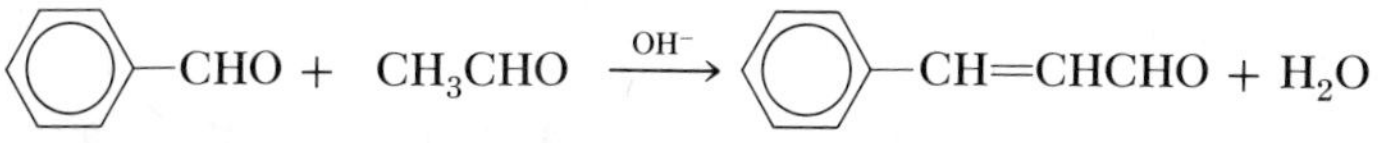

$$C_6H_5\text{—CHO} + CH_3CHO \xrightarrow{OH^-} C_6H_5\text{—CH=CHCHO} + H_2O$$

Benzaldehyde (excess) Acetaldehyde Cinnamaldehyde

An important reversible aldol condensation in biologic systems is the reversible cleavage of fructose diphosphate catalyzed by the enzyme **aldolase.**

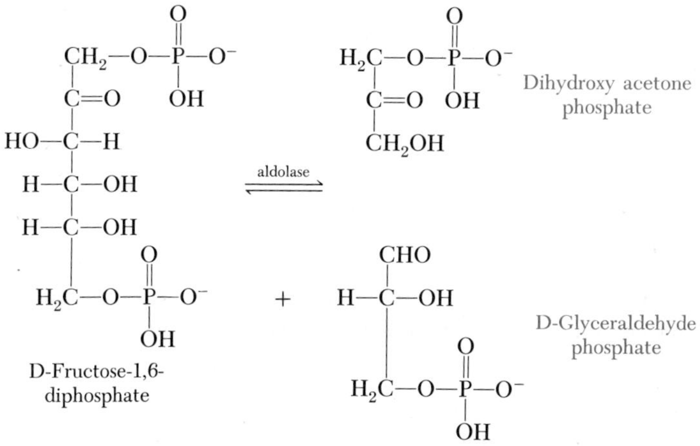

$$CH_2\text{—O—P(=O)(OH)—O}^- \text{—C=O—HO—C—H—H—C—OH—H—C—OH—}H_2C\text{—O—P(=O)(OH)—O}^- \underset{}{\overset{\text{aldolase}}{\rightleftharpoons}} H_2C\text{—O—P(=O)(OH)—O}^-\text{—C=O—}CH_2OH + CHO\text{—H—C—OH—}H_2C\text{—O—P(=O)(OH)—O}^-$$

D-Fructose-1,6-diphosphate; Dihydroxy acetone phosphate; D-Glyceraldehyde phosphate

Aldehydes which have *no* α-hydrogens cannot, of course, undergo the aldol condensation. However, in the presence of concentrated alkali, aldehydes of this type undergo an oxidation-reduction reaction. One molecule of the aldehyde is reduced to an alcohol, and a second molecule of the aldehyde is oxidized to an acid. Formaldehyde and benzaldehyde illustrate this kind of reaction. This reaction is known as a Cannizzaro reaction.

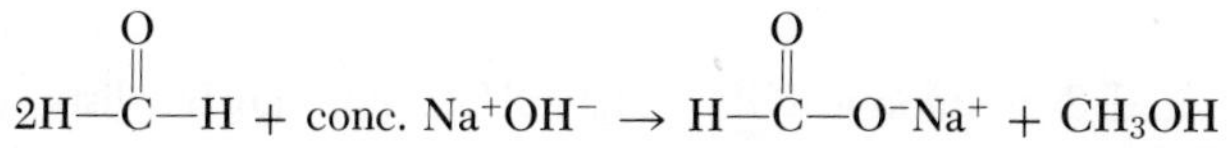

$$2H\text{—C(=O)—H} + \text{conc. } Na^+OH^- \rightarrow H\text{—C(=O)—O}^-Na^+ + CH_3OH$$

Formaldehyde Sodium formate Methanol

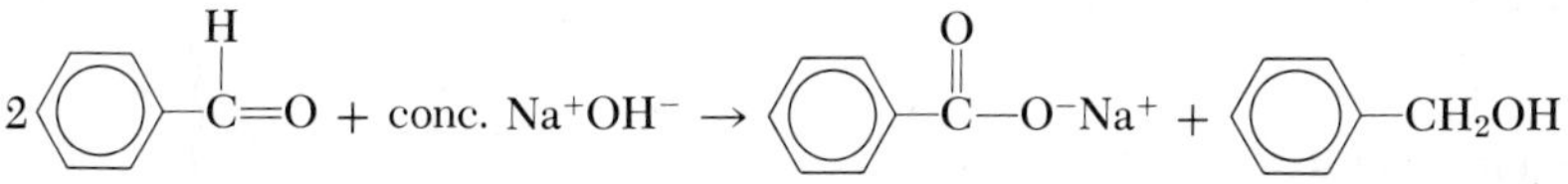

$$2\,C_6H_5\text{—C(H)=O} + \text{conc. } Na^+OH^- \rightarrow C_6H_5\text{—C(=O)—O}^-Na^+ + C_6H_5\text{—}CH_2OH$$

Benzaldehyde Sodium benzoate Benzyl alcohol

Halogen Substitution

In addition to the wide variety of reactions that occur at the carbonyl group of aldehydes and ketones, other kinds of reactions can also occur either in the alkyl group or in the aromatic ring of aldehydes and ketones. The most important of these reactions are halogen substitution in the alkyl group and electrophilic substitution in the aromatic ring.

Halogenation of alkyl containing aldehydes and ketones takes place at the α-hydrogen atoms in the presence of alkali. Halogenation may proceed stepwise if the amount of reagent is controlled, or may proceed with excess halogenating agent until all the α-hydrogen atoms have been replaced by halogen.

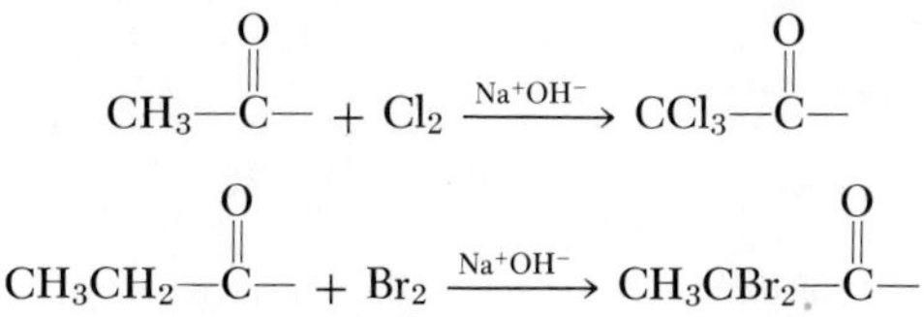

$$CH_3\text{—C(=O)—} + Cl_2 \xrightarrow{Na^+OH^-} CCl_3\text{—C(=O)—}$$

$$CH_3CH_2\text{—C(=O)—} + Br_2 \xrightarrow{Na^+OH^-} CH_3CBr_2\text{—C(=O)—}$$

When the carbonyl compound contains the grouping $CH_3{-}\overset{O}{\overset{\|}{C}}{-}$, a secondary reaction occurs after halogenation called a **haloform** reaction. This reaction involves cleavage of the group $CX_3{-}\overset{O}{\overset{\|}{C}}{-}$ (X = halogen) by base to give a haloform ($CHCl_3$, $CHBr_3$, CHI_3) and an acid. The reaction is illustrated below for acetophenone:

$$CH_3\overset{O}{\overset{\|}{C}}{-}C_6H_5 + 3I_2 + 3NaOH \rightarrow CI_3\overset{O}{\overset{\|}{C}}{-}C_6H_5 + 3H_2O + 3Na^+I^-$$

Acetophenone

$$CI_3\overset{O}{\overset{\|}{C}}{-}C_6H_5 \xrightarrow{NaOH} CHI_3\downarrow + C_6H_5{-}\overset{O}{\overset{\|}{C}}{-}O^-Na^+$$

Iodoform Sodium benzoate

Chloroform ($CHCl_3$) and bromoform ($CHBr_3$) are liquids, whereas iodoform (CHI_3) is a bright yellow crystalline solid (m.p. 119°C) which precipitates from solution. This type of cleavage reaction occurs only for the $CH_3{-}\overset{O}{\overset{\|}{C}}{-}$ grouping, since only this grouping can be converted to the $CX_3{-}\overset{O}{\overset{\|}{C}}{-}$ group that gives the haloform on cleavage. In most cases, sodium hydroxide and iodine are used, since the iodoform provides a visual means of detecting the reaction. Under these conditions the test is commonly referred to as the **iodoform test.** It is used to distinguish methyl ketones from other types of ketones. Thus, two ketone structures of molecular formula $C_5H_{10}O$, namely 2-pentanone ($CH_3\overset{O}{\overset{\|}{C}}CH_2CH_2CH_3$) and 3-pentanone ($CH_3CH_2\overset{O}{\overset{\|}{C}}CH_2CH_3$), can be distinguished very easily on the basis of the iodoform test. The 2-pentanone will give a positive iodoform test (will give a precipitate of CHI_3), whereas 3-pentanone will not give a precipitate of iodoform.

Aromatic Substitution Reactions

The carbonyl group is a *meta*-directing group and is also a deactivating group. Aromatic aldehydes and ketones, therefore, undergo electrophilic aromatic substitution reactions less easily than benzene, and more vigorous reaction conditions must be used to bring about ring substitution. The halogenation of a typical aromatic aldehyde and ketone is shown below. Similar reactions occur on nitration, sulfonation, and alkylation.

$$C_6H_5CHO + Br_2 \xrightarrow[\Delta]{Fe} m\text{-}BrC_6H_4CHO$$

Benzaldehyde *m*-Bromobenzaldehyde

$$C_6H_5COCH_3 + Br_2 \xrightarrow[\Delta]{Fe} m\text{-}BrC_6H_4COCH_3$$

Acetophenone *m*-Bromoacetophenone

USES OF IMPORTANT ALDEHYDES AND KETONES

As noted earlier, formaldehyde ($H-\overset{\overset{\displaystyle O}{\|}}{C}-H$) is a colorless gas that readily dissolves in water; a 40% solution is known as **formalin.** Formalin acts as a disinfectant and is used in embalming fluid and as a preservative of various tissues. Formaldehyde gas and the polymer paraformaldehyde are used extensively as insecticides, fumigating agents, and antiseptics. Large quantities of formaldehyde are used in the manufacture of synthetic resins and in the synthesis of other organic compounds.

Acetaldehyde is used in the production of acetic acid, ethyl acetate, synthetic rubber, and other organic compounds. Paraldehyde, a cyclic trimer of acetaldehyde, is more stable than acetaldehyde and serves as a source of the latter compound when heated. It is an effective hypnotic, or sleep producer, but has been replaced by other drugs since it has an irritating odor and an unpleasant taste.

Chloral hydrate is used medically as a sedative or hypnotic. Benzaldehyde is an important intermediate in the preparation of drugs, dyes, and other organic compounds. As noted earlier, it is also used in flavoring agents and perfumes. Cinnamaldehyde, C_6H_4—CH=CHCHO, is a constituent of the oil of cinnamon obtained from cinnamon bark.

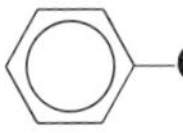

Another important aldehyde is **vanillin,** which may be obtained from the extracts of vanilla beans, and which also occurs in sugar beets, resins, and balsams. Vanillin may be prepared from the phenol derivative eugenol or guaiacol. Eugenol is first rearranged with alcoholic potassium hydroxide to form isoeugenol, which is then oxidized to vanillin. Vanillin is employed as a flavoring agent in many food products, such as candies, cookies, and ice cream.

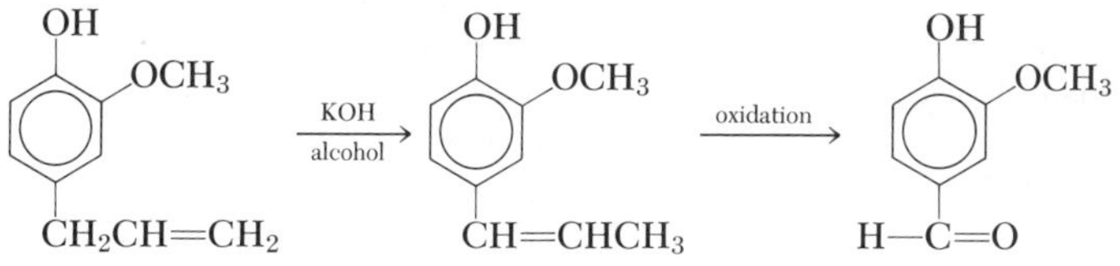

OH, OCH_3 / $CH_2CH{=}CH_2$ (Eugenol) $\xrightarrow[\text{alcohol}]{\text{KOH}}$ OH, OCH_3 / $CH{=}CHCH_3$ (Isoeugenol) $\xrightarrow{\text{oxidation}}$ OH, OCH_3 / $H-C{=}O$ (Vanillin)

Eugenol   Isoeugenol   Vanillin

Acetone, the most important ketone, is prepared by the oxidation of isopropyl alcohol, or by the butyl alcohol fermentation process from cornstarch. It is used as a solvent for cellulose derivatives, varnishes, lacquers, resins, and plastics.

Methyl ethyl ketone ($CH_3COCH_2CH_3$) is used by the petroleum industry in the dewaxing of lubricating oils. It is an excellent solvent for fingernail polish and is used as a polish remover.

Acetophenone is used as an intermediate in the preparation of other compounds and in the preparation of perfumes. If one of the hydrogen atoms on the methyl group is replaced by a chlorine atom, chloroacetophenone ($C_6H_5COCH_2Cl$) is formed. This compound is a lachrymator and is commonly used as tear gas.

Benzophenone is a colorless solid with a characteristically pleasant odor. It is used in the manufacture of perfume and soaps and in the preparation of other organic compounds.

An important aromatic ketone that is used extensively in the identification of amino acids and proteins is ninhydrin. This compound also represents one of the few stable hydrates known.

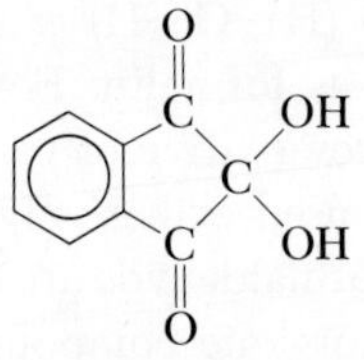

Ninhydrin

IMPORTANT TERMS AND CONCEPTS

acetal
acetaldehyde
acetone
aldol condensation
Benedict's solution
benzaldehyde
Cannizzaro reaction
carbonyl group
cyanohydrin
formaldehyde
formalin
haloform reaction
hemiacetal
hydrate
hydrazone
iodoform test
oxime
Tollen's reagent
vanillin

QUESTIONS

1. Draw a structural formula for each of the following compounds:
   (a) heptanal
   (b) 2-hexanone
   (c) *m*-nitrobenzaldehyde
   (d) cyclopentanone
   (e) 5-bromo-3-methylpentanal
   (f) *p*-chloroacetophenone
   (g) methyl cyclohexyl ketone
   (h) *p,p'*-dibromobenzophenone
   (i) 3-methylcyclohexanone
   (j) *trans*-3-pentenal

2. Name each of the following compounds.
   (a) $(CH_3)_2CHCHO$
   (b) 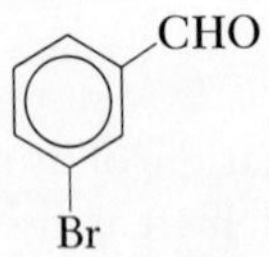 [benzene ring with CHO and Br in meta positions]
   (c) $CH_3CH_2\overset{O}{\overset{\|}{C}}CH(CH_3)_2$
   (d) CH_2BrCH_2CHO
   (e) 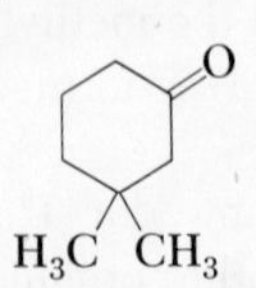 [cyclohexanone with two methyl groups (H_3C, CH_3) on carbon 4]
   (f) CCl_3CH_2CHO
   (g) 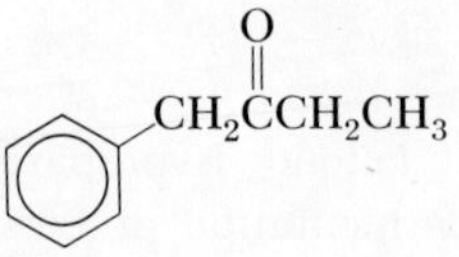 [benzene ring]$-CH_2\overset{O}{\overset{\|}{C}}CH_2CH_3$
   (h) 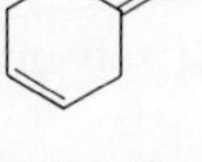 [cyclohexenone]
   (i) $CH_3CH{=}CHCHO$
   (j) $CH_3CH_2\overset{O}{\overset{\|}{C}}C(CH_3)_3$

3. What is the oxidizing agent in: (a) Fehling's solution; (b) Benedict's solution; (c) Tollen's reagent?

4. Write equations for the reactions of butyraldehyde with the following reagents:
   (a) $Ag(NH_3)_2^+$
   (b) $LiAlH_4$/ether
   (c) CH_3MgI/hydrolysis
   (d) phenylhydrazine
   (e) dilute Na^+OH^-

5. Write equations for the reactions of *p*-bromoacetophenone with the following reagents:
   (a) $NaOH + I_2$
   (b) hydroxylamine
   (c) phenyl magnesium bromide/hydrolysis
   (d) $LiAlH_4$/ether
   (e) $K_2Cr_2O_7/H_2SO_4$

6. Devise *simple* chemical tests that would enable you to distinguish one compound from the other in each of the following:
   (a) butanal and butanone
   (b) 2-hexanone and 3-hexanone
   (c) 3-pentanol and 3-pentanone

7. Write equations for the preparation of the following compounds. Give any necessary catalysts. More than one step may be necessary.
   (a) 1-methylcyclohexanol from cyclohexanol
   (b) 3-hexanone from 3-hexene
   (c) acetone from acetaldehyde
   (d) 1-phenylethanol from benzene
   (e) benzophenone from benzaldehyde
   (f) *m*-bromobenzoic acid from benzaldehyde
   (g) 3-pentanone from propionic acid
   (h) α-methyl styrene from acetophenone
   (i) benzoic acid from acetophenone
   (j) benzyl bromide from benzaldehyde

SUGGESTED READING

Brown: New Selective Reducing Agents. Journal of Chemical Education, Vol. 38, p. 173, 1961.
Cash: Nucleophilic Reactions at Trigonally Bonded Carbon. Journal of Chemical Education, Vol. 41, p. 108, 1964.
Gaylord: Reduction with Complex Metal Hydrides. Journal of Chemical Education, Vol. 34, p. 367, 1957.
Moore: The Art and the Science of Perfumery. Journal of Chemical Education, Vol. 37, p. 434, 1960.
Seelye and Turney: The Iodoform Reaction. Journal of Chemical Education, Vol. 36, p. 572, 1959.

CARBOXYLIC ACIDS AND ACID DERIVATIVES

CHAPTER 10

The *objectives* of this chapter are to enable the student to:

1. Recognize a carboxylic acid or a carboxylic acid derivative.
2. Write the names and structures of carboxylic acids and acid derivatives.
3. Explain the acidity of carboxylic acids relative to water and alcohols which also contain the hydroxyl group.
4. Deduce the effect of substituents on the acidity of carboxylic acids.
5. Describe the preparation of carboxylic acids and acid derivatives.
6. Explain the reactions of carboxylic acids.
7. Recognize the interconvertability of the different types of acid derivatives.
8. Recognize a saponification and transesterification reaction.
9. Recognize the important carboxylic acids, acid derivatives, and their current uses.

Organic compounds composed of the three elements carbon, hydrogen, and oxygen have been discussed in the previous chapters. These compounds included alcohols, ethers, aldehydes, and ketones. Further oxidation of primary alcohols or aldehydes eventually leads to the formation of another group of important organic compounds, **carboxylic acids,** which contain the carboxyl functional group $-\overset{O}{\overset{\|}{C}}-OH$. These acids are widely distributed in nature, especially in foodstuffs. Citric acid in citrus fruits, oxalic acid in fruits and vegetables, acetic acid in vinegar, amino acids in proteins, and fatty acids in fats and lipids are some typical examples of naturally occurring organic acids.

Important derivatives of organic acids that will be considered in this chapter include salts, esters, acid halides, acid anhydrides, and amides.

NOMENCLATURE

Carboxylic acids are defined as organic compounds which contain one or more carboxyl groups (—COOH) in the molecule. Some typical acids are illustrated in the following examples:

$\mathrm{HC(=O)OH}$ Formic acid $\mathrm{CH_3C(=O)OH}$ Acetic acid $\mathrm{C_6H_5C(=O)OH}$ Benzoic acid $\mathrm{HOC(=O)—C(=O)OH}$ Oxalic acid

$\mathrm{C_6H_4(C(=O)OH)_2}$ Phthalic acid $\mathrm{C_6H_4(COOH)_2}$ Terephthalic acid $\mathrm{HO—C(CH_2COOH)_2—COOH}$ 3-Hydroxy-3-carboxypentanedioic acid (citric acid)

Hydrogen, alkyl, and aryl groups may be attached to the carboxyl group, and more than one carboxyl group may be present in the molecule.

Since organic acids* occur so widely in nature either as the free acid or an acid derivative, they were among the earliest organic compounds investigated. Common names were given to many of these acids, and these names are still used frequently today. For the aliphatic acids, the IUPAC nomenclature is also used. This system adds the ending **-oic acid** to the stem name of the hydrocarbon with the corresponding number of carbon atoms. The carboxyl group is given the number **1**, and all the other usual IUPAC rules are then applied. The aromatic acids are named either by common names or as derivatives of the parent aromatic acid, benzoic acid. Some typical illustrations of the use of these nomenclature systems are shown below. The common name (for the aliphatic acids) is given in parentheses.

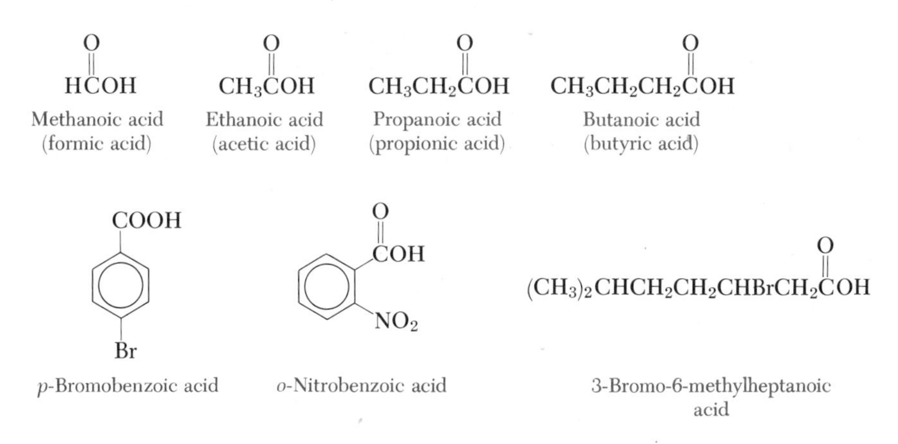

Methanoic acid (formic acid) Ethanoic acid (acetic acid) Propanoic acid (propionic acid) Butanoic acid (butyric acid)

p-Bromobenzoic acid *o*-Nitrobenzoic acid 3-Bromo-6-methylheptanoic acid

In some cases the position of substituents on the aliphatic chain is indicated by the Greek letters α, β, γ, δ, and so forth.

Di-acids, tri-acids, and so forth, are named in the IUPAC system by adding the ending **-dioic acid, -trioic acid,** and so forth, to the name of the hydrocarbon containing the same number of carbon atoms. In addition, common names are used for many of these compounds. Some examples are shown in the following compounds (the common name is in parentheses):

*The carboxyl group may be written as $\mathrm{—C(=O)—OH}$, $\mathrm{—COOH}$, or $\mathrm{—CO_2H}$.

$$\overset{\beta}{\underset{3}{HOCH_2}}\overset{\alpha}{\underset{2}{CH_2}}\underset{1}{CO_2H}$$

3-Hydroxypropanoic acid
(β-hydroxypropionic acid)

$$HO\overset{O}{\overset{\|}{C}}-\overset{O}{\overset{\|}{C}}OH$$

Ethanedioic acid
(oxalic acid)

$$HO\overset{O}{\overset{\|}{C}}CH_2\overset{O}{\overset{\|}{C}}OH$$

1,3-Propanedioic acid
(malonic acid)

$$HO\overset{O}{\overset{\|}{C}}(CH_2)_2\overset{O}{\overset{\|}{C}}OH$$

1,4-Butanedioic acid
(succinic acid)

$$HOOC(CH_2)_3COOH$$

1,5-Pentanedioic acid
(glutaric acid)

$$HO\overset{O}{\overset{\|}{C}}(CH_2)_4\overset{O}{\overset{\|}{C}}OH$$

1,6-Hexanedioic acid
(adipic acid)

PHYSICAL PROPERTIES OF CARBOXYLIC ACIDS

Carboxylic acids are, as the name indicates, acidic compounds. When dissolved in water, they can ionize slightly to donate a proton to a more basic substance, such as water. Most simple carboxylic acids are only slightly ionized in water, and these acids

$$\underset{\text{Carboxylic acid}}{R-\overset{O}{\overset{\|}{C}}-OH} + H_2O \rightleftharpoons \underset{\text{Carboxylate anion}}{R-\overset{O}{\overset{\|}{C}}-O^-} + H_3O^+$$

are fairly weak. The ionization constant, K_a, is a measure of the acid strength of these compounds. Two factors are important in determining the increased acidity of carboxylic acids relative to water and alcohol, both of which also contain the hydroxyl (—OH) group. First, the inductive effect of the carbonyl group facilitates the loss of the proton by decreasing the electron density of the O—H bond:

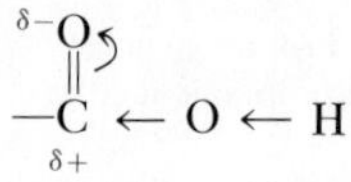

$$-\underset{\delta+}{\overset{\delta- O}{\overset{\|}{C}}} \leftarrow O \leftarrow H$$

Secondly, and more importantly, resonance stabilization of the resulting anion assists the loss of the proton, since energy is gained by formation of the delocalized ion:

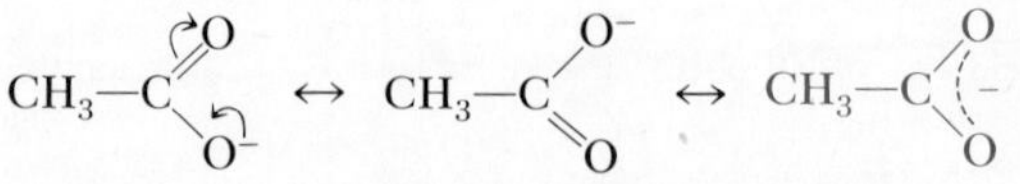

$$CH_3-C\begin{matrix}\nearrow O\\ \searrow O^-\end{matrix} \leftrightarrow CH_3-C\begin{matrix}\nearrow O^-\\ \searrow O\end{matrix} \leftrightarrow CH_3-C\begin{matrix}O\\ \vdots -\\ O\end{matrix}$$

For the unsubstituted aliphatic acids, the acidity varies with changes in the alkyl group. Substitution of the α-hydrogens by highly electronegative groups, such as the halogens, has a significant effect on the acidity. The effect is an additive one, and the acidity increases as the number of electron-attracting groups in the α-position increases. This effect is only large when the electron-attracting group is in the α-position. When this type of group is further removed from the carboxyl group (β, γ, δ, and so forth), the effect on the acidity of the acid is small.

Aromatic acids are very slightly more acidic than the aliphatic acids. Strongly electron-attracting substituents on the ring also increase the acidity of the aromatic acid, and the effect is generally about the same as in the aliphatic acids. Some typical acids

TABLE 10–1 IONIZATION CONSTANTS OF CARBOXYLIC ACIDS

ACID	STRUCTURE	K_a
Acetic acid	CH_3COOH	1.8×10^{-5}
Propionic acid	CH_3CH_2COOH	1.0×10^{-5}
n-Butyric acid	$CH_3CH_2CH_2COOH$	1.5×10^{-5}
Chloroacetic acid	CH_2ClCO_2H	1.5×10^{-3}
Dichloroacetic acid	$CHCl_2CO_2H$	5.0×10^{-2}
Trichloroacetic acid	CCl_3CO_2H	1×10^{-1}
α-Chloropropionic acid	$CH_3CHClCO_2H$	1.6×10^{-3}
β-Chloropropionic acid	$CH_2ClCH_2CO_2H$	8×10^{-5}
Benzoic acid	$C_6H_5CO_2H$	6×10^{-5}
p-Chlorobenzoic acid	$Cl-C_6H_4-CO_2H$	1×10^{-4}

and substituted acids are tabulated in Table 10–1 with their (approximate) K_a to illustrate the preceding points.

Like alcohols, the carboxyl group contains a hydroxyl group which is capable of forming hydrogen bonds with other acid molecules, or with other similar kinds of molecules, such as water. Consequently, the lower molecular weight carboxylic acids are water soluble, the borderline solubility being about 4 to 5 carbon atoms. Branching of the carbon chain increases water solubility, and some branched acids of 5 to 6 carbon atoms are soluble in water.

The carboxylic acids have abnormally high boiling points compared to other functional compounds of similar molecular weight. For example, propionic acid (mol. wt. 74) boils at 141°C, *n*-pentane (mol. wt. 74) boils at 35°C, and *n*-butanol (mol. wt. 74) boils at 118°C. The carboxylic acids have been found to consist largely of dimers; consequently, two hydrogen bonds per molecule are formed, thereby accounting for the high boiling points of this class of compounds.

```
       O---H—O
     //       \
R—C             C—R
     \        //
       O—H---O
```

Carboxylic acid dimer

The lower molecular weight aliphatic acids are liquids with sharp, unpleasant odors. As the molecular weight increases, the volatility decreases, and the higher acids ($>C_{10}$) are solids with little odor. The aliphatic acids in the C_{12} to C_{18} range are used in making soaps and candles (Table 10–2).

PREPARATION OF CARBOXYLIC ACIDS

Oxidation

Oxidation of primary alcohols and aldehydes (see reactions of aldehydes) gives carboxylic acids. This type of oxidation can be used to prepare both aliphatic and aromatic acids. Oxidizing agents, such as potassium dichromate and sulfuric acid or alkaline potassium permanganate, are generally used for these reactions as shown on the following page:

$$CH_3CH_2CH_2CH_2OH + K_2Cr_2O_7 \xrightarrow{H_2SO_4} CH_3CH_2CH_2\overset{\overset{\large O}{\|}}{C}OH$$

n-Butanol → Butyric acid

$$C_6H_5CHO + KMnO_4 \xrightarrow{KOH} C_6H_5CO_2^-K^+ \xrightarrow{H_2SO_4} C_6H_5CO_2H$$

Benzaldehyde → Potassium benzoate → Benzoic acid

Di-acids can be prepared similarly by oxidation of the corresponding diols or di-aldehydes. This type of oxidation method gives a carboxylic acid with the same number of carbon atoms as the original alcohol or aldehyde.

Benzoic acid and other aromatic acids can also be prepared by the oxidation of alkyl side-chains with either nitric acid or alkaline potassium permanganate. The alkyl group attached to the ring, regardless of its length, is degraded (broken down) until it is converted to the carboxyl group. Since the alkylbenzene can be easily prepared by alkylation and acylation reactions (see reactions of aromatic hydrocarbons), this is a useful preparative method for some aromatic acids. Similar types of oxidations of aliphatic hydrocarbons require more vigorous conditions, are not easily controlled, and are not

TABLE 10-2 Physical Properties of Acids and Acid Derivatives

Compound	Structure	M.P. °C	B.P. °C
Formic acid	$H\overset{\overset{O}{\|}}{C}OH$	+8	+101
Acetic acid	CH_3CO_2H	+17	+118
Propionic acid	$CH_3CH_2CO_2H$	−21	+141
n-Butyric acid	$CH_3CH_2CH_2CO_2H$	−4	+164
Isobutyric acid	$(CH_3)_2CHCO_2H$	−46	+153
n-Valeric acid	$CH_3(CH_2)_3CO_2H$	−34	+186
Methyl acetate	$CH_3CO_2CH_3$	−98	+57
Ethyl acetate	$CH_3CO_2CH_2CH_3$	−84	+77
Acetyl chloride	CH_3COCl	−112	+52
Acetamide	CH_3CONH_2	+82	+222
Acetic anhydride	$(CH_3CO)_2O$	−73	+140
Benzoic acid	$C_6H_5CO_2H$	122	+249
Benzoyl chloride	C_6H_5COCl	−1	+197
Benzamide	$C_6H_5CONH_2$	+130	+290
Phthalic anhydride	$C_6H_4(CO)_2O$	+130	+285
Methyl benzoate	$C_6H_5CO_2CH_3$	−12	+199

of extensive preparative value. Some typical examples are shown in the following equations:

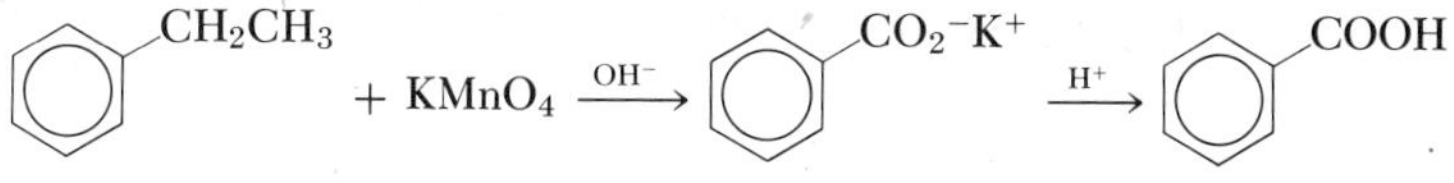

$$C_6H_5CH_2CH_3 + KMnO_4 \xrightarrow{OH^-} C_6H_5CO_2^-K^+ \xrightarrow{H^+} C_6H_5COOH$$

Ethylbenzene — Potassium benzoate — Benzoic acid

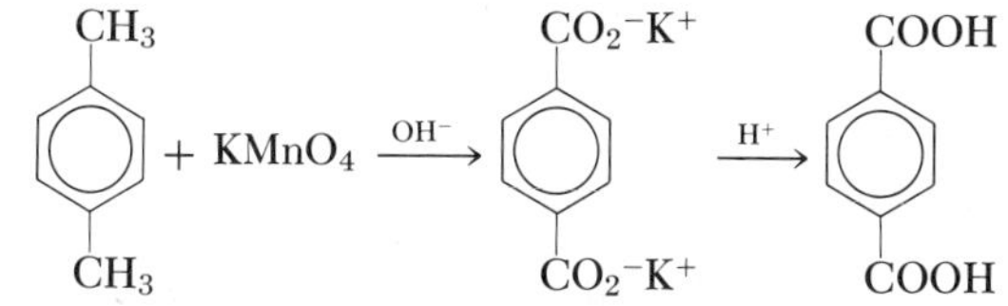

$$p\text{-}CH_3C_6H_4CH_3 + KMnO_4 \xrightarrow{OH^-} p\text{-}K^+\,{}^-O_2CC_6H_4CO_2^-K^+ \xrightarrow{H^+} p\text{-}HOOCC_6H_4COOH$$

p-Xylene — Potassium terephthalate — Terephthalic acid

Carbonation of Organometallic Reagents

When organometallic reagents, such as Grignard and lithium reagents, are treated with carbon dioxide (CO_2), a complex is formed that gives a carboxylic acid on hydrolysis. This reaction may be rationalized as the addition of the organometallic reagent across the carbonyl group (similar to the addition of organometallics to aldehydes and ketones), as shown below:

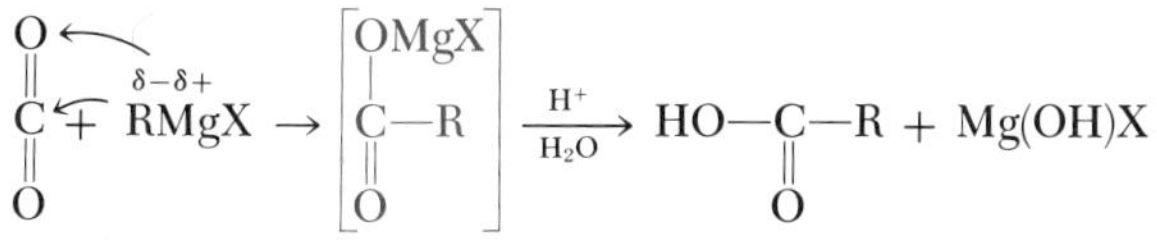

$$O{=}C{=}O + \overset{\delta-\ \delta+}{RMgX} \rightarrow \left[\begin{array}{c}OMgX\\ |\\ C{-}R\\ \|\\ O\end{array}\right] \xrightarrow[H_2O]{H^+} HO{-}\underset{\underset{O}{\|}}{C}{-}R + Mg(OH)X$$

Carbon dioxide — Complex — Carboxylic acid

Both aliphatic and aromatic carboxylic acids can be prepared from the appropriate alkyl or aryl halides, as illustrated in the following examples:

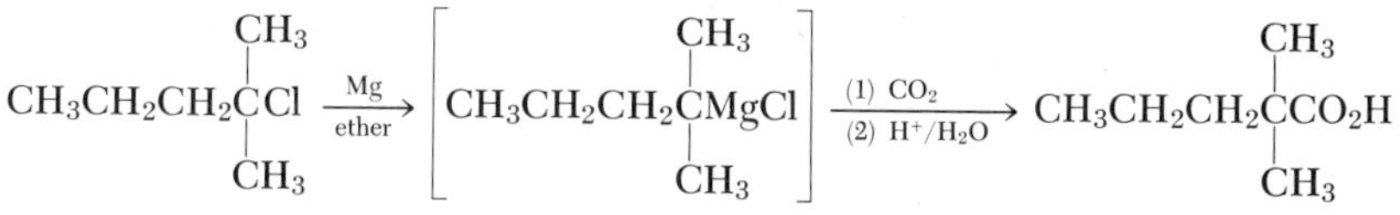

$$CH_3CH_2CH_2C(CH_3)_2Cl \xrightarrow[ether]{Mg} [CH_3CH_2CH_2C(CH_3)_2MgCl] \xrightarrow[(2)\ H^+/H_2O]{(1)\ CO_2} CH_3CH_2CH_2C(CH_3)_2CO_2H$$

2-Chloro-2-methylpentane — Grignard reagent — 2,2-Dimethylpentanoic acid

The acids prepared by this method contain one more carbon atom than the original alkyl or aryl halide.

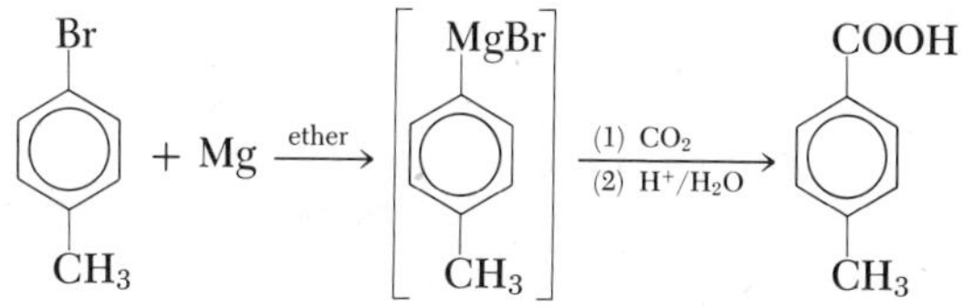

$$p\text{-}BrC_6H_4CH_3 + Mg \xrightarrow{ether} [p\text{-}BrMgC_6H_4CH_3] \xrightarrow[(2)\ H^+/H_2O]{(1)\ CO_2} p\text{-}HOOCC_6H_4CH_3$$

p-Bromotoluene — Grignard reagent — *p*-Methylbenzoic acid

Hydrolysis of Nitriles

The acid or base hydrolysis of nitriles affords another convenient method of preparing certain classes of carboxylic acids. The nitriles are conveniently prepared from alkyl halides via a nucleophilic displacement reaction:

$$R\,C{\equiv}N \xrightarrow[\text{base}]{\text{acid or}} R\overset{O}{\overset{\|}{C}}OH$$

Since nucleophilic displacement of aryl halides does not readily occur, the aromatic nitriles must be prepared by other methods (see reactions of diazonium compounds in the next chapter). Like the organometallic method described in the preceding section, this method of preparation gives a carboxylic acid with one more carbon atom than the original halide. Some typical examples of this method are illustrated below:

$$\underset{\text{1-Bromopentane}}{CH_3CH_2CH_2CH_2CH_2Br} + NaCN \rightarrow \underset{\text{Hexanenitrile}}{CH_3CH_2CH_2CH_2CH_2CN} + NaBr$$

$$\xrightarrow{H^+/H_2O} \underset{\text{Hexanoic acid}}{CH_3CH_2CH_2CH_2CH_2\overset{O}{\overset{\|}{C}}OH} + NH_4^+$$

$$\underset{\text{Benzonitrile}}{C_6H_5CN} + NaOH \xrightarrow{\Delta} \underset{\text{Sodium benzoate}}{C_6H_5\overset{O}{\overset{\|}{C}}O^-Na^+} + NH_3$$

$$\xrightarrow{H^+} \underset{\text{Benzoic acid}}{C_6H_5\overset{O}{\overset{\|}{C}}OH}$$

The nomenclature of nitriles is as follows: the common name is formed by dropping the **-ic** ending of the common name for the corresponding acid and adding the term **-onitrile** (the "o" is added for euphony). In the IUPAC system, the term **-nitrile** is added to the parent hydrocarbon name. Some representative examples are given below (the common name is in parentheses):

CH_3CN	$CH_3(CH_2)_3CN$	$C_6H_5C{\equiv}N$	$p\text{-}CH_3C_6H_4CN$
Ethanenitrile (acetonitrile)	Pentanenitrile (*n*-valeronitrile)	Benzonitrile	*p*-Toluonitrile

REACTIONS OF CARBOXYLIC ACIDS

Salt Formation

Like their inorganic analogues, carboxylic acids react with bases to form salts. The name of the salt is formed by naming the cation, followed by the name of the acid in which the **-ic** ending (of either the common or IUPAC name) has been changed to **-ate.** Many of these salts are water-soluble. Since these salts are the salts of weak acids, they are readily hydrolyzed, and the acid can be regenerated by treatment of the acid salt with a stronger acid such as HCl or H_2SO_4. Some typical examples are shown below:

$$CH_3CO_2H + NaOH \rightarrow CH_3CO_2^-Na^+ + H_2O$$

Acetic acid — Sodium acetate

$$C_6H_5\overset{O}{\overset{\|}{C}}OH + NaOH \rightarrow C_6H_5\overset{O}{\overset{\|}{C}}O^-Na^+ + H_2O$$

Benzoic acid — Sodium benzoate

$$\downarrow HCl$$

$$C_6H_5COOH$$

Benzoic acid

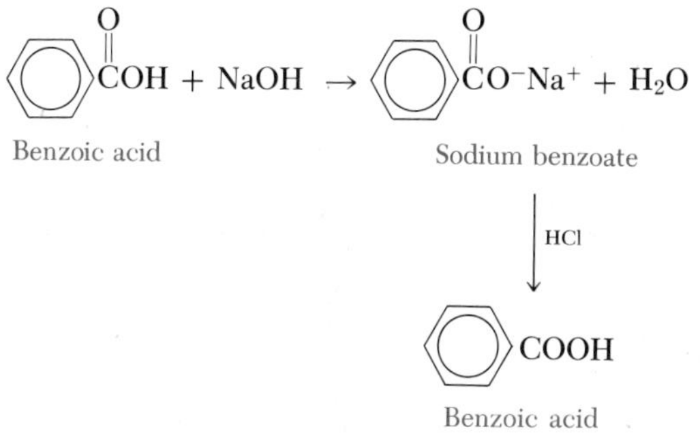

The solubility of these salts in water can be used to separate a carboxylic acid from a mixture of other functional groups. Treatment of the mixture with base converts the acidic components to water-soluble salts. Separation of the water layer effectively separates the salts from the remaining water insoluble components of the mixture. Treatment of the water layer with a strong acid regenerates the pure acid.

Conversion Into Functional Derivatives

Acid Halides (Acyl Halides).* The —OH group of a carboxylic acid can be replaced by a halogen atom to give a class of compounds called **acid halides,** RCOX (where X = halogen). This reaction is analogous to the replacement of the —OH group of alcohols and phenols by halogen to give alkyl or aryl halides. The same type of reagents used in replacing the hydroxyl group of alcohols can be used to form the acid halides; namely, reagents such as phosphorus trichloride, phosphorus pentachloride, and thionyl chloride (see reactions of alcohols). The hydrogen halides cannot be used to prepare the acid halides.

$$CH_3\overset{O}{\overset{\|}{C}}OH + SOCl_2 \rightarrow CH_3\overset{O}{\overset{\|}{C}}Cl + SO_2\uparrow + HCl\uparrow$$

Acetic acid — Thionyl chloride — Acetyl chloride

$$C_6H_5\overset{O}{\overset{\|}{C}}OH + PCl_5 \rightarrow C_6H_5\overset{O}{\overset{\|}{C}}Cl + POCl_3 + HCl\uparrow$$

Benzoic acid — Benzoyl chloride

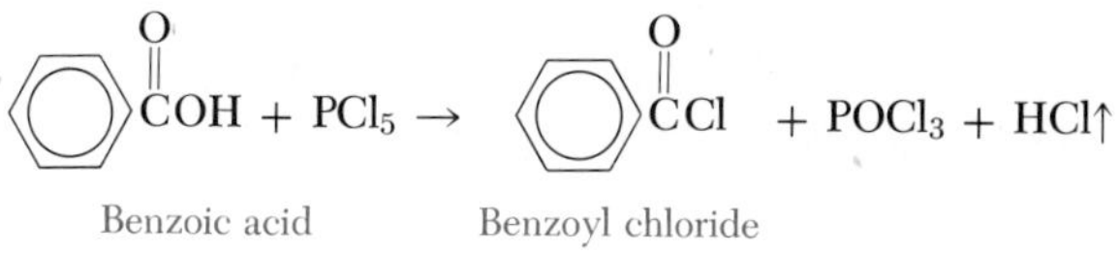

*The $R-\overset{O}{\overset{\|}{C}}-$ group is known as the acyl group (see acylation reactions). In many reactions of acids, this portion of the acid molecule remains unaffected in the chemical reaction. Many acid derivatives, particularly the halide derivatives, are named according to this group. The name is formed by changing the **-ic** ending of the common acid name to **-yl.**

Esters. Carboxylic acids react with alcohols in the presence of an acid to give a class of compounds called **esters,** RCO_2R' (see reactions of alcohols). The name of the ester is formed by changing the **-ic** ending of the acid (either the common or IUPAC name) to **-ate,** preceded by the name of the alkyl or aryl group derived from the alcohol. Some typical examples are shown below:

$$C_6H_5\overset{O}{\overset{\|}{C}}OH + CH_3OH \underset{}{\overset{H^+}{\rightleftharpoons}} C_6H_5\overset{O}{\overset{\|}{C}}OCH_3 + H_2O$$

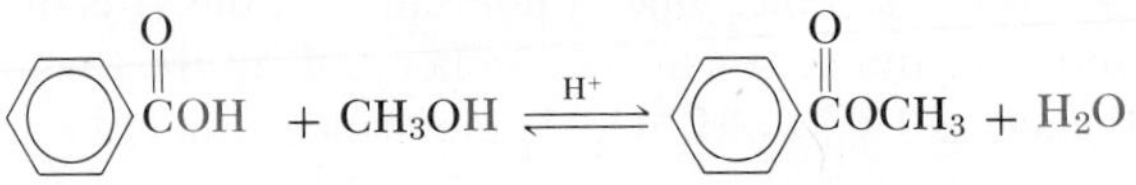

Benzoic acid — Methyl alcohol — Methyl benzoate

$$CH_3\overset{O}{\overset{\|}{C}}OH + CH_3CH_2OH \overset{H^+}{\rightleftharpoons} CH_3\overset{O}{\overset{\|}{C}}OCH_2CH_3 + H_2O$$

Acetic acid — Ethyl alcohol — Ethyl acetate

As noted in Chapter 6, phosphate esters are associated with almost all biologic reactions. Several important sugar phosphates are shown on page 217. Fats are esters of fatty acids and glycerol, and esters of this type are also found in nature (p. 227). Phosphatides in animal and vegetable cells and cephalins found in brain tissue also contain both phosphate and organic ester linkages (p. 232). Similarly, the important nucleotide ATP, which is the source of energy for many biologic reactions, contains phosphate ester linkages (p. 287).

Acid Anhydrides. Carboxylic acids undergo dehydration (loss of a mole of water) when treated with dehydrating agents to give a class of compounds called **acid anhydrides,** $R\overset{O}{\overset{\|}{C}}-O-\overset{O}{\overset{\|}{C}}-R'$. Anhydrides are named by adding the name anhydride after the acid from which the anhydride is derived. Cyclic anhydrides, containing five- or six-membered rings, are formed from the appropriate di-acids on dehydration. Some representative examples are shown below:

$$2CH_3\overset{O}{\overset{\|}{C}}-OH \xrightarrow{H_2SO_4} CH_3\overset{O}{\overset{\|}{C}}-O-\overset{O}{\overset{\|}{C}}-CH_3$$

Acetic acid — Acetic anhydride

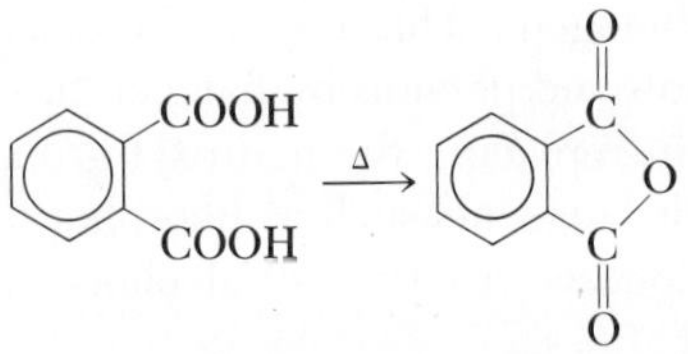

$$C_6H_4(COOH)_2 \xrightarrow{\Delta} C_6H_4(CO)_2O$$

Phthalic acid — Phthalic anhydride

Mixed anhydrides are formed *via* the elimination of one mole of water between an inorganic acid and an organic acid. Biologically, the inorganic acid is usually phosphoric acid. Anhydrides of this type are important in biologic reactions. For example, a mixed

$$R\overset{O}{\overset{\|}{C}}OH + HO-\overset{O}{\overset{\|}{\underset{OH}{\underset{|}{P}}}}-OH \rightleftharpoons R\overset{O}{\overset{\|}{C}}-O-\overset{O}{\overset{\|}{\underset{OH}{\underset{|}{P}}}}-OH + H_2O$$

Carboxylic acid — Phosphoric acid — Mixed anhydride

anhydride, 1,3-Di-PO_4^- glyceric acid is formed during carbohydrate metabolism. In this case the anhydride is formed from glyceraldehyde 3-PO_4 (see p. 322).

Amides. Carboxylic acids react with ammonia to give salts, which on heating give a class of compounds called **amides,** $RCONH_2$. Substituted amides, $RCONHR'$ or

$$CH_3\overset{O}{\overset{\|}{C}}OH + NH_3 \rightarrow CH_3CO_2^-NH_4^+ \xrightarrow{\Delta} CH_3\overset{O}{\overset{\|}{C}}NH_2 + H_2O$$

Acetic acid — Ammonium acetate — Acetamide

$RCONR_2'$, can be prepared by using amines (RNH_2 or R_2NH) in place of ammonia. The unsubstituted amides ($RCONH_2$) are named by dropping the **-ic** and **-oic** endings of the common and IUPAC names of the acid and adding the term **-amide.** When groups other than hydrogen are present on the nitrogen atom, their position and type is indicated by adding the prefix **N-** to indicate their position, followed by the name of the group attached to the nitrogen atom. Some typical examples are shown below:

$$H\overset{O}{\overset{\|}{C}}NH_2 \qquad CH_3\overset{O}{\overset{\|}{C}}NH_2 \qquad H\overset{O}{\overset{\|}{C}}N(CH_3)_2 \qquad C_6H_5\overset{O}{\overset{\|}{C}}\overset{H}{N}CH_2CH_3$$

Formamide — Acetamide — N,N-Dimethylformamide — N-Ethylbenzamide

The formation of the amide linkage catalyzed by an enzyme is of great biologic importance. For example, when two amino acids react to form a substituted amide, the carbonyl group of one amino acid reacts with the α-amino group of the other amino acid to form a **peptide.** Proteins are high-molecular weight compounds that contain many

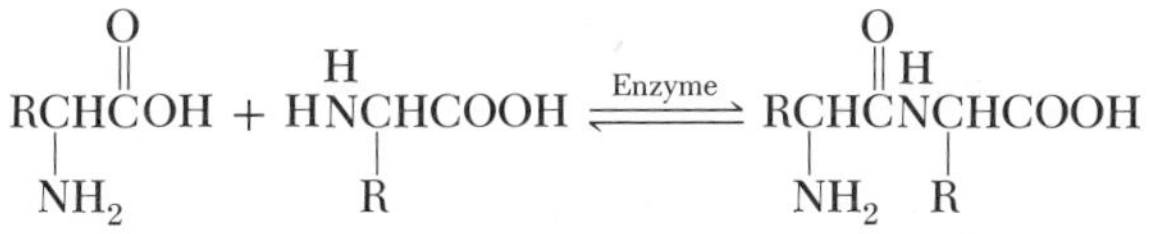

$$\underset{NH_2}{RCH}\overset{O}{\overset{\|}{C}}OH + H\overset{H}{N}\underset{R}{CH}COOH \xrightleftharpoons{\text{Enzyme}} \underset{NH_2}{RCH}\overset{O}{\overset{\|}{C}}\overset{H}{N}\underset{R}{CH}COOH$$

Peptide

amino acids linked by peptide (amide) bonds. These types of compounds will be discussed in more detail in Chapter 15.

SUBSTITUTION REACTIONS

In addition to the many reactions involving the hydroxyl group of the acid, reactions involving the alkyl group of the aliphatic acids or the aromatic ring of the aromatic acids are also possible.

The most important reaction in the alkyl chain is the halogenation reaction. Halogenation with phosphorus and halogen causes halogenation in the α-position of the acid, as shown below:

$$RCH_2CO_2H + X_2 \xrightarrow{P} RCHXCO_2H + HX\uparrow$$

X=Cl, Br

These α-halo acids are important intermediates in the preparation of amino acids (α-amino acids).

$$CH_3CH_2COOH + Br_2 \xrightarrow{P} CH_3\underset{\displaystyle Br}{\underset{|}{C}}HCO_2H \xrightarrow{NH_3} CH_3\underset{\displaystyle NH_2}{\underset{|}{C}}HCO_2H$$

Propionic acid — α-Bromopropionic acid — α-Aminopropionic acid (alanine)

Aromatic carboxylic acids undergo the usual electrophilic aromatic substitution reactions. The carboxyl group is a *meta*-director and deactivates relative to benzene. The bromination of benzoic acid represents a typical example of this kind of reaction. Nitration, sulfonation, alkylation, and acylation proceed similarly.

$$C_6H_5CO_2H + Br_2 \xrightarrow{Fe} m\text{-}BrC_6H_4CO_2H$$

Benzoic acid — *m*-Bromobenzoic acid

REACTIONS OF ACID DERIVATIVES

The acid derivatives undergo a wide variety of reactions, many of which involve the conversion of one acid derivative into another acid derivative. The ease of these types of conversions depends on the reactivity of the acid derivative and on the reaction conditions. The order of reactivity of the acid derivatives in the three most common types of reactions (hydrolysis, ammonolysis, and alcoholysis) is: acid halides > acid anhydrides > esters > amides. These three types of reactions are outlined below for the acid derivatives:

Acid Halides:

Hydrolysis $$R{-}\overset{\displaystyle O}{\overset{\|}{C}}{-}Cl + HOH \rightarrow R{-}\overset{\displaystyle O}{\overset{\|}{C}}{-}OH + HCl$$

Ammonolysis $$R{-}\overset{\displaystyle O}{\overset{\|}{C}}{-}Cl + 2NH_3 \rightarrow RCONH_2 + NH_4Cl$$

Alcoholysis $$R{-}\overset{\displaystyle O}{\overset{\|}{C}}{-}Cl + R'OH \rightarrow RCOOR' + HCl$$

Acid Anhydrides:

$$R{-}\overset{\displaystyle O}{\overset{\|}{C}}{-}O{-}\overset{\displaystyle O}{\overset{\|}{C}}{-}R + H_2O \rightarrow 2RCO_2H$$

$$R{-}\overset{\displaystyle O}{\overset{\|}{C}}{-}O{-}\overset{\displaystyle O}{\overset{\|}{C}}{-}R + 2NH_3 \rightarrow RCONH_2 + RCO_2^-NH_4^+$$

$$R{-}\overset{\displaystyle O}{\overset{\|}{C}}{-}O{-}\overset{\displaystyle O}{\overset{\|}{C}}{-}R + R'OH \rightarrow RCO_2R' + RCO_2H$$

Esters:

$$R-\overset{\overset{\Large O}{\|}}{C}-OR' + H_2O \xrightarrow{OH^-} R\overset{\overset{\Large O}{\|}}{C}O^- + R'OH$$ Called a saponification reaction

$$R-\overset{\overset{\Large O}{\|}}{C}-OR' + NH_3 \longrightarrow RCONH_2 + R'OH$$

$$R-\overset{\overset{\Large O}{\|}}{C}-OR' + R''OH \longrightarrow R-\overset{\overset{\Large O}{\|}}{C}-OR'' + R'OH$$ Transesterification reaction

Amides:

$$RCONH_2 + H_2O \xrightarrow{OH^-} RCO_2^- + NH_3$$

IMPORTANT ACIDS, ACID DERIVATIVES, AND THEIR USES

Formic acid occurs free in nature in small amounts, and its presence is made known in an unpleasant fashion. Anyone who has been bitten by an ant or brushed against stinging nettles has felt the irritating effect of formic acid injected under the skin. Formic acid is prepared commercially by heating powdered sodium hydroxide with carbon monoxide gas under pressure. The sodium formate produced is converted to formic acid by treatment with sulfuric acid.

$$\underset{\text{Carbon monoxide}}{CO + NaOH} \xrightarrow[\text{pressure}]{\Delta} \underset{\text{Sodium formate}}{H\overset{\overset{\Large O}{\|}}{C}O^-Na^+} \xrightarrow{H_2SO_4} \underset{\text{Formic acid}}{H\overset{\overset{\Large O}{\|}}{C}OH} + NaHSO_4$$

Formic acid is a colorless liquid with a sharp, irritating odor. It is a slightly stronger acid than most carboxylic acids and produces blisters when it comes in contact with the skin. It is used in the manufacture of esters, salts, plastics, and oxalic acid.

Acetic acid has been known for centuries as an essential component of vinegar. It is formed from the oxidation of ethyl alcohol that is produced from the fermentation of fruit juices in the preparation of the vinegar. Cider vinegar from fruit juices contains about 4% acetic acid in addition to flavoring and coloring agents from the fruit. White vinegar is prepared by diluting acetic acid to the proper concentration with water.

Acetic acid is produced commercially by the oxidation of acetaldehyde, which is obtained by a catalytic oxidation of ethylene. Commercially produced acetic acid is about 99.5% pure and is called "glacial acetic acid," since at temperatures below 17°C it freezes to an ice-like solid.

Lactic acid, $CH_3CHOHCOOH$, is an important hydroxy acid. It is formed when lactose (milk sugar) is fermented by lactobacillus bacteria. The taste of sour milk and buttermilk is due to the presence of lactic acid. Sour milk is often used in baking for its leavening effect on the dough. The lactic acid reacts with the sodium bicarbonate of baking soda to produce carbon dioxide throughout the dough.

In the process of muscular contraction, lactic acid is formed by the muscle tissues and released into the blood stream. In many of the so-called "cycles" involved in the oxidation of carbohydrates, lipids, and proteins to produce energy in the body, lactic acid is an essential component.

Oxalic acid occurs in the leaves of vegetables such as rhubarb and is one of the strongest naturally occurring acids. Its preparation is closely related to that of formic acid, since the sodium salt of oxalic acid is produced by heating sodium formate with alkali. Oxalic acid is used to remove iron stain from fabrics and from porcelain ware, and to bleach straw and leather goods.

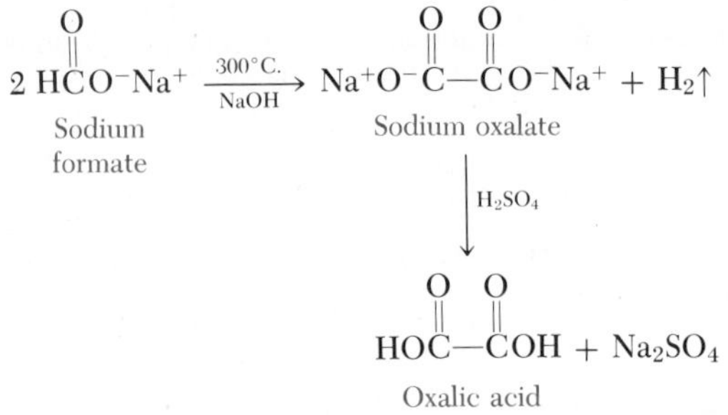

$$2\,\mathrm{HCO^-Na^+}\ (\text{Sodium formate}) \xrightarrow[\mathrm{NaOH}]{300^\circ\mathrm{C.}} \mathrm{Na^+O^-\overset{O}{\overset{\|}{C}}-\overset{O}{\overset{\|}{C}}O^-Na^+}\ (\text{Sodium oxalate}) + \mathrm{H_2}\uparrow$$

$$\text{Sodium oxalate} \xrightarrow{\mathrm{H_2SO_4}} \mathrm{HO\overset{O}{\overset{\|}{C}}-\overset{O}{\overset{\|}{C}}OH}\ (\text{Oxalic acid}) + \mathrm{Na_2SO_4}$$

Citric acid, is a normal constituent of citrus fruits. Large quantities are produced by a fermentation process from starch or molasses. It is commonly employed to impart a sour taste to food products and beverages. In fact, most of the citric acid produced annually is used by the food and soft drink industries.

Benzoic acid is a colorless solid, slightly soluble in hot water. It is used in the synthesis of organic compounds, and its sodium salt (sodium benzoate) is used as an antiseptic and food preservative. Several substituted benzoic acids are also important commercial materials. ***Para*-aminobenzoic acid** is classified as a vitamin. It is apparently necessary for the proper physiological functioning of chickens, mice, and bacteria, but as yet has not been proved essential in the diet of humans. Certain derivatives of this acid are used as local anesthetics. Procaine, or Novocain, is probably the most important local anesthetic derived from *p*-aminobenzoic acid. Saccharin, a derivative of *o*-sulfobenzoic acid, has a relative sweetness several hundred times that of sucrose. It is used as a sweetening agent when the sugar intake must be restricted. Recently another more potent sweetening agent named Sucaryl has been developed. This sweetening agent has properties similar to saccharin and is available in both liquid and solid form. Sucaryl salts as a class are called **cyclamates.** Recent data has shown that large doses of cyclamates induce cancer in rats. Consequently, the federal government has banned the use of cyclamates in all beverages and food products. At the present time, saccharin is the only artificial sweetener allowed in the U. S.

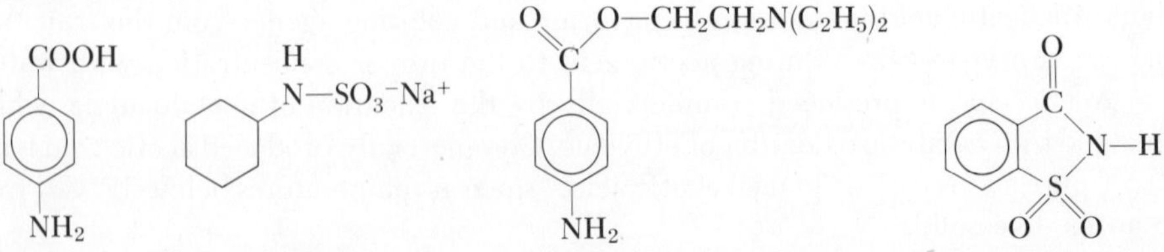

COOH / C_6H_4 / NH_2 — *p*-Aminobenzoic acid

C_6H_{11}–N(H)–$SO_3^-Na^+$ — Sucaryl

$O{=}C(O{-}CH_2CH_2N(C_2H_5)_2)$ / C_6H_4 / NH_2 — Procaine

C_6H_4 fused ring with C=O, N—H, SO_2 — Saccharin

Acid Derivatives

Organic acid derivatives, such as salts, are often used in place of the acids for many commercial uses. **Sodium acetate** is often used for its buffering effect in reducing the acidity of inorganic acids. It is also used in the preparation of soaps and pharmaceutical agents. **Lead acetate,** sometimes called sugar of lead because of its sweet taste, is used externally to treat poison ivy and certain skin diseases. **Paris green** is a complex salt that contains copper acetate and is used as an insecticide. Calcium propionate is added to bread to prevent molding. The calcium salt of lactic acid is sometimes used to supplement the calcium of the diet. Many infants receive additional iron in their diets from the salt, ferric ammonium citrate. Magnesium citrate has long been used as a saline cathartic. A solution of sodium citrate is employed to prevent the clotting of blood used in transfusions, and potassium oxalate will prevent the clotting of blood specimens drawn for analysis in the clinical laboratory. The sodium salt of salicylic acid is used extensively as an antipyretic or fever-lowering agent, and as an analgesic in the treatment of rheumatism and arthritis.

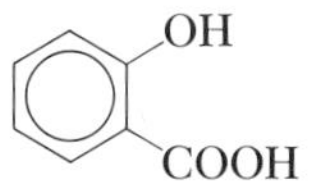

OH, COOH
Salicylic acid

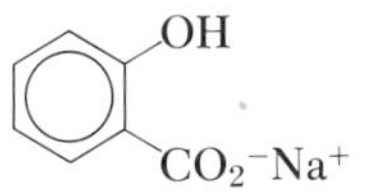

OH, $CO_2^-Na^+$
Sodium salicylate

Commercially, only a few of the esters are produced in large quantity. The two most important are ethyl acetate and butyl acetate, which are used as solvents, especially for nitrocellulose in the preparation of lacquers. Some of the higher molecular weight esters are plastics, whereas others are used in the production of plastics.

Many esters occur in free form in nature and are responsible for the odor of most flowers and fruits. The characteristic taste and odors of different esters find application in the manufacture of artificial flavoring extracts and perfumes. Synthetic esters that are commonly used as food flavors are amyl acetate for banana, octyl acetate for orange, ethyl butyrate for pineapple, amyl butyrate for apricot, isobutyl formate for raspberry, and ethyl formate for rum.

Other esters are commonly used as therapeutic agents in medicine. Ethyl acetate is employed as a stimulant and antispasmodic in colic and bronchial irritations. It is also applied externally in the treatment of skin diseases caused by parasites. Ethyl nitrite, when mixed with alcohol, is called elixir of niter and is used as a diuretic and antispasmodic. Amyl nitrite is used to lower blood pressure temporarily and causes relaxation of muscular spasms in asthma and in the heart condition known as angina pectoris. Glyceryl trinitrate, or nitroglycerin, is a vasodilator that has physiological action similar to amyl nitrate.

Methyl benzoate and methyl salicylate are two of the aromatic esters used in the manufacture of perfumes and flavoring agents. Methyl benzoate has the odor of new-mown hay, whereas methyl salicylate smells and tastes like wintergreen. One of the most important esters of salicylic acid is that formed with acetic acid. **Acetylsalicylic acid,** or aspirin, is the common antipyretic and analgesic drug. An increasing number of medications for the relief of simple headaches and the pain of rheumatism and arthritis contain aspirin combined with other pharmaceutical agents. For example, aspirin combined with antacid and buffering agents is absorbed more rapidly from the intestinal tract and should therefore afford more rapid relief of aches and pains. Several esters of *p*-aminobenzoic acid act as local anesthetics. Butesin is the butyl ester, whereas procaine is the diethyl amino ester of this acid.

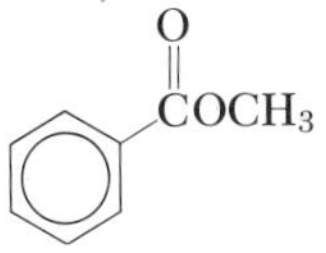

O, $COCH_3$
Methyl benzoate

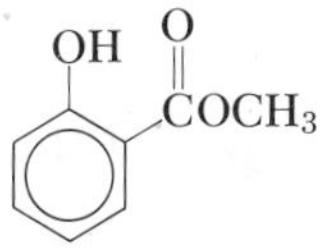

OH, O, $COCH_3$
Methyl salicylate

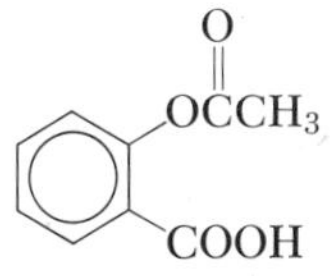

O, $OCCH_3$, COOH
Acetylsalicylic acid

Polyesters make up some of the most important commercial polymers. One of the most interesting and important types of vinyl plastics is the acrylic resins. They are polymers of acrylic acid esters, such as methyl acrylate and methyl methacrylate.

$H_2C{=}CHCOOH$	$H_2C{=}CHCO_2CH_3$	$H_2C{=}C(CH_3)CO_2CH_3$
Acrylic acid	Methyl acrylate	Methyl methacrylate

Polymerization of methyl methacrylate results in a clear, transparent plastic that can be used in place of glass. Lucite, Plexiglas, and Perspex are some of the trade names given this polymer. It can be formed into strong, flexible sheets that are highly transparent and lighter than glass, or it can be molded into transparent articles.

Another important polyester, Dacron, is prepared from terephthalic acid and ethylene glycol. The initial condensation produces an ester linkage, but the initial condensation product still has two functional groups left which can react further. This type of condensation continues until thousands of the ester units are contained in the chain. Polymers of this type will be considered in more detail in Chapter 12.

Phthalic anhydride is another very important organic compound that is usually produced commercially by the oxidation of *ortho*-xylene or naphthalene. Large quantities of this material are used as plasticizers of synthetic resins and in the manufacture of the glyptal type of weather-resistant protective coating.

The most important amide is **urea,** $H_2N{-}\overset{O}{\overset{\|}{C}}{-}NH_2$, which is a naturally occurring diamide of carbonic acid. The synthesis of this compound by Wöhler in 1828 was the keystone in the development of organic chemistry. Urea is prepared by the action of heat and pressure on carbon dioxide and ammonia.

$$CO_2 + 2NH_3 \xrightarrow[\text{pressure}]{\Delta} \underset{\text{Urea}}{H_2N\overset{O}{\overset{\|}{C}}NH_2} + H_2O$$

On heating, urea decomposes to ammonia and isocyanic acid. Alcohols react with isocyanic acid to form carbamates.

$$\underset{\text{Urea}}{H_2N\overset{O}{\overset{\|}{C}}NH_2} \xrightarrow{\Delta} \underset{\text{Ammonia}}{NH_3} + \underset{\text{Isocyanic acid}}{HN{=}C{=}O}$$

$$\underset{\text{Isocyanic acid}}{HN{=}C{=}O} + \underset{\text{Ethyl alcohol}}{HOCH_2CH_3} \rightarrow \underset{\text{Ethyl carbamate}}{H_2N\overset{O}{\overset{\|}{C}}OCH_2CH_3}$$

Meprobamate (Miltown) is 2-methyl-2-*n*-propyl-1,3-propanediol dicarbamate, a widely used tranquilizing agent.

$$\underset{\text{Miltown}}{(CH_3CH_2CH_2)(CH_3)C(CH_2O\overset{O}{\overset{\|}{C}}NH_2)_2}$$

One of the most important commercial uses of urea is the base catalyzed condensation of urea with substituted diethyl malonates to produce the sedatives known as the **barbiturates.**

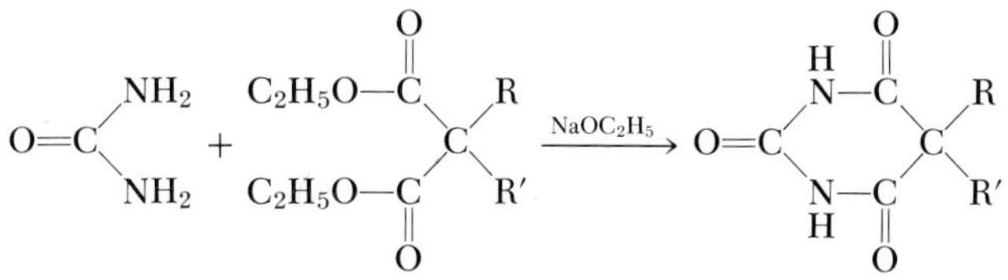

$$O{=}C(NH_2)_2 + (C_2H_5O{-}CO)_2CRR' \xrightarrow{NaOC_2H_5} O{=}C(NH{-}CO)_2CRR'$$

Urea — Substituted diethyl malonates — A barbiturate

An important polyamide, nylon, is prepared by the condensation of adipic acid and hexamethylenediamine. The initial reaction involves a neutralization reaction between the acid and the amine to give a nylon salt, which on heating at high temperatures produces nylon (see Chapter 12).

IMPORTANT TERMS AND CONCEPTS

acetic acid
acid anhydride
acid halide (acyl halide)
amide
amino acid
aspirin
benzoic acid
carboxyl group
citric acid
cyclamate
ester
lactic acid
nitrile
oxalic acid
peptide
saccharin
saponification reaction
sucaryl
transesterification
urea

QUESTIONS

1. Draw a structural formula for each of the following compounds:
   (a) potassium propionate
   (b) isobutyryl chloride
   (c) ethyl benzoate
   (d) succinic anhydride
   (e) diethyl phthalate
   (f) butyramide
   (g) sodium oxalate
   (h) *m*-bromobenzoic acid
   (i) methyl formate
   (j) isohexanoic acid

2. Name each of the following compounds:
   (a) $CH_3(CH_2)_3CH(CH_3)CO_2H$
   (b) CCl_3CO_2H
   (c) $CH_3CO_2^-K^+$
   (d) $CH_3CH_2CH_2CO_2CH_2CH_3$
   (e) $CH_3CH_2CH_2CONH_2$
   (f) CH_3COBr
   (g) $CH_2(CO_2CH_2CH_3)_2$
   (h) 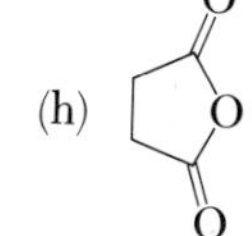
   (i) $(CF_3CO)_2O$
   (j) 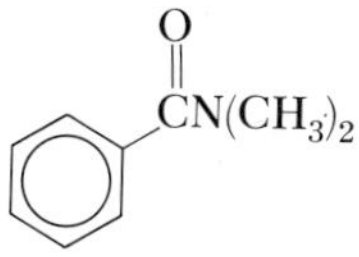 $C_6H_5{-}\overset{O}{\overset{\|}{C}}N(CH_3)_2$

3. Write equations for the reactions of *n*-butyric acid with the following reagents:
   (a) CH_3CH_2OH/H^+
   (b) $SOCl_2$
   (c) NH_3/heat
   (d) KOH
   (e) H_2SO_4/heat
   (f) PBr_3
   (g) $P + Br_2$
   (h) $LiAlH_4$/ether

4. Write equations for the reactions of benzoic acid with the following reagents:
   (a) $Ca(OH)_2$
   (b) PCl_5
   (c) Br_2/Fe
   (d) HNO_3/H_2SO_4
   (e) benzyl alcohol/H^+

5. Write equations for the preparation of the following compounds. Give any necessary catalysts. More than one step may be necessary.
   (a) *n*-butyric acid from *n*-butyl alcohol
   (b) ethyl propionate from propionic acid
   (c) benzoic acid from benzene
   (d) N-methylpropionamide from ethyl bromide
   (e) benzoyl chloride from benzene
   (f) *n*-propionic acid from ethyl iodide
   (g) α-bromobutyric acid from *n*-propyl iodide
   (h) terephthalic acid from *p*-bromotoluene
   (i) *p*-bromobenzoic acid from benzene
   (j) acetic anhydride from ethyl alcohol

6. Using equations, show how each of the following conversions can be carried out:
   (a) ethyl propionate to *n*-propionic acid
   (b) acetyl chloride to ethyl acetate
   (c) benzoyl chloride to N,N-dimethylbenzamide
   (d) succinic acid to 1,4-butanediol
   (e) *p*-bromobenzoyl chloride to *p*-bromobenzoic acid

SUGGESTED READING

Carter: The History of Barbituric Acid. Journal of Chemical Education, Vol. 28, p. 524, 1951.
Davidson: Acids and Bases In Organic Chemistry. Journal of Chemical Education, Vol. 19, p. 154, 1942.
On Sweetness and Sweeteners. Chemistry, Vol. 44, p. 21, 1971.

AMINES AND AMINE DERIVATIVES

CHAPTER 11

The *objectives* of this chapter are to enable the student to:

1. Distinguish primary, secondary, and tertiary amines.
2. Write the structures and names of the common alkyl and aromatic amines.
3. Explain the basicity of amines.
4. Describe the preparation of amines.
5. Recognize the more common reactions of amines.
6. Define a diazonium salt and explain its synthetic value in organic chemistry.
7. Recognize the important amines, amine derivatives, and their current uses.

Amines are organic derivatives of ammonia in which one or more of the hydrogen atoms have been replaced by an alkyl or aryl group. The characteristic functional group present in amines is called the **amino group** and is written as **—NH_2**. Like alcohols, amines are classified as primary, secondary and tertiary amines. In the case of amines the classification is determined by the *number* of alkyl or aryl groups attached to the nitrogen atom. Some representative examples are illustrated below:

CH_3NH_2	$CH_3N(H)$—C_6H_5	$(CH_3)_2NCH_2CH_3$
Primary amine	Secondary amine	Tertiary amine

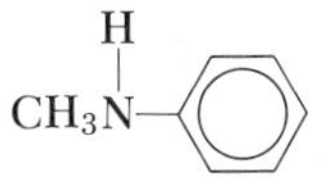

Another description that classifies amines as primary, secondary, or tertiary describes the nitrogen atom on the basis of its having two, one, or no —NH bonds.

Amines can be named as amino derivatives of hydrocarbons according to the IUPAC system—as amino alkanes for example. This system, however, is little used. The common name nomenclature system is most widely used. It consists of naming the alkyl or aryl groups attached to the nitrogen atom, using the appropriate prefixes if two or more identical substituents are attached to the nitrogen, followed by the word "amine." The following examples illustrate this system of nomenclature:

$CH_3CH_2NH_2$ — Ethyl amine
$(CH_3)_2NH$ — Dimethylamine
$(CH_3CH_2)_3N$ — Triethylamine
$(C_6H_5)_2NH$ — Diphenylamine
$C_6H_5NH_2$ — Aniline

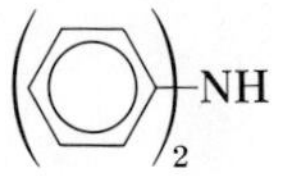

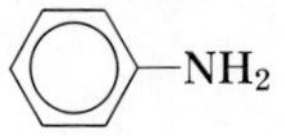

Simple aromatic amines are named as derivatives of the parent aromatic amine, **aniline.** The substituted anilines are named in the same manner as other substituted benzene derivatives. If there are two or more substituents, then numbers must be used to indicate the positions of the substituents.

$CH_3N(H)CH_2CH_3$ — Methyl ethyl amine
$C_6H_5N(H)CH_3$ — N-Methylaniline
$C_6H_5N(CH_3)_2$ — N,N-Dimethylaniline 
$O_2N{-}C_6H_4{-}NH_2$ — *p*-Nitroaniline

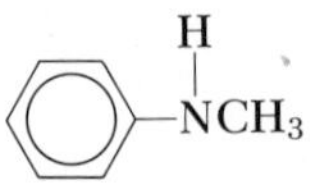

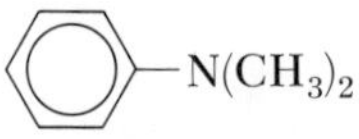

The nitrogen atom may also be part of a ring system, and both cyclic and aromatic types of amines are known, as illustrated below:

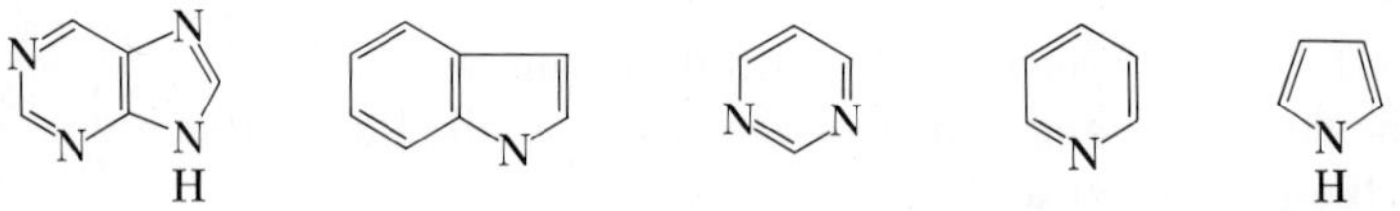

Purine   Indole   Pyrimidine   Pyridine   Pyrrole

PHYSICAL PROPERTIES OF AMINES

The lower molecular weight amines are gases that are soluble in water. As might be anticipated for derivatives of ammonia, the aqueous solutions of the water-soluble amines are alkaline. The volatile amines have unpleasant odors that combine the odor of ammonia with that of decayed fish. The higher molecular weight ($>C_6$) amines are insoluble in water and soluble in organic solvents, and the unpleasant odor decreases with decreasing volatility. The properties of some typical amines are listed in Table 11-1.

Amines occur widely in living cells. For example, nucleic acids contain various pyrimidine and purine derivatives (Chapter 16). Also niacin and some coenzymes also contain heterocyclic amines (p. 303 and 305). The amino group of amino acids is also of importance in peptide and protein formation as discussed in the last chapter under amides, and is also important in protein metabolism (see p. 347). These important biochemical applications of amines will be discussed more thoroughly in the last section of this text.

Other important compounds which we have seen previously are LSD and barbiturates (see Chapter 2).

Amines are basic like ammonia. Their base strength depends on the type of group (alkyl or aryl) attached to the nitrogen atom. Alkyl groups attached to the nitrogen atom increase the basicity of the amine relative to ammonia, and the alkyl amines are slightly stronger bases than ammonia. On the other hand, aromatic groups attached to the nitrogen atom decrease the basicity of the amine relative to ammonia. Aniline is a much weaker base than ammonia; diphenylamine is only weakly basic; and triphenylamine is essentially

TABLE 11-1 PHYSICAL PROPERTIES OF AMINES

COMPOUND	STRUCTURE	M.P. °C	B.P. °C	K_a
Methylamine	CH_3NH_2	−93	−6	4×10^{-4}
Ethylamine	$CH_3CH_2NH_2$	−81	+17	5×10^{-4}
n-Propylamine	$CH_3CH_2CH_2NH_2$	−83	+48	4×10^{-4}
Dimethylamine	$(CH_3)_2NH$	−92	+7	5×10^{-4}
Diethylamine	$(CH_3CH_2)_2NH$	−50	+56	9×10^{-4}
Trimethylamine	$(CH_3)_3N$	−117	+3	6×10^{-5}
Triethylamine	$(CH_3CH_2)_3N$	−115	+90	5×10^{-4}
Aniline	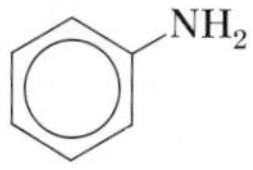 $C_6H_5NH_2$	−6	+184	4×10^{-10}
o-Toluidine	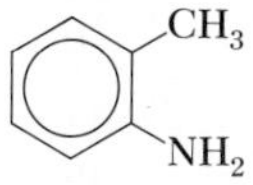 (CH_3, NH_2 ortho on benzene)	−15	+200	
m-Toluidine	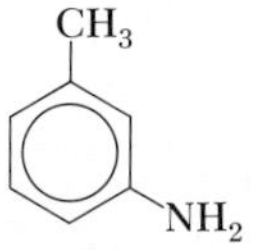 (CH_3, NH_2 meta on benzene)	−30	+203	
p-Toluidine	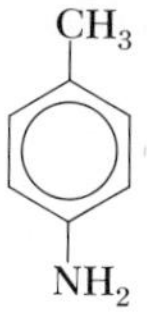 (CH_3, NH_2 para on benzene)	+44	+201	
Nitrobenzene	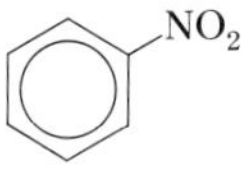 $C_6H_5NO_2$	+6	+211	

a neutral compound. The decreased basicity of aromatic amines is due to resonance delocalization of the lone pair of electrons on nitrogen with the aromatic ring. This type

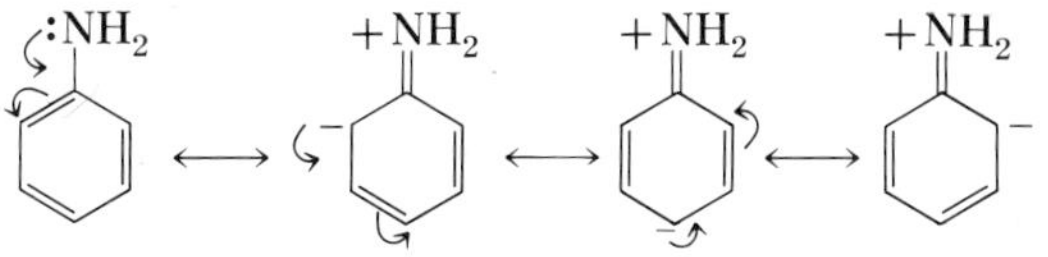

$:NH_2$ (aniline) ⟷ $+NH_2$ (negative charge ortho) ⟷ $+NH_2$ (negative charge para) ⟷ $+NH_2$ (negative charge ortho)

of resonance interaction makes the lone pair of nitrogen less available to associate with a proton.

The boiling points of primary and secondary amines are higher than might be expected on the basis of molecular weight. Primary and secondary amines can form hydrogen bonds, which accounts for their high boiling points. Tertiary amines, which cannot form hydrogen bonds, have boiling points similar to the hydrocarbons of corresponding molecular weights. The simple methyl amines and the corresponding hydrocarbons illustrate this point, as shown below:

CH_3NH_2 (C_2H_6, B.P. −88°C)
B.P. −6°C

$(CH_3)_2NH$ (C_3H_8, B.P. −42°C)
B.P. +7°C

$(CH_3)_3N$ (C_4H_{10}, B.P. 0°C)
B.P. +3°C

SALT FORMATION

Ammonia reacts with acids to form salts. The unshared pair of electrons on the nitrogen atom is used to form a new N—H bond, and the ammonia molecule is converted into the ammonium ion.

$$\underset{\text{Ammonia}}{H{-}\ddot{N}(H){-}H} + \underset{\text{Hydrogen chloride}}{H^+Cl^-} \rightarrow \underset{\text{Ammonium chloride}}{\left[H{-}N(H)_2{-}H \right]^+ Cl^-}$$

The organic derivatives of ammonia, the amines, behave similarly. Aqueous mineral acids and carboxylic acids convert amines into salts. The amine salts are typical ionic saltlike compounds.

$$\begin{matrix} R{-}\ddot{N}H_2 & & R\overset{+}{N}H_3 & & R\ddot{N}H_2 \\ R_2\ddot{N}H & \xrightarrow{HX} & R_2\overset{+}{N}H_2\ X^- & \xrightarrow{OH^-} & R_2\ddot{N}H + H_2O + X^- \\ R_3\ddot{N} & & R_3\overset{+}{N}H & & R_3\ddot{N} \end{matrix}$$

They are nonvolatile solids, generally soluble in water and insoluble in nonpolar organic solvents. The difference in solubility behavior between amines and their salts can be used to separate amines from a mixture of other nonbasic organic compounds. Treatment of such a mixture with aqueous dilute hydrochloric acid converts the amine into a water-soluble salt that dissolves in the water layer. Separation of the aqueous layer, followed by the addition of hydroxide (a stronger base than amines), liberates the free amine.

In addition to simple salt formation with acids, tertiary amines also react with alkyl halides to form a type of compound called a **quaternary ammonium** salt.

$$\underset{\text{Amine}}{R_3\ddot{N}} + \underset{\text{Alkyl halide}}{R'{-}X} \rightarrow \underset{\text{Quaternary ammonium salt}}{\left[R{-}N(R')(R){-}R \right]^+ X^-} \xrightarrow[H_2O]{Ag_2O} \underset{\text{Quaternary ammonium hydroxide}}{\left[R{-}N(R')(R){-}R \right]^+ OH^-} + AgX\downarrow$$

This type of reaction is another illustration of nucleophilic displacement reactions of alkyl halides (see reactions of alkyl halides). The amine is the nucleophile in this case. In the product, the substituted ammonium ion is again formed. Treatment of the quaternary ammonium salt with aqueous silver oxide converts the salt into a quaternary ammonium hydroxide. These hydroxides are very strong bases and are comparable in base strength to the alkali metal hydroxides (NaOH, KOH).

PREPARATION OF AMINES

Reduction Methods

The reduction of compounds containing the nitro (—NO_2) group affords a convenient path to primary amines, particularly the aromatic amines.

$$\underset{\text{Nitro compound}}{R{-}NO_2} + \underset{\text{Reducing agent}}{[H]} \rightarrow \underset{\text{Amine}}{RNH_2}$$

The aliphatic nitro compounds are more difficult to prepare than the aromatic nitro compounds, and this method is not generally used for preparing aliphatic amines. The aromatic nitro compounds, however, can be easily prepared by nitration of the appropriate aromatic hydrocarbon. Reduction can be carried out catalytically (H_2 + Pt), but the more normal reducing agents used are a metal and a mineral acid, as shown below:

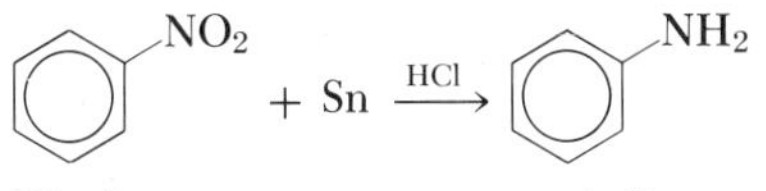

$$\underset{\text{Nitrobenzene}}{C_6H_5NO_2} + Sn \xrightarrow{HCl} \underset{\text{Aniline}}{C_6H_5NH_2}$$

$$\underset{m\text{-Nitrotoluene}}{CH_3C_6H_4NO_2} + Fe \xrightarrow{HCl} \underset{m\text{-Toluidine}}{CH_3C_6H_4NH_2}$$

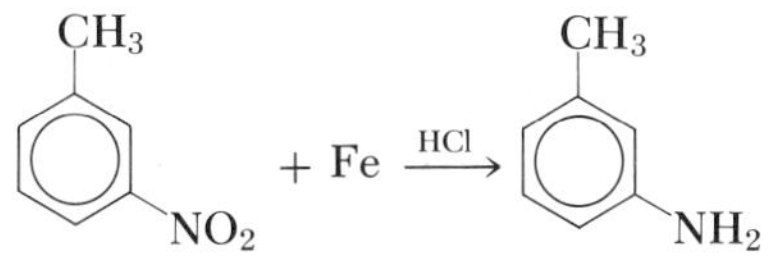

Reduction of nitriles also gives primary amines. Since the nitriles can be conveniently prepared from alkyl halides (see reactions of alkyl halides), this method is quite useful for preparing aliphatic amines. However, aromatic nitriles cannot be easily prepared by halide displacement, and this reaction is not often used for aryl halides. This method complements the reduction of nitro compounds. The normal reducing agents for this kind of reduction are $LiAlH_4$/ether, Na/CH_3CH_2OH, and H_2/Ni.

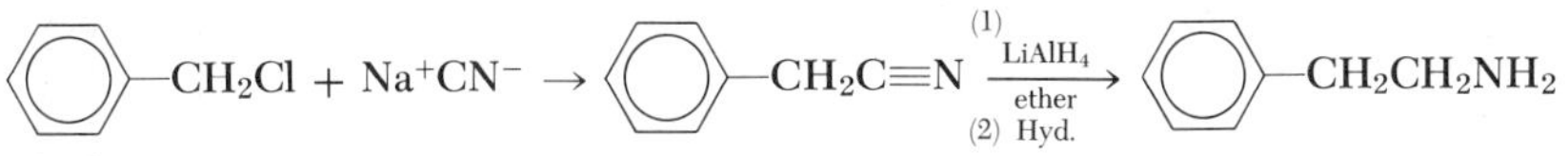

$$\underset{\text{Benzyl chloride}}{C_6H_5{-}CH_2Cl} + Na^+CN^- \rightarrow \underset{\text{Benzyl cyanide (phenylacetonitrile)}}{C_6H_5{-}CH_2C{\equiv}N} \xrightarrow[\text{ether (2) Hyd.}]{(1)\ LiAlH_4} \underset{\beta\text{-Phenylethylamine}}{C_6H_5{-}CH_2CH_2NH_2}$$

Amides also give amines on reduction with lithium aluminum hydride in ether. The type of amine formed on reduction depends on the type of amide reduced. Substituted amides provide a convenient path to secondary and tertiary amines.

$$\underset{\text{N-Methylacetamide}}{CH_3C(=O)NHCH_3} + LiAlH_4 \xrightarrow{\text{ether}} \underset{\text{Methyl ethyl amine}}{CH_3CH_2NHCH_3}$$

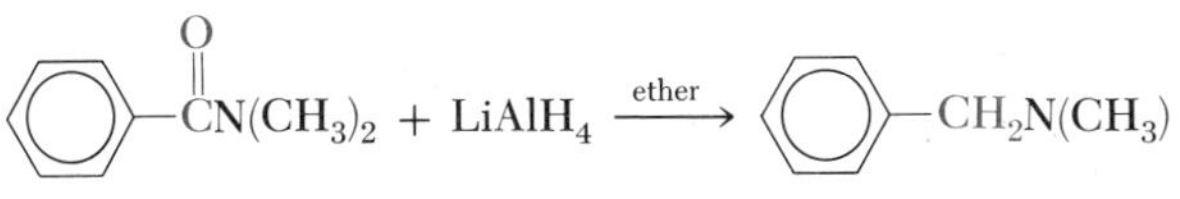

$$\underset{\text{N,N-Dimethylbenzamide}}{C_6H_5{-}C(=O)N(CH_3)_2} + LiAlH_4 \xrightarrow{\text{ether}} \underset{\text{Benzyl dimethylamine}}{C_6H_5{-}CH_2N(CH_3)}$$

Alkylation of Alkyl Halides

The use of ammonia or amines in a nucleophilic displacement reaction with alkyl halides (see reactions of alkyl halides) provides a classical route to the aliphatic amines, as shown on the next page.

$$R{-}X + \ddot{N}H_3 \rightarrow [R{-}\overset{+}{N}H_3]X^- \xrightarrow{NaOH} R\ddot{N}H_2 + Na^+X^- + H_2O$$

$X = Cl, Br, I$

Unfortunately, the alkylated product (RNH_2) also can react with the alkyl halide to produce some secondary amine, which in turn can give tertiary amine, which in turn can give a quaternary ammonium salt.

$$\underset{\text{Alkyl halide}}{R{-}X} + \ddot{N}H_3 \rightarrow \underset{\text{Primary amine}}{R\ddot{N}H_2} \xrightarrow{R-X} \underset{\text{Secondary amine}}{R_2\ddot{N}H} \xrightarrow{R-X} \underset{\text{Tertiary amine}}{R_3\ddot{N}} \xrightarrow{R-X} \underset{\text{Quaternary ammonium salt}}{R_4{}^+NX^-}$$

In practice, a mixture of all four products results. Use of a large excess of ammonia gives the primary amine in largest quantity. Although of use commercially, the formation of mixtures of this type limits this method for general laboratory use where a pure amine is desired.

An important biologic quaternary ammonium compound is acetylcholine, which is

$$\underset{\text{Acetylcholine}}{CH_3COOCH_2CH_2\overset{+}{N}(CH_3)_3} \xrightarrow{\text{cholinesterase}} CH_3CO_2{}^- + \underset{\text{Choline}}{HOCH_2CH_2N^+(CH_3)_3}$$

present in nerve cells and which aids in transmission of nerve impulses in the body. When a nerve cell is stimulated, acetylcholine is released. The released acetylcholine in turn stimulates an adjacent nerve cell, which in turn releases acetylcholine. This process continues until the nerve impulse reaches the brain. After the nerve impulse is complete, the enzyme cholinesterase immediately hydrolyses the acetylcholine to acetate and choline. If cholinesterase is inhibited by a chemical from carrying out this hydrolysis reaction, or if the stimulation process of acetylcholine is blocked, then the nervous system is completely disrupted and paralysis and death quickly result. The concept behind the use of nerve gases such as diisopropylfluorophosphate (p. 145) is to inhibit cholinesterase, thereby resulting in disruption of the nervous system.

REACTIONS OF AMINES

Salt Formation

As noted earlier in this chapter, amines, both aliphatic and aromatic, form salts with mineral acids. The free amines can be regenerated by treatment of the salt with a strong base.

$$\underset{\text{Aniline}}{C_6H_5{-}\ddot{N}H_2} + HCl \rightarrow \underset{\text{Anilinium chloride (aniline hydrochloride)}}{C_6H_5{-}\overset{+}{N}H_3Cl^-}$$

$$(CH_3)_2\ddot{N}H + HBr \rightarrow \underset{\text{Dimethylamine hydrobromide}}{(CH_3)_2\overset{+}{N}H_2Br^-}$$

Alkylation

As noted in the preparation of amines, alkylation of amines occurs readily. Since the alkylation is difficult to control, this reaction is most useful in preparing quaternary ammonium salts.

$$(CH_3)_3\ddot{N} + CH_3I \rightarrow (CH_3)_4N^+I^- \xrightarrow{Ag_2O} (CH_3)_4N^+OH^- + AgI\downarrow$$

Acylation

Carboxylic acid amides ($RCONH_2$) are prepared via the reaction of ammonia with an acid halide or an acid anhydride. Similar types of acylation reactions occur when primary and secondary amines react with acid halides and acid anhydrides. Since tertiary amines have no available N—H bond, they cannot be acylated.

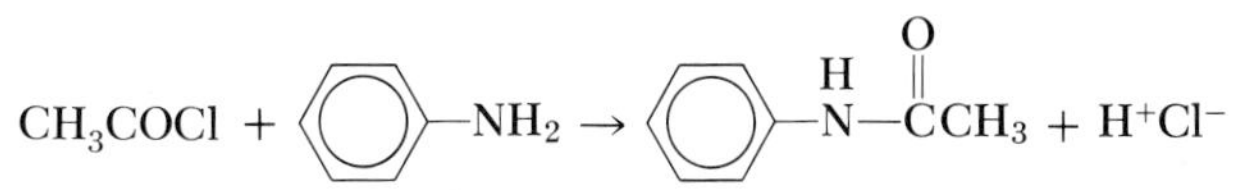

$$CH_3COCl + C_6H_5\text{—}NH_2 \rightarrow C_6H_5\text{—}\overset{H}{N}\text{—}\overset{O}{\overset{\|}{C}}CH_3 + H^+Cl^-$$

Acetyl chloride Aniline Acetanilide

$$CH_3COCl + CH_3\ddot{N}H_2 \rightarrow CH_3\text{—}\overset{O}{\overset{\|}{C}}\text{—}\overset{H}{N}CH_3 + H^+Cl^-$$

Acetyl chloride Methyl amine N-Methyl acetamide

In addition to being acylated by carboxylic acid derivatives, primary and secondary amines also react with sulfonic acid halides to form sulfonamides, as illustrated below for benzene sulfonyl chloride (the acid chloride of benzene sulfonic acid):

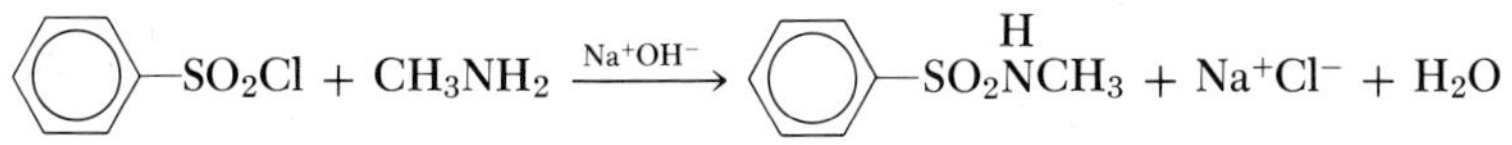

$$C_6H_5\text{—}SO_2Cl + CH_3NH_2 \xrightarrow{Na^+OH^-} C_6H_5\text{—}SO_2\overset{H}{N}CH_3 + Na^+Cl^- + H_2O$$

Benzene sulfonyl chloride Methyl amine N-Methyl benzenesulfonamide

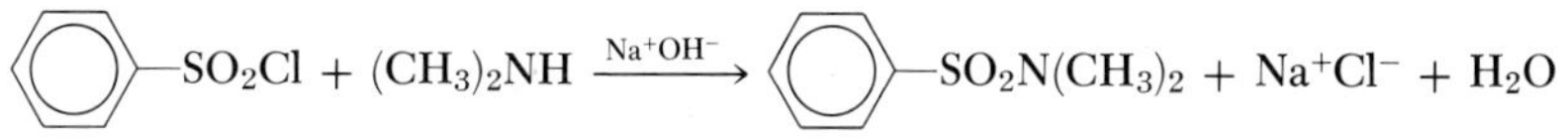

$$C_6H_5\text{—}SO_2Cl + (CH_3)_2NH \xrightarrow{Na^+OH^-} C_6H_5\text{—}SO_2N(CH_3)_2 + Na^+Cl^- + H_2O$$

Benzene sulfonyl chloride Dimethyl amine N,N-Dimethyl benzenesulfonamide

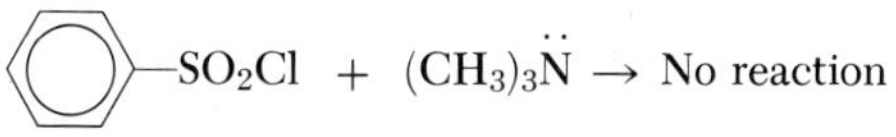

$$C_6H_5\text{—}SO_2Cl + (CH_3)_3\ddot{N} \rightarrow \text{No reaction}$$

Benzene sulfonyl chloride Trimethyl amine

Again, tertiary amines do not react. The sulfonamide derivative formed from the primary amine still has one NH bond left, and in the presence of excess sodium hydroxide will form a water-soluble sodium salt.

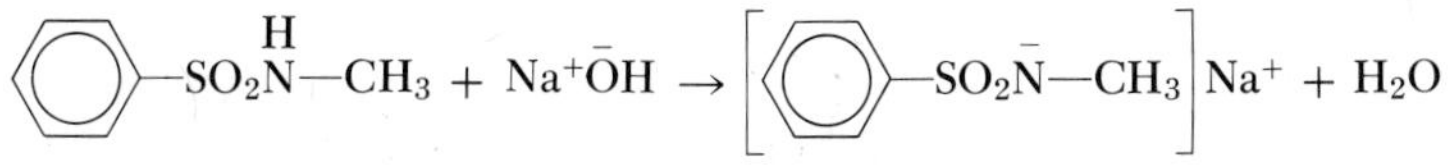

$$C_6H_5\text{—}SO_2\overset{H}{N}\text{—}CH_3 + Na^+\bar{O}H \rightarrow \left[C_6H_5\text{—}SO_2\bar{N}\text{—}CH_3\right]Na^+ + H_2O$$

Water-insoluble Water-soluble

The water-insoluble sulfonamide derivative of a secondary amine, $(CH_3)_2NH$, has no NH bond and cannot form an aqueous soluble salt. These reactions can be used to distinguish a primary, secondary, or tertiary amine, and they form the basis for the "Hinsberg test for amines."

Sulfanilamide is the parent compound of the important class of chemotherapeutic agents known as the sulfa drugs. The majority of the drugs are prepared by an acylation reaction of a sulfanilamide derivative, as shown below:

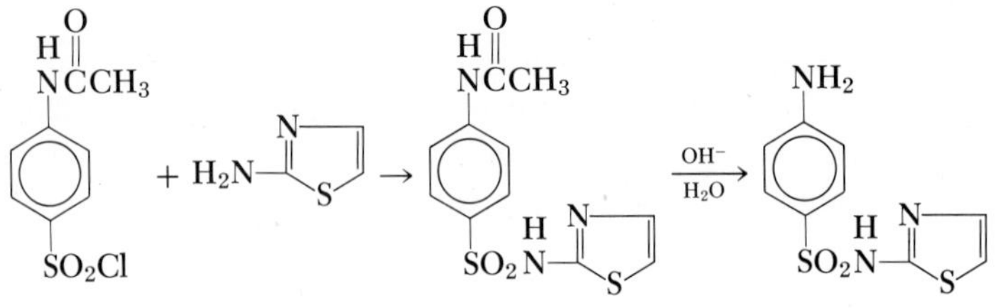

2-Aminothiazole

Sulfathiazole

The structure of the sulfa drug can be varied by varying the structure of the amine being condensed with the sulfonyl chloride (see reactions of aromatic hydrocarbons).

Aromatic Substitution Reactions

The amino group ($-NH_2$) and amino derivatives ($-N(H)C(=O)R$, and $-N(H)R$, and $-N(H)R_2$) are *ortho-para* directing groups. They are also activating groups, and compounds containing these functional groups undergo electrophilic aromatic substitution with ease. In practice, the acylated derivative is usually employed, as this group, ($-N(H)C(=O)R$), is less of an activator than the amine group itself, and the substitution reaction is more easily controlled. The free amine can be regenerated by hydrolysis of the acylated compound. This type of reaction is illustrated below for the bromination of aniline:

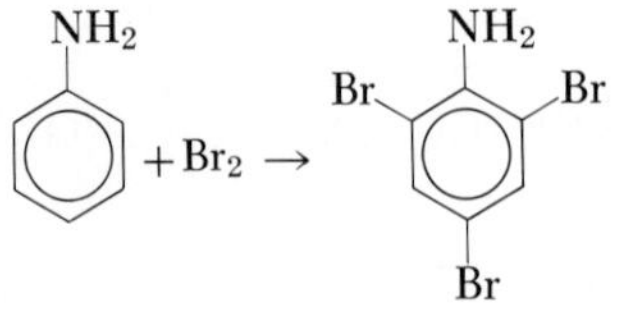

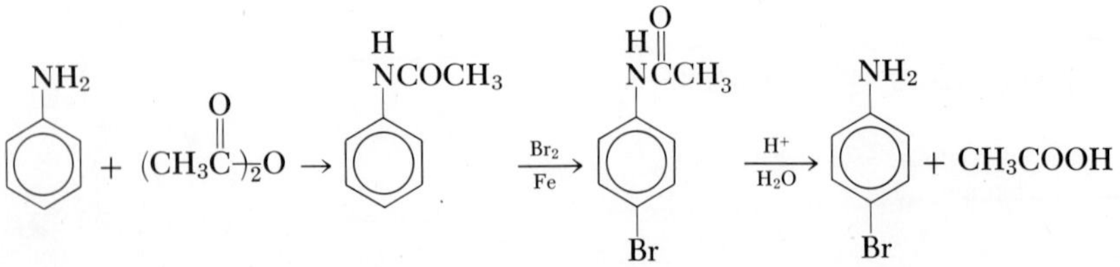

Aniline Acetic anhydride Acetanilide Main product *p*-Bromoaniline

Nitration, sulfonation, and Friedel-Crafts reactions proceed similarly.

Reaction with Nitrous Acid

Primary amines react with nitrous acid (HONO) to give a **diazonium** salt, and this type of reaction is known as **diazotization.** Since nitrous acid is not a stable acid, it is generated *in situ* from sodium nitrite and hydrochloric acid.

$$R\ddot{N}H_2 + NaNO_2 + HCl \xrightarrow{0°C} [R\overset{+}{N}{\equiv}N:]Cl^-$$

A diazonium salt

Diazonium salts are generally not isolated, as they decompose easily and are explosive in the dry state. Aliphatic diazonium salts (R = alkyl) are usually not stable even at 0°C and are not as useful as the aromatic diazonium salts in synthesis. The aromatic diazonium compounds have a reasonable stability if kept at low temperatures (0° to 5°C), and undergo a variety of displacement and coupling reactions.

The main type of displacement reaction of aromatic diazonium salts is one which involves displacement of molecular nitrogen by a nucleophile. Nucleophiles, such as CN^-, I^-, Br^-, Cl^-, H_2O, and ROH, are effective in this type of reaction. Displacement by these various types of nucleophiles is illustrated below for benzene diazonium chloride:

$$C_6H_5\ddot{N}H_2 + NaNO_2 + HCl \xrightarrow{0°C} [C_6H_5\overset{+}{N}{\equiv}N]Cl^-$$

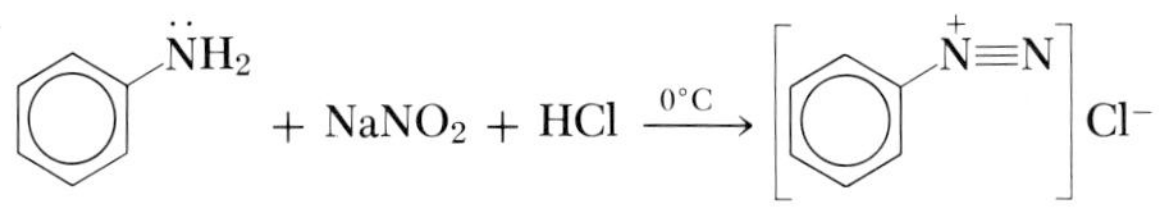

Aniline — Benzenediazonium chloride

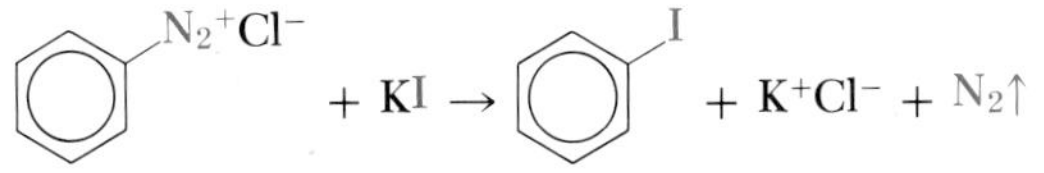

$$C_6H_5N_2^+Cl^- + KI \rightarrow C_6H_5I + K^+Cl^- + N_2\uparrow$$

Iodobenzene

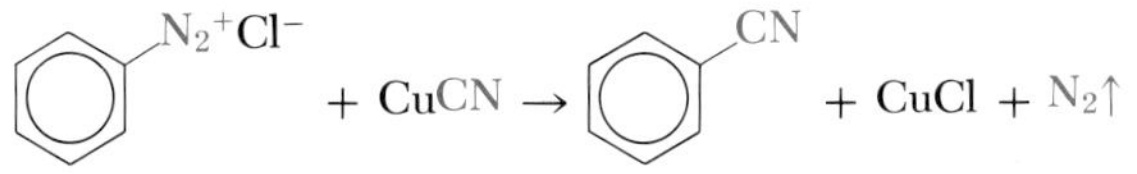

$$C_6H_5N_2^+Cl^- + CuCN \rightarrow C_6H_5CN + CuCl + N_2\uparrow$$

Benzonitrile

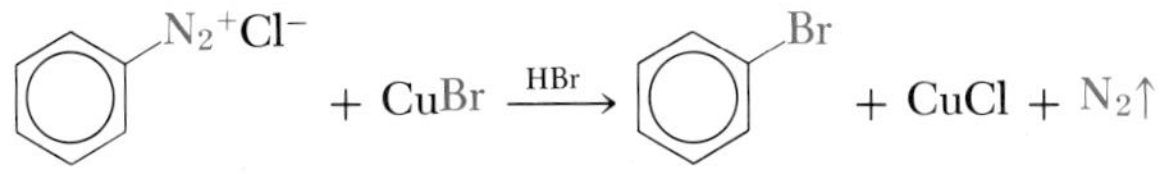

$$C_6H_5N_2^+Cl^- + CuBr \xrightarrow{HBr} C_6H_5Br + CuCl + N_2\uparrow$$

Bromobenzene

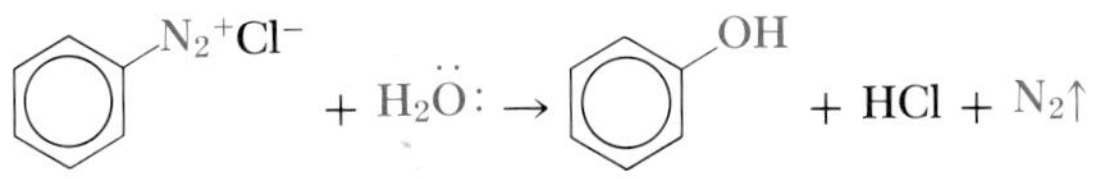

$$C_6H_5N_2^+Cl^- + H_2\ddot{O}: \rightarrow C_6H_5OH + HCl + N_2\uparrow$$

Phenol

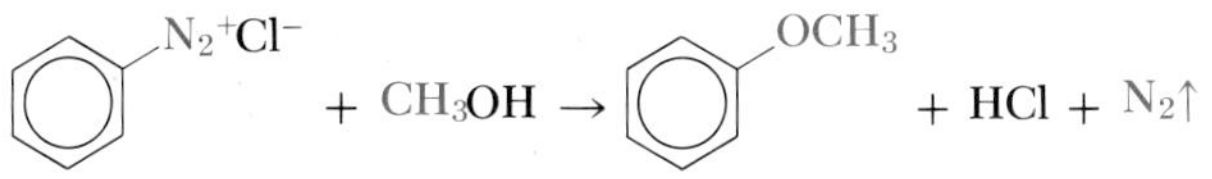

$$C_6H_5N_2^+Cl^- + CH_3OH \rightarrow C_6H_5OCH_3 + HCl + N_2\uparrow$$

Anisole

$$C_6H_5N_2^+Cl^- + CuCl \rightarrow C_6H_5Cl + CuCl + N_2\uparrow$$

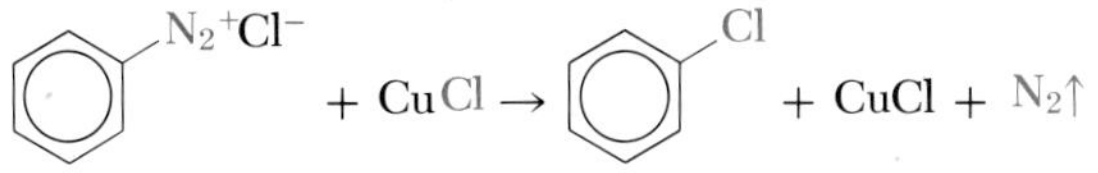

Chlorobenzene

This type of displacement reaction is particularly useful for introducing cyano and iodo groups into an aromatic ring, as these functional groups cannot usually be introduced by nucleophilic displacement or electrophilic substitution reactions. This type of process is illustrated on the next page for the preparation of *ortho*-iodotoluene from *ortho*-toluidine.

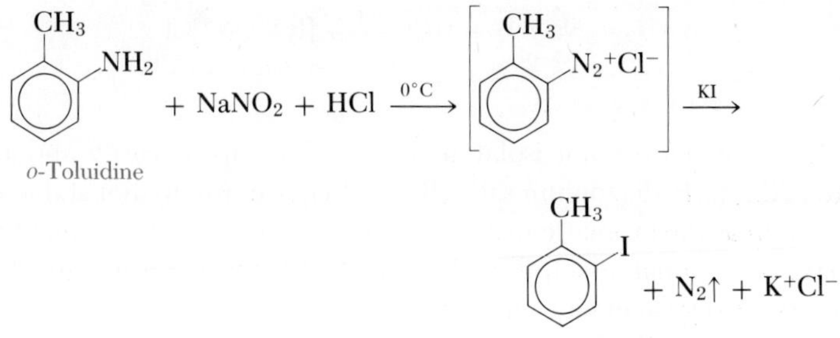

$$o\text{-Toluidine} + NaNO_2 + HCl \xrightarrow{0°C} [\text{o-}CH_3C_6H_4N_2^+Cl^-] \xrightarrow{KI} o\text{-}CH_3C_6H_4I + N_2\uparrow + K^+Cl^-$$

o-Toluidine

o-Iodotoluene

The diazo group can also be replaced by hydrogen (—H) if the diazonium salt is heated with hypophosphorous acid, as shown below:

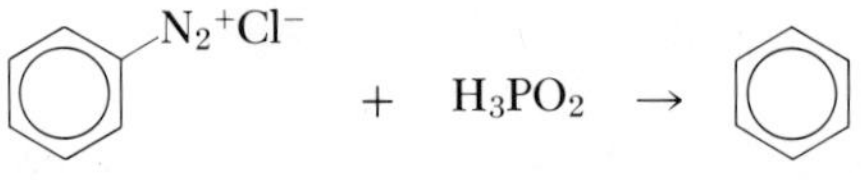

$$C_6H_5N_2^+Cl^- + H_3PO_2 \rightarrow C_6H_6$$

Benzenediazonium chloride Hypophosphorous acid Benzene

The removal of the amino group (and hence the nitro group from which it was formed) by this type of reaction can be used advantageously in synthesis. For example, 1,3,5-tribromobenzene can be prepared by the following sequence of reactions:

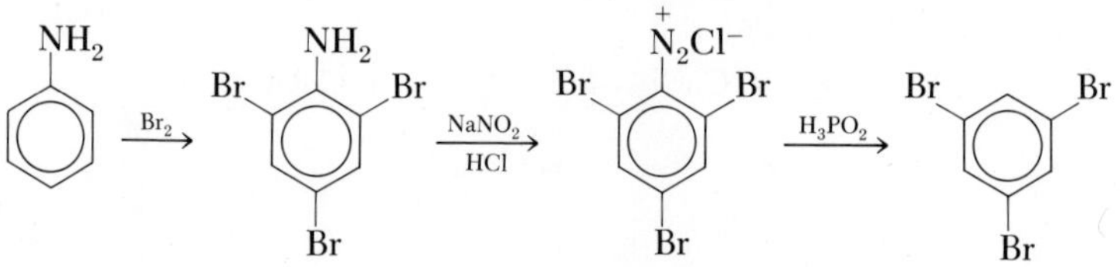

$$C_6H_5NH_2 \xrightarrow{Br_2} 2,4,6\text{-}Br_3C_6H_2NH_2 \xrightarrow[HCl]{NaNO_2} 2,4,6\text{-}Br_3C_6H_2\overset{+}{N_2}Cl^- \xrightarrow{H_3PO_2} 1,3,5\text{-}Br_3C_6H_3$$

Note that the bromines in the final product are situated **meta** to each other. If one had attempted to brominate benzene to obtain this compound, the wrong isomer would have been obtained, since bromine is an **ortho-para** directing group. Hence, in many syntheses, the amino group is used to orient the substituents into the desired positions on the aromatic ring, and when this has been accomplished, the amino group is then removed.

Another important reaction of diazonium salts is their **coupling** reaction in alkaline or neutral solution with reactive aromatic compounds such as phenols or aromatic amines. The coupling takes place between the diazonium salt and the *para* position of the aromatic compound, and the nitrogen is retained. The (—N=N—) linkage is known as the **azo** group, and azo compounds of this type are generally colored. Related compounds containing the azo group make up the important group of dyestuffs known as azo dyes.

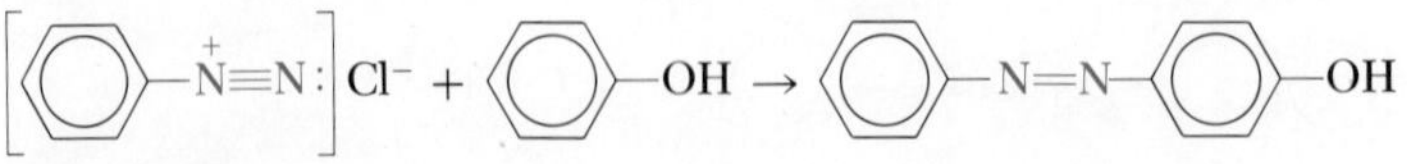

$$[C_6H_5\overset{+}{N}\equiv N:]Cl^- + C_6H_5OH \rightarrow C_6H_5N{=}NC_6H_4OH$$

Benzene diazonium chloride Phenol *p*-Hydroxyazobenzene

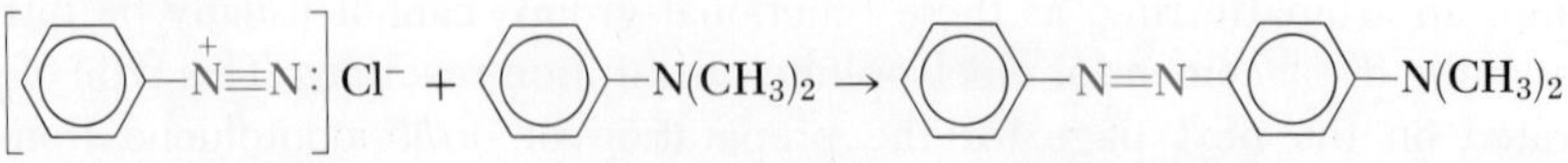

$$[C_6H_5\overset{+}{N}\equiv N:]Cl^- + C_6H_5N(CH_3)_2 \rightarrow C_6H_5N{=}NC_6H_4N(CH_3)_2$$

N,N-Dimethyl aniline *p*-Dimethylaminoazobenzene

IMPORTANT AMINES, AMINE DERIVATIVES AND THEIR USES

Dimethylamine

When allowed to react with nitrous acid, followed by reduction of the product by hydrogen, dimethyl hydrazine, $(CH_3)_2NNH_2$, is formed. Hydrazines of this type have been used as rocket propellants.

Di- and trimethyl amines are essential in the preparation of quaternary ammonium types of anion exchange resins. In general, amines are utilized in the preparation of dyes, drugs, herbicides, fungicides, soaps, disinfectants, insecticides, and photographic developers, and as choline chloride for use in animal feeds.

Among the aromatic amines, aniline is used to synthesize other important organic compounds used as dyes and dye intermediates, antioxidants, and drugs. Other derivatives of aromatic amines, such as *p*-aminosulfonic acid and *p*-toluidine, are used as dye intermediates.

Several derivatives of the aromatic amines have medicinal properties and are used as drugs. **Acetanilide** was used for many years as an antipyretic and analgesic drug. The toxicity of this compound in the body resulted in a search for similar compounds that were less toxic. **Phenacetin,** or **acetophenetidin,** which is closely related to acetanilide, was found to be less toxic and to possess the beneficial effects of acetanilide. Sulfanilamide and other sulfa drugs have been noted earlier for their therapeutic effects.

Another type of important organic compounds that contains amino groups is **amino acids.** Amino acids may be considered as organic acids containing an amino group. Two simple amino acids are illustrated below.

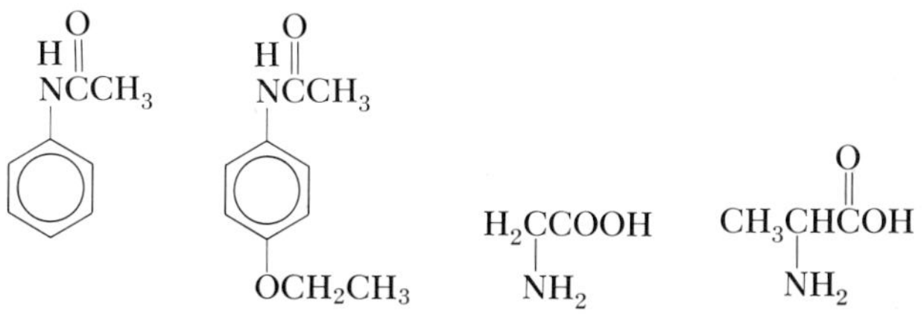

$C_6H_5NHCOCH_3$	$CH_3CH_2OC_6H_4NHCOCH_3$	$H_2C(NH_2)COOH$	$CH_3CH(NH_2)COOH$
Acetanilide	Phenacetin	Glycine	Alanine

Amino acids are the fundamental units in the protein molecule and will be discussed later in detail under the chemistry of proteins.

IMPORTANT TERMS AND CONCEPTS

acetylcholine
amino group
aniline
azo group
coupling reaction
diazonium salt
diazotization
Hinsberg test for amines
phenacetin
primary amine
quaternary ammonium salt
secondary amine
sulfanilamide
tertiary amine

QUESTIONS

1. Draw a structural formula for each of the following compounds:
   (a) diethylamine
   (b) tri-*n*-propyl amine
   (c) aniline
   (d) tetraethylammonium hydroxide
   (e) *m*-nitroaniline
   (f) isohexyl amine
   (g) benzyl amine
   (h) *o*-toluidine
   (i) cyclobutyl amine
   (j) *p*-bromo-N,N-dimethylaniline

2. Give each of the following compounds an appropriate name:

(a) 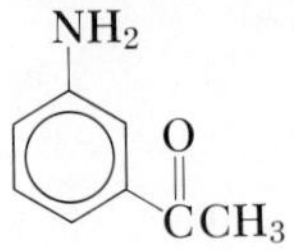 benzene ring with NH_2 and $\overset{O}{\overset{\|}{C}}CH_3$ (meta)

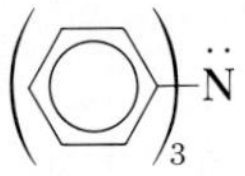

(b) $CH_3CH_2\overset{H}{N}CH_2CH_2CH_3$

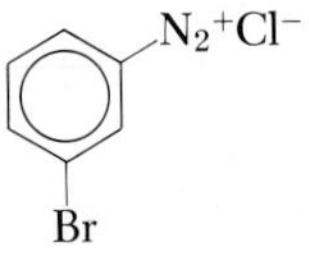

(c) 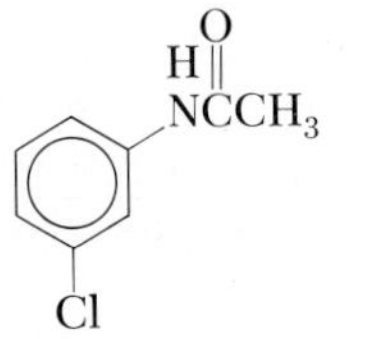 benzene ring with $\overset{H}{N}\overset{O}{\overset{\|}{C}}CH_3$ and Cl (meta)

(d) $(C_6H_5)_3\ddot{N}$

(e) benzene ring with $N_2^+Cl^-$ and Br (meta)

3. The fishy odor of a solution of dimethylamine in water is lost when an equimolar amount of hydrochloric acid is added. *Explain.*

4. Write equations for the reactions of *n*-butyl amine with each of the following reagents:
   (a) acetyl chloride
   (b) HI
   (c) benzene sulfonyl chloride/OH^-
   (d) ethylene oxide
   (e) excess CH_3I followed by Ag_2O

5. Write equations for the reactions of *p*-bromoaniline with the following reagents:
   (a) H_2SO_4
   (b) acetic anhydride
   (c) $NaNO_2/HCl$
   (d) $NaNO_2/HCl$ followed by N,N-dimethylaniline
   (e) benzene sulfonyl chloride/OH^-

6. Compare the structural features of novacaine (p. 178) with those of acetylcholine. Suggest an explanation of its use as a local anesthetic.

7. Write equations for the preparation of the following compounds. Give any necessary catalysts. More than one step may be necessary.
   (a) aniline hydroiodide from nitrobenzene
   (b) acetanilide from nitrobenzene
   (c) *n*-butyl amine from *n*-butyric acid
   (d) *p*-bromobenzonitrile from aniline
   (e) 1,6-hexanediamine from 1,4-butanediol
   (f) *p*-methylaniline from benzene
   (g) *n*-butyl amine from *n*-pentanoic acid
   (h) *p*-nitroaniline from benzene
   (i) *p*-fluorobromobenzene from aniline
   (j) 1,3,5-tribromobenzene from aniline

8. Devise a simple chemical test that would enable you to distinguish one compound from the other in each of the following:
   (a) $CH_3(CH_2)_4CH_2NH_2$ and $CH_3(CH_2)_3CH_2NHCH_3$
   (b) $(CH_3)_3N$ and $CH_3CH_2NHCH_3$

9. Outline a diagnostic test (with equations) that could be used to distinguish between *n*-propyl amine, di-*n*-propyl amine, and tri-*n*-propyl amine.

10. Arrange the following compounds in order of decreasing base strength:

NH_3, CH_3NH_2, $C_6H_5NH_2$ (aniline), CH_3COOH, CH_3CH_2OH, H_2O, p-$NO_2C_6H_4NH_2$ (benzene ring with NH_2 and NO_2 para)

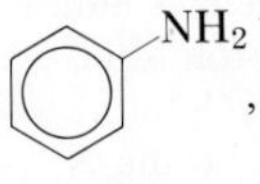

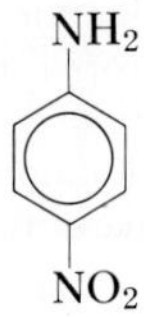

SUGGESTED READING

Amundsen: Sulfanilamide and Related Chemotherapeutic Agents. Journal of Chemical Education, Vol. 19, p. 167, 1942.

Barron, Jarvik, and Sterling: Hallucinogenic Drugs. Scientific American, Vol. 210, No. 4, p. 29, 1964.

Plummer and Yorkman: Antihypertensive and Diuretic Agents. Past, Present, and Future. Journal of Chemical Education, Vol. 37, p. 179, 1960.

Ray: Alkaloids. The World's Pain Killers. Journal of Chemical Education, Vol. 37, p. 451, 1960.

POLYMERS AND POLYMERIZATION REACTIONS

CHAPTER 12

The *objectives* of this chapter are to enable the student to:

1. Define a polymer or macromolecule.
2. Define and recognize common monomer units.
3. Distinguish the difference between condensation and addition polymers.
4. Deduce that polymerization reactions represent the same reactions studied from a monofunctional viewpoint, but that the degree of reaction is different.
5. Recognize the common terms polyester, nylon, Bakelite, polypropylene, polystyrene, polyethylene, Teflon, and polyacrylate and to associate these terms with a repeating structural unit.
6. Explain that a polymer may have different configurations and that these configurations influence the properties of the polymer.

A polymer (or macromolecule) is a large molecule of high molecular weight (from a few thousand to several million) that is composed of small components called monomers. Many macromolecules occur in nature, such as proteins, nucleic acids, polysaccharides, starch, and cellulose, and these polymers will be discussed in more detail in the biochemistry section of the text. The main concern of this chapter will be man-made (synthetic) polymers that have achieved widespread commercial use.

The monomers used in the preparation of polymers are usually fairly simple monofunctional or bifunctional compounds that have previously been encountered in this portion of the text. In most cases the reactions which yield polymers can be readily understood from the material which has been studied by students thus far in organic chemistry. The main difference between the organic reactions which have previously been studied and polymer-forming reactions is in the degree of reaction. Most of the reactions studied thus far have involved a reaction between two reactants to yield a relatively simple product. Polymerization reactions, however, are reactions which are functionally capable of proceeding indefinitely and which could in theory give a compound of infinite molecular weight. Consequently, in a polymerization reaction the initially formed intermediate or product is capable of further reaction and simple 1:1

products do not result. For example, in the free-radical polymerization of ethylene, shown below, the intermediate radical can add to a second molecule of ethylene to form a new radical, which in turn can add to a third molecule to form another radical, and this process can proceed until a reaction (called a *termination reaction*) occurs to interrupt this chain process.

$$\underset{\text{Radical}}{R\cdot} + \underset{\text{Ethylene}}{CH_2{=}CH_2} \rightarrow RCH_2CH_2\cdot \xrightarrow{CH_2=CH_2} RCH_2CH_2CH_2CH_2\cdot$$

$$\downarrow nCH_2{=}CH_2$$

$$\underset{\text{Polyethylene}}{R(CH_2CH_2)_{n+2}R} \xleftarrow[\text{termination step}]{R\cdot} R[CH_2CH_2]_{n+1}CH_2CH_2\cdot$$

Note that the ethylene unit is repeated numerous times and the molecular weight of the polymer depends upon the value of n. In most polymerizations, reaction conditions are adjusted to make n large so that high-molecular weight materials are obtained. Polymers which contain the same repeating unit, such as polyethylene, are called **homopolymers.** If two different monomers are used to form the polymer, a **copolymer** is obtained. For example, butadiene and styrene can be copolymerized to form a synthetic rubber used in automobile tires. Molecular weights of 25,000 to 500,000 can be attained by this type of copolymerization. If the two monomers are fed into the polymerization

$$\underset{\text{1,3-Butadiene}}{CH_2{=}CH{-}CH{=}CH_2} + \underset{\text{Styrene}}{C_6H_5CH{=}CH_2} \xrightarrow{R\cdot} \underset{\text{SBR rubber}}{R{+}CH_2{-}\overset{H}{\underset{H}{C}}{=}\overset{H}{C}{-}CH_2\underset{C_6H_5}{CH}CH_2{+}_nR}$$

reaction without any special precautions, a **random copolymer** is obtained in which no definite sequential pattern of monomer units is found. Special reactions, however, can be used to put large blocks of units of the same monomer back to back in the repeating chain. Copolymers of this type are called **block polymers** and have different physical properties than random copolymers.

Polymerization reactions can be conveniently divided into two types, namely, **condensation polymers** and **addition polymers.** In a condensation polymerization reaction two molecules are joined (condensed) together and a small molecule, such as water or an alcohol, is removed or eliminated in this reaction. In order for a condensation polymerization to form high-molecular weight materials, the condensation reaction must occur over and over again. Consequently, the monomers employed in a condensation polymerization must have two or more functional groups which can enter into the reaction responsible for building the polymer chain. Most of the chemical reactions involved in a condensation polymerization are well known reactions, such as esterification and amide formation, which have been encountered earlier in this treatment of organic chemistry.

An addition polymer has the same ratio of elements in the polymer repeating unit as was present in the monomer, since the monomeric units are merely added together to form the polymer, as shown earlier in the formation of polyethylene. Until recently, most addition polymers were generally formed from monomers which contained carbon-carbon double bonds. However, addition polymerizations involving other types of unsaturation, such as carbon-oxygen double bonds or carbon-nitrogen double bonds, are currently being investigated by polymer chemists and show promise of yielding useful polymers.

Another distinguishing feature between condensation and addition polymers is the type of reaction mechanism which produces them. Generally, condensation polymers involve a stepwise reaction which consumes the monomer, but the molecular weight buildup is slow or is accomplished in a second step. On the other hand, addition polymers occur by a chain mechanism in which consumption of the monomer is slow, but rapid formation of high-molecular weight polymers is attained. These differences will become more evident as representative examples of these two types of polymers are considered.

CONDENSATION POLYMERS

The type of bifunctional compounds which are condensed together to form a condensation polymer will determine the functional group linkage in the repeating unit of the polymer. Thus, if an acid and an alcohol are employed, a polyester will result. If an acid and an amine are condensed, a polyamide results, as shown below. Note that a molecule of water is eliminated in the condensation reaction.

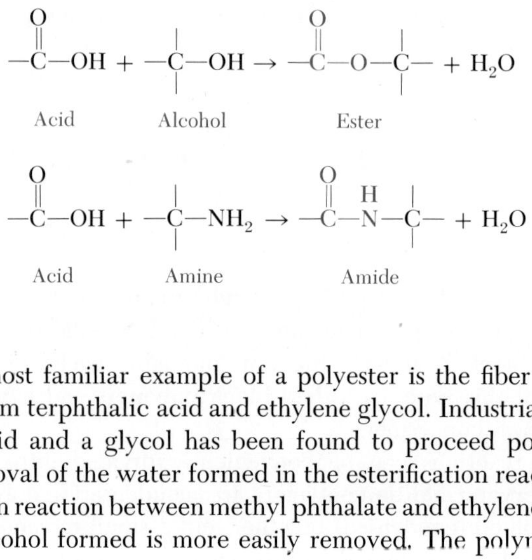

$$\underset{\text{Acid}}{-\overset{\overset{O}{\|}}{C}-OH} + \underset{\text{Alcohol}}{-\overset{|}{\underset{|}{C}}-OH} \rightarrow \underset{\text{Ester}}{-\overset{\overset{O}{\|}}{C}-O-\overset{|}{\underset{|}{C}}-} + H_2O$$

$$\underset{\text{Acid}}{-\overset{\overset{O}{\|}}{C}-OH} + \underset{\text{Amine}}{-\overset{|}{\underset{|}{C}}-NH_2} \rightarrow \underset{\text{Amide}}{-\overset{\overset{O}{\|}}{C}-\overset{H}{N}-\overset{|}{\underset{|}{C}}-} + H_2O$$

Probably the most familiar example of a polyester is the fiber **Dacron,** which is a polyester formed from terphthalic acid and ethylene glycol. Industrially, direct esterification between an acid and a glycol has been found to proceed poorly because of the difficulty in the removal of the water formed in the esterification reaction. Consequently, the transesterification reaction between methyl phthalate and ethylene glycol is employed, since the methyl alcohol formed is more easily removed. The polymerization is carried

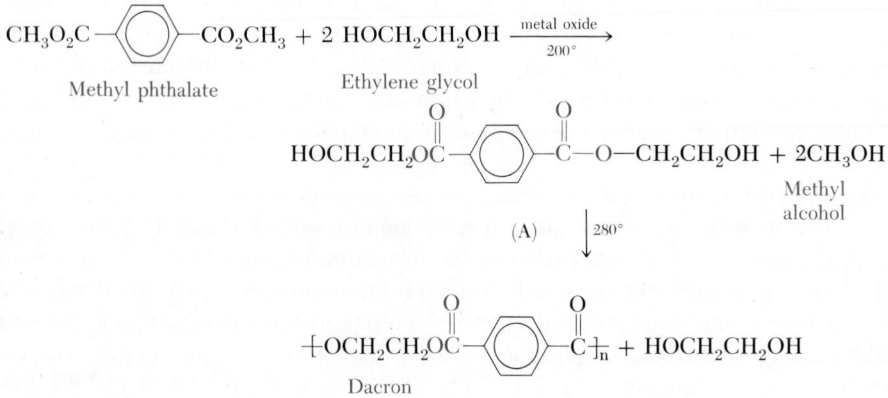

$$\underset{\text{Methyl phthalate}}{CH_3O_2C-C_6H_4-CO_2CH_3} + 2\ \underset{\text{Ethylene glycol}}{HOCH_2CH_2OH} \xrightarrow[200^\circ]{\text{metal oxide}}$$

$$\underset{\text{(A)}}{HOCH_2CH_2O\overset{\overset{O}{\|}}{C}-C_6H_4-\overset{\overset{O}{\|}}{C}-O-CH_2CH_2OH} + \underset{\text{Methyl alcohol}}{2CH_3OH}$$

$$\downarrow 280^\circ$$

$$\underset{\text{Dacron}}{\left[OCH_2CH_2O\overset{\overset{O}{\|}}{C}-C_6H_4-\overset{\overset{O}{\|}}{C}\right]_n} + HOCH_2CH_2OH$$

out in several steps. First, ester interchange is carried out at 200°C to give methyl alcohol (which is removed by distillation) and a new monomer (A). After removal of the methanol is completed, the temperature is raised to 280°C, and polymerization occurs to give Dacron and ethylene glycol which is also removed by distillation. Dacron fiber, also called Terylene or Teron, can be set in permanent creases and finds widespread commercial use in combination with wool in wash-and-wear fabrics. When cast into a film (Mylar, Cronar), a product of high tensile strength is obtained and Mylar is currently used extensively in making magnetic recording tapes.

The best known type of polyamides are the nylons, which were developed as a substitute for silk during World War II. The nylons are generally prepared by the copolymerization of a dibasic acid and a diamine. The different types of nylons are distinguished by the use of numbers. These numbers refer to the number of carbon atoms in the diamine and in the dibasic acid. For example, **Nylon 66** is prepared by a two-stage process from hexamethylenediamine (six carbons) and adipic acid (six carbons). When stoichiometric amounts of these two monomers are allowed to react, a 1:1 salt is formed. Further treatment of this salt at 270°C and 250 psi yields Nylon 66.

$$\underset{\text{Hexamethylenediamine}}{H_2N(CH_2)_6NH_2} + \underset{\text{Adipic acid}}{HO\overset{O}{\overset{\|}{C}}(CH_2)_4\overset{O}{\overset{\|}{C}}OH} \longrightarrow H_2N(CH_2)_6\overset{+}{N}H_3\overset{-}{O}-\overset{O}{\overset{\|}{C}}(CH_2)_4\overset{O}{\overset{\|}{C}}OH$$

$$\xrightarrow{\text{heat, pressure}} \underset{\text{Nylon 66}}{\left[\overset{H}{N}(CH_2)_6\overset{H}{N}-\overset{O}{\overset{\|}{C}}(CH_2)_4\overset{O}{\overset{\|}{C}}\right]_n} + nH_2O$$

This particular nylon has high strength because of its ability to form intermolecular hydrogen bonds. Consequently, it shows good resistance to breakage by stretching or abrasion and finds widespread use in carpet fabrics and machine parts, such as gear assemblies, which require no lubrication.

Other nylons, such as Nylon 610 from hexamethylenediamine and sebacic acid and Nylon 6 from ε-caprolactam, are also commercially useful.

$$\underset{\varepsilon\text{-Caprolactam}}{\text{(cyclic } \overset{O}{\overset{\|}{C}}-NH\text{)}} + H_2O \longrightarrow H_3\overset{+}{N}(CH_2)_5\overset{O}{\overset{\|}{C}}-O^- \xrightarrow[\text{pressure}]{\text{heat}} \underset{\text{Nylon 6}}{\left[(CH_2)_5\overset{O}{\overset{\|}{C}}-\overset{H}{\overset{|}{N}}\right]_n}$$

CROSS-LINKED POLYMERS

Simple polyesters and polyamides are long linear molecules without linkages interconnecting the individual strands of the polymer. If the individual strands can be connected so that a network of polymers results, this process is called *cross-linking* and the resultant polymers are called *cross-linked polymers*. For example, if a dibasic acid is copolymerized with a trihydroxylalcohol, such as glycerol, the third hydroxyl function is available for forming a connection between the individual strands of the polymers and crosslinking can occur. Polyester polymers with cross-links are called **alkyd resins.** For example, the alkyd resin **Glyptal** is commercially prepared from glycerol and phthalic anhydride. Generally, cross-linked polymers such as Glyptal are rigid, insoluble materials that do not flow nor soften when heated. When heated to their melting point, polymers of this type, called **thermosetting polymers,** undergo a permanent change and

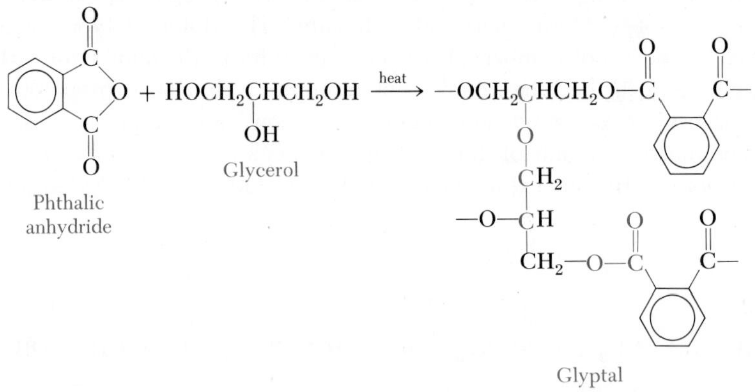

Phthalic anhydride + $HOCH_2CHCH_2OH$ (with OH on the central carbon; Glycerol) $\xrightarrow{\text{heat}}$ Glyptal

Phthalic anhydride

Glycerol

Glyptal

set to a solid which cannot be remelted. In contrast to this behavior, **thermoplastic polymers** soften and flow when heated and can be remelted many times without change.

The resin **Bakelite** is a copolymer of phenol and formaldehyde. The resultant thermosetting polymers find extensive use in heat and electrical resistant materials such as electric appliance handles and electrical switches.

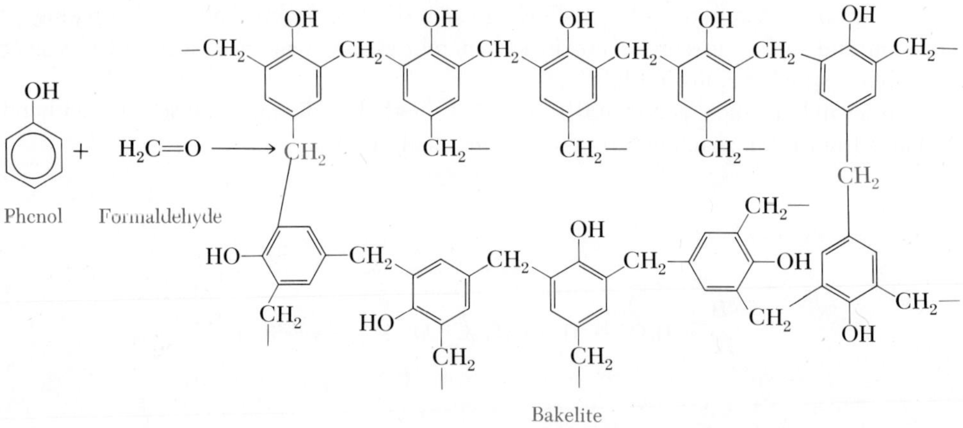

Phenol + $H_2C{=}O$ (Formaldehyde) $\longrightarrow$ Bakelite

Phenol

Formaldehyde

Bakelite

ADDITION POLYMERS

Addition polymers are formed by some sort of a chain mechanism. These polymerizations may occur by anionic, cationic, or free-radical mechanisms depending on the type of monomer employed. Each of these mechanistic classes are characterized by initiation, propagation, and termination steps. For example, in the free-radical (peroxide) polymerization of an olefinic compound these steps may be outlined as shown on the next page.

Initiation (Formation of the reactive intermediate, in this case a free-radical)

$$\underset{\text{Benzoyl peroxide}}{C_6H_5\overset{O}{\overset{\|}{C}}-O-O-\overset{O}{\overset{\|}{C}}-C_6H_5} \xrightarrow{\text{heat}} \underset{\text{Benzoyl radical}}{2\ C_6H_5\overset{O}{\overset{\|}{C}}-O\cdot} \rightarrow \underset{\text{Phenyl radical}}{2\ C_6H_5\cdot} + 2\ CO_2$$

$$C_6H_5\cdot + CH_2{=}CH_2 \rightarrow C_6H_5CH_2CH_2\cdot$$

The initiator radical (phenyl radical in this example) adds to the unsaturated monomer in the initiation step to generate the monomer free-radical.

Propagation

$$C_6H_5CH_2CH_2\cdot + CH_2{=}CH_2 \rightarrow C_6H_5CH_2CH_2CH_2CH_2\cdot$$

$$\downarrow n\ CH_2{=}CH_2$$

$$C_6H_5(CH_2CH_2)_{n+1}CH_2CH_2\cdot$$

In the propagation step consecutive addition of the monomer occurs to build up the growing chain. The value of n determines the molecular weight of the polymer.

Termination

Termination interrupts the growing chain and stops the polymerization reaction. Coupling of two free-radicals may result in termination of the chain. Chain transfers (*via* hydrogen abstraction) may also terminate the chain reaction. In fact, chain transfer agents are commonly added to regulate the molecular weight (the value of n) of the polymer.

Coupling

$$2\ C_6H_5(CH_2CH_2)_nCH_2CH_2\cdot \rightarrow 2\ C_6H_5(CH_2CH_2)_nCH_2CH_2CH_2CH_2(CH_2CH_2)_nC_6H_5$$

Chain-Transfer

$$C_6H_5(CH_2CH_2)_nCH_2CH_2\cdot + RSH \rightarrow C_6H_5(CH_2CH_2)_nCH_2CH_3 + RS\cdot$$

These three types of steps are present in all chain type reactions. Although in most cases the actual mechanisms may appear to be more complex than outlined previously, they merely involve either more complex steps to generate the initiator or more variety of termination steps.

Polypropylene is an addition polymer obtained by polymerization of propylene with a trialkylaluminum titanium tetrachloride (Ziegler-Natta) catalyst. This highly ordered polymer has a high melting point (165 to 170°C) and high tensile strength and is easily converted to molded objects and fibers.

$$\underset{\text{Propylene}}{CH_3CH{=}CH_2} \xrightarrow[TiCl_4]{R_3Al} \underset{\text{Polypropylene}}{\left[\overset{CH_3}{\overset{|}{C}H}-CH_2\right]_n}$$

Another important and well known addition polymer is polytetrafluoroethylene (**Teflon**). Since pure tetrafluoroethylene will polymerize explosively under free-radical conditions, special catalyst systems have been developed to control this polymerization. Teflon has excellent thermal stability over a wide temperature range, is chemically

$$\underset{\text{Tetrafluoroethylene}}{nCF_2{=}CF_2} \xrightarrow{\text{catalyst}} \underset{\text{Teflon}}{+\!CF_2{-}CF_2\!+_n}$$

unreactive, and shows excellent low-friction properties. It is used in electrical insulators, in valve packings and gaskets in which chemical inertness is required, and in a wide variety of antistick applications, such as nonlubricated bearings and nonstick frying pans.

Another familiar and important addition polymer is *poly(vinyl chloride)* which is prepared by the polymerization of vinyl chloride. Poly(vinyl chloride), PVC, is used in sewer pipes to replace cast-iron pipes, in a variety of transparent items such as plastic raincoats, in phonograph records, and in insulation for electric wire.

$$\underset{\text{Vinyl chloride}}{nCH_2{=}CHCl} \xrightarrow{\text{catalyst}} \underset{\text{PVC}}{+\!CH_2CHCl\!+_n}$$

When vinyl chloride is copolymerized with vinylidine chloride, the polymers used in *Saran Wrap* are obtained. This copolymer has low moisture transmission and forms a tough film which is particularly useful in food packaging.

$$\underset{\text{Vinyl chloride}}{CH_2{=}CHCl} + \underset{\text{Vinylidene chloride}}{CH_2{=}CCl_2} \xrightarrow{\text{catalyst}} \underset{\text{Saran wrap}}{+\!CH_2CCl_2CH_2CHCl\!+_n}$$

Other important addition polymers are *polystyrene*, which is used in the form of a foam (Styrofoam) as a heat-insulating material and in many novelty items, *poly (methyl methacrylate)*, which is used as a clear transparent plastic (Plexiglass, Lucite) in many

$$\underset{\text{Styrene}}{C_6H_5CH{=}CH_2} \xrightarrow{\text{catalyst}} \underset{\text{Polystyrene}}{+\!\underset{C_6H_5}{\underset{|}{C}}HCH_2\underset{C_6H_5}{\underset{|}{C}}HCH_2\!+_n}$$

$$\underset{\text{Methyl methacrylate}}{CH_2{=}C(CH_3)CO_2CH_3} \xrightarrow{\text{catalyst}} \underset{\text{Poly(methyl methacrylate)}}{+\!CH_2\overset{CH_3}{\overset{|}{\underset{CO_2CH_3}{\underset{|}{C}}}}\!+_n}$$

automobile accessories, and *polyacrylonitrile*, which is the major constituent of acrylic fibers (Orlon, Acrilan).

$$\underset{\text{Acrylonitrile}}{CH_2{=}CHCN} \xrightarrow{\text{catalyst}} \underset{\text{Polyacrylonitrile}}{+\!CH_2\underset{CN}{\underset{|}{C}}H\!+_n}$$

CONFIGURATION OF POLYMERS

When a substituted vinyl monomer ($CH_2{=}CHR$) is polymerized, three possible configurations of the R-group (in relation to the backbone of the polymer) are possible as shown below.

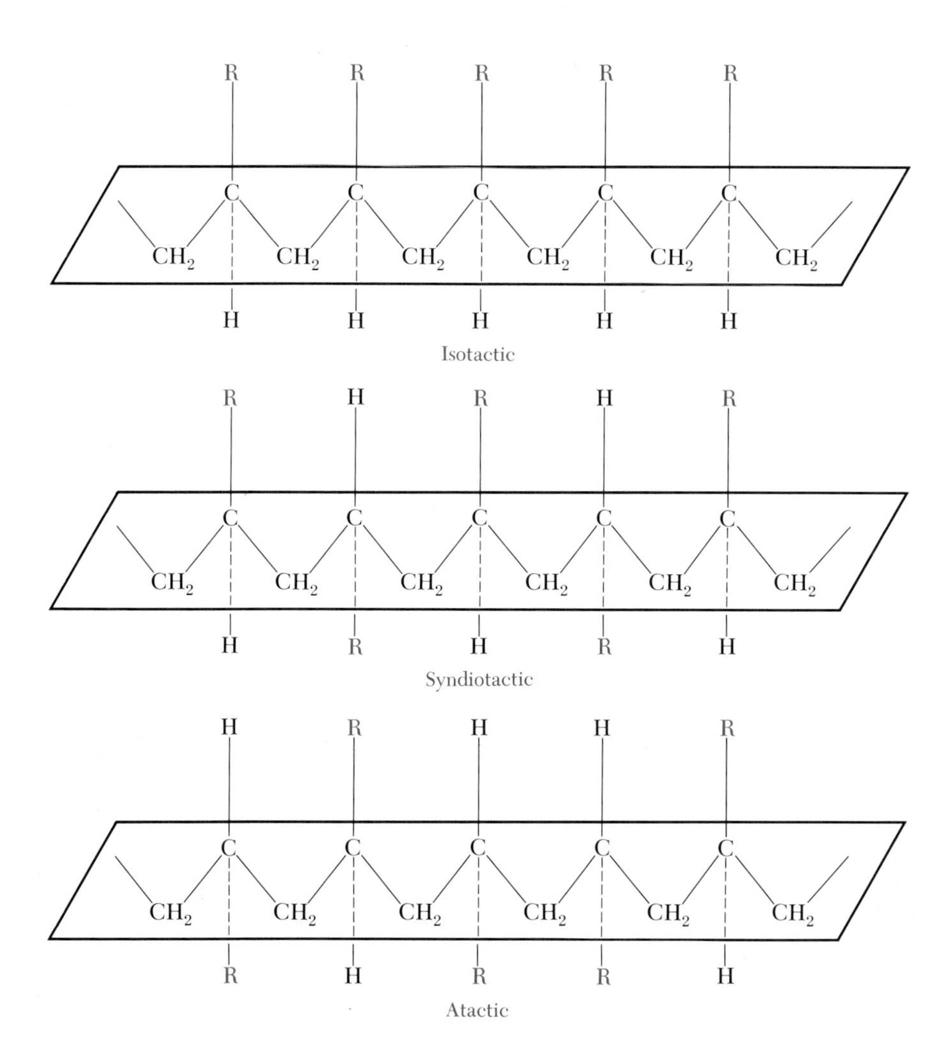

When all the "R's" are above (or below) the plane of the backbone of the polymer chain, the polymer is called **isotactic.** When the R's alternate above and below the backbone, the polymer is called **syndiotactic.** Where there is a random arrangement of the R groups, the polymer is called **atactic.** Most free-radical polymerizations yield atactic polymers, whereas most ionic polymerizations yield either isotactic or syndiotactic polymers. In general, isotactic and syndiotactic polymers are crystalline and exhibit good mechanical properties. In contrast, atactic polymers are generally noncrystalline and show weak mechanical properties. It wasn't until the discovery of the Ziegler-Natta catalyst system that polymer chemists were able to control the configuration of the resultant polymers and hence the properties of the polymers. With the Ziegler-Natta catalyst either isotactic or syndiotactic polymers can be produced depending on the exact type of catalyst use. For this important discovery, Ziegler and Natta were awarded the Nobel Prize.

IMPORTANT TERMS AND CONCEPTS

addition polymer
atactic polymer
condensation polymer
copolymer
cross-linking
homopolymer
isotactic polymer
macromolecule
monomer
Orlon
polyamide
polyethylene
polymer
polymerization reaction
polypropylene
Styrofoam
syndiotactic polymer
thermoplastic polymer
thermosetting polymer

QUESTIONS

1. How does an addition polymer differ from a condensation polymer?

2. What functional group(s) is contained in *each* of the following polymers?
   (a) Nylon 6
   (b) Dacron
   (c) Orlon
   (d) Glyptal
   (e) Plexiglass

3. Kel-F is an addition polymer of chlorotrifluoroethylene. Draw the repeating unit of this polymer.

4. Outline the preparation of Nylon 610.

5. What is the repeating unit in *each* of the following polymers?
   (a) Nylon 66
   (b) Dacron
   (c) Polystyrene
   (d) Polyacrylonitrile
   (e) Lucite
   (f) Teflon
   (g) Orlon
   (h) Saran Wrap
   (i) PVC
   (j) Polypropylene

SUGGESTED READING

Bruck: Thermally Stable Polymeric Materials. Journal of Chemical Education, Vol. 42, p. 18, 1965.
Fisher: New Horizons in Elastic Polymers. Journal of Chemical Education, Vol. 37, p. 369, 1960.
Heckert: Synthetic Fibers. Journal of Chemical Education, Vol. 30, p. 166, 1953.
Mayo: Contribution of Vinyl Polymerization to Organic Chemistry. Journal of Chemical Education, Vol. 36, p. 157, 1959.
McGrew: Structure of Synthetic High Polymers. Journal of Chemical Education, Vol. 35, p. 178, 1958.
Moncrieff: Linear Polymerization and Synthetic Fibers. Journal of Chemical Education, Vol. 31, p. 233, 1954.
Mondano: Plastics in the Electronic Industries. Journal of Chemical Education, Vol. 31, p. 383, 1954.
Price: The Effect of Structure on Chemical and Physical Properties of Polymers. Journal of Chemical Education, Vol. 42, p. 13, 1965.

CARBOHYDRATES

CHAPTER 13

The *objectives* of this chapter are to enable the student to:

1. Distinguish between the d and l forms of an optically active compound.
2. Describe the structure of α-D(+) glucose and its relation to α-D-glucopyranose.
3. Recognize the α and β isomers of the pyranose and furanose forms of sugars.
4. Describe the glycoside or acetal linkage between two monosaccharides.
5. Write the formula for sucrose and explain why it does not reduce Benedict's solution.
6. Explain why ascorbic acid is a good reducing agent.
7. Illustrate the structure of a polysaccharide using several molecules of glucose.
8. Recognize the difference between the structures of cellulose and glycogen.

In the preceding section we have considered the chemical properties and reactions of organic compounds, as well as the importance of the possible application of these compounds to biological systems. With this organic chemistry background, we are better equipped to approach a consideration of the important section on biochemistry. In the early chapters of the section, the organic chemistry of carbohydrates, lipids, and proteins will be considered. These compounds not only make up the three major types of foods but are also the essential organic constituents of the body.

The carbohydrates will be discussed in the present chapter, since they possess the simplest chemical structure. As a class of compounds, they include simple sugars, starches, and celluloses. Simple sugars such as glucose, fructose, and sucrose are constituents of many fruits and vegetables. Starches are the storage form of carbohydrates in plants, and cellulose is the main supporting structure of trees and plants. About 75 per cent of the solid matter of plants consists of carbohydrates. In microorganisms and in the cells of higher animals carbohydrates play a fundamental role, serving as an important source of energy for the cell and as a storage form of chemical energy. Specific carbohydrates function as structural units in cell walls and membranes, and in cellular components responsible for function and growth.

OPTICAL ACTIVITY

Stereoisomerism is a common phenomenon in organic chemistry. Examples of **structural** and **geometric isomers** were considered on p. 28 and 68 in Chapters 2 and 4. Optical isomerism is frequently encountered in organic compounds of biochemical interest and is essential in a study of the composition, properties, and reactions of carbohydrates.

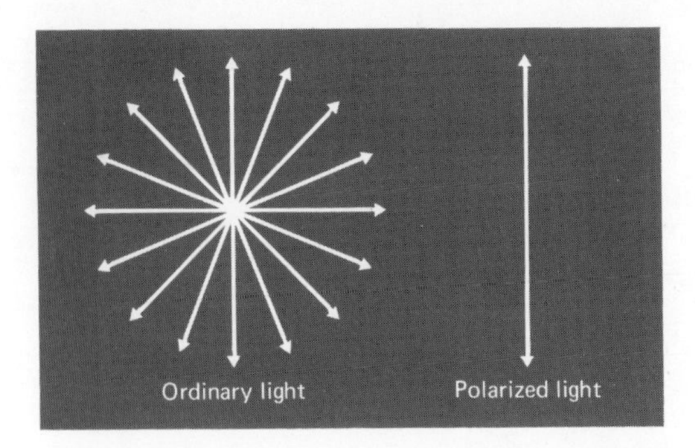

FIGURE 13-1 Rays of ordinary light coming toward the observer and vibrating in all planes at right angles to the path of the light, compared to the rays of plane polarized light vibrating in only one plane.

Many organic molecules, including the carbohydrates, exhibit the phenomenon of **optical activity.** Any optically active compound possesses the property of rotating a plane of polarized light. Ordinary light may be thought of as radiant energy propagated in the form of wave motion whose vibrations take place in all directions at right angles to the path of the beam of light. If a light ray is considered as perpendicular to the plane of the page and passing through it, the vibrations may be represented as spokes on a wheel (Fig. 13–1). Certain minerals, such as Iceland spar and tourmaline, and polaroid sheets or discs (properly oriented crystals embedded in a transparent plastic) allow only light vibrating in a single plane to pass through their crystals. When ordinary light is passed through a Nicol prism, which consists of two pieces of tourmaline cemented together, the resulting beam is traveling in one direction and in one plane, and is called **plane polarized** or just **polarized** light (Fig. 13–1).

Carbohydrates in solution show the property of optical rotation, i.e., a beam of polarized light is rotated when it passes through the solution. The extent to which the beam is rotated, or the angle of rotation, is determined with an instrument called a **polarimeter.** A simple polarimeter is shown in Figure 13–2.

The two Nicol prisms are arranged in such a manner that the light readily passes through them to illuminate the eyepiece uniformly. A cylindrical cell containing a solution of an optically active substance is then placed between the two Nicol prisms. Since the solution rotates the plane polarized light, it will not pass through the second Nicol prism and the field of the eyepiece will become dark. The analyzing Nicol prism is then rotated until the field of the eyepiece is again uniformly illuminated, and the number of degrees through which the prism is rotated as well as the extent of rotation can be determined. A substance whose solution rotates the plane of polarized light to the right is said to be **dextrorotatory**; one whose solution rotates the light to the left is called **levorotatory.**

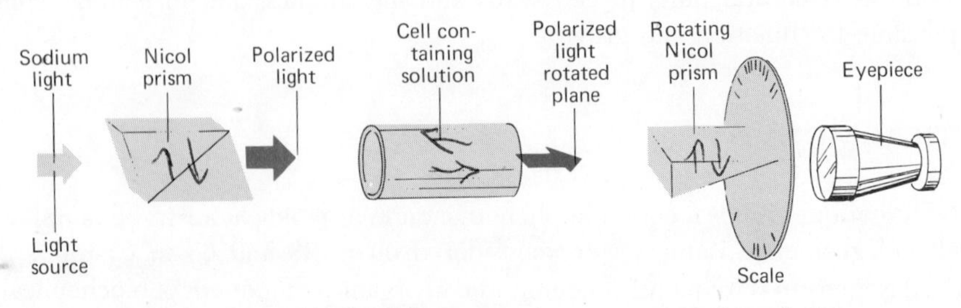

 FIGURE 13-2 A diagrammatic sketch of the essential components of a polariscope.

The rotation is designated d- for dextrorotatory or l- for levorotatory; for example, d-lactic acid and l-lactic acid.

To standardize experimental work with the polarimeter, the term **specific rotation** has been adopted. The specific rotation $[\alpha]_D^{20^\circ}$ of a substance is the rotation in angular degrees produced by a column of solution 1 decimeter long whose concentration is 1 gram per ml. The terms $]_D^{20^\circ}$ refer to the temperature of the solution, 20°C, and the source of plane polarized light, which is the monochromatic sodium light of wavelength 5890 to 5896 Å corresponding to the D line in the yellow part of the spectrum. The specific rotation of an optically active substance in solution is ordinarily expressed as:

$$[\alpha]_D^{20^\circ} = \frac{100\alpha}{lc}$$

where α is the observed rotation in degrees, l is the length of the tube in decimeters, and c is the concentration of the solution in grams per 100 ml. If the specific rotation of a sugar is known, its concentration in solution may readily be determined by use of this equation.

Van't Hoff and La Bel independently advanced the same theory to explain the fundamental reason for the optical activity of a compound. They postulated that the presence of an **asymmetric carbon atom** in a compound was responsible for the optical activity. An asymmetric carbon atom was defined as one which had four different groups attached to it. A simple compound such as lactic acid contains one asymmetric carbon atom, which is marked with an asterisk in the following illustration:

```
      COOH                 COOH
       |                    |
  H—C°—OH             HO—C°—H
       |                    |
      CH3                  CH3
 d-Lactic acid        l-Lactic acid
```

The d and l forms of lactic acid are mirror images of each other, rotating the plane of polarized light an equal extent in opposite directions. These two compounds, which are identical in composition and differ only in spatial configuration, are known as **stereoisomers.** A molecular model, constructed as shown in Figure 13–3, in which the asymmetric carbon atom is the central sphere and is joined to four other spheres representing a hydrogen atom, a hydroxyl group, a methyl group, and a carboxyl group, may help to explain optical activity. This model emphasizes the asymmetric carbon atom and the fact that the d and l forms of lactic acid cannot be superimposed upon one another. One compound, however, resembles the other by being its **mirror image;** for if one model is held before a mirror, the image in the mirror corresponds to the other model. Such

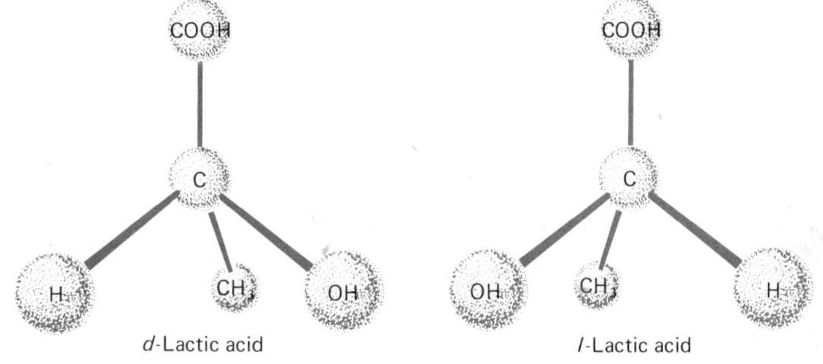

FIGURE 13-3 The spatial relationship of the groups attached to the asymmetric carbon atom of the d and l forms of lactic acid.

mirror image isomers as d- and l-lactic acid are an enantiomeric pair, and the d and l forms are **enantiomers.** Although the presence of an asymmetric carbon atom is most frequently responsible for the optical activity of a compound, any condition that removes the elements of symmetry that make mirror images identical produces optical activity. A molecule such as tartaric acid with two asymmetric carbon atoms that are alike (attached to the same four kinds of groups) may exist in a form in which there is a plane of symmetry and each half molecule is a mirror image of the other. This is known as the *meso* form and exhibits an optical activity of zero. The two forms of tartaric acid representing the *meso* form are shown as follows:

```
        O                  O
       //                 //
      C—OH               C—OH
      |                  |
   H—C—OH            HO—C—H
------+------------------+--------
   H—C—OH            HO—C—H
      | O                | O
      |//                |//
      C—OH               C—OH
```

meso-Tartaric acid

A compound having two like asymmetric carbon atoms, therefore, has only three optical isomers, the d-, and l-, and the *meso* form. Before Van't Hoff and La Bel postulated the presence of asymmetric carbon atoms, Pasteur recognized two forms of crystals of tartaric acid and carefully separated them by hand. One form was dextrorotatory in solution, the other levorotatory.

Lactic acid from different natural sources exhibits differences in optical activity. For example, the lactic acid involved in the contraction of muscle tissue in the body is the dextro form, whereas the levo form may be isolated from the fermentation products of cane sugar. When milk sours, the lactic acid that is formed consists of an equal mixture of the d and l forms and does not rotate the plane of polarized light. In general, when a compound that exhibits optical activity is synthesized in the laboratory, a mixture of equal parts of the dextro and levo forms results. Such a mixture is called a **racemic mixture.** Reactions carried out in the body or in the presence of microorganisms often produce optically active isomers, since the reactions are catalyzed by enzymes which themselves are optically active. The enzyme reactions are often specific for the d or l component of a racemic mixture.

As naturally occurring optically active compounds such as the carbohydrates and the amino acids are studied, it will be seen that the isomeric form, and therefore the structure, is closely related to the physiological activity of these substances.

COMPOSITION

Carbohydrates are composed of carbon, hydrogen, and oxygen, and the hydrogen and oxygen are usually in the proportion of two to one, the same as in water. The name carbohydrate (which signifies hydrate of carbon) is based on this relationship of hydrogen and oxygen. The term is misleading, however, because water does not exist as such in a carbohydrate. The classic definition of carbohydrates stated that they were compounds of C, H, and O in which the H and O were in the same proportion as in water. But a compound such as acetic acid, $C_2H_4O_2$, fits this definition and yet is not classed as a carbohydrate, whereas an important carbohydrate such as deoxyribose, $C_5H_{10}O_4$, a constituent of DNA (deoxyribonucleic acid) found in every cell, does not fit the definition. Carbohydrates are now defined as derivatives of polyhydroxyaldehydes or polyhydroxyketones. A sugar that contains an aldehyde group is called an **aldose,** and one that contains a ketone group is termed a **ketose.**

CLASSIFICATION

The simplest carbohydrates are known as **monosaccharides,** or simple sugars. Monosaccharides are derivatives of straight-chain polyhydric alcohols and are classified according to the number of carbon atoms in the chain. A sugar with two carbon atoms is called a diose; with three, a triose; with four, a tetrose; with five, a pentose; and with six, a hexose. The ending -ose is characteristic of sugars. When two monosaccharides are linked together by splitting out a molecule of water, the resulting compound is called a **disaccharide.** The combination of three monosaccharides results in a **trisaccharide,** although the general term for carbohydrates composed of two to five monosaccharides is **oligosaccharide.** Polymers composed of several monosaccharides are called **polysaccharides.**

Carbohydrates which will be considered in this chapter may be classified as follows:

I. Monosaccharides
- Trioses—$C_3H_6O_3$
  - Aldose—Glyceraldehyde
  - Ketose—Dihydroxyacetone
- Pentoses—$C_5H_{10}O_5$
  - Aldoses—Arabinose, Xylose, Ribose
- Hexoses—$C_6H_{12}O_6$
  - Aldoses—Glucose, Galactose
  - Ketoses—Fructose
- Ascorbic acid

II. Disaccharides—$C_{12}H_{22}O_{12}$
- Sucrose (glucose + fructose)
- Maltose (glucose + glucose)
- Lactose (glucose + galactose)

III. Polysaccharides
- Hexosans
  - Glucosans—Starch, Glycogen, Dextrin, Cellulose

IV. Mucopolysaccharides
- Hyaluronic acid
- Chondroitin sulfate
- Heparin

Trioses

The trioses are important compounds in muscle metabolism, and are the basic sugars to which all monosaccharides are referred. The definition of a simple sugar may readily be illustrated by the use of the trioses. The polyhydric alcohol from which they are derived is glycerol. Oxidation on the end carbon atom produces the aldose sugar known as glyceraldehyde; oxidation on the center carbon yields the keto triose, dihydroxyacetone. It can be seen from the formula of glyceraldehyde that one asymmetric carbon atom

H
H—C—OH
H—C—OH
H—C—OH
H

Glycerol
(polyhydric alcohol)

→

H
C=O
H—C—OH
H—C—OH
H

Glyceraldehyde (aldose)

H
H—C—OH
C=O
H—C—OH
H

Dihydroxyacetone (ketose)

is present. Therefore this sugar can exist in two forms, one of which rotates plane polarized light to the right, the other to the left. Originally the two forms were designated d and l for the dextro and levo rotation of polarized light. Modern terminology employs the D and L, written in small capital letters, for structural relationships, and a (+) and (−) for direction of rotation.

```
       H                    H                 H
       |                    |                 |
      C=O                  C=O               C=O
       |                    |                 ¦
  H—C*—OH             HO—C*—H            H▬C◄OH
       |                    |                 ¦
     CH2OH                CH2OH             CH2OH
```

D (+) Glyceraldehyde   L (−) Glyceraldehyde   Perspective formula

Fischer projection formula

The isomeric forms of sugars are often represented as the **Fischer projection formula.** The asymmetric carbon atom * of glyceraldehyde would represent the central sphere as in the model in Figure 13–3, with the H and OH groups projecting in front of the plane of the paper and the aldehyde and primary alcohol group projecting behind. As ordinarily written, the horizontal bonds are understood to be in front of the plane and the vertical bonds behind the plane of the paper. The **perspective formula** emphasizes the position of the groups using dotted lines to connect those behind the plane and heavy wedges to represent groups in front of the plane. The Fischer projection formulas are always written with the aldehyde or ketone groups (the most highly oxidized groups) at the top of the structure; therefore, all monosaccharides with the hydroxyl group on the right of the carbon atom next to the bottom primary alcohol group are related to D-glyceraldehyde and are called D-sugars.

In like manner, if the hydroxyl group on the carbon atom next to the end primary alcohol group is on the left, it is related to L-glyceraldehyde and is an L-sugar. The direction of rotation of polarized light cannot be ascertained from the formula, but must be determined experimentally, and is designated (+) for dextrorotatory and (−) for levorotatory sugars.

Only two optical isomers of aldotriose exist, since it contains only one asymmetric carbon atom. A sugar such as a tetrose with two structurally different carbon atoms would exhibit a total of four different isomers, and as each new asymmetric carbon atom is added to the structure the number of isomers is doubled. The total number of isomers of a sugar is 2^n, where n is the number of different asymmetric carbon atoms. For example, pentoses contain three asymmetric carbon atoms and can form eight isomers; hexoses contain four asymmetric carbon atoms and can exist as 16 different isomers.

Pentoses

The pentoses are sugars whose molecules contain five carbon atoms and three asymmetric carbon atoms. They occur in nature combined in polysaccharides from which the monosaccharides may be obtained by hydrolysis with acids. Arabinose is obtained from gum arabic and the gum of the cherry tree, and xylose is obtained by hydrolysis of wood, corn cobs, or straw. Ribose and deoxyribose are constituents of the ribose nucleic acids, RNA, and deoxyribose nucleic acids, DNA, that are essential components of the cytoplasm and nuclei of cells.

Hexoses

The hexoses are by far the most important monosaccharides from a nutritional and physiological standpoint. The bulk of the carbohydrates used as foods consist of hexoses free or combined in disaccharides and polysaccharides. Glucose, fructose, and galactose are the hexoses commonly occurring in foods, whereas mannose is a constituent of a vegetable polysaccharide. Glucose, also called **dextrose,** is the normal sugar of the blood and tissue fluids and is utilized by the cells as a source of energy. Fructose often occurs free in fruits and is the sweetest sugar of all the monosaccharides. Galactose is a constituent of **milk sugar** and is found in brain and nervous tissue. All these monosaccharides are D-sugars. The hydroxyl group on the carbon next to the primary alcohol is on the right. Fructose is a ketose sugar and the others are aldose sugars.

Although glucose and galactose are represented as simple aldehyde structures, this form does not explain all the reactions they undergo. Both of these aldoses, for example, do not give a positive Schiff test for aldehyde. Also, when a glucose solution is allowed to stand, a change in its specific rotation may be observed.

```
       H                                          H
       |                                          |
       C=O             CH2OH                      C=O
       |                 |                        |
    H—C—OH               C=O                   H—C—OH
       |                 |                        |
   HO—C—H            HO—C—H                   HO—C—H
       |                 |                        |
    H—C—OH            H—C—OH                  HO—C—H
       |                 |                        |
    H—C—OH            H—C—OH                   H—C—OH
       |                 |                        |
      CH2OH             CH2OH                    CH2OH
   D-Glucose          D-Fructose              D-Galactose
   (aldose)           (ketose)                (aldose)
```

Freshly prepared aqueous solutions of crystalline glucose often yield a specific rotation as high as +113 degrees, whereas glucose crystallized from pyridine exhibits a specific rotation as low as +19 degrees. On standing, both of these solutions change their rotation until an equilibrium value of +52.5 degrees is reached. This change in rotation is called **mutarotation.** Since the specific rotation of an organic compound is related to its structure, as is its melting point, boiling point, and other properties, it may be suspected that glucose exists in two different isomeric forms. This has been shown to be true and is explained by the existence of an **intramolecular bridge structure** involving carbon atoms 1 and 5. It can be seen from these formulas that the free aldehyde group no longer exists and a new asymmetric carbon is produced. To indicate the position of the hydroxyl group on the first carbon and to distinguish between the two new isomers, the α-isomer has the OH on the right and the β-isomer on the left as shown. When the α- or β-isomer is dissolved in water, an equilibrium mixture of 37 per cent α and 63 per cent β, with a specific rotation of 52.5 degrees, is formed. This intramolecular bridge

```
    H   OH                 H                 HO   H
     \ /                   |                   \ /
      C-------|            C=O                  C-------|
      |       |            |                    |       |
   H—C—OH     |         H—C—OH               H—C—OH     |
      |       |            |                    |       |
  HO—C—H      O  ⇄    HO—C—H        ⇄      HO—C—H      O
      |       |            |                    |       |
   H—C—OH     |         H—C—OH               H—C—OH     |
      |       |            |                    |       |
   H—C--------|         H—C—OH               H—C--------|
      |                    |                    |
     CH2OH                CH2OH                CH2OH
  α-D-Glucose          D-Glucose,           β-D-Glucose
     +113°           straight chain            +19°
                          form
```

structure and the phenomenon of mutarotation are common to all aldohexoses, and since the structure contains an additional asymmetric carbon atom, the number of possible isomers is doubled.

A further projection of the structure results from the random motion of the open chain which allows the alcoholic hydroxyl group on carbon-5 of the aldohexose molecule to approach the aldehyde group on the end of the molecule. Ring formation could then result from the formation of a hemiacetal between the aldehyde and hydroxyl group (see p. 158). The hemiacetal formation in an aldohexose molecule may be represented as follows:

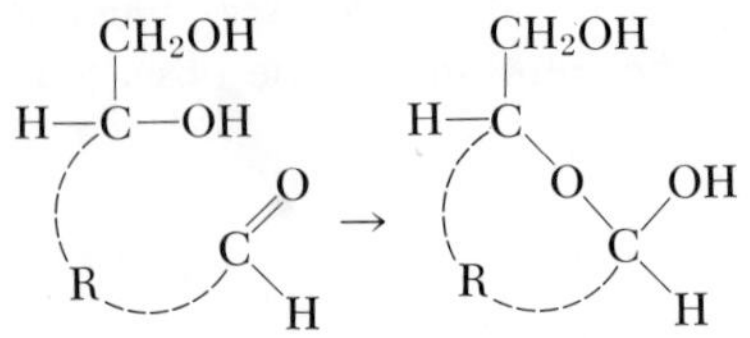

CH_2OH, H—C—OH, R, C, O, H → CH_2OH, H—C, R, O, C, OH, H

Haworth suggested that the sugars be represented as derivatives of the heterocyclic rings pyran and furan.

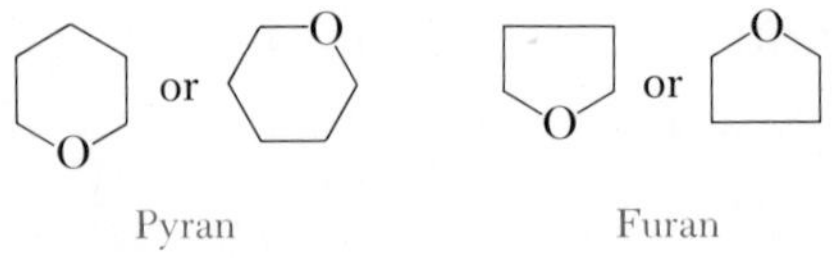

O or O   O or O

Pyran   Furan

The relation between the straight chain structure of glucose and Haworth's **glucopyranose** may best be understood by writing glucose in a chain as shown in A. The chain is folded and a rotation of groups occurs around carbon-5 to bring the primary alcohol group (carbon-6) into the proper spatial relation to the other groups (B). The hemiacetal is then formed between the aldehyde on carbon-1 and the OH group on carbon-5 to form α-D-glucopyranose, shown in C.

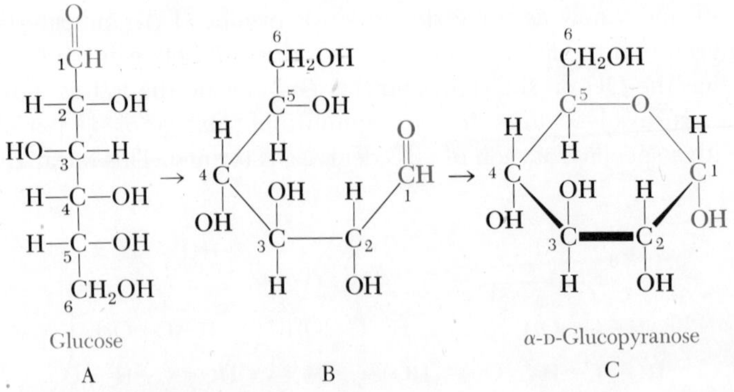

Glucose
A

B

α-D-Glucopyranose
C

The heavy lines represent the base of a space model in which the five carbons and one oxygen are in the same plane perpendicular to the plane of the paper. The thick bonds of the ring extend toward the reader, whereas the thin bonds of the ring are behind the plane of the paper. Groups which are ordinarily written on the right of the oxide ring structure appear below the plane of the pyranose ring, and those to the left of the carbon chain appear above the plane. The glucopyranose structure, C, has the OH on carbon-1 below the plane of the ring and would therefore be an α-isomer. As mentioned earlier (p. 35), the chair form of the pyranose ring structure is more stable than the boat form configuration and the α-D-glucopyranose formula is readily apparent in this form.

Monosaccharides such as the pentoses and fructose, whose oxide rings enclose four carbon atoms, are written as derivatives of furan as shown:

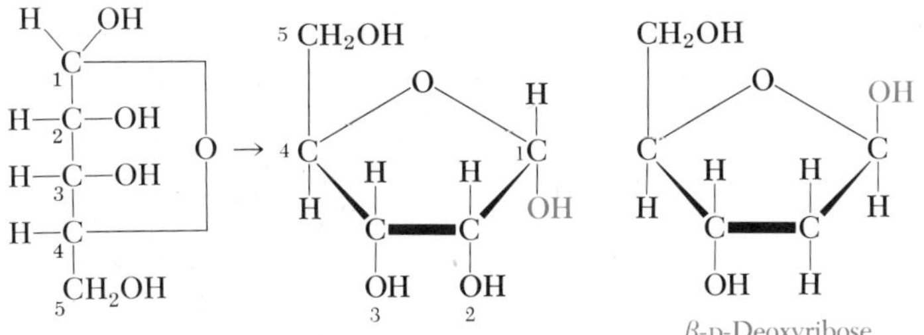

α-D-Ribose

β-D-Deoxyribose

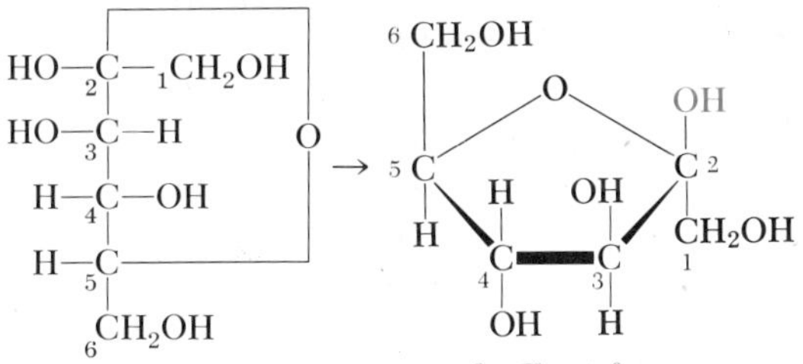

β-D-Fructofuranose

The oxide ring structure of fructose is represented as shown in the above structure, since the α and β forms of the ketoses are derived from the corresponding aldoses containing one less carbon atom by replacing the hydrogen on the first carbon with CH_2OH. This representation results in the correct model of β-D-fructofuranose. It can readily be observed in the preceding structures that the OH group on carbon-1 (carbon-2 in ketofuranoses) indicates the α or β form of the sugar. The α-isomers have the OH extending below the plane of the ring, whereas in the β-isomers the OH extends above the plane.

REACTIONS OF CARBOHYDRATES

Dehydration

When aldohexoses or aldopentoses are heated with strong acids, they are dehydrated to form furfural derivatives. Pentoses yield furfural (p. 104) whereas aldohexoses form hydroxymethyl furfural.

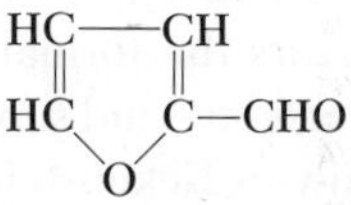

Furfural

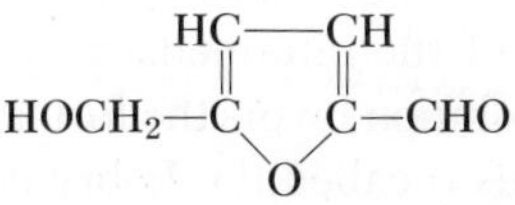

Hydroxymethyl furfural

The furfural derivatives formed in this reaction combine with α-naphthol to give a purple color. This color is the basis of the Molisch test, a general test for carbohydrate. Furfural reacts with orcinol to yield a green color which constitutes Bial's test for pentoses.

Acetal or Glycoside Formation

When monosaccharides are treated with an alcohol in a strong acid solution, they form **glycosides.** The hemiacetal structure reacts with alcohols or an alcoholic hydroxyl group to form an acetal or glycoside (see p. 158).

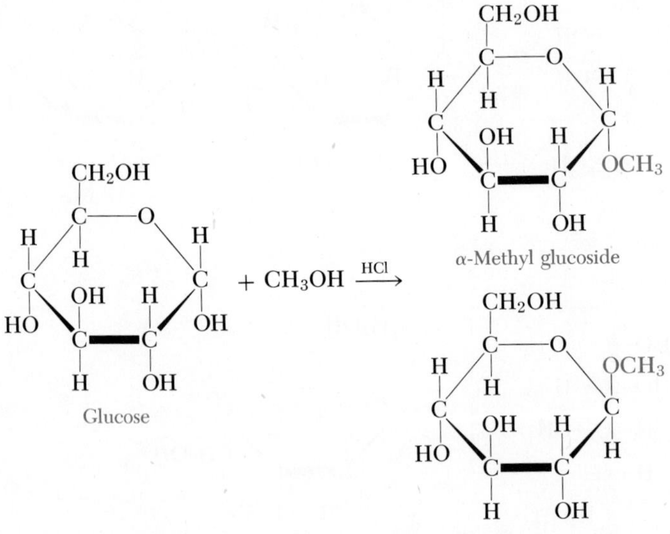

$$\text{Glucose} + CH_3OH \xrightarrow{HCl} \alpha\text{-Methyl glucoside} \quad / \quad \beta\text{-Methyl glucoside}$$

Glucose

α-Methyl glucoside

β-Methyl glucoside

The position of the methyl group below or above the plane of the ring indicates α- or β-methyl glucoside in that order. This is a very important reaction since many of the disaccharides and polysaccharides are glycosides in which one of the alcoholic hydroxyl groups in the second monosaccharide reacts with the hemiacetal in the first, as shown below:

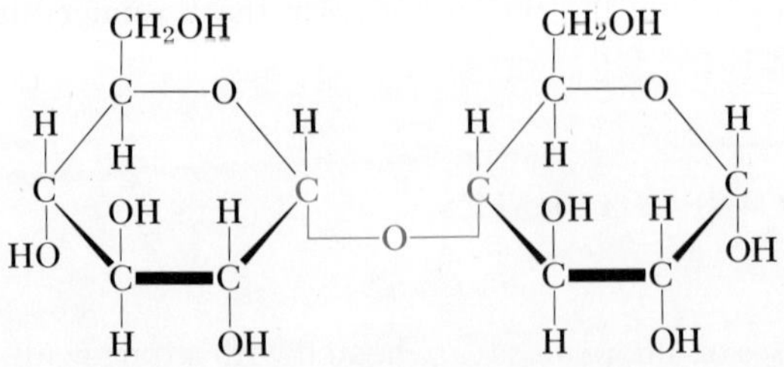

α-1-4 Linked disaccharide

When glucose, fructose, or mannose is added to a saturated solution of $Ba(OH)_2$ and allowed to stand, it forms the same intermediate **enediol.** The loss of asymmetry in the second carbon atom of the intermediate enediol favors the formation of the other two sugars. The alkaline solution favors the formation of enediols and suppresses the formation of ring structures. This is called the **Lobry de Bruyn-von Eckenstein transformation** and may be illustrated as shown on the following page.

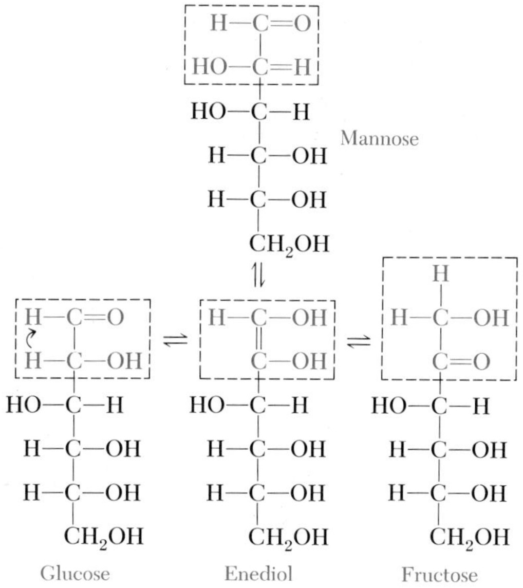

```
                  H—C=O
                    |
                 HO—C=H
                    |
                 HO—C—H
                    |          Mannose
                  H—C—OH
                    |
                  H—C—OH
                    |
                    CH2OH
                    ⇅
                                        H
                                        |
  H—C=O        ⇌    H—C—OH    ⇌    H—C—OH
    |                 ||               |
  H—C—OH              C—OH             C=O
    |                 |                |
 HO—C—H            HO—C—H           HO—C—H
    |                 |                |
  H—C—OH            H—C—OH           H—C—OH
    |                 |                |
  H—C—OH            H—C—OH           H—C—OH
    |                 |                |
    CH2OH             CH2OH            CH2OH
  Glucose           Enediol          Fructose
```

Oxidation

In Chapter 9 one of the important reactions of carbonyl compounds was the oxidation to carboxylic acids. Sugars that contain *free or potential aldehyde or ketone groups* in the hemiacetal type structure are oxidized in alkaline solution by Cu^{+2} and Ag^{+}. The reaction of aldehydes in Benedict's, Fehling's, or the silver mirror test has already been described (p. 154). Sugars that undergo oxidation in these reactions are called **reducing sugars.**

All the reducing sugars that are capable of reducing Cu^{+2} to Cu^{+} are oxidized in the reaction described in the preceding section. If the aldehyde group of glucose is oxidized to a carboxyl group by a weak oxidizing agent, such as NaOBr, gluconic acid is formed. Oxidation of the primary alcohol group, either by chemical agents or enzymes, produces glucuronic acid. Further oxidation with concentrated HNO_3 converts both end groups into carboxyl groups as in saccharic acid.

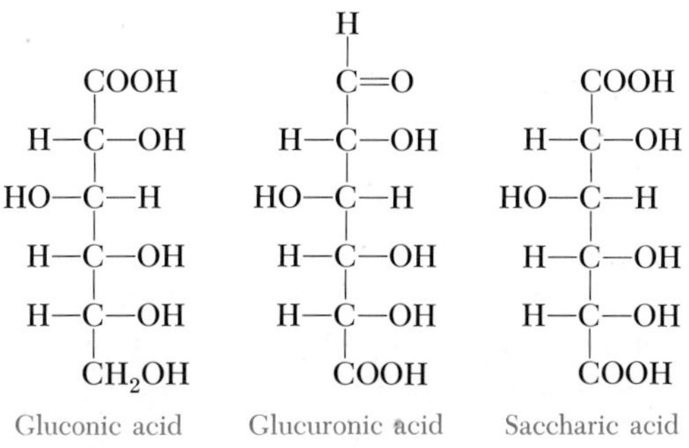

```
                         H
                         |
     COOH                C=O              COOH
     |                   |                |
   H—C—OH              H—C—OH           H—C—OH
     |                   |                |
  HO—C—H              HO—C—H           HO—C—H
     |                   |                |
   H—C—OH              H—C—OH           H—C—OH
     |                   |                |
   H—C—OH              H—C—OH           H—C—OH
     |                   |                |
     CH2OH               COOH             COOH
 Gluconic acid     Glucuronic acid   Saccharic acid
```

Glucuronic acid combines with drugs and toxic compounds in the body, and the conjugated glucuronides are excreted in the urine. Oxidation of galactose with concentrated HNO_3 produces mucic acid. This acid crystallizes readily and the reaction is used as a test for the presence of galactose.

Ascorbic Acid (Vitamin C)

Ascorbic acid is an enediol of a hexose sugar acid. The reduced or enediol form is readily oxidized to form **dehydroascorbic acid.** Both forms are biologically active, however, treatment with a weak alkali opens the oxide ring and produces an inactive molecule.

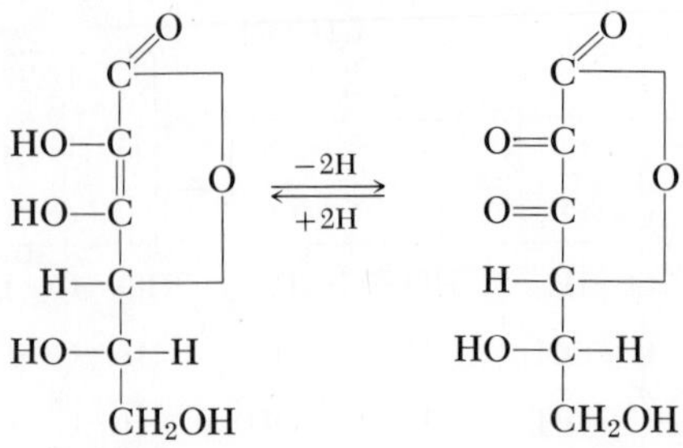

```
          O                                O
         //                               //
        C ─────┐                         C ─────┐
        |      |                         |      |
   HO ─ C      |        −2H         O ═ C      |
        ‖      O      ⇌                  |      O
   HO ─ C      |        +2H         O ═ C      |
        |      |                         |      |
    H ─ C ─────┘                     H ─ C ─────┘
        |                                |
   HO ─ C ─ H                       HO ─ C ─ H
        |                                |
       CH2OH                            CH2OH
  L-Ascorbic acid              Dehydroascorbic acid
```

A deficiency of ascorbic acid in the diet results in the disease known as **scurvy.** As early as 1720, citrus fruits were used as a cure for scurvy. The fact that all British ships were later required to carry stores of lime juice to prevent scurvy on long voyages led to the use of the terms "limey" for sailors, "lime juicers" for ships, and "lime house district" for the section of town in which sailors lived. Early symptoms of scurvy are loss of weight, anemia, and fatigue. As the disease progresses, the gums become swollen and bleed readily, and the teeth loosen. The bones become brittle and hemorrhages develop under the skin and in the mucous membrane. Extreme scurvy is not commonly seen today, although many cases of subacute, or latent, scurvy are recognized. Symptoms such as sore receding gums, sores in the mouth, tendency to fatigue, lack of resistance to infections, defective teeth, and pains in the joints are indicative of **subacute scurvy.**

Man, monkeys, and guinea pigs are the only species that are known to be susceptible to the lack of ascorbic acid. Other animals possess the ability to synthesize this vitamin. The richest sources of ascorbic acid are citrus fruits, tomatoes, and green leafy vegetables. A large percentage of the ascorbic acid in foods is destroyed or lost in cooking. Prolonged boiling and the addition of sodium bicarbonate to maintain the green color of vegetables can destroy 70 to 90 per cent of the vitamin C content.

Ascorbic acid may function in oxidation or reduction processes in the body since it is a powerful reducing agent. The adrenal cortex contains appreciable amounts of ascorbic acid, which may function in the synthesis of steroid hormones in the adrenal gland. Ascorbic acid is also thought to be involved in hydroxylation reactions and in electron transport in the microsomal region of the cell. The biochemistry of ascorbic acid deficiency in the body is not as yet well understood. Linus Pauling (p. 260) has suggested that the common cold can be prevented or treated with large doses of ascorbic acid. Considerable controversy has arisen regarding this suggestion since neither the cause of the common cold nor the action of ascorbic acid in the body is clearly understood.

Fermentation

The enzyme mixture called **zymase** present in common bread yeast will act on some of the hexose sugars to produce alcohol and carbon dioxide. The fermentation of glucose may be represented as follows:

$$\underset{\text{Glucose}}{C_6H_{12}O_6} \xrightarrow{\text{zymase}} \underset{\text{Ethyl alcohol}}{2C_2H_5OH} + 2CO_2$$

The common hexoses (with the exception of galactose) ferment readily, but pentoses are not fermented by yeast. Disaccharides must first be converted into their monosaccharide constituents by other enzymes present in yeast before they are susceptible to fermentation by zymase.

There are many other types of fermentation of carbohydrates besides the common alcoholic fermentation. When milk sours, the lactose of milk is converted into lactic acid by a fermentation process. Citric acid, acetic acid, butyric acid, and oxalic acid may all be produced by special fermentation processes.

Ester Formation

Esters formed between the hydrogen atom of a hydroxyl group of phosphoric acid and the hydroxyl group of a monosaccharide are common (p. 121), and several of these phosphorylated sugars are encountered in carbohydrate metabolism (Chapter 20). In the metabolic reactions in the body the location of the hydroxyl group or groups on the compounds to be phosphorylated is controlled by enzymes specific for that reaction. The organic chemist utilizes phosphoryl group donors in nonaqueous systems and chemically blocks other potentially reactive groups in order to phosphorylate the desired hydroxyl group in a compound.

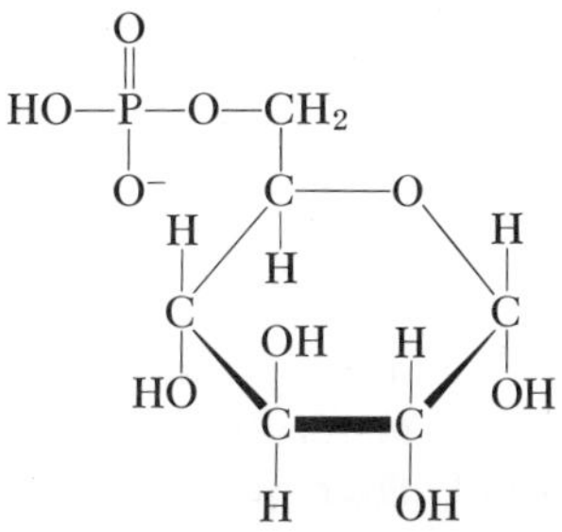

α-D-Glucose-6-phosphate

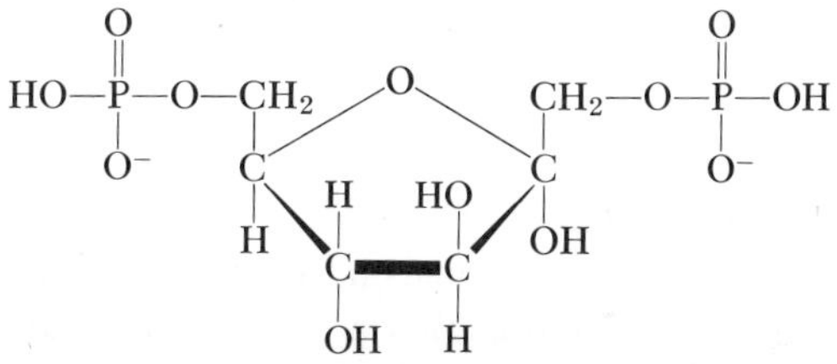

α-D-Fructose-1, 6-diphosphate

DISACCHARIDES

A disaccharide is composed of two monosaccharides whose combination involves the splitting out of a molecule of water. The acetal linkage is always made from the aldehyde group of one of the sugars to a hydroxyl or ketone group of the second. In the structure of the individual disaccharides, such as sucrose, lactose, and maltose, the exact location of the hemiacetyl linkage and the isomeric forms involved in the linkage are obviously involved in the proof of the structure of the disaccharide. Information concerning the monosaccharides formed by hydrolysis, the reduction products, the change in optical rotation on hydrolysis, and the reaction as a reducing sugar is used in the proof of the structure. In order to reduce Benedict's solution, disaccharides must have a potential aldehyde or ketone group that is not involved in the acetal linkage between the two sugars.

Sucrose

Sucrose is commonly called **cane sugar** and is the ordinary sugar that is used for sweetening purposes in the home. It is found in many plants such as sugar beets, sorghum cane, the sap of the sugar maple, and sugar cane. Commercially it is prepared from sugar cane and sugar beets.

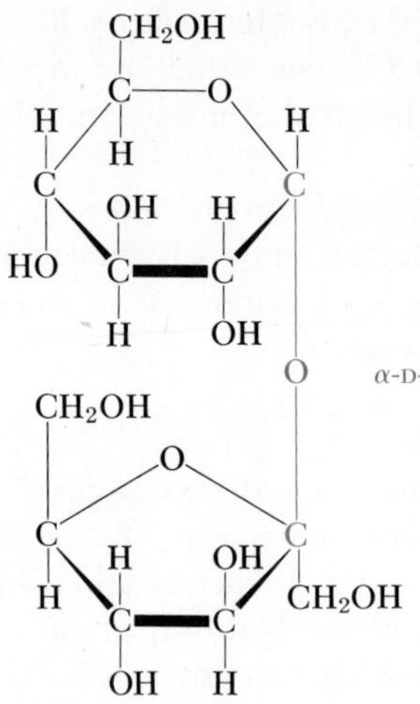

α-D-Glucopyranose-β-D-fructofuranoside

Sucrose

Sucrose is composed of a molecule of glucose joined to a molecule of fructose in such a way that the linkage involves the reducing groups of both sugars (carbon-1 of glucose and carbon-2 of fructose). It is the only common mono- or disaccharide that will not reduce Benedict's solution. When sucrose is hydrolyzed, either by the enzyme sucrase or by an acid, a molecule of glucose and a molecule of fructose are formed. The fermentation of sucrose by yeast is possible, since the yeast contains the two enzymes sucrase and zymase. The sucrase first hydrolyzes the sugar, and then the zymase ferments the monosaccharides to form alcohol and carbon dioxide.

Lactose

The disaccharide present in milk is lactose, or **milk sugar.** It is synthesized in the mammary glands of animals from the glucose in the blood. Commercially, it is obtained from milk whey and is used in infant foods and special diets. Lactose, when hydrolyzed by the enzyme lactase or by an acid, forms a molecule of glucose and a molecule of galactose. Lactose will reduce Benedict's solution, but is not fermented by yeast. From its reducing properties, it is obvious that the linkage between its constituent monosaccharides does not involve both potential aldehyde groups (carbon-1 of galactose is connected to carbon-4 of glucose). This linkage also illustrates the formation of an ether linkage by dehydration, as discussed on page 130.

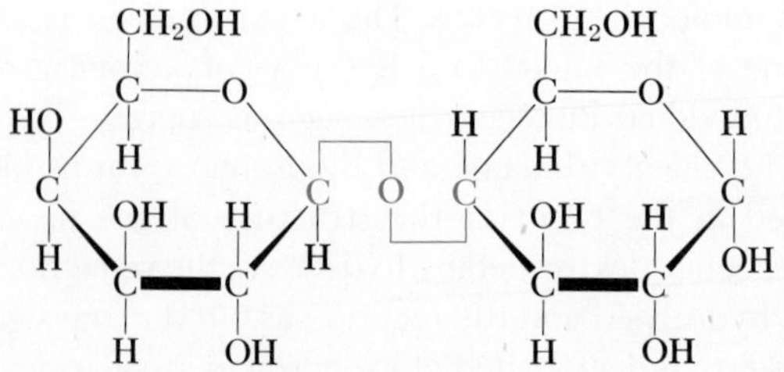

β-D-Galactopyranosyl-α-D-glucopyranose

Lactose

Maltose

Maltose is present in germinating grains. Since it is obtained as a product of the hydrolysis of starch by enzymes present in malt, it is often called **malt sugar.** It is also formed in the animal body by the action of enzymes on starch in the process of digestion.

Commercially, it is made by the partial hydrolysis of starch by acid in the manufacture of corn syrup. Maltose reduces Benedict's solution and is fermented by yeast. On hydrolysis it forms two molecules of glucose.

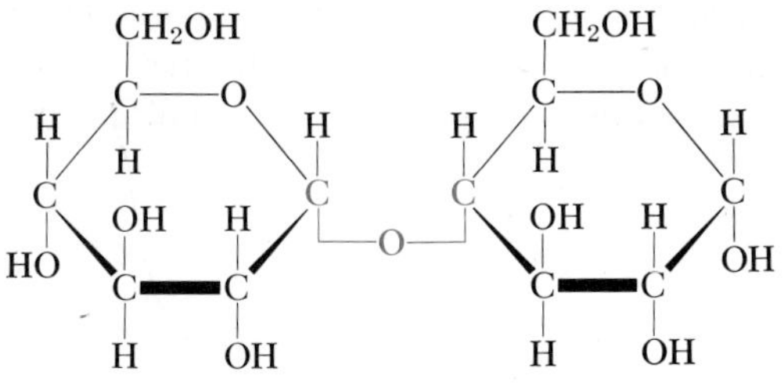

α-D-Glucopyranosyl-α-D-glucopyranose
Maltose

Cellobiose

Cellobiose is similar to maltose since it is composed of two glucose molecules joined in a carbon-1 to carbon-4 linkage. In contrast to maltose, the glucose in cellobiose is in the β form.

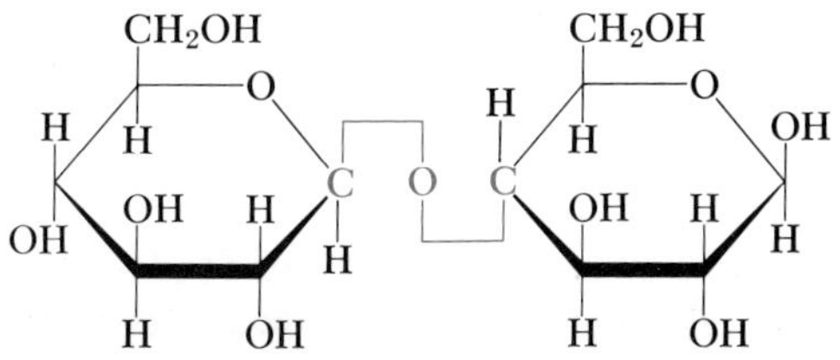

Cellobiose

POLYSACCHARIDES

The polysaccharides are complex carbohydrates that are made up of many monosaccharide molecules and therefore possess a high molecular weight. They differ from the simple sugars in many ways. They do not have a sweet taste, are usually insoluble in water, and when dissolved by chemical means form colloidal solutions because of their large molecules. Although most polysaccharides have a terminal monomer present as a reducing sugar, the contribution of this portion to the properties of the molecule decreases as the size of the polymer increases. Most polysaccharides, therefore, do not behave as reducing sugars, although most oligosaccharides will reduce Benedict's reagent.

There are polysaccharides formed from pentoses or from hexoses, and there are also mixed polysaccharides. Of these, the most important are composed of the hexose glucose and are called **hexosans,** or more specifically, **glucosans.** As in a disaccharide, whenever two molecules of a hexose combine, a molecule of water is split out. For this reason, a hexose polysaccharide may be represented by the formula $(C_6H_{10}O_5)_x$. The x represents the number of hexose molecules in the individual polysaccharide. Because of the complexity of the molecules, the number of glucose units in any one polysaccharide is still an estimate. In addition, the molecular weight values obtained for polysaccharides from different plant and animal sources show considerable variation. For this reason the molecular weights of the polysaccharides described in the following section are only approximations.

Starch

From a nutritional standpoint, starch is the most important polysaccharide. It is made up of glucose units and is the storage form of carbohydrates in plants. It consists of two types of polysaccharides: **amylose,** composed of a chain of glucose molecules connected

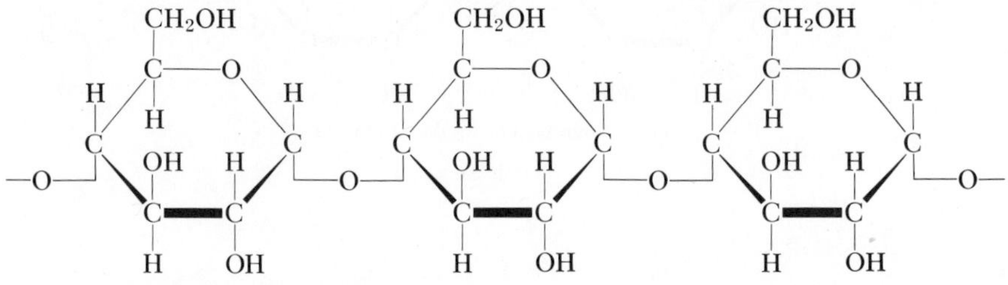

Amylose

by α-1,4 linkages, and **amylopectin,** which is a branched chain or polymer of glucose with both α-1,4 and α-1,6 linkages. The repeating structure of glucose molecules in amylose is usually represented as glucopyranose units as shown in the accompanying diagram. Amylose has a molecular weight of about 50,000, compared to about 300,000 for amylopectin. The branching of the glucose chain in amylopectin occurs about every 24 to 30 glucose molecules.

Starch will not reduce Benedict's solution and is not fermented by yeast. When starch is hydrolyzed by enzymes or by an acid, it is split into a series of intermediate compounds possessing small numbers of glucose units. The product of complete hydrolysis is the free glucose molecule. A characteristic reaction of starch is the formation of a blue compound with iodine. This test is often used to follow the hydrolysis of starch, since the color changes from blue through red to colorless with decreasing molecular weight:

starch →	amylodextrin →	erythrodextrin →	achroodextrin →	maltose →	glucose
blue	blue	red	colorless	colorless with iodine	

Dextrins

Dextrins are found in germinating grains, but are usually obtained by the partial hydrolysis of starch. Those formed from amylose have straight chains, whereas those derived from amylopectin exhibit branched chains of glucose molecules. The larger branched chain molecules give a red color with iodine and are the erythrodextrins. They are soluble in water and have a slightly sweet taste. Large quantities of dextrins are used in the manufacture of adhesives because they form sticky solutions when wet. An example of their use is the mucilage on the back of postage stamps.

Glycogen

Glycogen is the storage form of carbohydrate in the animal body and is often called animal starch. It is found in liver and muscle tissue. It is soluble in water, does not reduce Benedict's solution, and gives a red-purple color with iodine. The glycogen molecule is similar to the amylopectin molecule in that it has branched chains of glucose with α-1,4 and α-1,6 linkages that occur about every 12 to 18 glucose molecules. The branched

Branched chain of glucose molecules in amylopectin and glycogen

chain structure common to both glycogen and amylopectin is shown above and represented in Figure 13–4.

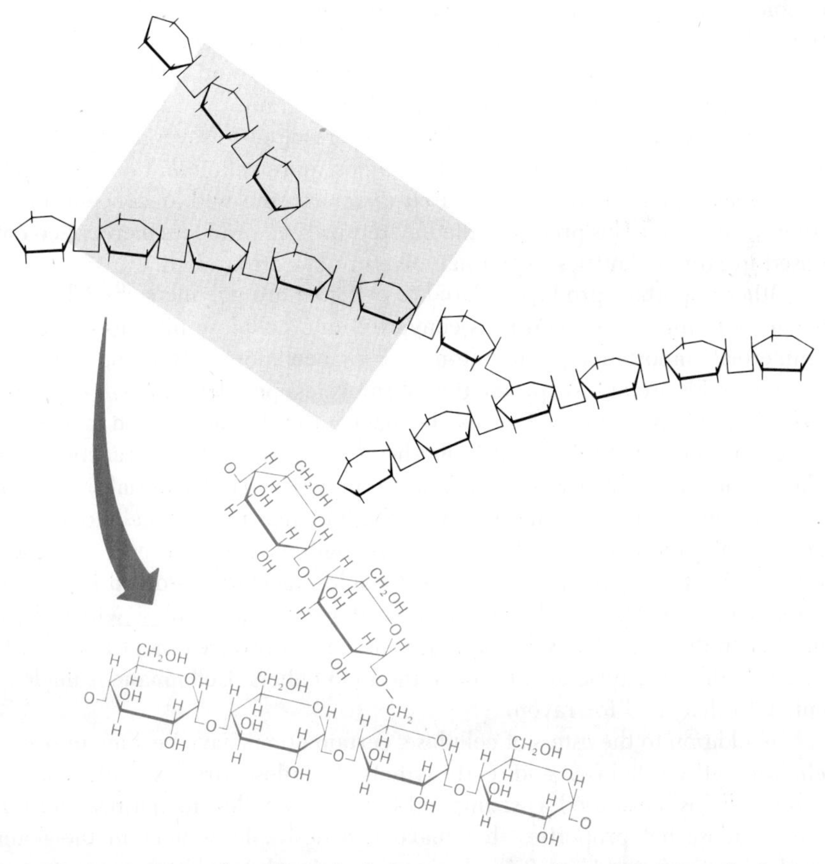

FIGURE 13-4 Representation of the branched chain structure of amylopectin and glycogen. The shaded portion is illustrated in detail as follows.

The molecular weight of glycogen is greater than that of amylopectin, and often exceeds 5,000,000. When glycogen is hydrolyzed in the animal body, it forms glucose to help maintain the normal sugar content of the blood.

Cellulose

Cellulose is a polysaccharide that occurs in the framework, or supporting structure, of plants. It is composed of a straight chain polymer of glucose molecules similar in structure to that pictured for amylose. The major difference concerns the linkage of the glucose molecules. In amylose the linkage is α-1,4, as in maltose, whereas in cellulose the linkage is β-1,4, which occurs in **cellobiose.** The maltose type of structure is hydrolyzed by enzymes and serves as a source of dietary carbohydrate. In contrast, the

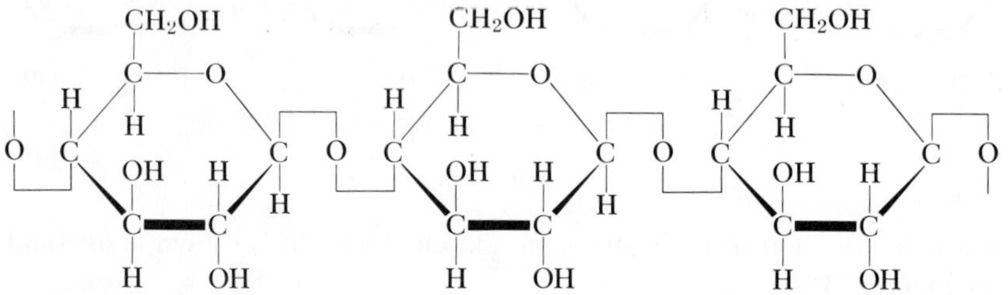

Cellulose

cellobiose structure is insoluble in water, will not reduce Benedict's solution and is not attacked by enzymes present in the human digestive tract. The molecular weight of cellulose has been estimated to be between 200,000 and 2,000,000. The structure of cellulose is represented in the accompanying diagram.

The chemical treatment of cellulose has resulted in several important commercial products. For example, cotton, which is almost pure cellulose, takes on a silk-like luster and increases in strength when treated under tension with a concentrated solution of sodium hydroxide. This process, called mercerization, produces mercerized cotton, which is used in large quantities in the manufacture of cotton cloth.

Other important products related to cellulose are polymers (see Chapter 12). When treated with nitric acid, cotton is converted into cellulose nitrates, which are esters of commercial importance. Gun-cotton is a nitrocellulose containing about 13 per cent nitrogen and is used in the production of smokeless powder and high explosives. Another nitrocellulose is pyroxylin, which can be made into celluloid, motion picture film, artificial leather, and lacquers for automobile finishes. Cellulosc treated with acetic anhydride yields cellulose acetate, which is a thermoplastic polymer. This material finds applications in the manufacture of safety motion picture films, plastics, and cellulose acetate fiber or yarn. It is also fabricated into films for wrapping food products and to serve as sandwiches between sheets of glass in safety glass. Cellulose acetate dissolved in a volatile solvent is used in fingernail polish. **Rayon** is produced by treating cellulose with sodium hydroxide and carbon disulfide. The solution that results from this treatment is forced through fine holes into dilute sulfuric acid to make the rayon fibers. **Cellophane** is made by a process similar to that used for rayon.

In addition to the esters of cellulose, certain ethers have become important. Methylcellulose, ethylcellulose, and carboxymethylcellulose are examples of these ethers. Methylcelluloses are used as sizing and finish for textiles, for pastes, and for cosmetics. Ethylcellulose has properties that make it a desirable adjunct in the manufacture of plastics, coatings, and films. It is soluble in organic solvents but very resistant to the action of alkalis. Carboxymethylcellulose is used as a protective colloid, a sizing agent for textiles, and as a builder in the manufacture of synthetic detergents.

MIXED POLYSACCHARIDES

Heparin

Heparin is a polysaccharide that possesses anticoagulant properties. It prevents the clotting of blood by inhibiting the conversion of prothrombin to thrombin. Thrombin acts as a catalyst in converting plasma fibrinogen into the fibrin clot. The structure of heparin is still uncertain but it contains a repeating unit of α-1,3 linked glucuronic acid and glucosamine, with sulfate groups on some of the hydroxyl and amino groups.

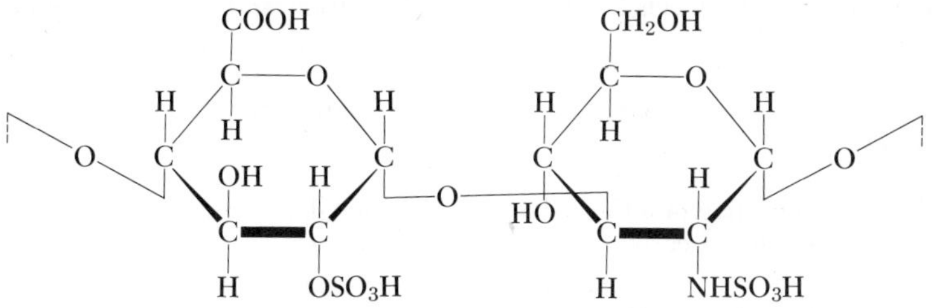

D-Glucuronic acid-2-sulfate D-Glucosamine sulfate

α-1,3 linked repeating unit of heparin

Hyaluronic Acid

A structural polysaccharide found in higher animals is the mucopolysaccharide hyaluronic acid. It is an essential component of the ground substance, or intercellular cement, of connective tissue. Hyaluronic acid has a high viscosity and a molecular weight in the millions. The molecule consists of repeating units of D-glucuronic acid and N-acetyl-D-glucosamine joined in a β-1,3 linkage.

Chondroitin Sulfate

Chondroitin sulfates are structural polysaccharides found in the ground substance and cartilage of mammals. Its structure is similar to hyaluronic acid except that galactosamine sulfate replaces the acetyl glucosamine.

IMPORTANT TERMS AND CONCEPTS

acetal
aldose
asymmetric carbon atom
disaccharides
enantiomers
fructofuranose
glucopyranose
hemiacetal
hexose
ketose
monosaccharides
mutarotation
optical activity
pentose
polarized light
polysaccharides
reducing sugar
starch

QUESTIONS

1. Write the formula for the aldose, glyceraldehyde. Would this compound exhibit optical activity in solution? Explain.

2. Explain the difference between the *meso* form of an optically active compound and a racemic mixture.

3. Write the formulas for D-fructose and L-fructose. Star the asymmetric carbon atoms in each molecule.

4. Write the formula for lactic acid. Would this compound be classified as a carbohydrate? Explain.

5. What is (1) an aldose, (2) a hexose, (3) a pentose, (4) a ketose, and (5) a disaccharide?

6. Explain fully what is meant by (1) D (+) glucose, and (2) L (−) fructose.

7. How does the phenomenon of mutarotation complicate the representation of the formulas for carbohydrates? Explain with an example.

8. Explain the relation between hemiacetal ring formation and Haworth's representation of glucopyranose.

9. How are the α- and β- isomers of the pyranose and furanose ring forms of the sugars indicated in the structures?

10. Write an equation to illustrate the formation of an acetal or a glucoside between β-D-galactopyranose and the OH on carbon-4 of α-D-glucopyranose.

11. When a reducing sugar reacts with Benedict's solution, what other products are formed besides Cu_2O? Why is the formation of Cu_2O important in the test?

12. Write the formula for dihydroxyacetone. Use the formula to illustrate the formation of dihydroxyacetone-3-phosphate.

13. L-Gulonic acid, an intermediate in the formation of ascorbic acid, is formed from D-glucuronic acid by reducing the aldehyde group to a primary alcohol group. Write the formula for L-gulonic acid.

14. Write the formula for sucrose and use it to explain why sucrose will not reduce Benedict's solution.

15. Explain the significance of the Lobry de Bruyn-von Eckenstein transformation.

16. Write a partial polysaccharide structure that illustrates both α-1,4 and α-1,6 linkages between the monosaccharides.

17. Explain why polysaccharides such as starch, glycogen, and cellulose are not considered as reducing sugars.

18. Compare starch, dextrins, glycogen, and cellulose as to size of molecule, chemical composition, color with iodine, and importance.

19. How would you account for the great difference in properties between starch and cellulose?

20. What happens when starch is hydrolyzed?

SUGGESTED READING

Browne: The Origins of Sugar Manufacture in America. Journal of Chemical Education, Vol. 10, p. 323, 1933.

Heuser: The High Points in the Development of Cellulose Chemistry since 1876. Journal of Chemical Education, Vol. 29, p. 449, 1952.

Hudson: Emil Fischer's Discovery of the Configuration of Glucose. Journal of Chemical Education, Vol. 18, p. 353, 1941.

McCord and Getchell: Cotton. Chemical & Engineering News, November 14, 1960, p. 106.

Mowery: Criteria for Optical Activity in Organic Molecules. Journal of Chemical Education, Vol. 46, p. 269, 1969.

Pauling: Vitamin C and the Common Cold. San Francisco, W. H. Freeman and Company, 1970.

Slocum, Sugarman, and Tucker: The Two Faces of D and L Nomenclature. Journal of Chemical Education, Vol. 48, p. 597, 1971.

Vennos: Construction and Uses of an Inexpensive Polarimeter. Journal of Chemical Education, Vol. 46, p. 459, 1969.

LIPIDS

CHAPTER 14

The *objectives* of this chapter are to enable the student to:

1. Illustrate the formation of a triglyceride from glycerol and three molecules of a fatty acid.
2. Write an equation illustrating the process of saponification.
3. Describe the separation of lipids by thin-layer and gas-liquid chromatography.
4. Distinguish between phospholipids and glycolipids.
5. Recognize the sterol nucleus in the steroid hormones.
6. Distinguish between the structures of estrone, testosterone, and cortisone.
7. Describe the relation between vitamin A and Δ^{11} *cis*-retinal.
8. Describe the formation of vitamin D_2 from ergosterol.

The common fats and oils compose a class of organic compounds of biological importance. The general term for this type of material is lipid. **Lipids** are characterized by the presence of fatty acids or their derivatives and by their solubility in fat solvents such as acetone, alcohol, ether, and chloroform. Lipids are essential constituents of practically all plant and animal cells. In the human body they are concentrated in cell membranes and in brain and nervous tissue. Repeated extraction of specimens of body tissue with hot fat solvents will invariably yield a mixture of lipid components.

Chemically, lipids are composed of five main elements: carbon, hydrogen, oxygen, and occasional nitrogen and phosphorus. At present there is no generally accepted method of classification of lipids. Some schemes divide them into simple lipids, compound lipids, and steroids, but a more practical classification may be that followed in this chapter:

Fats—esters of fatty acids with glycerol.
Phospholipids—compounds that contain phosphorus, fatty acids, glycerol, and a nitrogenous compound.
Sphingolipids—compounds that contain a fatty acid, phosphoric acid, choline, and an amino alcohol, sphingosine.
Glycolipids—composed of a carbohydrate, a fatty acid, and an amino alcohol.
Steroids—high molecular weight cyclic alcohols.
Fat soluble vitamins—vitamins A, D, E, and K.
Waxes—esters of fatty acids with alcohols other than glycerol.

FATTY ACIDS

Since all fats are esters of fatty acids and glycerol, it may be well to consider the composition and properties of these substances before discussing lipids in general. Fatty acids, although not lipids themselves, are sometimes classified as derived lipids, since they are constituents of all the above types of lipids. The fatty acids that occur in nature almost always have an even number of carbon atoms in their molecules. They are usually straight chain organic acids that may be saturated or unsaturated. Some of the important fatty acids that occur in natural fats are listed in Table 14–1.

In the series of saturated fatty acids, those up to and including capric acid are liquid at room temperature. The most important saturated fatty acids are **palmitic** and **stearic acids.** They are components of the majority of the common animal and vegetable fats.

Unsaturated fatty acids are characteristic constituents of oils. **Oleic acid,** which contains one double bond, is the most common unsaturated fatty acid. Its formula is written:

$$CH_3(CH_2)_7CH{=}CH(CH_2)_7COOH$$

Ricinoleic acid is an unsaturated fatty acid characterized by the presence of a hydroxyl group and is found in castor oil. Its formula is as follows:

$$CH_3(CH_2)_5CHOHCH_2CH{=}CH(CH_2)_7COOH$$

The structure of fatty acids suggested by x-ray analysis is that of a zigzag configuration with the carbon-carbon bond forming a 109° angle.

TABLE 14-1 SOME IMPORTANT FATTY ACIDS OCCURRING IN NATURAL FATS

NAME	FORMULA	CARBON ATOMS	POSITION OF DOUBLE BONDS	OCCURRENCE
Saturated				
Butyric	C_3H_7COOH	4		Butter fat
Caproic	$C_5H_{11}COOH$	6		Butter fat
Caprylic	$C_7H_{15}COOH$	8		Coconut oil
Capric	$C_9H_{19}COOH$	10		Palm kernel oil
Lauric	$C_{11}H_{23}COOH$	12		Coconut oil
Myristic	$C_{13}H_{27}COOH$	14		Nutmeg oil
Palmitic	$C_{15}H_{31}COOH$	16		Animal and vegetable fats
Stearic	$C_{17}H_{35}COOH$	18		Animal and vegetable fats
Arachidic	$C_{19}H_{39}COOH$	20		Peanut oil
Unsaturated				
Palmitoleic (1 =)*	$C_{15}H_{29}COOH$	16	Δ9†	Butter fat
Oleic (1 =)	$C_{17}H_{33}COOH$	18	Δ9	Olive oil
Linoleic (2 =)	$C_{17}H_{31}COOH$	18	Δ9, 12	Linseed oil
Linolenic (3 =)	$C_{17}H_{29}COOH$	18	Δ9, 12, 15	Linseed oil
Arachidonic (4 =)	$C_{19}H_{31}COOH$	20	Δ5, 8, 11, 14	Lecithin

*Number of double bonds.

†Δ9 indicates a double bond between carbon 9 and 10, Δ12 between carbon 12 and 13, and so forth.

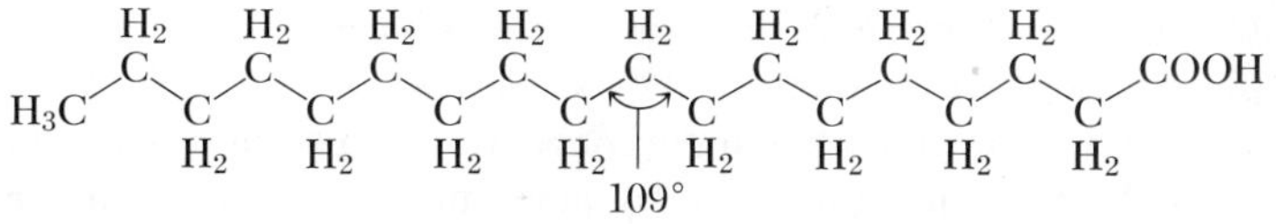

```
    H2   H2   H2   H2   H2   H2   H2   H2
    C    C    C    C    C    C    C    C    COOH
H3C   C    C    C    C    C    C    C    C
      H2   H2   H2   H2   H2   H2   H2   H2
                        109°
```

Stearic acid

Unsaturated fatty acids exhibit a type of geometrical isomerism, discussed on page 69, known as ***cis-trans* isomerism.** The *cis* configuration is found in nature, and the *cis*-form of unsaturated fatty acids is less stable than the *trans*-form. Oleic acid exists in the *cis* configuration; the *trans* configuration of the fatty acid is called elaidic acid.

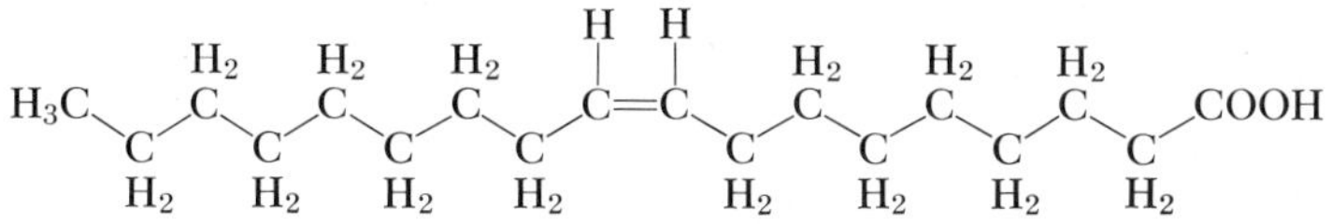

```
                              H  H
       H2   H2   H2           |  |     H2   H2   H2
H3C    C    C    C           C=C      C    C    C     COOH
    C    C    C    C              C    C    C    C
    H2   H2   H2   H2             H2   H2   H2   H2
```

Oleic acid (*cis*-form)

```
    H2   H2   H2   H2     H
    C    C    C    C      |      H2   H2   H2
H3C   C    C    C     C=C       C    C    C    COOH
      H2   H2   H2    |      C    C    C    C
                      H      H2   H2   H2   H2
```

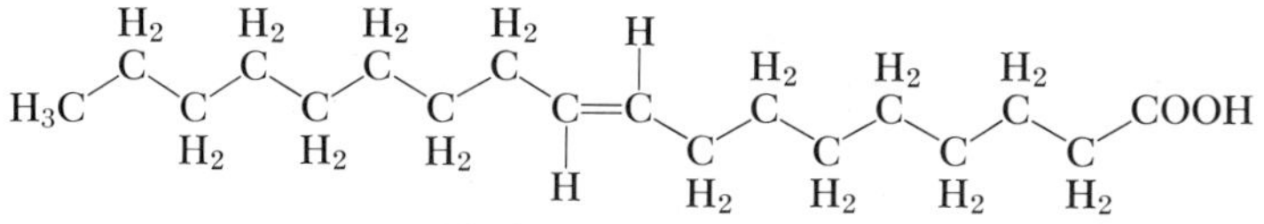

Elaidic acid (*trans*-form)

The **prostaglandins** are cyclic fatty acids found in seminal plasma and other tissues. They are derived from polyunsaturated fatty acids such as arachidonic. A typical example of the prostaglandins is in the compound PGE_2.

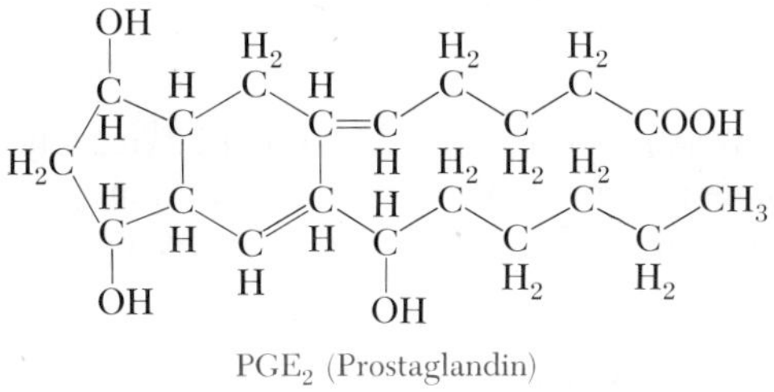

```
      OH
      |      H2          H2    H2
      C  H   C  H        C     C
     /H   C      C=C       C      COOH
 H2C      |      |   H  H2 H2 H2
     \H   C      C  H   C    C     CH3
      C  H   C   H   C    C     C
      |      H       |    H2    H2
      OH             OH
```

PGE_2 (Prostaglandin)

These compounds are involved in the control of lipid metabolism (Chapter 21) and are thought to depress the action of cyclic-3′,-5′-AMP (p. 335).

From a nutritional standpoint, the three most commonly occurring fatty acids in edible animal and vegetable fats are palmitic, stearic, and oleic acids.

FATS

From a chemical standpoint fats are esters of fatty acids and glycerol (p. 126). This combination of 3 molecules of fatty acid with 1 molecule of glycerol may be illustrated as shown in the accompanying diagram.

Tristearin is called a **simple glyceride** because all the fatty acids in the fat molecule are the same. Other examples of simple glycerides would be tripalmitin and triolein. In most naturally occurring fats, different fatty acids are found in the same molecule. These are called **mixed glycerides** and may contain both saturated and unsaturated fatty acids. The glycerides are classed as **neutral lipids** since their molecules are not charged.

Both fats and oils are esters of fatty acids and glycerol. In general, fats are solid at room temperature and are characterized by a relatively high content of saturated fatty acids. Oils are liquids that contain a high concentration of unsaturated fatty acids. A fat that contains short chain saturated fatty acids may also exist as a liquid at room temperature.

Most of the common animal fats are glycerides that contain saturated and unsaturated fatty acids. Since the saturated fatty acids predominate, these fats are solid at room

$$\begin{array}{ccccc}
\mathrm{H} & & & & \mathrm{H}\quad \mathrm{O} \\
| & & \| & & |\qquad \| \\
\mathrm{H{-}C{-}OH} & & \mathrm{HO{-}C{-}C_{17}H_{35}} & & \mathrm{H{-}C{-}O{-}C{-}C_{17}H_{35}} \\
| & & \mathrm{O} & & |\qquad \mathrm{O} \\
| & & \| & & |\qquad \| \\
\mathrm{H{-}C{-}OH} & + & \mathrm{HO{-}C{-}C_{17}H_{35}} & \rightarrow & \mathrm{H{-}C{-}O{-}C{-}C_{17}H_{35}} + 3H_2O \\
| & & \mathrm{O} & & |\qquad \mathrm{O} \\
| & & \| & & |\qquad \| \\
\mathrm{H{-}C{-}OH} & & \mathrm{HO{-}C{-}C_{17}H_{35}} & & \mathrm{H{-}C{-}O{-}C{-}C_{17}H_{35}} \\
| & & & & | \\
\mathrm{H} & & & & \mathrm{H} \\
\text{Glycerol} & & \text{3 molecules of stearic acid} & & \text{Tristearin, a fat}
\end{array}$$

temperature. Beef fat, mutton fat, lard, and butter are important examples of animal fats. Butter fat is readily distinguished from other animal fats because of its relatively high content of short chain fatty acids.

Glycerides that are found in vegetables usually exist as oils rather than fats. Vegetable oils such as olive oil, corn oil, cottonseed oil, and linseed oil are characterized by their high content of oleic, linoleic, and linolenic acids. Coconut oil, like butter fat, contains a relatively large percentage of short chain fatty acids.

Reactions of Fats

Glycerol Portion. When glycerol or a liquid containing glycerol is heated with a dehydrating agent, **acrolein** is formed. Acrolein has a very pungent odor and is sometimes formed by the decomposition of glycerol in the fat of frying meats.

$$\begin{array}{lcl}
\mathrm{CH_2OH} & & \mathrm{H{-}C{=}O} \\
| & & \quad | \\
\mathrm{CHOH} & + \xrightarrow[\mathrm{KHSO_4}]{\mathrm{O_2}} & \quad \mathrm{CH} + 2H_2O \\
| & & \quad \| \\
\mathrm{CH_2OH} & & \quad \mathrm{CH_2} \\
\text{Glycerol} & & \text{Acrolein}
\end{array}$$

The formation of acrolein is often used as a test for fats, since all fats yield glycerol when they are heated.

Rancidity. Many fats develop an unpleasant odor and taste when they are allowed to stand in contact with air at room temperature. The two common types of rancidity are **hydrolytic** and **oxidative.** Hydrolytic changes in fats are the result of the action of enzymes or microorganisms producing free fatty acids. If these acids are of the short chain variety, as is butyric acid, the fats develop a rancid odor and taste. This type of rancidity is common in butter.

The most common type of rancidity is the oxidative type. The unsaturated fatty acids in fats undergo oxidation at the double bonds. The combination with oxygen results in the formation of peroxides, volatile aldehydes, ketones, and acids.

Heat, light, moisture, and air are factors that accelerate oxidative rancidity. The prevention of rancidity of lard and vegetable shortenings that are used in the manufacture of crackers, pretzels, pastries, and similar food products has long been an important problem. Modern packaging has helped considerably in this connection, although a more important contribution has been the development of "**antioxidants.**" These compounds usually contain phenolic groups in their structure, i.e., tocopherol, or vitamin E (p. 241), which effectively inhibits the autoxidation of unsaturated fatty acids. The majority of the vegetable shortenings on the market as well as certain brands of lard are protected from rancidity by the addition of antioxidants.

Hydrogenation. It has already been stated that the main difference between oils and fats is the number of unsaturated fatty acids in the molecule. Vegetable oils may be converted into solid fats by the addition of hydrogen to the double bonds of the unsaturated fatty acids. This process has already been illustrated and discussed on page 75.

Hydrolysis. Fats may be hydrolyzed to form free fatty acids and glycerol by the action of acid, alkali, superheated steam, or the enzyme lipase. In hydrolysis of a fat, the 3 water molecules (that were split out when the 3 fatty acid molecules combined with 1 glycerol molecule in an ester linkage to make the fat molecule) are replaced with the resultant splitting of the fat into glycerol and fatty acids. Commercially, fats are a cheap source of glycerol for use in the manufacture of high explosives and pharmaceuticals. For this purpose the fat is hydrolyzed with superheated steam and a reagent which contains naphthalene, oleic acid, and sulfuric acid. The advantage of this method is that glycerol is readily separated from the fatty acids.

Saponification

Hydrolysis by an alkali is called **saponification,** and produces glycerol and salts of the fatty acids that are called soaps. In the laboratory, fats are usually saponified by an alcoholic solution of an alkali. The fats are more soluble in hot alcohol and the reaction is therefore more rapid. **Soaps** may be defined as metallic salts of fatty acids. The saponification of a fat may be represented as follows:

$$\begin{array}{l} CH_2{-}O{-}C({=}O){-}C_{17}H_{35} \\ | \\ CH{-}O{-}C({=}O){-}C_{17}H_{35} \\ | \\ CH_2{-}O{-}C({=}O){-}C_{17}H_{35} \\ \text{Tristearin} \end{array} + 3NaOH \rightarrow \begin{array}{l} CH_2OH \\ | \\ CHOH \\ | \\ CH_2OH \\ \text{Glycerol} \end{array} + \begin{array}{c} 3C_{17}H_{35}C({=}O){-}ONa \\ \text{Sodium stearate} \\ \text{(soap)} \end{array}$$

Sodium salts of fatty acids are known as **hard soaps,** whereas potassium salts form **soft soaps.** The ordinary cake soaps used in the home are sodium soaps. Yellow laundry soap contains resin, which increases the solubility of soap and its lathering properties and has some detergent action. White laundry soap in the form of bars, soap chips, or powdered soap contains sodium silicate and a water-softening agent such as sodium carbonate or sodium phosphate. Tincture of green soap, commonly used in hospitals, is a solution of potassium soap in alcohol. When sodium soaps are added to hard water, the calcium and magnesium salts present replace sodium to form insoluble calcium and magnesium soaps. The familiar soap curd formed in hard water is due to these **insoluble soaps.**

Detergents. These compounds are a mixture of the sodium salts of the sulfuric acid esters of lauryl and cetyl alcohols. They may be used in hard water because they do not form insoluble compounds with calcium and magnesium.

Extensive research on new detergents and emulsifying agents has resulted in the development of several hundred products possessing almost any desired property. The chemical reaction involved in the formation of detergents and the alteration in structure that results in biodegradable detergents is outlined on page 98. At present over four billion pounds of detergents are sold each year with synthetic detergents, called **syndets,** outselling soap by more than four to one.

It has been commonly observed that soaps and detergents are emulsifying agents that will convert a mixture of oil and water into a permanent emulsion. The **cleansing power** of soaps and detergents is related to their action as emulsifying agents and to their ability to lower surface tension. By emulsifying the grease or oily material that holds the dirt on the skin or clothing, one can rinse off the particles of grease and dirt with water. The ability of soaps and detergents to break or stabilize oil and water emulsion has been given the name **"detergency."** This property may be illustrated with a diagram (Fig. 14–1).

The hydrocarbon portion of the soap molecule tends to dissolve in the oil droplets, while the carboxyl group is strongly attracted to the aqueous phase. As a result of this phenomenon, each oil droplet is negatively charged and tends to repel other oil droplets, resulting in a stable emulsion. **Detergents** are composed of a hydrophilic group similar to the carboxyl group and a hydrocarbon chain.

Analysis of Lipids

For many years the chemical analysis of lipid mixtures has been most difficult. Determination of the **saponification number** yielded a rough measure of molecular weight, the **iodine number** was used for the content of unsaturated fatty acids, the **Reichert-Meissl number** for the content of volatile fatty acids, and the **acetyl number** for the amount of hydroxy fatty acids. With the exception of the iodine number that is discussed in detail on page 77, these determinations have been replaced by more sensitive methods. Thin-layer chromatography and gas-liquid chromatography are presently the methods of choice for the analysis of lipids.

Thin-layer chromatography is carried out on a thin, uniform layer of silica gel spread on a glass plate and activated by heating in an oven (100° to 250°C). Samples of lipid

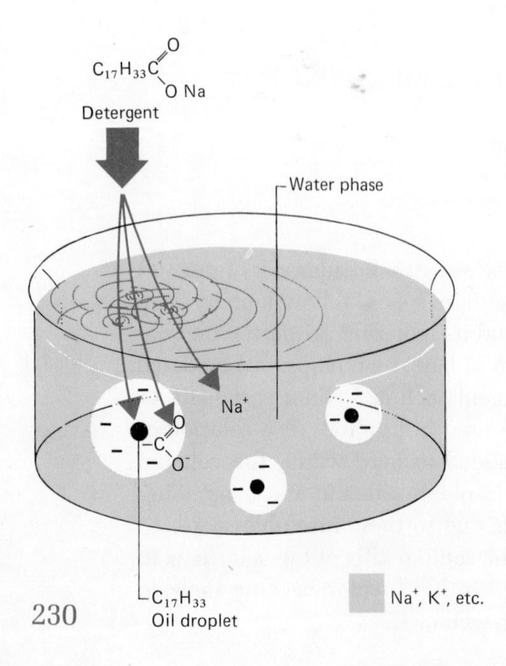

FIGURE 14–1 A diagram illustrating the action of a soap or detergent in stabilizing an oil and water emulsion.

material in the proper solvent are spotted along one edge of the plate with micropipettes. After evaporation of the solvent, the plates are placed vertically in a covered glass tank which contains a layer of suitable solvent on the bottom. Within a few minutes the lipids are separated by the solvent rising through the thin layer carrying the spots to different locations on the silica gel by a combination of adsorption on the gel and varying distribution in the solvent system. The plates are removed, dried, and sprayed with various detection agents to visualize the lipid components. An example of the thin-layer chromatographic separation of several of the phospholipids is shown in Figure 14–2. This technique is very sensitive and can be made quantitative by removing the spots and measuring the concentration of the component by gas-liquid chromatography.

Gas-liquid chromatography is another powerful tool of the lipid chemist. Any substance that is volatile or can be made into a volatile derivative, for example fatty acids being converted into their methyl ester, can be separated and analyzed by this technique. The volatile substance is injected into a long column which contains a nonvolatile liquid on a finely divided inert solid. The column is heated and the volatile material is carried through the tube by an inert gas such as helium. Separation depends on the difference in vapor pressure and the partition coefficients of the components in the nonvolatile liquid. As the fractionated components reach the end of the column, they pass over a detection device that is extremely sensitive to differences in organic material carried by the gas, and it records the changes in the gas flow as peaks on a recorder chart. By the use of

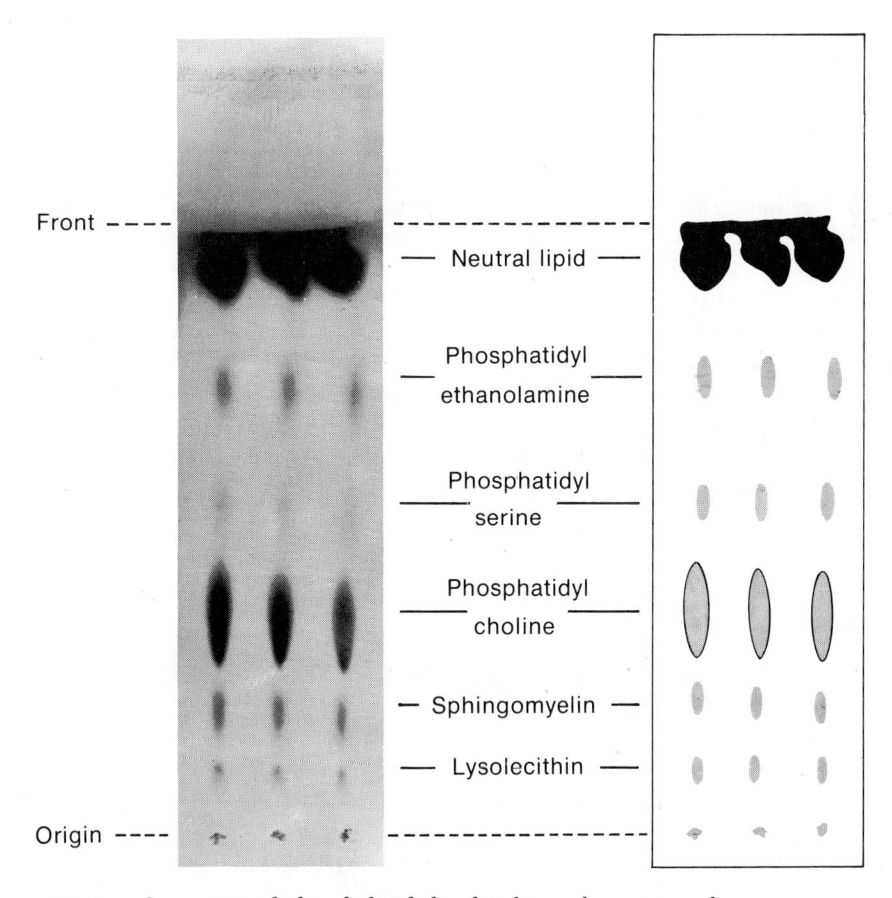

FIGURE 14–2 Separation of phospholipids by thin layer chromatography.

helium gas alone as the control, and known lipid components as standards to determine the position and area under the peaks, quantitative analysis of lipids can be achieved.

PHOSPHOLIPIDS

The phospholipids are found in all animal and vegetable cells. They are composed of glycerol, fatty acids, phosphoric acid, and a nitrogen-containing compound. More specifically, they are esters of **phosphatidic acid** with choline, ethanolamine, serine, or inositol (hexahydroxycyclohexane).

Phosphatidyl Choline, or Lecithins

The lecithins are esters of phosphatidic acid and choline.

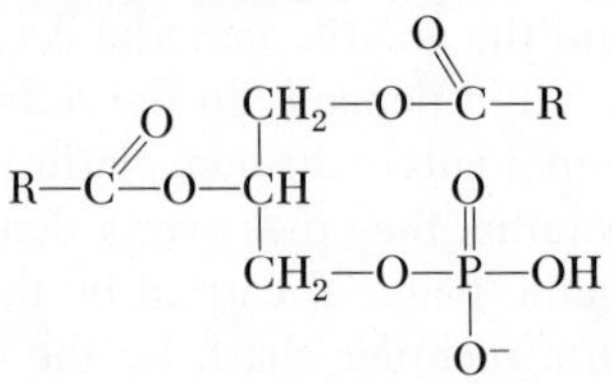

$$CH_2{-}O{-}C({=}O){-}R$$
$$R{-}C({=}O){-}O{-}CH$$
$$CH_2{-}O{-}P({=}O)(O^-){-}OH$$

L-α-Phosphatidic acid

$$CH_2{-}O{-}C({=}O){-}R$$
$$R{-}C({=}O){-}O{-}CH$$
$$CH_2{-}O{-}P({=}O)(O^-){-}O{-}CH_2CH_2\overset{+}{N}(CH_3)_3$$

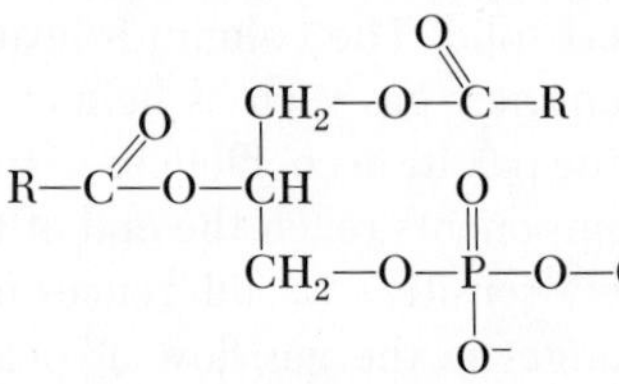

α-Lecithin

The formula is written with the fatty acid on the left side of the central, or β-carbon, to indicate optical activity and an asymmetric carbon atom. Naturally occurring phosphatides have the L form and may contain at least five different fatty acids; however, the β-carbon usually is attached to an unsaturated fatty acid. In addition, the formula indicates that lecithin exists in the dissociated state since phosphoric acid is a fairly strong acid and choline is a strong base. **Choline** is a quaternary ammonium compound whose basicity in aqueous solution is similar to that of KOH (p. 188).

The lecithins are constituents of brain, nervous tissue, and egg yolk. From a physiological standpoint they are important in the transportation of fats from one tissue to another and are essential components of the protoplasm of all body cells. In industry lecithin is obtained from soybeans and finds wide application as an emulsifying agent.

If the oleic acid on the central carbon atom of lecithin is removed by hydrolysis, the resulting compound is called **lysolecithin.** Disintegration of the red blood cells, or hemolysis, is caused by intravenous injection of lysolecithin. The venom of snakes such as the cobra contains an enzyme capable of converting lecithins into lysolecithins, which accounts for the fatal effects of the bite of these snakes. A few insects and spiders produce toxic effects by the same mechanism.

Phosphatidyl Ethanolamine, or Cephalins

The cephalins are found in brain tissue and are essentially mixtures of phosphatidyl ethanolamine and phosphatidyl serine.

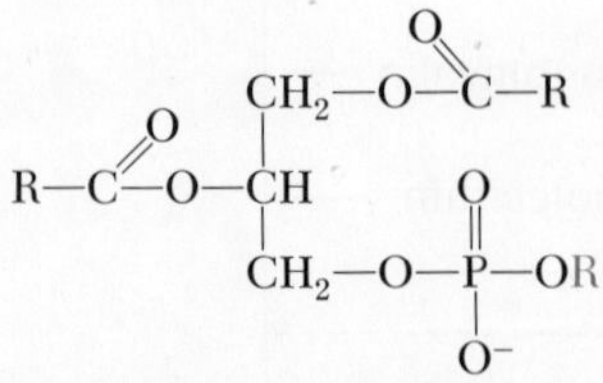

$$CH_2{-}O{-}C({=}O){-}R$$
$$R{-}C({=}O){-}O{-}CH$$
$$CH_2{-}O{-}P({=}O)(O^-){-}OR^1$$

Basic structure

For Phosphatidyl serine,

R^1 is $-CH_2CH(COOH)NH_2$

For Phosphatidyl ethanolamine,

R^1 is $-CH_2CH_2NH_2$

The cephalins are involved in the blood-clotting process and are therefore essential constituents of the body.

Other types of phospholipids related to the lecithins and cephalins are the phosphatidyl inositols and the plasmologens, which are derivatives of phosphatidyl ethanolamine. These compounds are not completely characterized at present but are found in brain, heart, and liver tissue.

SPHINGOLIPIDS

Sphingomyelins

The sphingomyelins differ chemically from the lecithins or cephalins. The common structure in both sphingolipids and glycolipids is **ceramide,** which is composed of sphingosine and a fatty acid (R).

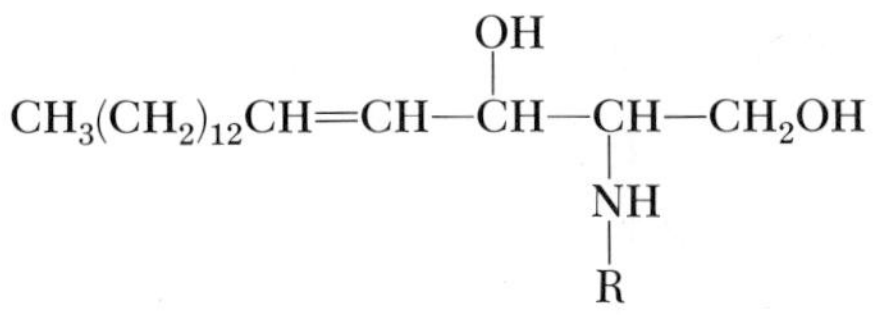

$$CH_3(CH_2)_{12}CH{=}CH{-}\underset{}{\overset{OH}{\overset{|}{C}}}H{-}\underset{\underset{\underset{R}{|}}{NH}}{\underset{|}{C}}H{-}CH_2OH$$

Ceramide

$$\text{Ceramide}{-}O{-}\overset{\overset{O}{\|}}{\underset{\underset{O^-}{|}}{P}}{-}O{-}CH_2CH_2\overset{+}{N}(CH_3)_3$$

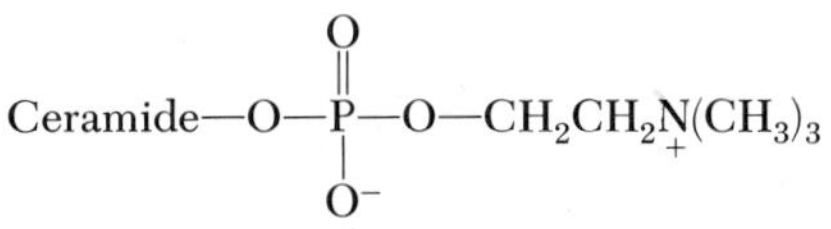

Sphingomyelin

The sphingomyelins are found in large amounts in brain and nervous tissue and are essential constituents of the protoplasm of cells.

GLYCOLIPIDS

The glycolipids are compound lipids that contain a carbohydrate.

Cerebrosides

These lipids are often called **cerebrosides** because they are found in brain and nervous tissue. They are composed of a carbohydrate and a ceramide. There are four different cerebrosides, each containing a different fatty acid. **Kerasin** contains lignoceric acid ($C_{23}H_{47}COOH$), **phrenosin** a hydroxy lignoceric acid, **nervon** an unsaturated lignoceric acid with one double bond called nervonic acid, and **oxynervon,** a hydroxy derivative of nervonic acid. The carbohydrate in these lipids is usually galactose, although glucose is sometimes present. The structure of a typical cerebroside may be represented as follows:

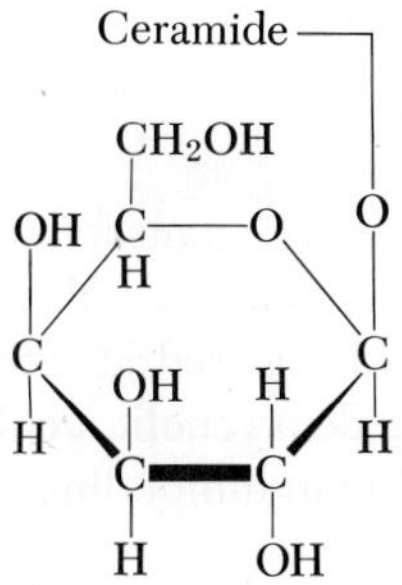

Ceramide
CH₂OH
OH C—O O
H
C C
OH H
H H
C—C
H OH

Cerebroside

STEROIDS

The steroids are derivatives of cyclic alcohols of high molecular weight that occur in all living cells. The lipid material from tissue that is not saponifiable by alkaline hydrolysis contains the steroids. The parent hydrocarbon compound for all the steroids is the cyclopentanophenanthrene nucleus, also called the sterol nucleus. This structure is an integral part of the cholesterol molecule which may be used to illustrate the lettering system for the rings and the number system for the carbon atoms.

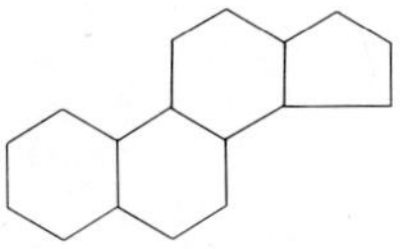

Sterol nucleus 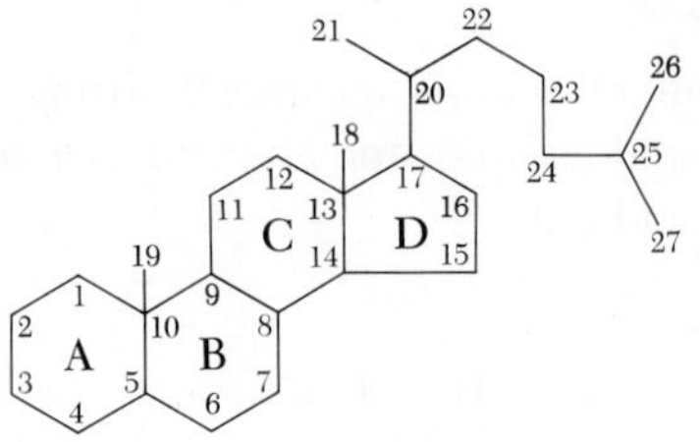   Cholesterol structure designation

The most common sterol is **cholesterol,** which is found in brain and nervous tissue and in gallstones. The structure of cholesterol is shown as follows (each ring is completely saturated and where there is a double bond in the ring it is specifically designated):

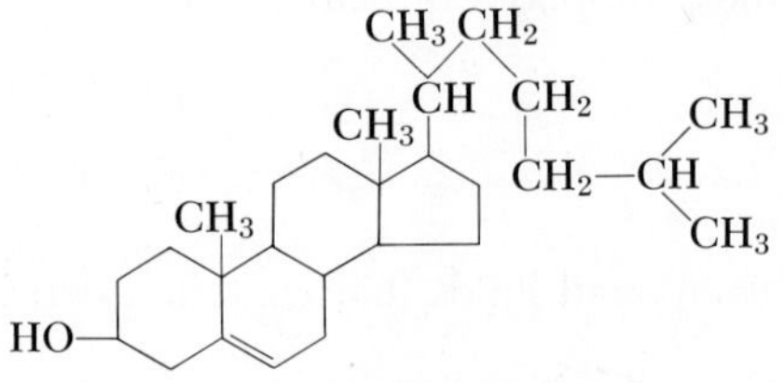

Cholesterol

Cholesterol reacts with acetic anhydride and sulfuric acid in a dry chloroform solution to yield a green color. This is called the **Liebermann-Burchard reaction** and is the basis for both qualitative detection and quantitative methods for cholesterol. Cholesterol is the precursor of bile acids, sex hormones, hormones of the adrenal cortex, and vitamin D, which will be discussed in the following sections.

BILE SALTS

The bile salts are natural emulsifying agents found in the bile, a digestive fluid formed by the liver. Cholesterol and bile pigments are also important constituents of the bile. Bile is stored in the gall bladder and released at intervals to assist in the digestion and absorption of fats. **Cholic acid** and **deoxycholic acid** are the major bile acids that are combined with glycine or taurine by an amide linkage to form bile salts such as glycocholate or taurocholate.

Basic structure

Cholic acid: R^1 is OH and R^2 is OH
Deoxycholic acid: R^1 is OH and R^2 is H
Glycocholic acid: R^1 is NH_2CH_2COOH (Glycine)
Taurocholic acid: R^1 is $NH_2CH_2CH_2SO_3H$ (Taurine)

Chenodeoxycholic acid with hydroxyl groups at positions 3 and 7 in rings A and B, and **lithocholic acid** with a single hydroxyl group at position 3, are also found in human bile.

HORMONES OF THE ADRENAL CORTEX

The cortex of the adrenal gland is an endocrine gland (Fig. 14–3) which produces a group of hormones with important physiological functions. If the gland exhibits decreased function, as in Addison's disease, electrolyte and water balance are abnormal, carbohydrate and protein metabolisms are adversely affected, and the patient is more sensitive to cold and stress. Typical steroid hormones of the gland are represented as follows:

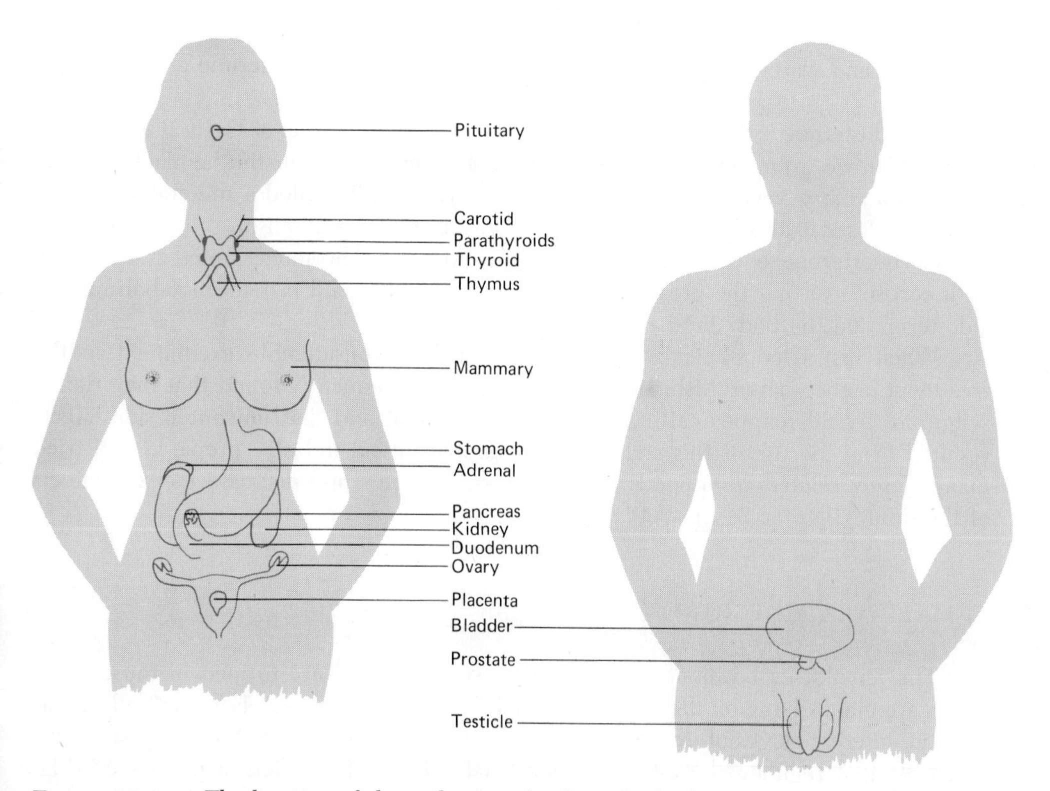

FIGURE 14-3 The location of the endocrine glands in the body.

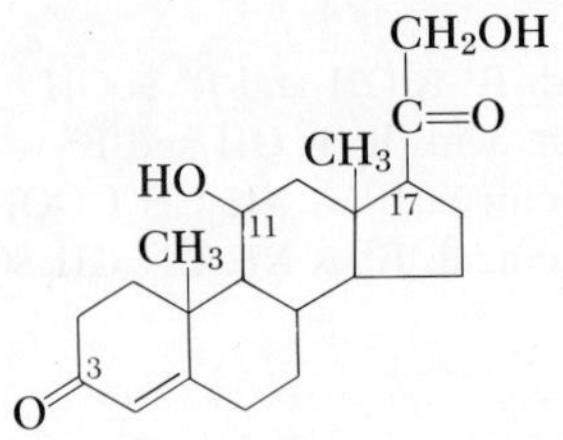

Corticosterone

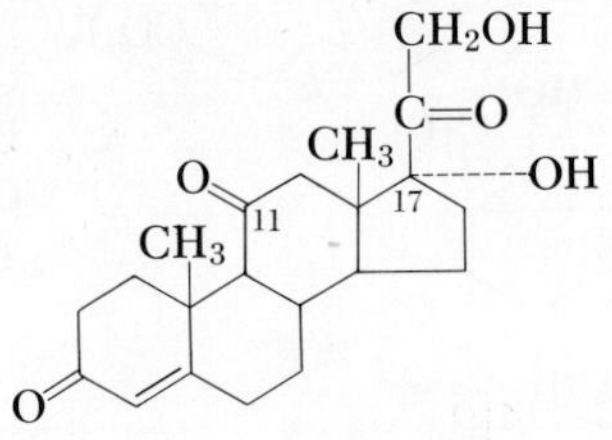

11-Dehydro-17-hydroxycorticosterone
(compound E, cortisone)

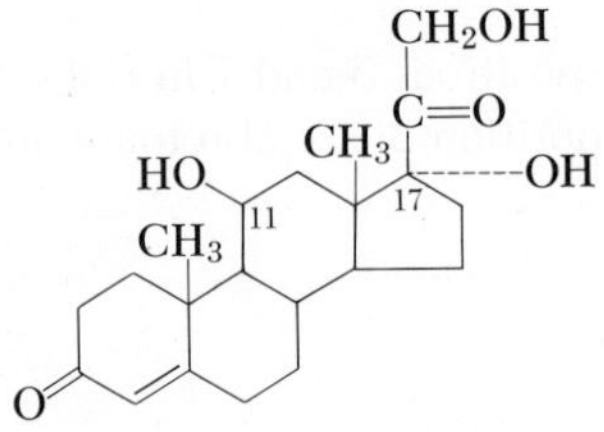

17-Hydroxycorticosterone
(hydrocortisone, cortisol)

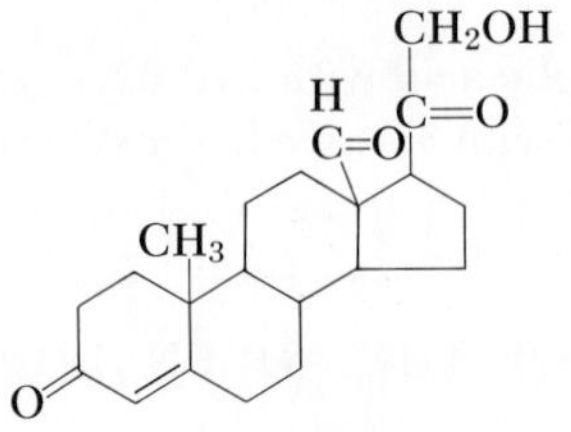

Aldosterone
(aldehyde form)

Corticosterone was the original name of the first adrenal cortical hormone, which accounts for the naming of other hormones as derivatives of this compound. Three major types of adrenal cortical hormones illustrate the relation of structure to physiological activity.

1. Compounds containing an oxygen on the C-11 position (C—OH, or C=O) exhibit greatest activity in carbohydrate and protein metabolism. Examples are **corticosterone, cortisone,** and **cortisol.**

2. Hormones without an oxygen on the C-11 position have their greatest effect on electrolyte and water metabolism. Examples are **11-deoxycorticosterone** and **11-deoxycortisol.**

3. **Aldosterone** is the only compound without a methyl group at C-18. It is replaced by an aldehyde group that can exist in the aldehyde form or in the hemiacetal form. Aldosterone has a very potent effect on electrolytes and is called a **mineralocorticoid.** In higher doses it also acts on carbohydrate and protein metabolism.

Corticosterone, cortisol, and aldosterone are the major hormones found in the blood, with **cortisol** exerting the greatest effect on carbohydrate and protein metabolism, and aldosterone on the body fluid electrolytes.

When first tried clinically, **cortisone** stimulated considerable excitement in the treatment of rheumatoid arthritis. However, it was subsequently found that the original symptoms would reappear after a period of treatment and that unwanted side effects resulted from the use of this steroid. The pharmaceutical industry prepared and tried many closely related compounds to increase the potency and decrease the side effects of the drug (Chapter 24, p. 384).

FEMALE SEX HORMONES

The female sex hormones are steroid in structure and are formed in the ovaries, which are glands lying on the sides of the pelvic cavity (Fig. 14–3). Follicles and corpus lutea in different stages of development are located in the cortex of the ovary and form hormones that regulate the **estrus** or **menstrual cycle** and function in pregnancy (Fig. 14–4). Follicular hormones are also responsible for the development of the secondary sexual characteristics that occur at puberty.

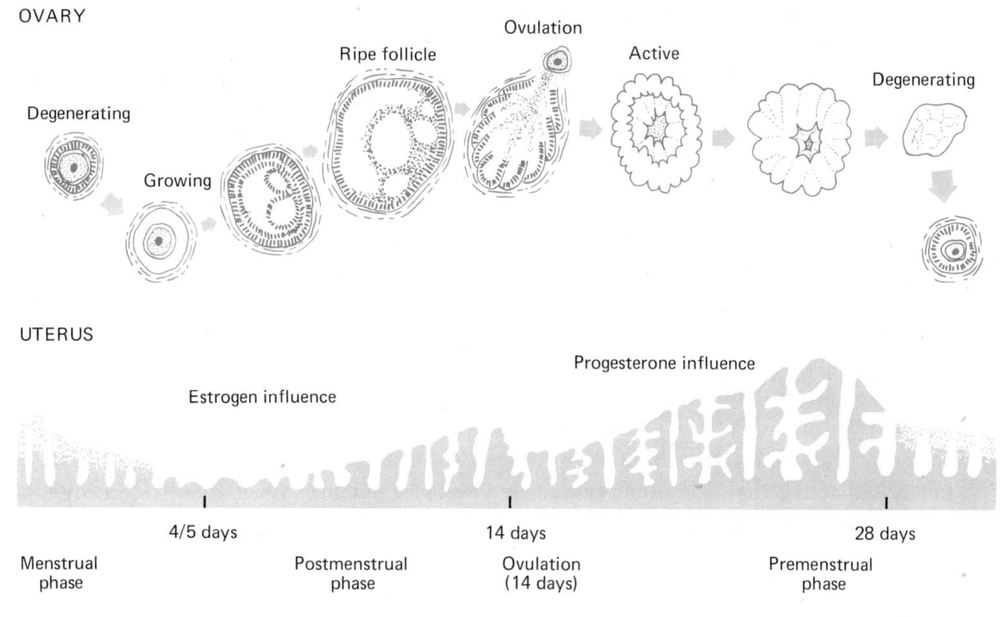

FIGURE 14-4 The sequence of events in the menstrual cycle.

Hormones of the Follicle

The liquid within the follicle contains at least two hormones, known as **estrone** and **estradiol.** Estrone (theelin) was the first hormone to be isolated from the follicular liquid, but estradiol (dihydrotheelin) is more potent than estrone and may be the principal hormone.

Estrone Estradiol

These two compounds are excreted in the urine in increased amounts during pregnancy.

The Hormone of the Corpus Luteum

The hormone produced by the corpus luteum is called **progesterone.** In the body progesterone is converted into **pregnanediol** by reduction before it is excreted in the urine. These two compounds are similar to the estrogens in chemical structure.

Progesterone Pregnanediol

The main function of progesterone is the preparation of the uterine endometrium for implantation of the fertilized ovum. If pregnancy occurs, this hormone is responsible for the retention of the embryo in the uterus and for the development of the mammary glands prior to lactation. In the normal menstrual cycle the administration of progesterone inhibits ovulation, a property used in the development of "the pill" (p. 384).

MALE SEX HORMONES

The male sex hormones are produced by the testes, which are two oval glands located in the scrotum of the male (Fig. 14–3). Small glands in the testes form spermatozoa, which are capable of fertilizing a mature ovum. Between the cells that manufacture spermatozoa are the **interstitial cells,** which produce a hormone called **testosterone.** This hormone is probably converted into other compounds such as **androsterone** before being excreted in the urine. **Dehydroandrosterone** has also been isolated from male urine but is much less active than the other two hormones. The male sex hormones, or **androgens,** have structures similar to the estrogens.

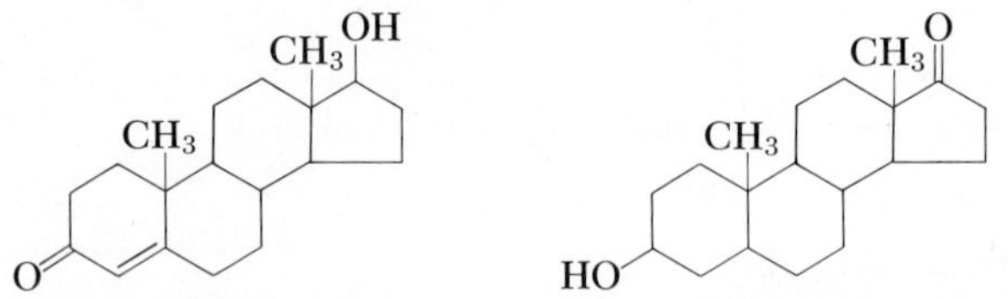

Testosterone

Androsterone

The main function of the androgens in man is the development of masculine sexual characteristics, such as deepening of the voice, the growth of a beard, and distribution of body hair at puberty. They also control the function of the glands of reproduction (seminal vesicles, prostate, and Cowper's gland).

FAT SOLUBLE VITAMINS

After vitamin C deficiency was related to **scurvy** and vitamin B to **beriberi,** it was noted in experiments prior to 1920 that certain animal fats such as butter and cod liver oil were capable of promoting growth in young rats which were fed a purified diet. These fat soluble vitamins were first collectively called vitamin A, but now include vitamins A, D, E, and K.

Vitamin A

Vitamin A is closely related to the carotenoid pigments, alpha, beta, and gamma carotene and cryptoxanthin, which are polyunsaturated hydrocarbons. The carotene pigment, beta carotene, has an all-*trans* structure and is an active precursor of the vitamin.

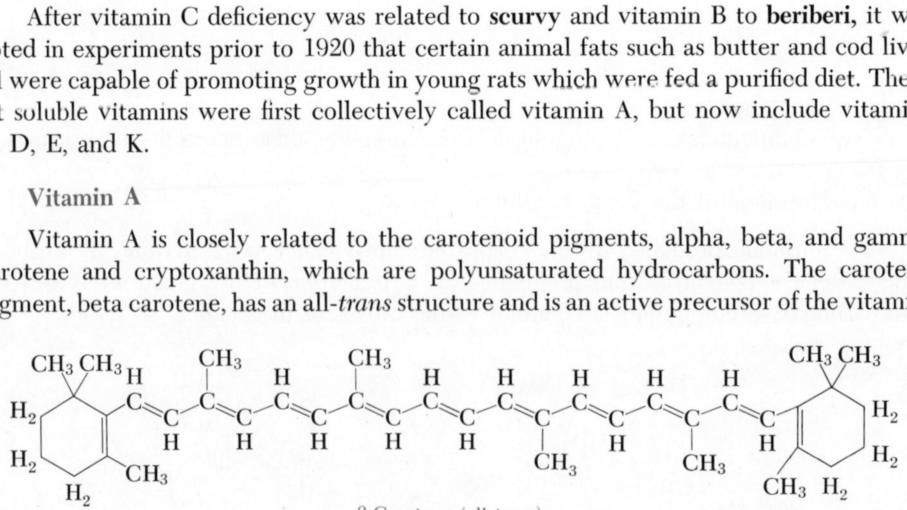

β-Carotene (all-*trans*)

Vitamin A represents half the beta carotene molecule with the ends oxidized to primary alcohol groups.

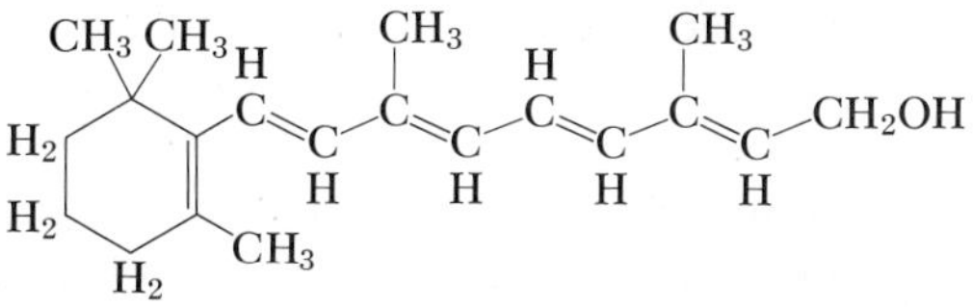

Vitamin A (all-*trans*)

The carotene pigments and cryptoxanthine can be converted into vitamin A in the animal body. The vitamin is soluble in fat and fat solvents and is a liquid at room temperature.

A diet deficient in vitamin A will not support growth, and the deficiency adversely affects the epithelial cells of the mucous membrane of the eye, the respiratory tract, and the genitourinary tract. The process in which the mucous membrane hardens and dries up is known as **keratinization.** The eye is first to show the effect of a deficiency and one of the first symptoms of the lack of the vitamin is **night blindness.** Later the eyes develop a disease called **xerophthalmia.** This disease is characterized by inflamed eyes and eyelids. The eyes ultimately become infected, and when this infection involves the cornea and lens, sight is permanently lost. A continued deficiency of vitamin A results in extensive infection in the respiratory tract, the digestive tract, and the urinary tract. Vitamin A deficiency also causes sterility, since it affects the lining of the genital tract. It is therefore necessary for normal reproduction and lactation. More recently the vitamin has been found essential for the synthesis of mucopolysaccharides which form the ground substance of structural tissue and for the maintenance of the stability of cellular membranes and the membranes of subcellular particles such as lysosomes and mitochondria.

Fish liver oils are potent sources of vitamin A. Eggs, liver, milk and dairy products, green vegetables, and tomatoes are good food sources of the vitamin. The body has the ability to store vitamin A in the liver when it is present in the food in excess of the body requirements. Infants obtain a store of the vitamin in the first milk (colostrum) of the mother, which is ten to one hundred times as rich in vitamin A as ordinary milk.

The function of vitamin A in the visual process involves the formation of **rhodopsin,** a visual pigment which is a complex composed of retinine and a protein (opsin). Retinine has been identified as vitamin A-aldehyde and is now called **retinal,** which may exist in the *cis* or all-*trans* form. The relation between rhodopsin, retinal, and vitamin A and the visual cycle is shown in the following outline:

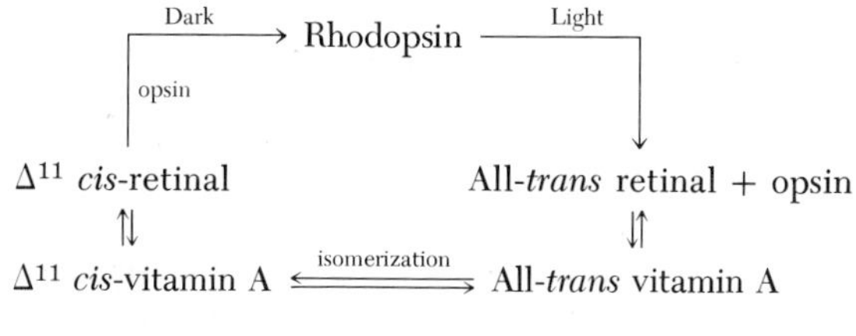

Dark → Rhodopsin — Light

opsin

Δ^{11} *cis*-retinal   All-*trans* retinal + opsin

Δ^{11} *cis*-vitamin A ⇌ (isomerization) All-*trans* vitamin A

All-*trans* retinal is vitamin A with the primary alcohol group oxidized to an aldehyde. The structure of Δ^{11} *cis*-retinal is shown as follows:

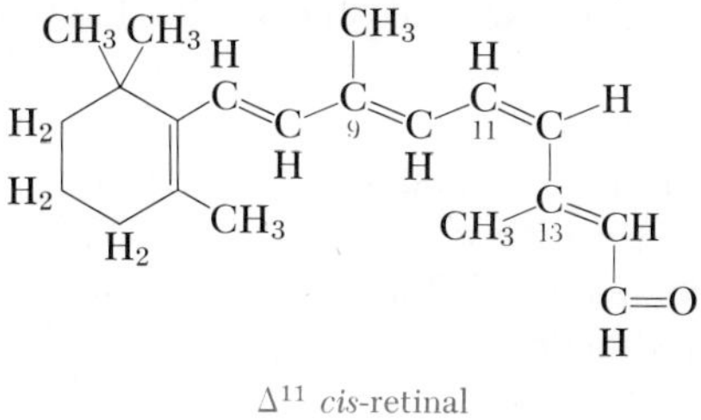

Δ^{11} *cis*-retinal

When light strikes rhodopsin, isomerization of Δ^{11} *cis*-retinal to the all-*trans* retinal occurs, and the complex splits into the protein opsin and all-*trans* retinal. The latter compound is reduced to all-*trans* vitamin A. In the regeneration of the visual pigment rhodopsin, the all-*trans* vitamin A is first isomerized to Δ^{11} *cis*-vitamin A, then oxidized to Δ^{11} retinal which combines with opsin to form rhodopsin. The time required to adapt to dim light when coming from a brightly lighted room is the time needed to regenerate a normal concentration of the visual pigment.

There is a loss of vitamin A during the regeneration of rhodopsin after exposure to light. Since this supply of vitamin must come from the blood, the normal rate of rhodopsin synthesis is therefore dependent on the vitamin A concentration in the blood.

Vitamin D

Chemical Nature. Several compounds with vitamin D activity exist, although only two of them commonly occur in antirachitic drugs and foods. These two compounds are produced by the irradiation of ergosterol and 7-dehydrocholesterol with ultraviolet light. Ergosterol is a sterol that occurs in yeast and molds, whereas 7-dehydrocholesterol is found in the skin of animals. Irradiated ergosterol is called **calciferol,** or vitamin D_2; irradiated 7-dehydrocholesterol is called vitamin D_3. The structures of the two forms of vitamin D are very similar.

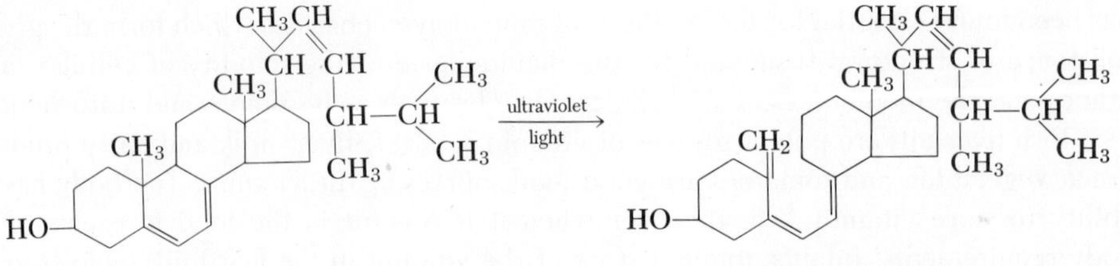

Ergosterol —ultraviolet light→ Vitamin D_2 (calciferol)

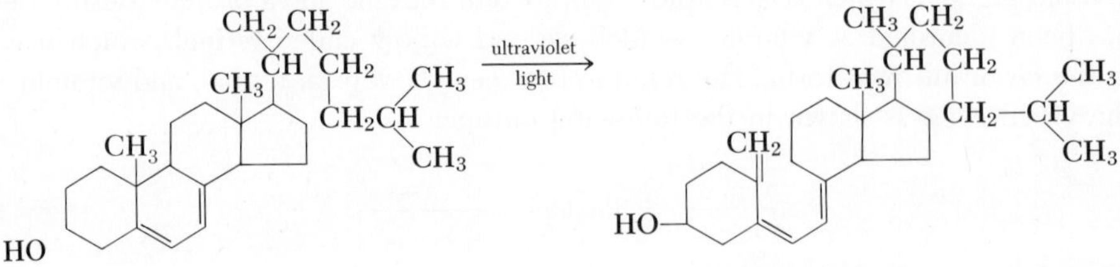

7-Dehydrocholesterol —ultraviolet light→ Vitamin D_3

The lack of vitamin D in the diet of infants and children results in an abnormal formation of the bones, a disease called **rickets.** Calcium, phosphorus, and vitamin D are all involved in the formation of bones and teeth. Bowed legs, a "rachitic rosary" of the ribs, an abnormal formation of the ribs known as "pigeon breast," and poor tooth development are common signs of vitamin D deficiency in small children. Rickets does not occur in adults after bone formation is complete, although the condition of **osteomalacia** may occur in women after several pregnancies. In osteomalacia, the bones soften and abnormalities of the bony structure may occur.

The fish liver oils are the most potent sources of vitamin D; fish such as sardines, salmon, and herring are the richest food sources. The ultraviolet rays in sunlight form vitamin D by irradiation of 7-dehydrocholesterol in the skin. Children who play outdoors in the summer materially increase the vitamin D content of their bodies. Milk in particular and other foods have their vitamin D content increased by the addition of small amounts of irradiated ergosterol.

The main function of vitamin D in the body is to increase the utilization of calcium and phosphorus in the normal formation of bones and teeth. The exact mode of action of the vitamin is not known, although it increases calcium and phosphorus absorption from the intestine, stimulates the activity of the enzyme phosphatase, and is essential for normal growth.

Vitamin E

Vitamin E is chemically related to a group of compounds called **tocopherols.** Alpha-, beta-, and gamma-tocopherol have vitamin E activity, but alpha-tocopherol is the most potent.

```
            CH3
             |
             C     O      CH3          CH3             CH3             CH3
          // \   /  \   /               |               |               |
   CH3—C       C       C—(CH2)3—CH—(CH2)3—CH—(CH2)3—CH—CH3
         |      ||       |
    HO—C       C       CH2
          \\  /  \   /
             C     CH2
             |
            CH3
```

Alpha-tocopherol

The other two tocopherols differ only in the number and position of the CH_3 groups on the aromatic ring. Beta is a 1,4-di-CH_3, and gamma a 1,2-di-CH_3, derivative. Vitamin E is stable to heat but is destroyed by oxidizing agents and ultraviolet light. Oxidative rancidity of fats rapidly destroys the potency of the vitamin.

It has been known for several years that when animals such as rabbits and rats are maintained on a vitamin E deficient diet, muscular dystrophy, creatinuria, and anemia develop in rabbits, and changes in reproductive organs and function occur in rats. The richest source of vitamin E is wheat germ oil. Corn oil, cottonseed oil, egg yolk, meat, and green leafy vegetables are good sources of this vitamin.

The tocopherols are excellent antioxidants and prevent the oxidation of several substances in the body, including unsaturated fatty acids and vitamin A. As an antioxidant, vitamin E may protect mitochondrial systems in the cell from irreversible oxidation by lipid peroxides. It may also protect lung tissue from damage by oxidants present in smog-contaminated atmospheres.

Vitamin K

Vitamin K is a derivative of 1,4-naphthoquinone, as is illustrated in the formula for vitamin K:

```
                O
       H        ||
       C        C
    //   \    /   \
  HC       C        C—CH3                                                  CH3
   |       ||       ||                                                      |
  HC       C        C—CH2—CH=C—(CH2)3—CH—(CH2)3—CH—(CH2)3—CH
    \\   /   \    /              |                |               |               |
       C        C               CH3             CH3             CH3             CH3
       H        ||
                O
```

Vitamin K

The 2-methyl-1,4-napthoquinones and naphthohydroquinones possess vitamin activity. Vitamin K is fat soluble and therefore is soluble in ordinary fat solvents. It is stable to heat, but is destroyed by alkalies, acids, oxidizing agents, and light.

A diet lacking in vitamin K will cause an increase in the clotting time of blood. This condition produces hemorrhages under the skin and in the muscle tissue. The abnormality in the clotting mechanism is due to a reduction in the formation of **prothrombin,** one of the factors in the normal process.

Rich sources of vitamin K_1 are alfalfa, spinach, cabbage, and kale. The vitamin K_2 series is present in bacteria. The bacteria present in putrefying fish meal are capable of synthesizing vitamin K_2 and are potent sources.

The vitamin is essential in the synthesis of prothrombin by the liver, but the exact mechanism of synthesis is as yet unknown. It is also thought to be involved in the metabolic reactions and electron transport systems in the mitochondria of cells. The role of **prothrombin** in the clotting of blood is shown in the following sequence:

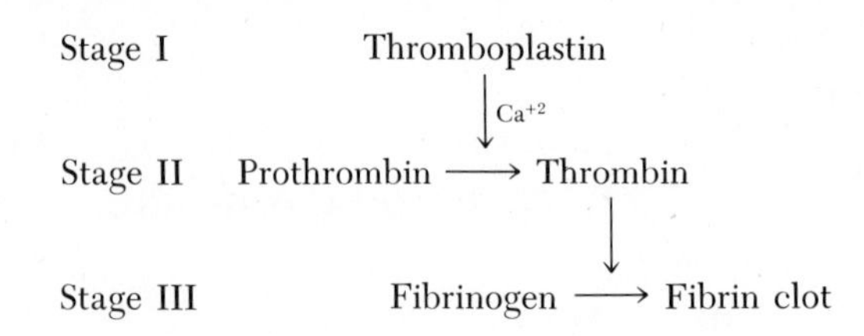

Although the clotting process is represented as two simple equations, it involves a large series of factors and reactions in which prothrombin is converted to the enzyme thrombin, which in turn converts plasma **fibrinogen** to the **fibrin clot.**

WAXES

Waxes are simple lipids that are esters of fatty acids and high molecular weight alcohols. Fatty acids such as myristic, palmitic, and carnaubic are combined with alcohols that contain from 12 to 30 carbon atoms. Common, naturally occurring waxes are **beeswax, lanolin, spermaceti,** and **carnauba wax.** Beeswax is found in the structural part of the honeycomb. Lanolin, from wool, is the most important wax from a medical standpoint, since it is widely used as a base for many ointments, salves, and creams. Spermaceti, obtained from the sperm whale, is used in cosmetics, some pharmaceutical products, and in the manufacture of candles. Carnauba wax is obtained from the carnauba palm and is widely used in floor waxes and in automobile and furniture polishes.

IMPORTANT TERMS AND CONCEPTS

acrolein
calciferol
cephalin
cerebroside
cholesterol
Δ^{11} *cis*-retinal
detergents
estrone
lecithin
rancidity
saponification
steroids
testosterone
tocopherols
triglyceride
unsaturated fatty acid

QUESTIONS

1. Name and write the formulas for the three most commonly occurring fatty acids.

2. Are most commonly occurring fats composed of simple glycerides or mixed glycerides? Explain.

3. Illustrate the formation of an ester linkage between glycerol and three molecules of butyric acid. How would you name the product?

4. Write the formula for a triglyceride that would exist as a solid at room temperature and for one that would exist as an oil.

5. Explain how you would test for the presence of glycerol in the laboratory and why this test can be used as a general test for fats.

6. What type of rancidity occurs in common shortenings? How is this prevented? Explain.

7. What chemical reaction occurs when butter becomes rancid? What agent initiates this reaction?

8. Write an equation illustrating the process of saponification. Name all compounds in the equation.

9. How do insoluble soaps, soft soaps, and hard soaps differ from each other?

10. Discuss the advantages and disadvantages of the new detergents and emulsifying agents.

11. Explain the cleansing power of soaps and detergents.

12. What analytical procedure would you employ to separate a mixture of phospholipids and neutral lipids into its components? Explain.

13. Write the formula for a typical phosphatidyl ethanolamine. What function does this compound serve in the body?

14. Explain the relation between the structure of the adrenal cortical steroids and their physiological function.

15. What is the major structural difference between aldosterone and the other steroid hormones of the adrenal cortex?

16. Write the structure for estrone and indicate how it differs from the male sex hormones.

17. What is the relation between Δ^{11} *cis*-retinal and vitamin A?

18. Compare vitamins D_2 and D_3 as to source and structure.

SUGGESTED READING

Beyler: Some Recent Advances in the Field of Steroids. Journal of Chemical Education, Vol. 37, p. 497, 1960.
Cayle: Enzymes as Detergent Additives. Journal of Chemical Education, Vol. 46, p. 780, 1969.
Dowling: Night Blindness. Scientific American, Vol. 215, No. 4, p. 78, 1966.
Frank: Carotenoids. Scientific American, Vol. 194, No. 1, p. 80, 1956.
Gaucher: An Introduction to Chromatography. Journal of Chemical Education, Vol. 46, p. 729, 1969.
Herndon: The Structure of Choleic Acids. Journal of Chemical Education, Vol. 44, p. 724, 1967.
Hubbard and Kropf: Molecular Isomers in Vision. Scientific American, Vol. 216, No. 6, p. 64, 1967.
Johnson and Williams: Action of Sight upon the Visual Pigment Rhodopsin. Journal of Chemical Education, Vol. 47, p. 736, 1970.
Kushner and Hoffman: Synthetic Detergents. Scientific American, Vol. 185, No. 4, p. 26, 1951.
Pike: Prostaglandins. Scientific American, Vol. 225, No. 5, p. 84, 1971.
Ruchelman: Gas Chromatography: Medical Diagnostic Aid. Chemistry, Vol. 43, No. 11, p. 14, 1970.
Sarett: The Hormones. Journal of Chemical Education, Vol. 37, p. 184, 1960.
Solubilization Pegged as a Key to Detergency. Chemical & Engineering News, Aug. 29, 1960, p. 34.
Young: Visual Cells. Scientific American, Vol. 233, No. 4, p. 80, 1970.

PROTEINS

CHAPTER 15

The *objectives* of this chapter are to enable the student to:

1. Differentiate between proteins, carbohydrates, and fats on the basis of their elementary composition.
2. Explain why naturally occurring amino acids are called alpha amino acids.
3. Recognize a sulfur-containing, an aromatic, and a heterocyclic amino acid.
4. Graphically illustrate the pK_1, the pK_2, and the isoelectric point of an amino acid like glycine.
5. Illustrate the formation and hydrolysis of a dipeptide.
6. Describe the separation of amino acids by ion exchange chromatography.
7. Describe the analysis of the insulin molecule by Sanger.
8. Recognize the difference between polypeptides similar to oxytocin and vasopressin.
9. Distinguish between the primary, secondary, and tertiary structures of a protein.
10. Describe how to determine the amount of protein in a solution.
11. Describe three methods of precipitating a protein out of a solution.
12. Write the equation for the precipitation of a protein by a heavy metal.

The chemistry of proteins is more complex than that of carbohydrates and lipids. Proteins were recognized as essential nitrogen-containing constituents of protoplasm by Mulder in 1839. He named them **proteins,** from a Greek word meaning "of prime importance." Proteins are fundamental constituents of all cells and tissues in the body and are present in all fluids except the bile and urine. They are also essential constituents of the diet required for the synthesis of body tissue, enzymes, certain hormones, and protein components of the blood.

Proteins are made by plant cells by a process starting with photosynthesis from carbon dioxide, water, nitrates, sulfates and phosphates. The complicated process of synthesis has not as yet been completely elucidated. Animals can synthesize only a limited amount of protein from inorganic sources and are mainly dependent on plants or other animals for their source of dietary protein. Proteins are used in the body for growth of new tissue, for maintenance of existing tissue, and as a source of energy. When used for energy, they are broken down by oxidation to form simple substances such as water, carbon dioxide, sulfates, phosphates, and simple nitrogen compounds that are excreted from the body. These same products are formed in decaying plant and animal matter. The simple nitrogen compounds such as amino acids and urea are converted into ammonia,

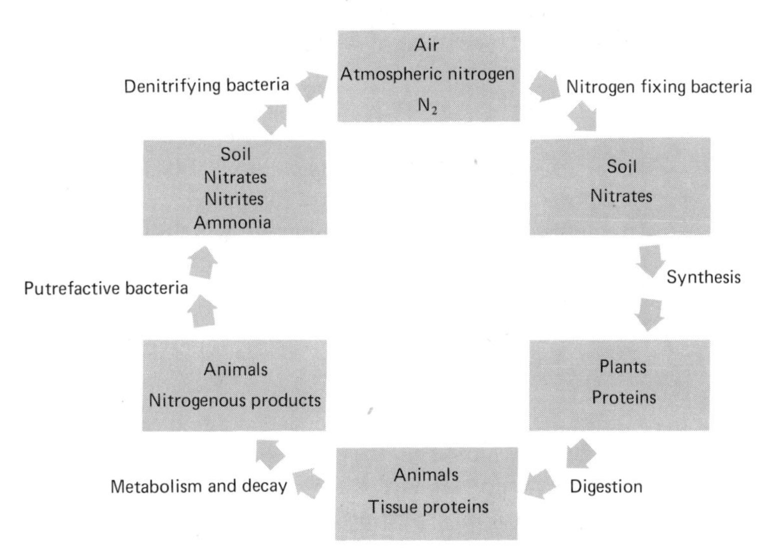

FIGURE 15-1 A simple diagram showing the events occurring in the nitrogen cycle.

nitrites, and nitrates. The growing plants then use these inorganic compounds to form new proteins, and the cycle is completed (Fig. 15-1).

ELEMENTARY COMPOSITION

The five elements that are present in most naturally occurring proteins are **carbon, hydrogen, oxygen, nitrogen,** and **sulfur.** There is a wide variation in the amount of sulfur in proteins. Gelatin, for example, contains about 0.2 per cent, in contrast to 3.4 per cent in insulin.

Other elements, such as phosphorus, iodine, and iron, may be essential constituents of certain specialized proteins. Casein, the main protein of milk, contains phosphorus, an element of utmost importance in the diet of infants and children. Iodine is a basic constituent of the protein in the thyroid gland and is present in sponges and coral. Hemoglobin of the blood, which is necessary for the process of respiration, is an iron-containing protein. Most proteins show little variation in their elementary composition; the average content of the five main elements is as follows:

Element	*Average Per Cent*
Carbon	53
Hydrogen	7
Oxygen	23
Nitrogen	16
Sulfur	1

The relatively high content of nitrogen differentiates proteins from fats and carbohydrates.

MOLECULAR WEIGHT

Protein molecules are very large, as indicated by the approximate formula for oxyhemoglobin:

$$C_{2932}H_{4724}N_{828}S_8Fe_4O_{840}$$

The molecular weight of oxyhemoglobin would thus be about 68,000. The common protein egg albumin has a molecular weight of about 34,500. In general, protein molecules have weights that vary from 34,500 to 50,000,000. Their extremely large size can readily be appreciated when they are compared with the molecular weight of a fat such as tripalmitin, which is 807, of glucose, which is 180, or of inorganic ions such as sodium and chloride (Fig. 15-2).

AMINO ACIDS

In addition to their large size, protein molecules are also very complicated. Like any complex molecule, they may be broken down by hydrolysis into smaller molecules whose structure is more easily determined. As will be illustrated later, the hydrolysis involves the simple splitting of an amide group. Common reagents used for the hydrolysis of proteins are acids (HCl and H_2SO_4), bases (NaOH), and enzymes (proteases). The simple molecules that are formed by the complete hydrolysis of a protein are called **amino acids.**

Before considering the properties and reactions of proteins, it may be well to study the individual amino acids. An amino acid is essentially an organic acid that contains an amino group. If a hydrogen is replaced by an amino group on the carbon atom that is next to the carboxyl group in acetic acid, CH_3COOH, the simple amino acid **glycine** will be formed.

$$\begin{array}{l} CH_2COOH \\ | \\ NH_2 \end{array}$$

The chemical name for this amino acid would be aminoacetic acid. The carbon atom next to the carboxyl group is called the alpha (α) carbon, the next beta (β), the next one gamma (γ), the next delta (δ), and the fifth from the carboxyl group is called the

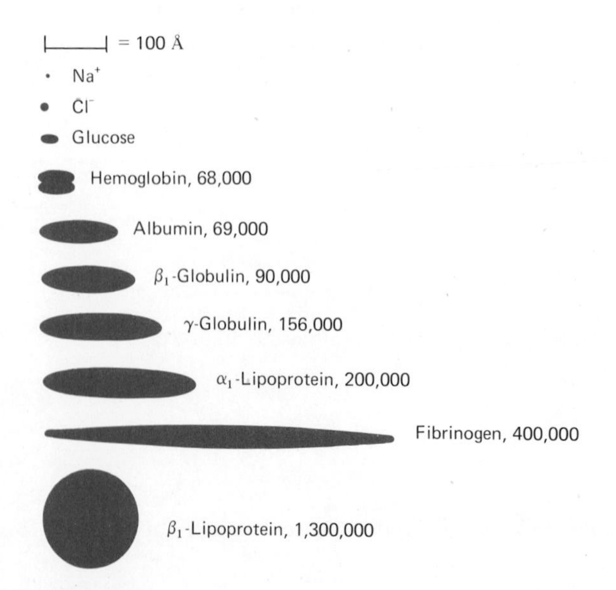

FIGURE 15-2 Relative dimensions of various protein molecules. (After Oncley, J. L.: Conference on the Preservation of the Cellular and Protein Components of Blood, published by the American Red Cross, Washington, D. C.)

epsilon (ε) carbon atom. Since all the amino acids have an amino group attached to the alpha carbon atom, they are known as **alpha amino acids.**

The amino acids are divided into groups according to their chemical structure. Examples of each group are given in the following classification. The common name for the amino acid is followed by the chemical name in parentheses, after which the abbreviation used in sequence and structure models is given. Example:

Glycine (aminoacetic acid) Gly

I. Aliphatic amino acids

A. With 1 amino and 1 carboxyl group:

Glycine (amino-acetic acid) Gly

$$\underset{\displaystyle NH_2}{\overset{|}{CH_2}}COOH$$

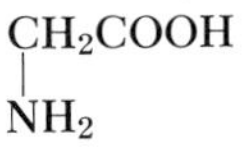

Alanine (α-aminopropionic acid) Ala

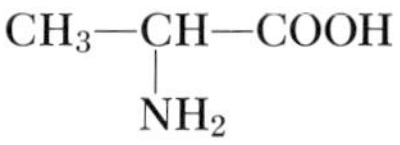

$$CH_3—\underset{\displaystyle NH_2}{\underset{|}{CH}}—COOH$$

Valine (α-aminoisovaleric acid) Val

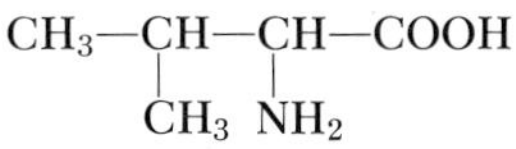

$$CH_3—\underset{\displaystyle CH_3}{\underset{|}{CH}}—\underset{\displaystyle NH_2}{\underset{|}{CH}}—COOH$$

Leucine (α-aminoisocaproic acid) Leu

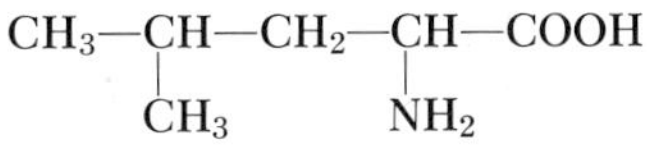

$$CH_3—\underset{\displaystyle CH_3}{\underset{|}{CH}}—CH_2—\underset{\displaystyle NH_2}{\underset{|}{CH}}—COOH$$

Isoleucine (α-amino-β-methylvaleric acid) Ileu

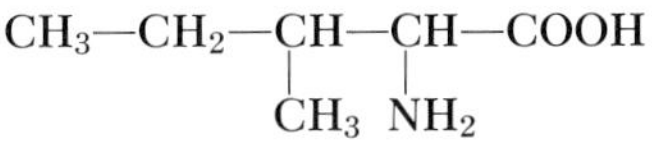

$$CH_3—CH_2—\underset{\displaystyle CH_3}{\underset{|}{CH}}—\underset{\displaystyle NH_2}{\underset{|}{CH}}—COOH$$

B. With 1 amino, 1 carboxyl, and 1 hydroxyl group:

Serine (α-amino-β-hydroxypropionic acid) Ser

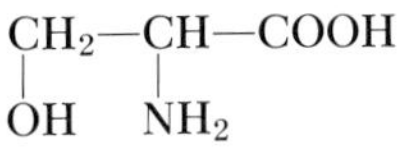

$$\underset{\displaystyle OH}{\underset{|}{CH_2}}—\underset{\displaystyle NH_2}{\underset{|}{CH}}—COOH$$

Threonine (α-amino-β-hydroxybutyric acid) Thr

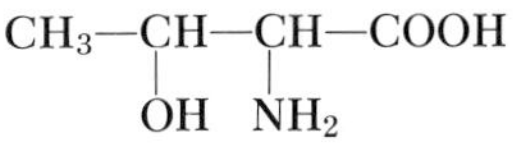

$$CH_3—\underset{\displaystyle OH}{\underset{|}{CH}}—\underset{\displaystyle NH_2}{\underset{|}{CH}}—COOH$$

C. With 1 amino and 2 carboxyl groups or their amides:

Aspartic acid (α-aminosuccinic acid) Asp

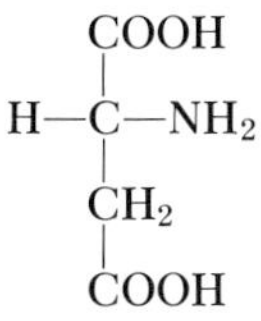

$$\begin{array}{c} COOH \\ | \\ H—C—NH_2 \\ | \\ CH_2 \\ | \\ COOH \end{array}$$

Asparagine (aspartic acid amide) $AspNH_2$

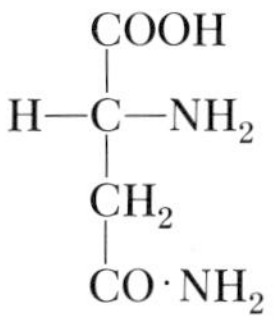

$$\begin{array}{c} COOH \\ | \\ H—C—NH_2 \\ | \\ CH_2 \\ | \\ CO\cdot NH_2 \end{array}$$

Glutamic acid (α-aminoglutaric acid) Glu

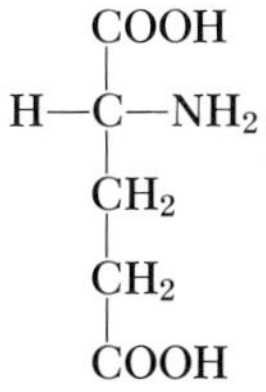

$$\begin{array}{c} COOH \\ | \\ H—C—NH_2 \\ | \\ CH_2 \\ | \\ CH_2 \\ | \\ COOH \end{array}$$

Glutamine (glutamic acid amide) $GluNH_2$

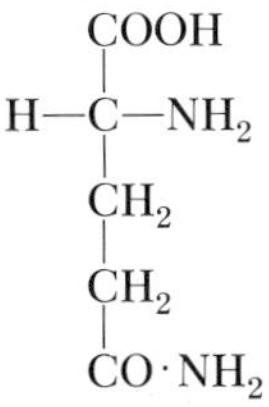

$$\begin{array}{c} COOH \\ | \\ H—C—NH_2 \\ | \\ CH_2 \\ | \\ CH_2 \\ | \\ CO\cdot NH_2 \end{array}$$

D. With 2 amino and 1 carboxyl group:
Lysine (α, ε-diaminocaproic acid) Lys

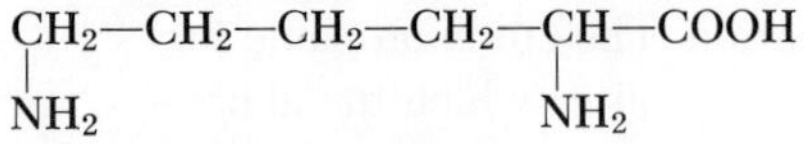

$$\underset{\displaystyle NH_2}{CH_2}-CH_2-CH_2-CH_2-\underset{\displaystyle NH_2}{CH}-COOH$$

Arginine (α-amino-δ-guanidovaleric acid) Arg

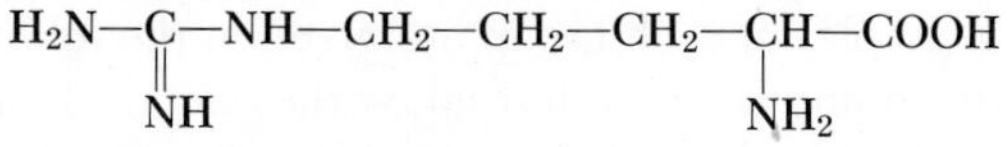

$$H_2N-\underset{\displaystyle NH}{\overset{}{C}}-NH-CH_2-CH_2-CH_2-\underset{\displaystyle NH_2}{CH}-COOH$$

E. Sulfur-containing amino acids:
Cysteine (α-amino-β-thiolpropionic acid) Cys

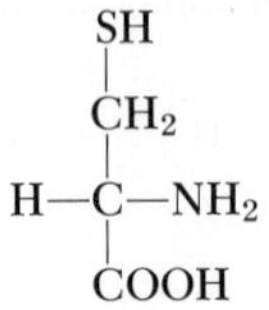

SH
|
CH_2
|
$H-C-NH_2$
|
COOH

Cystine [di(α-amino-β-thiopropionic acid)] Cys—Cys

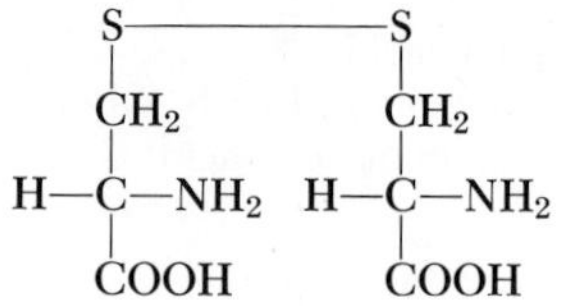

S————————S
| |
CH_2 CH_2
| |
$H-C-NH_2$ $H-C-NH_2$
| |
COOH   COOH

Methionine (α-amino-γ-methylthiobutyric acid) Met

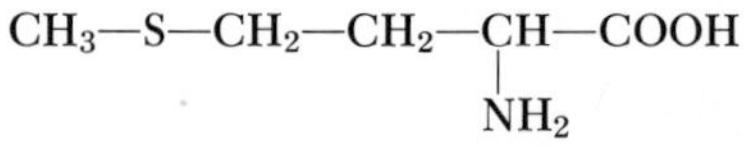

$$CH_3-S-CH_2-CH_2-\underset{\displaystyle NH_2}{CH}-COOH$$

II. Aromatic amino acids
Tyrosine (α-amino-β-parahydroxyphenylpropionic acid) Tyr

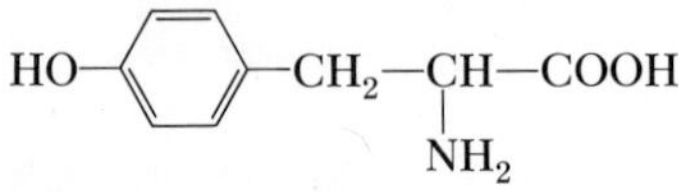

$$HO-C_6H_4-CH_2-\underset{\displaystyle NH_2}{CH}-COOH$$

Phenylalanine (α-amino-β-phenylpropionic acid) Phe

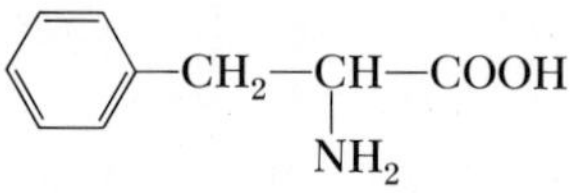

$$C_6H_5-CH_2-\underset{\displaystyle NH_2}{CH}-COOH$$

III. Heterocyclic amino acids
Tryptophan (α-amino-β-indolpropionic acid) Trp

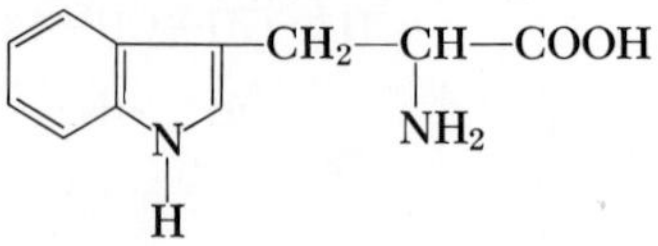

(indole ring, N—H) $-CH_2-\underset{\displaystyle NH_2}{CH}-COOH$

Histidine (α-amino-β-imidazolylpropionic acid) His

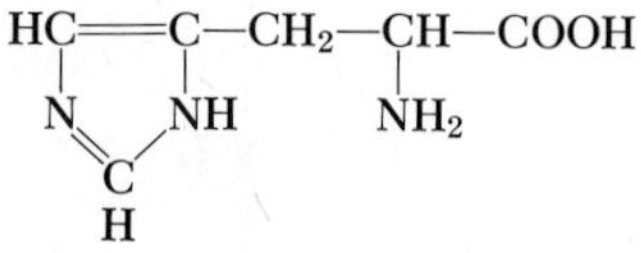

$HC{=}C-CH_2-\underset{\displaystyle NH_2}{CH}-COOH$
N   NH
 C
 H

Proline (pyrrolidine-2-carboxylic acid) Pro

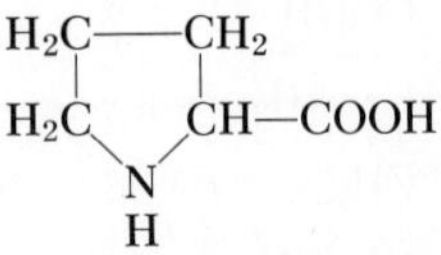

H_2C——CH_2
| |
H_2C $CH-COOH$
  N
  H

Hydroxyproline (4 hydroxypyrrolidine-2-carboxylic acid) Hypro

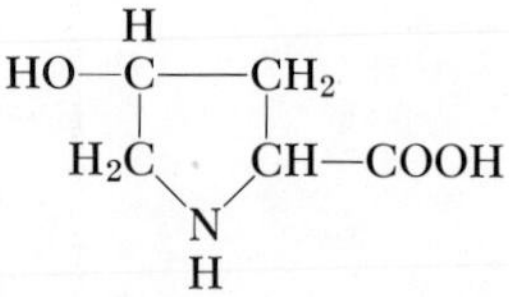

    H
HO—C——CH_2
   | |
  H_2C $CH-COOH$
     N
     H

Optical Activity of Amino Acids

All amino acids except glycine contain an asymmetric carbon atom in their formulas. For this reason they may exist in the D or L form. Using alanine as an example of a simple amino acid, we may compare the D and L forms to those of glyceraldehyde and lactic acid:

$$\begin{array}{c} \text{H} \\ \text{C=O} \\ | \\ \text{HO—C—H} \\ | \\ \text{CH}_2\text{OH} \end{array}$$

L-Glyceraldehyde

$$\begin{array}{c} \text{H} \\ \text{C=O} \\ | \\ \text{H—C—OH} \\ | \\ \text{CH}_2\text{OH} \end{array}$$

D-Glyceraldehyde

$$\begin{array}{c} \text{O} \\ /\!\!/ \\ \text{C—OH} \\ | \\ \text{HO—C—H} \\ | \\ \text{CH}_3 \end{array}$$

L-Lactic acid

$$\begin{array}{c} \text{O} \\ /\!\!/ \\ \text{C—OH} \\ | \\ \text{H—C—OH} \\ | \\ \text{CH}_3 \end{array}$$

D-Lactic acid

$$\begin{array}{c} \text{O} \\ /\!\!/ \\ \text{C—OH} \\ | \\ \text{H}_2\text{N—C—H} \\ | \\ \text{CH}_3 \end{array}$$

L-Alanine

$$\begin{array}{c} \text{O} \\ /\!\!/ \\ \text{C—OH} \\ | \\ \text{H—C—NH}_2 \\ | \\ \text{CH}_3 \end{array}$$

D-Alanine

Naturally occurring amino acids from plant and animal sources have the L configuration and would be designated L(+) or L(−), depending on their rotation of plane polarized light. D-alanine and D-glutamic acid have been obtained from microorganisms, especially from their cell walls.

Amphoteric Properties of Amino Acids

Amino acids behave both as weak acids and as weak bases, since they contain at least one carboxyl and one amino group. Substances that ionize as both acids and bases in aqueous solution are called **amphoteric.** An example would be glycine, in which both the acidic and basic groups are ionized in solution to form dipolar ions or **zwitterions.**

$$\begin{array}{l} \text{CH}_2\text{COO}^- \\ | \\ \text{NH}_3^+ \end{array}$$

The glycine molecule is electrically neutral, since it contains an equal number of positive and negative ions. The zwitterion form of glycine would therefore be isoelectric, and the pH at which the zwitterion does not migrate in an electric field is called the **isoelectric point.** Amphoteric compounds will react with either acids or bases to form salts. This is best illustrated by use of the zwitterion form of the amino acid.

$$\begin{array}{c} \text{CH}_3\text{—CH—COO}^- \\ | \\ \text{NH}_3^+ \end{array} + \text{HCl} \rightarrow \begin{array}{c} \text{CH}_3\text{—CH—COOH} \\ | \\ \text{NH}_3^+\text{Cl}^- \end{array}$$

$$\begin{array}{c} \text{CH}_3\text{—CH—COO}^- \\ | \\ \text{NH}_3^+ \end{array} + \text{NaOH} \rightarrow \begin{array}{c} \text{CH}_3\text{—CH—COO}^-\ \text{Na}^+ \\ | \\ \text{NH}_2 \end{array} + \text{H}_2\text{O}$$

From these equations it can be seen that the addition of a H^+ to an isoelectric molecule results in an increased positive charge (NH_3^+), since the acid represses the ionization of the carboxyl group. Conversely, the addition of a base to an isoelectric molecule results in an increased negative charge (COO^-), since the base represses the ionization of the amino group. Since proteins are composed of amino acids, they are amphoteric substances with specific isoelectric points and are able to neutralize both acids and bases. This property of proteins is responsible for their buffering action in blood and other fluids.

Titration of Amino Acids

Since amino acids serve as buffers in blood and other body fluids, their dipolar ionic nature should result in at least two dissociation constants when they react with acid or base. The **Henderson-Hasselbalch equation** for a simple buffer represents the dissociation constant, or pK, as the pH at which equal concentrations of the salt and acid forms of the buffer exist in solution.

$$pH = pK + \log\frac{[salt]}{[acid]} = pK + \log\frac{1}{1} = pK$$

The simple amino acid glycine may be used as an example of amino acids or proteins as buffers. When glycine in solution is titrated with acid or base, the molecule changes from the zwitterion form to a form dissociating as either a charged amino or carboxyl group.

$$\underset{\text{Acid solution}}{\underset{NH_3^+}{CH_2{-}COOH}} \underset{pK_1}{\overset{K_1}{\rightleftharpoons}} \overset{+}{H} + \underset{\substack{\text{Zwitterion}\\\text{Isoelectric point}\\pH=6.0}}{\underset{NH_3^+}{CH_2{-}COO^-}} + OH^- \underset{pK_2}{\overset{K_2}{\rightleftharpoons}} \underset{\text{Basic solution}}{\underset{NH_2}{CH_2COO^-}}$$

$$pH = 2.4 \qquad\qquad pH = 9.6$$

A mixture of the zwitterion and glycine in acid solution would compose a buffer whose pK_1 would equal the pH of the solution when equal quantities of the two forms were present in the solution. This point pK_1 at a pH of 2.4 also corresponds to the point of half neutralization of the carboxyl group. When base is added to glycine, the pK_2, or point of half neutralization of the amino group, corresponds to a pH of 9.6. The titration of glycine with acid or base shown in Figure 15–3 emphasizes the two buffering regions and the pK values that correspond to the buffer in acid solution or in alkaline solution.

The monoamino monocarboxylic acids each exhibit two pK values and act as buffers in two pH regions. The pH of the isoelectric point can be calculated by dividing the sum of the two pK values by 2 (glycine: $2.4 + 9.6 = 12.0 \div 2 = 6.0$). More complex amino acids such as aspartic acid and lysine have three pK values and can exist in four ionized forms. The ionization of aspartic acid may be represented as follows:

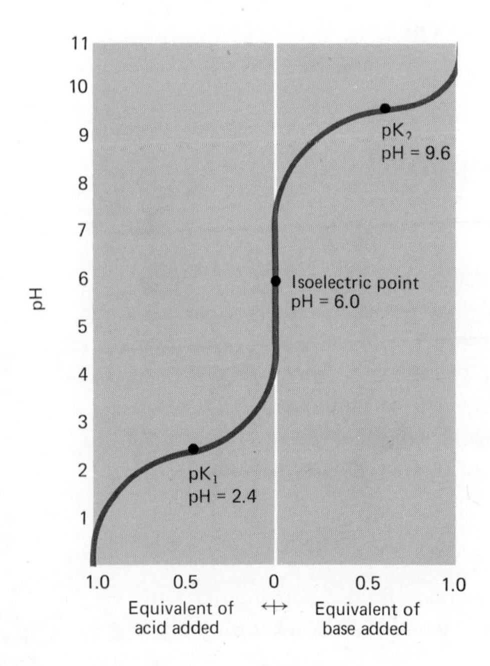

FIGURE 15–3 Titration curve for glycine.

$$\begin{array}{c} COOH \\ | \\ HC{-}NH_3^+ \\ | \\ CH_2 \\ | \\ COOH \end{array} \underset{}{\overset{K_1}{\rightleftharpoons}} \begin{array}{c} COO^- \\ | \\ HC{-}NH_3^+ \\ | \\ CH_2 \\ | \\ COOH \end{array} \underset{}{\overset{K_2}{\rightleftharpoons}} \begin{array}{c} COO^- \\ | \\ HC{-}NH_3^+ \\ | \\ CH_2 \\ | \\ COO^- \end{array} \underset{}{\overset{K_3}{\rightleftharpoons}} \begin{array}{c} COO^- \\ | \\ HC{-}NH_2 \\ | \\ CH_2 \\ | \\ COO^- \end{array}$$

$$pK_1 = 2.0 \qquad pK_2 = 4.0 \qquad pK_3 = 9.8$$

The titration of 20 ml of 0.1M aspartic acid in the form of the hydrochloride (form at extreme left in diagram) with 0.1M NaOH is shown in Figure 15–4. Aspartic acid may therefore act as a buffer in three different pH regions.

Reactions of Amino Acids

The fact that amino acids can ionize as both weak acids and weak bases and contain amino groups and carboxyl groups suggests a very reactive molecule. Many of the common reactions of organic chemistry may be applied to amino acids. The synthesis of amino acids from aldehydes by the addition of NH_4CN, the so-called Strecker reaction, is shown on page 156.

Reaction with Nitrous Acid. This is the basis of the Van Slyke method for the determination of free primary amino groups (see p. 190).

$$\underset{\displaystyle COOH}{R{-}\underset{|}{C}H{-}NH_2} + NaNO_2 + HCl \rightarrow [\underset{\displaystyle COOH}{R{-}\underset{|}{C}H}{-}\overset{+}{N}{\equiv}N]Cl^- \rightarrow \underset{\displaystyle COOH}{R{-}\underset{|}{C}H}{-}OH + N_2 + NaCl + H_2O$$

The nitrogen gas that is liberated in the reaction is collected and its volume measured. One half of this nitrogen comes from the amino acid and is used as a measure of the free amino nitrogen.

Reaction with 1-Fluoro-2,4-dinitrobenzene (FDNB). This compound, also called **Sanger's reagent,** reacts with the free amino group of an amino acid, as would an alkyl halide, to form a yellow colored dinitrophenylamino acid, DNP-amino acid.

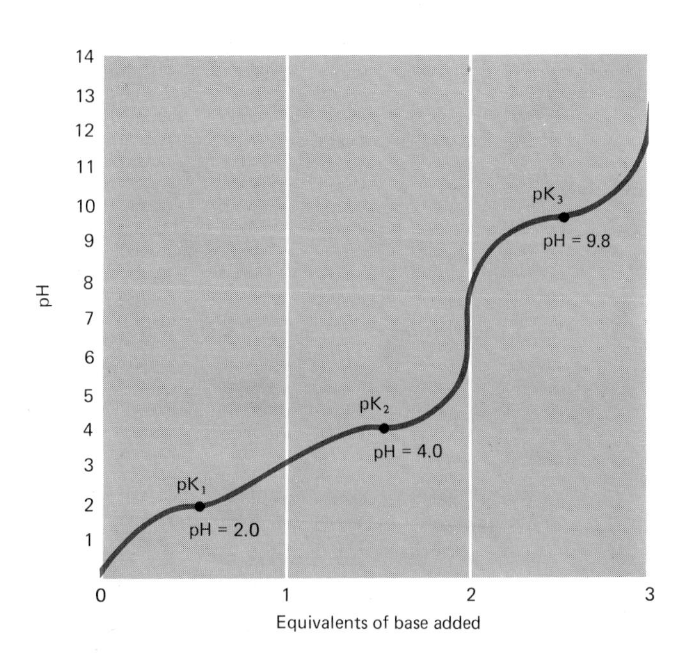

FIGURE 15-4 Titration curve for aspartic acid.

$$R{-}CH(COOH){-}NH_2 + F{-}C_6H_3(NO_2)_2 \rightarrow R{-}CH(COOH){-}NH{-}C_6H_3(NO_2)_2 + HF$$

Dinitrophenylamino acid,
or DNP-amino acid

This reaction will be found to be very important in the determination of protein structure, since the reagent reacts with the free amino group of the terminal amino acid in a protein and thus identifies the end amino acid in the structure.

Reaction with Ninhydrin. Amino acids react with ninhydrin (triketohydrindene hydrate) to form CO_2, NH_3, and an aldehyde. The amount of CO_2 that is liberated in the reaction can be used as a quantitative measure of the amino acid and is specific for an amino acid or compound with a free carboxyl group adjacent to an amino group. The NH_3 that is formed in the reaction combines with a molecule of reduced and a molecule of oxidized ninhydrin and forms a blue-colored compound. This compound may be measured colorimetrically for the quantitative determination of amino acids. The reaction is complex and may be represented by two main processes:

$$C_6H_4(CO)_2C(OH)_2 + R{-}CH(NH_2){-}COOH \rightarrow C_6H_4(CO)_2CH{-}NH_2 + CO_2 + R{-}CH{=}O$$

$$C_6H_4(CO)_2CH{-}NH_2 + (HO)_2C(CO)_2C_6H_4 \rightarrow C_6H_4(CO)_2CH{-}N{=}C(CO)_2C_6H_4 + 3H_2O$$

Blue-colored compound

Color Reactions of Specific Amino Acids. Certain amino acids, whether in the free form as in protein hydrolysates or combined in proteins, give specific color reactions that aid in their detection and determination. The **Millon test** depends on the formation of a red-colored mercury complex with tyrosine, whether free or in proteins, whereas tryptophan reacts with glyoxylic acid to produce a violet color in the **Hopkins-Cole test.** In the **Sakaguchi reaction,** the guanidino group in arginine forms a red color with α-naphthol and sodium hypochlorite, and cysteine and proteins that contain free sulfhydryl groups yield a red color with sodium nitroprusside. Both cystine and cysteine, free and in proteins, form a black precipitate of PbS in the **unoxidized sulfur test.**

POLYPEPTIDES

Amino acids are joined together in a polypeptide molecule by the peptide linkage. The **peptide linkage** is an amide linkage (p. 175) between the carboxyl group of one amino acid and the amino group of another amino acid, with the splitting out of a molecule of water. This type of linkage may be illustrated by the union of a molecule of alanine and a molecule of glycine:

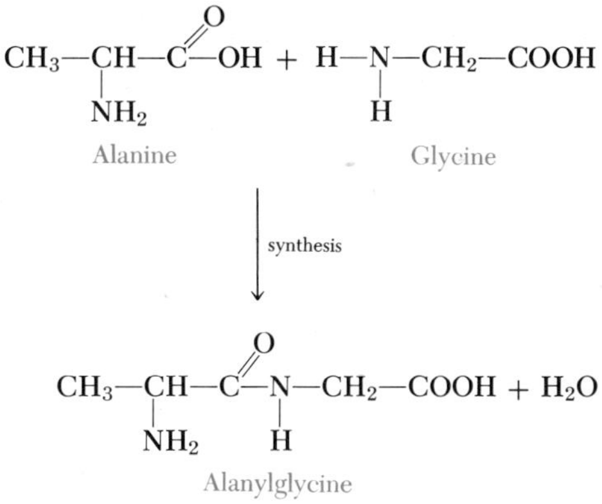

$$CH_3-\underset{\underset{NH_2}{|}}{CH}-\overset{\overset{O}{\mathbin{/\!/}}}{C}-OH + H-\underset{\underset{H}{|}}{N}-CH_2-COOH$$

Alanine Glycine

$$\xrightarrow{\text{synthesis}}$$

$$CH_3-\underset{\underset{NH_2}{|}}{CH}-\overset{\overset{O}{\mathbin{/\!/}}}{C}-\underset{\underset{H}{|}}{N}-CH_2-COOH + H_2O$$

Alanylglycine

The compound alanylglycine, which results from this linkage, is called a **dipeptide.** The union of three amino acids would result in a **tripeptide,** and the combination of several amino acids by the peptide linkage would be called a **polypeptide.** Since each amino acid has lost a water molecule when it joins to two other amino acids in a polypeptide, the remaining compound is called an **amino acid residue.** Proteins may be considered as complex polypeptides. A polypeptide chain illustrating the primary structure of a protein may be represented as follows:

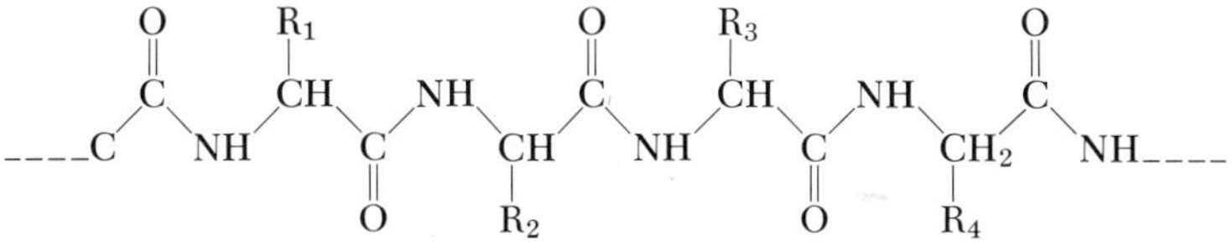

$$----C-\overset{\overset{O}{\|}}{C}-NH-\overset{\overset{R_1}{|}}{CH}-\underset{\underset{O}{\|}}{C}-NH-\underset{\underset{R_2}{|}}{CH}-\overset{\overset{O}{\|}}{C}-NH-\overset{\overset{R_3}{|}}{CH}-\underset{\underset{O}{\|}}{C}-NH-\underset{\underset{R_4}{|}}{CH_2}-\overset{\overset{O}{\|}}{C}-NH----$$

The R groups are the side chains of the specific amino acids in the polypeptide chain.

The Biuret Test

When a few drops of very dilute copper sulfate solution are added to a strongly alkaline solution of a peptide or protein, a violet color is produced. This is a general test for proteins and is given by peptides that contain two or more peptide linkages. **Biuret** is formed by heating urea and has a structure similar to the peptide structure of proteins:

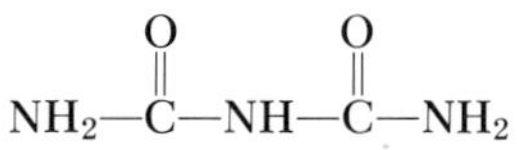

$$NH_2-\overset{\overset{O}{\|}}{C}-NH-\overset{\overset{O}{\|}}{C}-NH_2$$

Biuret

SEPARATION AND DETERMINATION OF AMINO ACIDS

Prior to 1950, it was thought that unraveling the combinations of the 23 different amino acids which form proteins of molecular weight from 34,500 to 50,000,000 was inconceivable. Proteins were subjected to hydrolysis and several amino acids could be separated and determined in the mixture. Amino acid sequence determination was made possible by the development of the techniques of chromatography.

Paper Chromatography

Several types of chromatographic techniques have been developed for the analysis of mixtures of molecules such as amino acids. **Paper chromatography** is relatively simple and produces excellent separation, detection, and quantitation of the individual amino acids. The technique of ascending paper chromatography is often used. When a strip of filter paper is held vertically in a closed glass cylinder with its lower end dipped in a mixture of water and an organic solvent such as butyl alcohol, phenol, or collidine, the mixture of water and organic solvent moves up the paper. If a solution containing a mixture of amino acids is added as a small spot just above the solvent level, the individual amino acids will be affected by the water phase and the organic phase as they move up the paper. A solvent partition will occur, and each amino acid will be carried to a particular location on the paper. This location depends on many factors, including the pH, the temperature, the concentration of the solvents, and the time of chromatography. After drying the paper, it is sprayed with a solution of ninhydrin, which yields a blue to purple color with each amino acid. By comparison with known amino acids, separation of the amino acids from a hydrolysate of a protein or polypeptide fragment may be achieved by the proper choice of conditions and solvents. This separation is often improved by a second chromatographic run during which different solvents are used and the dried paper from the first run is turned 90 degrees from the direction of the first migration. The results of a two-dimensional paper chromatography separation are shown in Figure 15–5 A.

A recent extension of the technique of paper chromatography involves thin-layer chromatography (see Chapter 14, p. 230). Instead of a strip of paper, the amino acids or other molecules to be separated are carried by the solvent mixture through a thin layer of cellulose powder, silica gel, and other adsorbents located on a glass plate, a plastic film, or a specially processed paper backing. The separation is much more rapid and a wide variety of reagents and dyes may be used as detection sprays. The speed of migration, sensitivity, and versatility of thin-layer chromatography make this technique a valuable addition to the tools of research in biochemistry.

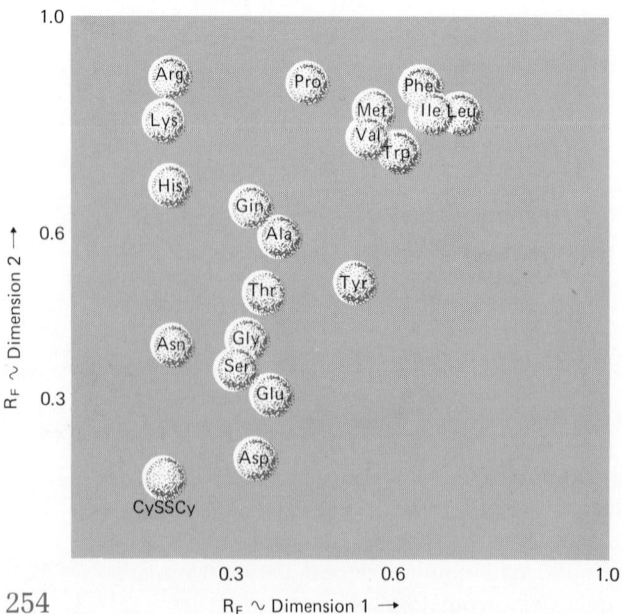

FIGURE 15–5A A schematic representation of a two-dimensional paper chromatogram. The solvent for dimension 1 is *n*-butanol-acetic acid-water (250 : 60 : 250 vol. per vol.) and for dimension 2, phenol-water-ammonia (120: 20: 0.3 per cent). Each solvent front moves with an R_F equal to 1 in each dimension. (After White et al.: Principles of Biochemistry. 4th ed. New York, McGraw-Hill, Inc., 1968, p. 114.)

Ion Exchange Chromatography

Columns of starch, cellulose powder, and alumina gels have been used to separate amino acids, but it has been difficult to isolate amino acids with similar properties from each other. More successful separations and quantitative determinations of amino acids in mixtures are achieved with **ion exchange chromatography.** Ion exchange resins are insoluble synthetic resins containing acidic or basic groups, such as $—SO_3H$ or $—OH$. A sulfonated polystyrene resin may be used as a cation exchange resin by the addition of Na ions to produce $—SO_3Na$ groups on the surface of the resin. Basic amino acids react with a cation exchange resin as follows:

$$\text{ResinSO}_3{}^-\text{Na}^+ + \underset{\text{Cl}^-}{\text{NH}_3{}^+\text{R}} \rightarrow \text{ResinSO}_3{}^-\text{NH}_3{}^+\text{R} + \text{Na}^+\text{Cl}^-$$

The resin is placed in a column or long glass tube, and the mixture of amino acids, which are dissolved in a small volume of buffer, is placed on top of the column. The column is washed with a buffer solution, and, as the amino acids pass down the column, the basic amino acids react with the $—SO_3{}^-Na^+$ groups, replacing Na^+, and are slowed in their passage. Glutamic and aspartic acids are least affected by the column and come off in the first buffer. They are followed by the neutral amino acids and finally the basic amino acids.

A more efficient removal of the different amino acids from the column is achieved by gradually increasing the pH and concentration of the eluting buffer to force the basic amino acids off the column. The elution of amino acids from their attachment to the resin by the use of sodium citrate buffer may be represented as follows:

$$\text{ResinSO}_3{}^-\text{NH}_3{}^+\text{R} + \text{Na}^+\text{citrate}^- \rightarrow \text{ResinSO}_3{}^-\text{Na}^+ + \text{NH}_3{}^+\text{Rcitrate}^-$$

By proper choice of resin, buffers, length of column, temperature, and elution rates, and collecting the eluted amino acids in fraction collectors that are coupled to automatic analyzing instruments, it is possible to obtain a quantitative amino acid analysis of a protein hydrolysate in a few hours. The **elution pattern** of representative amino acids from a cation exchange resin is illustrated in Figure 15–5 B.

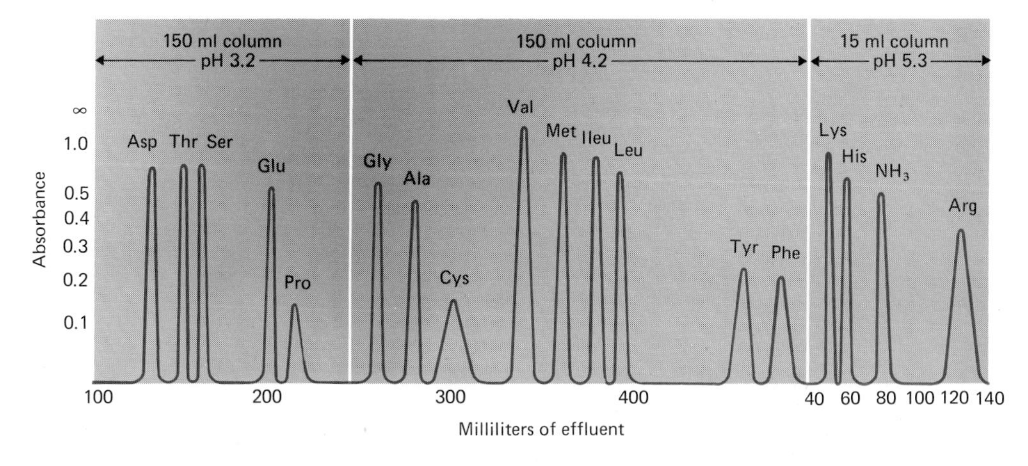

FIGURE 15-5*B* Chromatographic fractionation of a synthetic mixture of amino acids on columns of Amberlite IR-120. (After Moore et al.: Anal. Chem., *30:* 1186, 1958.)

Analysis of the Insulin Molecule by Sanger

Sanger's efforts to establish the structure of insulin were made possible by the advances in the methods of protein chemistry, especially the technique of chromatography. In addition, he found that a dinitrophenyl (DNP) group could be attached to free amino groups to form a yellow compound, as was discussed earlier. This DNP group remained attached to the amino acid residue even after hydrolysis was used to split the peptides, therefore making it possible to identify the terminal residue. Employing the DNP method, he first established that each insulin molecule contained two amino acid residues with free amino groups. He concluded, therefore, that insulin consists of two chains. The two chains were held together by the disulfide —S—S— bonds of cystine residues, which could be broken by mild oxidation. The two intact chains were obtained, and it was proven that one contained 21 amino acids whereas the other contained 30. Each chain was hydrolyzed with acid into smaller pieces and the amino acids identified by chromatography.

Using the DNP method, Sanger identified glycine as the amino acid on the amino end of the short chain and phenylalanine as the amino acid with the free amino group on the end of the long chain. After carefully breaking the chains into smaller peptides by hydrolysis with weak acid, the content and sequence of amino acids in each peptide was identified by the DNP reaction and paper chromatography. For example, Sanger found fragments of DNP gly-ileu, DNP gly-ileu-val, and DNP gly-ileu-val-glu, which gave him a good start on the amino end of the short-chain. A fragment DNP phe-val-asp-glu identified the beginning of the long chain. An example of the method used to establish the sequence of 12 amino acids in the long chain will illustrate the tedious complexity of the problem. The following fragments were obtained by hydrolysis and identified by DNP method and chromatography:

Dipeptides

ser-his leu-val glu-ala
his-leu val-glu

Tripeptides

ser-his-leu
leu-val-glu
val-glu-ala
ala-leu-tyr
tyr-leu-val

Higher Peptides

ser-his-leu-val-glu
val-glu-ala-leu leu-val-$CySO_3H$-gly

Deduced Sequence

-ser-his-leu-val-glu-ala-leu-tyr-leu-val-$CySO_3H$-gly-

In this fashion Sanger was able to determine the complete arrangement of amino acid residues in each part of the molecule.

The final difficult task was to determine the pairing of the half-cystine residues that joined the two chains. It was found that the shorter chain contained one disulfide linkage and the two chains were held together by two other disulfide bonds, as shown in Figure 15–6.

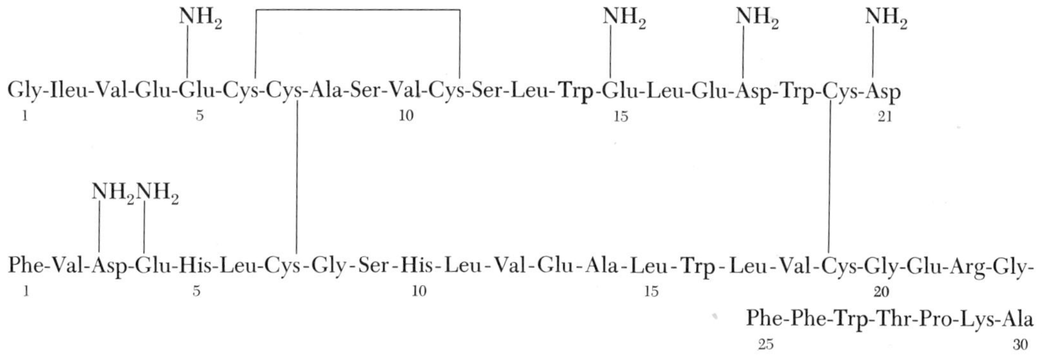

```
                NH2     ┌────────────┐                         NH2       NH2       NH2
                 |      |            |                          |         |         |
Gly-Ileu-Val-Glu-Glu-Cys-Cys-Ala-Ser-Val-Cys-Ser-Leu-Trp-Glu-Leu-Glu-Asp-Trp-Cys-Asp
1               5          |         10                     15                  |  21
                           |                                                    |
           NH2NH2          |                                                    |
            |  |           |                                                    |
Phe-Val-Asp-Glu-His-Leu-Cys-Gly-Ser-His-Leu-Val-Glu-Ala-Leu-Trp-Leu-Val-Cys-Gly-Glu-Arg-Gly-
1               5              10                 15                 20
                                                          Phe-Phe-Trp-Thr-Pro-Lys-Ala
                                                          25                       30
```

FIGURE 15-6 Amino acid sequence in beef insulin.

Amino Acid Sequence in Peptides and Proteins

After Sanger laid the foundation for the attack on the amino acid sequence of a protein molecule, other workers studied peptides and proteins. The naturally occurring tripeptide glutathione was shown to have a sequence as follows:

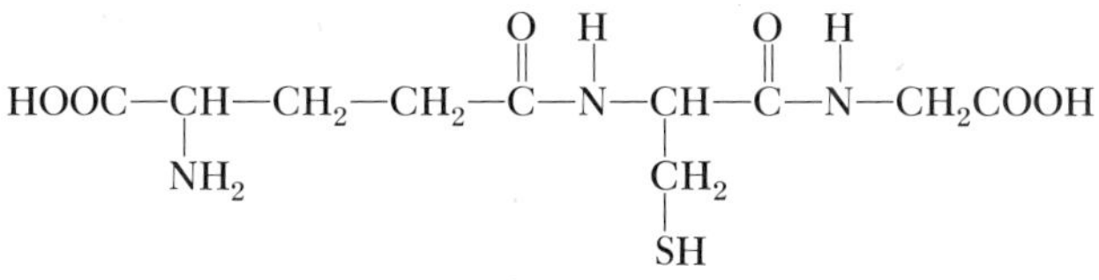

$$\mathrm{HOOC{-}\underset{\displaystyle NH_2}{CH}{-}CH_2{-}CH_2{-}\overset{\displaystyle O}{\overset{\|}{C}}{-}\overset{\displaystyle H}{\overset{|}{N}}{-}\underset{\displaystyle \underset{\displaystyle SH}{CH_2}}{CH}{-}\overset{\displaystyle O}{\overset{\|}{C}}{-}\overset{\displaystyle H}{\overset{|}{N}}{-}CH_2COOH}$$

Glutathione (α-glutamyl-cysteinyl-glycine)

This compound in the reduced or thiol form (as shown) is essential for the normal function of the red blood cells. Vincent du Vigneaud established the exact structure and sequence in two peptide hormones, **oxytocin** and **vasopressin,** that are elaborated by the posterior lobe of the pituitary gland. Oxytocin causes contraction of smooth muscles and is used in obstetrics to initiate labor. Vasopressin constricts vessels, raising blood pressure and affecting water and electrolyte balance. Each peptide contained eight amino acids, with the disulfide bridge of cystine across four of the amino acids.

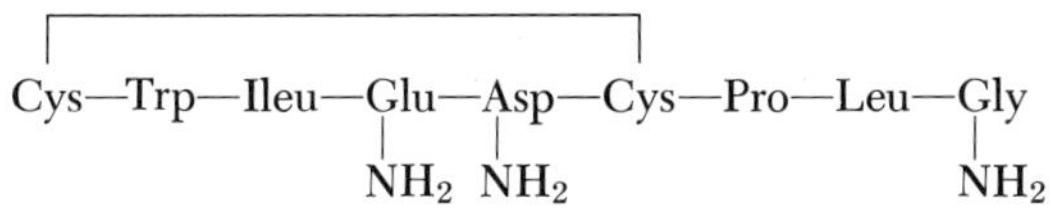

```
┌──────────────────────────┐
Cys—Trp—Ileu—Glu—Asp—Cys—Pro—Leu—Gly
              |    |                |
             NH2  NH2              NH2
```

Oxytocin

```
┌──────────────────────────┐
Cys—Trp—Phe—Glu—Asp—Cys—Pro—Arg—Gly
             |    |                |
            NH2  NH2              NH2
```

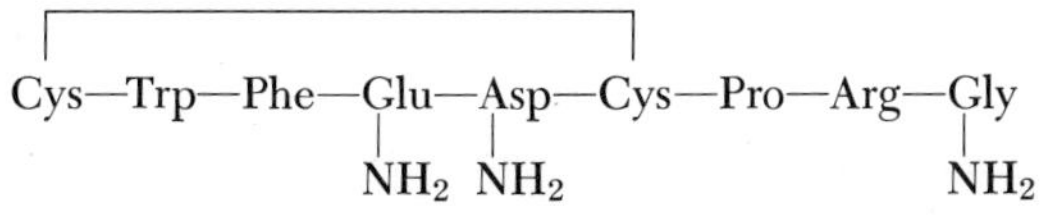

Vasopressin

The presence of two different amino acids in such a small peptide results in very different physiological activity. Du Vigneaud also succeeded in synthesizing these two molecules from amino acids and demonstrated the similar hormone activity of the synthetic peptides.

The amino acid sequence of the **adrenocorticotropic hormone, ACTH,** containing 39 amino acids, has been established. Larger protein molecules, such as the enzyme

ribonuclease (Fig. 15–7), with 124 amino acids, the α and β polypeptide chains of **hemoglobin** (141 and 146 amino acid residues, respectively), and the **tobacco mosaic virus** protein with 158 amino acid residues, have also been characterized.

The synthesis of polypeptides such as those just described is difficult and becomes more time-consuming as the molecule increases in size. Recently, an automated solid-phase synthesis technique has been developed by Merrifield. The synthesis is carried out in a single reaction vessel by an instrument programmed to add reagents and remove products at timed intervals. Briefly, the amino acid that will form the C-terminal end of the polypeptide is first attached to an insoluble resin particle. The second amino acid with its amino group blocked is added and a peptide bond formed between the NH_3 group of the first amino acid and the COOH group of the added amino acid. The blocking group is removed and the process is repeated until the desired polypeptide attached to the resin particle has been synthesized. Utilizing this technique both polypeptide chains of the insulin molecule were synthesized, and, more recently, the total synthesis of the enzyme pancreatic ribonuclease was achieved. The synthesis of such a hormone and enzyme in the laboratory may have far-reaching effects in medical therapy.

STRUCTURE OF PROTEINS

The chemical, physical, and biologic properties of specific proteins depend on the structure of the molecule as it exists in the native state. Proteins range in complexity from a simple polypeptide, such as vasopressin, with biologic activity, to a globular protein such as myoglobin, whose molecule involves cross linkages, helix formation, and folding and conformational forces.

Primary Structure

The amino acid sequence determinations have established the exact structure of the polypeptide chain in simple proteins. The bond distances and bond angles of a typical

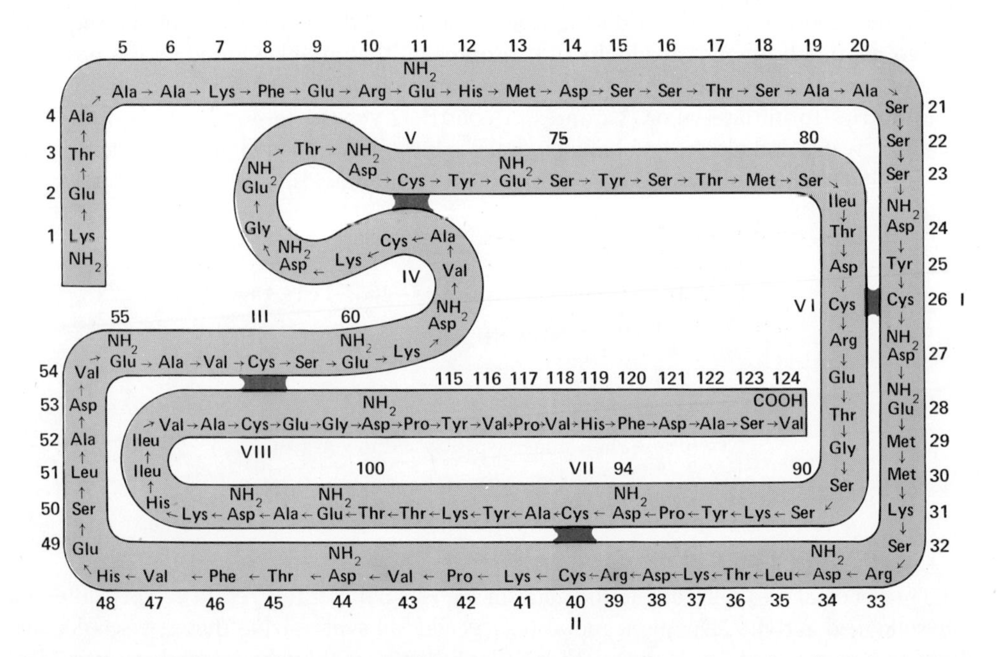

FIGURE 15-7 The complete amino acid sequence of enzyme ribonuclease. Standard three-letter abbreviations are used to indicate individual amino acid residues. (After Smyth et al.: J. Biol. Chem., *238:* 227, 1963.)

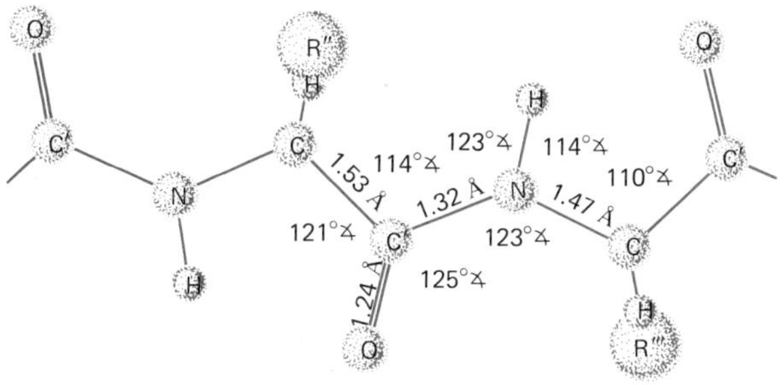

Figure 15-8 Dimensions of the fully extended polypeptide chain. (After Pauling and Corey: Adv. Protein Chem., *12:* 133, 1957.)

polypeptide chain are shown in Figure 15–8. The peptide linkage joining amino acids to produce a polypeptide is considered the **primary structure** of a protein.

Secondary Structure

If only peptide bonds were involved in protein structure, the molecules would consist of long polypeptide chains coiled in random shapes. Most **native proteins,** however, are either fibrous or globular in nature, and consist of polypeptide chains joined together or held in definite folded shapes by hydrogen bonds. This influence of hydrogen bonding on the protein molecule is often called the **secondary structure** of the protein. Although **hydrogen bonds** may be formed between several groups on the peptide chains, the most common bonding occurs between the carbonyl and amide groups of the peptide chain backbone, as shown:

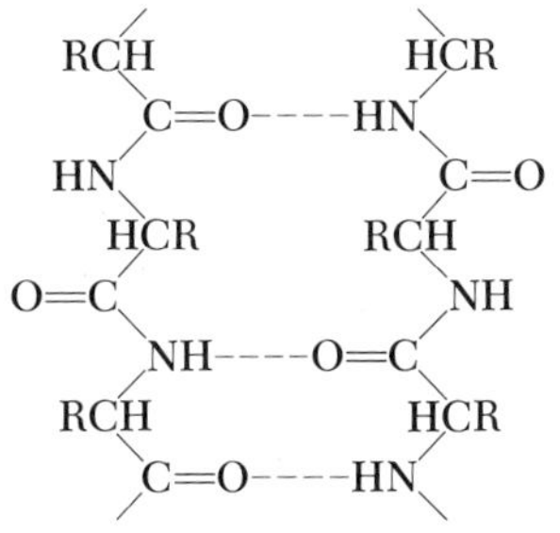

```
    RCH          HCR
       C=O----HN
    HN            C=O
      HCR      RCH
  O=C             NH
      NH----O=C
    RCH           HCR
       C=O----HN
```

This phenomenon explains the joining together of parallel polypeptide chains, which is evident in several fibrous proteins. A further structure that involves hydrogen bonds and accounts for the stereochemistry and the proper bond lengths and angles in the protein molecule is the **α-helix** proposed by Pauling (Fig. 15–9). The α-helix structure consists of a chain of amino acid units wound into a spiral which is held together by hydrogen bonds between a carbonyl group of one amino acid and the imido group of an amino acid residue further along the chain (Fig. 15–10). Each amino acid residue is 1.5 Å from the next amino acid residue, and the helix makes a complete turn for each 3.6 residues. The helix may be coiled in a right-handed or left-handed direction, but the right-handed helix is the most stable.

Tertiary Structure

The polypeptide chains of globular proteins are more extensively folded or coiled than those of fibrous proteins. This results from the activity of several types of bonds that hold the structure in a more complex and rigid shape. These bonds are responsible for the **tertiary structure** of proteins, and they exert stronger forces than hydrogen bonds

FIGURE 15-9 Linus Pauling (1901–) Professor of Chemistry, Stanford University. Winner of the 1954 Nobel Prize for Chemistry and the 1962 Nobel Prize for Peace.

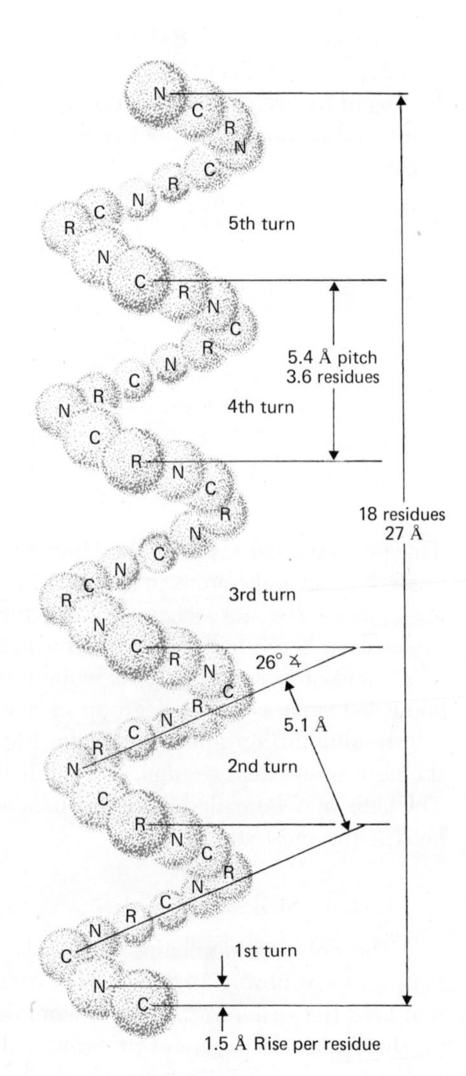

FIGURE 15-10 Representation of a polypeptide chain as an α-helical configuration. (After Pauling and Corey: Proc. Intern. Wool Textile Research Conf., *B*, 249, 1955.)

in holding together polypeptide chains or folds of individual chains. A strong **covalent bond** is formed between two cysteine residues, resulting in the disulfide bond. **Salt linkages,** or **ionic bonds,** may be formed between the basic amino acid residues of lysine and arginine and the dicarboxylic amino acids such as aspartic and glutamic. Also, there are many examples of **hydrophobic bonding** that result from the close proximity of aromatic groups or of like aliphatic groups on amino acid residues. Examples of these bonds may be seen in Figure 15–11.

Several types of physical measurements, including sedimentation in a high intensity centrifugal field, osmotic pressure, light scattering, optical rotation, and x-ray analysis, have been used to obtain information concerning the tertiary structure of proteins.

Sedimentation of Proteins

Measurement of the rate of sedimentation of proteins in a centrifugal field of high intensity was made possible by the development of the **ultracentrifuge** by Svedberg in the 1920's. A protein is usually centrifuged, and, if a single molecular species is used, the protein molecules will form a single boundary moving through the solvent away from the center of rotation. The size, shape, and especially the molecular weight of a protein determine its sedimentation rate. The **sedimentation constant, s,** is characteristic of a protein molecule.

X-Ray Diffraction Analysis

A three-dimensional picture of the protein molecule, **myoglobin,** has been developed from x-ray diffraction data by Kendrew and his co-workers. Myoglobin has a molecular weight of 17,000 and consists of a single polypeptide chain with 153 amino acid residues. It crystallizes readily from muscle extracts of the sperm whale. When x-rays strike an atom, they are diffracted (reflected) in proportion to the number of extranuclear electrons in the atom. The heavier atoms with a high atomic number, therefore, produce more diffraction than lighter atoms. A crystal, such as a protein crystal of myoglobin, when bombarded by a beam of monochromatic x-rays, yields a photographic pattern of the electron density of the atoms in the molecule. From a large series of electron density photographs in different planes, a three-dimensional picture of myoglobin was constructed. The model of the myoglobin that resulted from these studies is shown in Figure 15–12. The structural resolution achieved by Kendrew indicated that the major portion of the polypeptide chain was in the form of the right-handed helix proposed by Pauling.

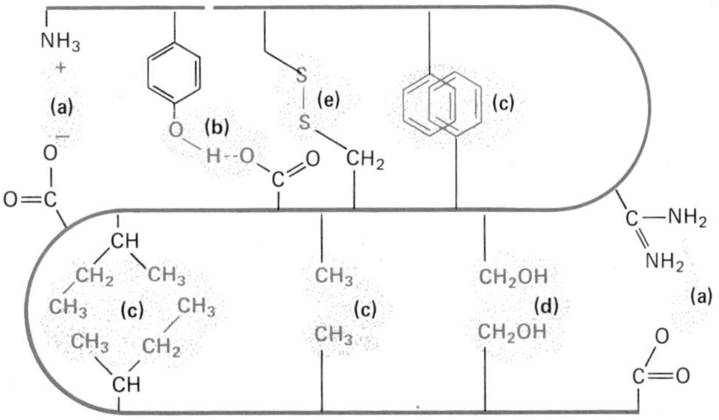

FIGURE 15–11 Some types of noncovalent bonds which stabilize protein structure: (*a*) Electrostatic interaction; (*b*) hydrogen bonding between tyrosine residues and carboxyl groups on side chains; (*c*) hydrophobic interaction of nonpolar side chains caused by the mutual repulsion of solvent; (*d*) dipole-dipole interaction; (*e*) disulfide linkage, a covalent bond. (After Anfinsen: The Molecular Basis of Evolution. New York, John Wiley and Sons, 1959, p. 102.)

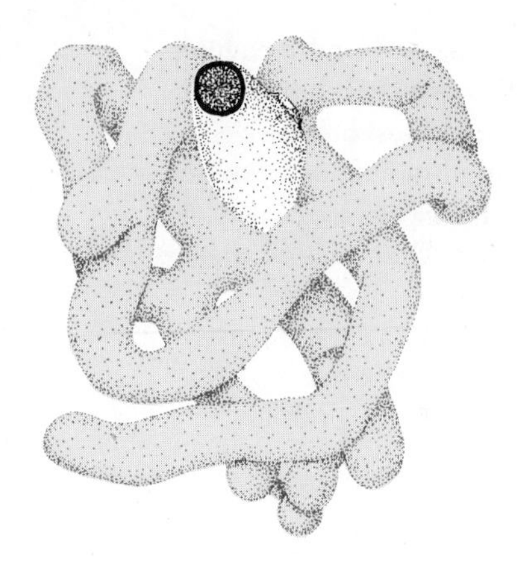

FIGURE 15-12 Model of the myoglobin molecule, derived from the 6 Å Fourier synthesis. The heme group is a dark gray disk (center top). (After Kendrew: Science, *139:* 1261, 1963.)

Quaternary Structure

This level of protein structure involves the polymerization, or degree of aggregation, of protein units. The hemoglobin molecule is a good example of subunit structure in proteins. **Native hemoglobin** has a molecular weight of 68,000 in a neutral solution. If the solution is diluted, made acid, or 4M with urea, the molecular weight changes to 34,000. This dissociation is due to the four polypeptide chains which exist in hemoglobin as two pairs of α and β chains. The enzyme **phosphorylase a,** which will be discussed under carbohydrate metabolism, contains four subunits which are inactive as catalysts until they are joined as a tetramer. **Insulin** is another example of a protein hormone containing subunits, and there are several proteins that are split into subunits when their disulfide bonds are converted to sulfhydryl groups. Several enzyme molecules have been separated into isoenzymes by physical techniques such as the process of electrophoresis. Isoenzymes apparently have the same activity and molecular weight as the parent enzyme but may be synthesized in different organs in the body. For example, lactic acid dehydrogenase is an enzyme with five isoenzymes whose syntheses take place in the heart and in liver tissue. Isoenzymes are becoming increasingly important in diagnosis of disease states.

CLASSIFICATION OF PROTEINS

Proteins are most often classified on the basis of their chemical composition or solubility properties. Of the three main types, **simple proteins** are classified by solubility properties, **conjugated proteins** by the non-protein groups, and **derived proteins** by the method of alteration.

Simple proteins such as **protamines** in the form of salmine and sturine from fish sperm, **histones** in the form of the globin in hemoglobin, and **albumins** in the form of egg albumin and serum albumin are all soluble in water, and protamines and histones contain a high proportion of basic amino acids. The **globulins** as lactoglobulin in milk are insoluble in water but soluble in dilute salt solutions; the **glutelins** as glutenin in wheat are insoluble in water and dilute salt solutions, but are soluble in dilute acid or alkaline solutions; **prolamines** as zein in corn and gliadin in wheat are soluble in 70 to 80 per cent ethyl alcohol; whereas the **albuminoids** as keratin in hair, horn, and feathers are insoluble in all the solvents mentioned above and can be dissolved only by hydrolysis.

Conjugated proteins include **nucleoproteins,** which consist of a basic protein such as histones or protamines combined with nucleic acid. They are found in cell nuclei and mitochondria. **Phosphoproteins** as casein in milk and vitellin in egg yolk are proteins linked to phosphoric acid; **glycoproteins** are composed of a protein and a carbohydrate and occur in mucin in saliva and mucoids in tendon and cartilage, whereas **chromoproteins** such as hemoglobin and cytochromes consist of a protein combined with a colored compound. **Lipoproteins** are proteins combined with lipids such as fatty acids, fats, and lecithin, and are found in serum, brain, and nervous tissue.

Derived proteins are an indefinite type of protein produced, for example, by partial hydrolysis, denaturation, and heat, and are represented by proteoses, peptones, metaproteins, and coagulated proteins.

DETERMINATION OF PROTEINS

It is often desirable to know the protein content of various foods and biological material. The analysis of the protein content of such material is based on its nitrogen content. Since the average nitrogen content of proteins is 16 per cent, the protein content of a substance may be obtained by multiplying its nitrogen value by the factor $100/16 = 6.25$. For example, if a certain food contained 2 per cent nitrogen on analysis, its protein content would equal 2 times 6.25, or 12.5 per cent.

Since the determination of nitrogen requires considerable equipment and is time consuming, several colorimetric methods for the quantitative estimation of protein have been developed. Methods based on the biuret test and the ninhydrin test are commonly used when the total protein content of many specimens is required. The aromatic amino acids, especially tryptophan and tyrosine, in proteins absorb ultraviolet light at a wave length of 280 nm. The measurement of light absorption at 280 nm is a convenient method for determining the amount of protein in a solution or in an eluate from a chromatograph column.

DENATURATION OF PROTEINS

Denaturation of a protein molecule causes changes in the structure that result in marked alterations of the physical properties of the protein. When in solution, proteins are readily denatured by standing in acids or alkalies, shaking, heating, reducing agents, detergents, organic solvents, and exposure to x-rays and light. Some of the effects of **denaturation** are loss of biological activity, decreased solubility at the isoelectric point, increased susceptibility to hydrolysis by proteolytic enzymes, and increased reactivity of groups that had been masked by the folding of chains in the native protein. Examples of the last mentioned are the uncovered SH groups of cysteine and the OH groups of tyrosine.

The cleavage of several hydrogen bonds and of several possible disulfide bonds often results in the loss of biological activity by denaturation. In some proteins, the native configuration is so stable that denaturation changes are reversible. Hemoglobin, for example, can be reversibly denatured.

PRECIPITATION OF PROTEINS

One of the most important characteristics of proteins is the ease with which they are precipitated by certain reagents. Many of the normal functions in the body are essentially precipitation reactions; for example, the clotting of blood or the precipitation of casein by rennin during digestion. Since animal tissues are chiefly protein in nature, reagents that precipitate protein will have a marked toxic effect if introduced into the body. Bacteria, which are mainly protein, are effectively destroyed when treated with suitable precipitants. Many of the common poisons and disinfectants act in this way. The following paragraphs contain a brief summary of the most common methods of protein precipitation.

By Heat Coagulation

When most protein solutions are heated, the protein becomes insoluble and precipitates, forming coagulated protein. Many protein foods coagulate when they are cooked. Tissue proteins and bacterial proteins are readily coagulated by heat. Routine examinations of urine specimens for protein are made by heating the urine in a test tube to coagulate any protein that might be present.

By Alcohol

Alcohol coagulates all proteins except the prolamines. A 70 per cent solution of alcohol is commonly used to sterilize the skin, since it effectively penetrates the bacteria. A 95 per cent solution of alcohol is not effective because it merely coagulates the surface of the bacteria and does not destroy them.

By Concentrated Inorganic Acids

Proteins are precipitated from their solutions by concentrated acids such as hydrochloric, sulfuric, and nitric acid. Casein, for example, is precipitated from milk as a curd when acted on by the hydrochloric acid of the gastric juice.

By Salts of Heavy Metals

Salts of heavy metals, such as mercuric chloride and silver nitrate, precipitate proteins. Since proteins behave as zwitterions, they will ionize as negative charges in neutral or alkaline solutions. The reaction with silver ions may be illustrated as follows:

$$\underset{\text{Protein}}{\underset{\displaystyle NH_2}{R-\overset{|}{CH}}-COO^-} + Ag^+ \rightarrow \underset{\text{Silver proteinate}}{\underset{\displaystyle NH_2}{R-\overset{|}{CH}}-COOAg}$$

These salts are used for their disinfecting action and are toxic when taken internally. A protein solution such as egg white or milk, when given as an antidote in cases of poisoning with heavy metals, combines with the metallic salts. The precipitate that is formed must be removed by the use of an emetic before the protein is digested and the heavy metal is set free to act on the tissue protein. A silver salt such as Argyrol is used in nose and throat infections, and silver nitrate is used to cauterize wounds and to prevent gonorrheal infection in the eyes of newborn babies.

By Alkaloidal Reagents

Tannic, picric, and tungstic acids are common alkaloidal reagents that will precipitate proteins from solution. When in acid solution the protein as a zwitterion ionizes as a positive charge. It will therefore react with picric acid as shown:

$$\underset{\text{Protein}}{\underset{\displaystyle NH_3^+}{R-\overset{|}{CH}}-COOH} + \text{picric acid} \rightarrow \underset{\text{Protein picrate}}{\underset{\displaystyle NH_3\text{—picrate}}{R-\overset{|}{CH}}-COOH}$$

Tannic and picric acids are sometimes used in the treatment of burns. When a solution of either of these acids is sprayed on extensively burned areas, it precipitates the protein to form a protective coating; this excludes air from the burn and prevents the loss of water. In an emergency, strong tea may be used as a source of tannic acid for the treatment of severe burns. Many other therapeutic agents have been used in the treatment of burns, the most recent being penicillin. Nevertheless, considerable quantities of tannic and picric acid preparations are still employed for this purpose.

By Salting Out

Most proteins are insoluble in a saturated solution of a salt such as ammonium sulfate. When it is desirable to isolate a protein from a solution without appreciably altering its chemical nature or properties, the protein may be precipitated by saturating the solution with $(NH_4)_2SO_4$. After filtration, the excess $(NH_4)_2SO_4$ is usually removed by dialysis. This salting out process finds wide application in the isolation of biologically active proteins.

IMPORTANT TERMS AND CONCEPTS

alpha amino acid
alpha helix
amino acid sequence
aromatic amino acids
chromatography
heterocyclic amino acids
isoelectric point
isoenzymes
oxytocin
peptide linkage
pK of amino acids
polypeptides
primary structure
secondary structure
tertiary structure
zwitterions

QUESTIONS

1. List the five main elements present in proteins with their average content. Using this list, how would you differentiate proteins from carbohydrates and fats?

2. What products are formed by complete hydrolysis of proteins? Which chemical agents would be used for complete hydrolysis?

3. Write the formula for an alpha amino acid containing three carbon atoms. How would you name this amino acid?

4. Write the formula and name of an amino acid that contains a heterocyclic ring.

5. Write the formula of an amino acid as a zwitterion and use the structure to explain the isoelectric point.

6. Illustrate with equations two reactions that involve the amino group of an amino acid.

7. Given the dipeptide alanylglycine as an unknown, explain the procedures you would employ to:
   a. Prove the identity of the two amino acids.
   b. Determine which amino acid in the dipeptide contained the free amino group.

8. Why is the biuret reagent used in the quantitative analysis of protein and the ninhydrin reagent used to spray amino acid spots in chromatography? Explain.

9. Briefly explain the application of chromatography and Sanger's reagent to the analysis of the insulin molecule.

10. Outline the amino acid sequence of vasopressin. How could the position of the disulfide bridge in the molecule be established?

11. How does the primary structure differ from the tertiary structure of a protein? Explain.

12. Briefly explain the meaning and significance of S, the Svedberg unit.

13. What is the relationship between isoenzymes and protein structure? Explain.

14. Describe two methods that can be used to determine the amount of protein in a solution.

15. Why is a protein solution used as an antidote in cases of poisoning with heavy metals? Explain.

16. Why are preparations containing tannic or picric acids sometimes used in the treatment of burns?

17. Illustrate with equations the precipitation of proteins with tannic acid and with silver salts.

SUGGESTED READING

Day and Ritter: Errors in Representing Structure of Proteins and Nucleic Acids. Journal of Chemical Education, Vol. 44, p. 761, 1967.
Delwiche: The Nitrogen Cycle. Scientific American, Vol. 223, No. 3, p. 136, 1970.
Doty: Proteins. Scientific American, Vol. 197, No. 3, p. 173, 1957.
Kendrew: Myoglobin and the Structure of Proteins. Science, Vol. 139, p. 1259, 1963.
Li: The ACTH Molecule. Scientific American, Vol. 209, No. 1, p. 46, 1963.
Merrifield: The Automatic Synthesis of Proteins. Scientific American, Vol. 218, No. 3, p. 56, 1968.
Perutz: The Hemoglobin Molecule. Scientific American, Vol. 211, No. 5, p. 64, 1964.
Stein and Moore: Chromatography. Scientific American, Vol. 184, No. 3, p. 35, 1951.
Stein and Moore: The Structure of Proteins. Scientific American, Vol. 204, No. 2, p. 81, 1961.
Vedvick and Coates: Hemoglobin: A Simple "Backbone" Type of Molecular Structure. Journal of Chemical Education, Vol. 48, p. 537, 1971.

NUCLEIC ACIDS

CHAPTER 16

The *objectives* of this chapter are to enable the student to:

1. Describe the structure of the products of complete hydrolysis of a nucleoprotein.
2. Distinguish between the components in DNA and RNA.
3. Illustrate hydrogen bonding and antiparallel chains in the DNA molecule.
4. Illustrate the tetranucleotide portion of one chain of DNA.
5. Distinguish between the different types of RNA molecules.
6. Describe the function and importance of DNA in the cell.

Nucleic acids were first isolated from cell nuclei and were named after their source. They were thought to be fairly simple groups that were conjugated with proteins to form nucleoproteins. These proteins are characterized by their content of **basic amino acids** such as **arginine** and **lysine.** We know now that nucleic acids are polymers of large molecular weight with nucleotides as the repeating unit. Deoxyribonucleic acid (DNA) is present in the nucleus and ribonucleic acid (RNA) in the cytoplasm of all living cells. DNA and RNA are essentially responsible for the transmission of genetic information and the synthesis of protein by the cell, respectively. Progressive hydrolysis of a nucleoprotein would yield the protein and nucleic acid and its components as shown:

$$\begin{array}{c} \text{Nucleoprotein} \rightarrow \text{Nucleic acid} + \text{Protein} \\ \downarrow \\ \text{Nucleotides} \\ \downarrow \\ \text{Nucleosides} + H_3PO_4 \\ \downarrow \\ \text{Purines and Pyrimidines} + \text{Pentose} \end{array}$$

More specific information concerning the hydrolysis products of DNA and RNA is necessary before the structure of these nucleic acids is considered.

DNA	RNA
Adenine	Adenine
Guanine	Guanine
Cytosine	Cytosine
Uracil	Thymine
D-2-Deoxyribose	D-Ribose
H_3PO_4	H_3PO_4

THE PYRIMIDINE AND PURINE BASES

The heterocyclic rings that form the nucleus for both the pyrimidine and purine bases have already been described (p. 107 and 108). The pyrimidine bases found in nucleic acids include cytosine, uracil, and thymine, which are represented as follows:

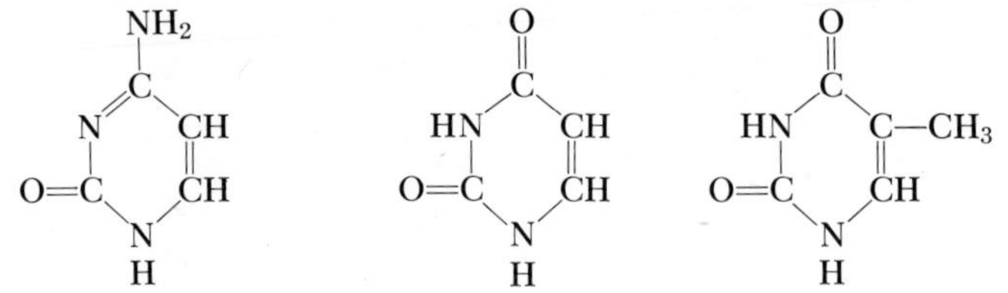

Cytosine (2-oxy-4-amino pyrimidine)

Uracil (2,4-dioxy pyrimidine)

Thymine (5-methyl uracil)

Both DNA and RNA contain the purines adenine and guanine.

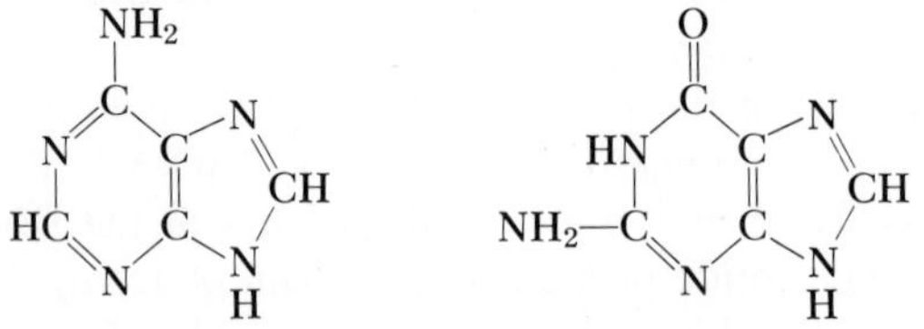

Adenine (6-amino purine)

Guanine (2-amino-6-oxy purine)

THE PENTOSE SUGARS

The pentose sugars that are essential components of the nucleic acids have already been considered on pages 210 and 213. On hydrolysis RNA yielded β-D-ribose, whereas DNA contained β-2-deoxy-D-ribose. Both pentose sugars occur in the furanose form.

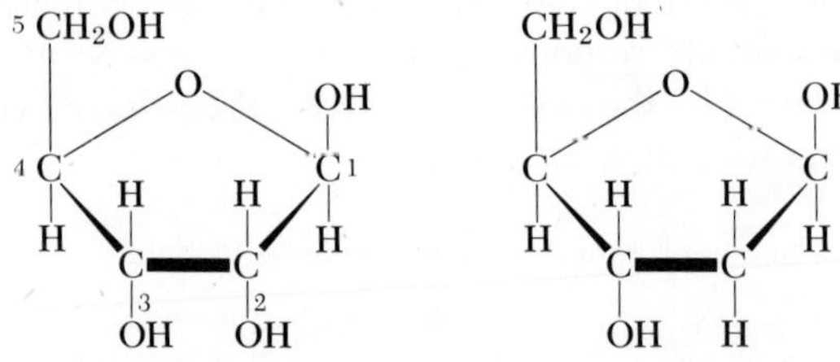

β-D-Ribose

β-2-deoxy-D-Ribose

NUCLEOSIDES

When a purine or pyrimidine base is combined with β-D-ribose or β-2-deoxy-D-ribose, the resultant molecule is called a **nucleoside.** The linkage of the two components is from the nitrogen of the base (position 1 in pyrimidines, 9 in purines) to carbon 1 of the pentose sugars. The formation of a nucleoside involves a reaction between the pyrimidine or purine and ribose phosphate. A nucleophilic displacement reaction occurs (p. 142), for example, in which the entering pyrimidine group takes its position on the opposite face of the ribose molecule from which the phosphate has left. Important examples of nucleosides are cytidine and adenosine:

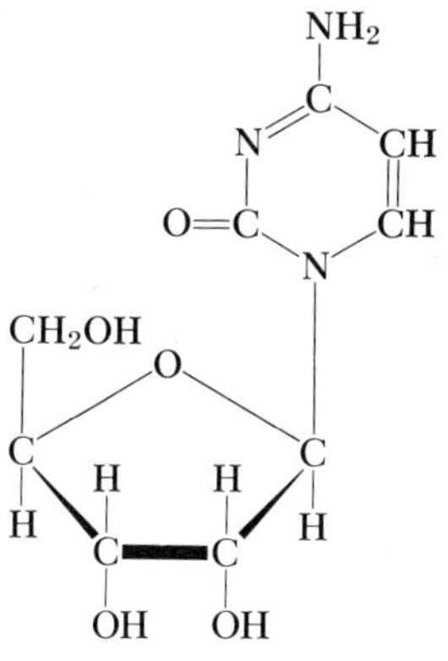

Cytidine

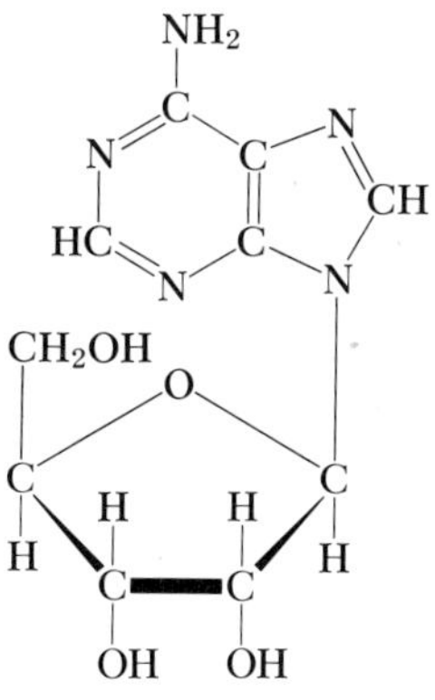

Adenosine

NUCLEOTIDES

When a phosphoric acid is attached to a hydroxyl group of the pentose sugar in the nucleoside by an ester linkage, the result is a **nucleotide.** The esterification may occur on the 2′, 3′, or 5′ hydroxyl of ribose and the 3′ or 5′ hydroxyl of deoxyribose. In the nucleotides, the carbon atoms of ribose or deoxyribose are designated by prime numbers to distinguish them from the atoms in the purine or pyrimidine bases. Yeast nucleic acid contains four mononucleotides: adenylic acid, guanylic acid, cytidylic acid, and uridylic acid. All these compounds include D-ribose in their structure and are named as acids because of the ionizable hydrogens of the phosphate group. Thymus nucleic acid yields nucleotides on hydrolysis that contain β-2-deoxy-D-ribose and thymidylic acid instead of uridylic acid. These structures may be represented by adenylic acid and uridylic acid, two compounds of prime importance in muscle metabolism and carbohydrate metabolism:

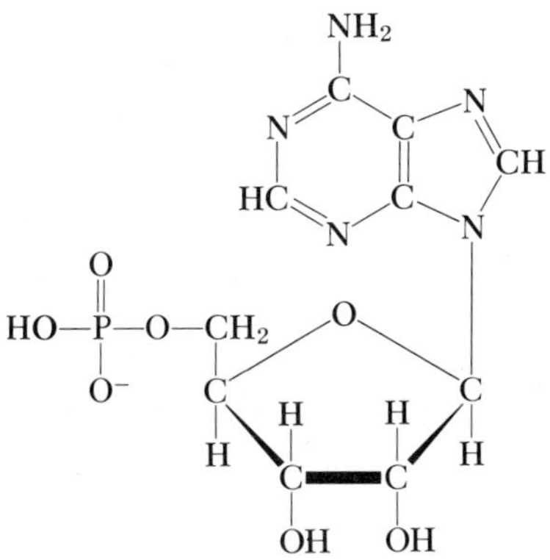

Adenylic acid
(adenosine-5′-monophosphate, AMP)

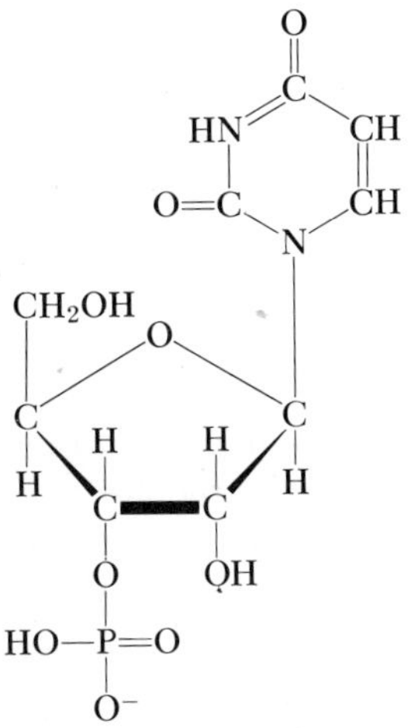

Uridylic acid
(uridine-3′-monophosphate, UMP)

In addition to the nucleotides that are integral components of yeast and thymus nucleic acids, several nucleotides and their derivatives occur free in the tissues and are essential constituents of tissue metabolism. For example, adenylic acid, also designated adenosine monophosphate (AMP), is found in muscle tissue. The biochemist commonly uses abbreviations such as AMP and, for uridine monophosphate, UMP, to designate nucleotides. Adenylic acid may also exist as the diphosphate (adenosine diphosphate, ADP) or as the triphosphate (adenosine triphosphate, ATP) (p. 287). These two compounds are sources of high energy phosphate bonds and are involved in many metabolic reactions.

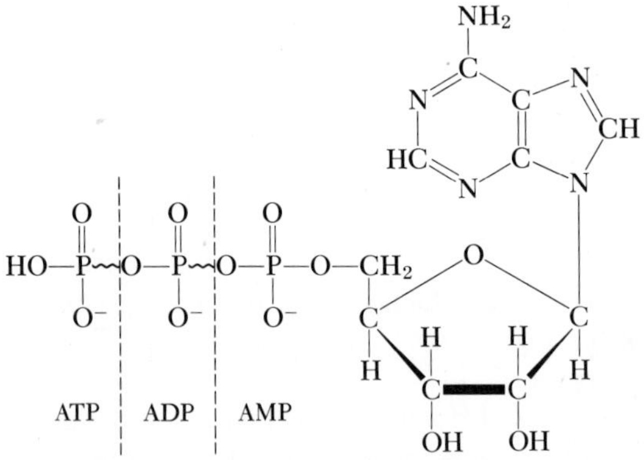

ATP ADP AMP

Adenosine triphosphate

Nucleotides also combine with vitamins to form coenzymes, which will be discussed in Chapter 19.

NUCLEIC ACIDS

The Structure of DNA

The structure of DNA has been shown to consist of chains of nucleotides linked together with phosphate groups that connect carbon atom 3 of one sugar molecule to the number 5 carbon of the next sugar. A portion of the chain containing one each of the four nucleotides of DNA will be represented, although apparently the nucleotides can occur in random order in the chain. This structure is on page 271. The 3,5 linkage from one deoxyribose molecule to another is characteristic in DNA and is the predominant linkage in RNA. This structure and those for nucleosides and nucleotides illustrate the importance of the acetal linkage in biologic compounds (p. 212). The molecular weights of DNA preparations from various sources exhibit a range from 6 to 16 million and illustrate the large size of the polymer. A series of hydrolytic experiments utilizing weak acids, bases, and enzymes has established a 1 : 1 ratio between cytosine and guanine nucleotides and a similar 1 : 1 ratio between adenine and thymine nucleotides. A shorthand representation of this unwieldy structure follows the DNA formula on the following page.

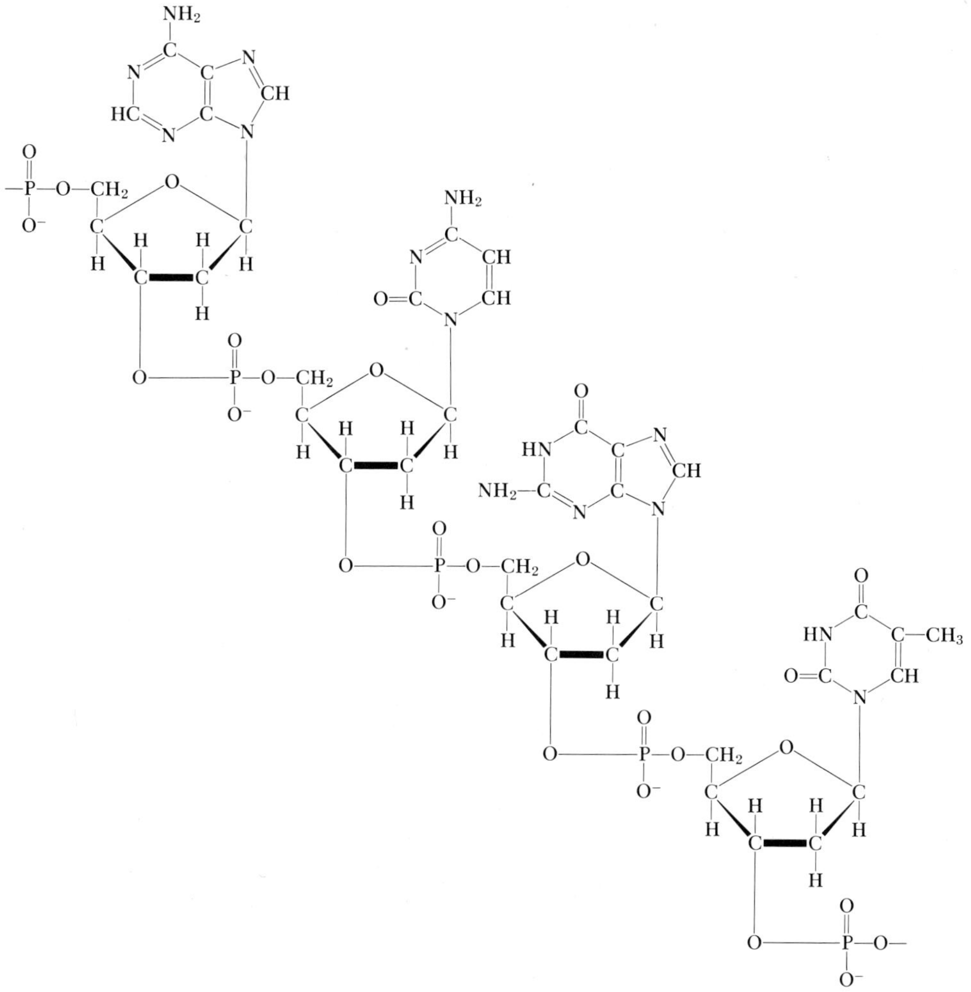

Tetranucleotide portion of one chain of DNA

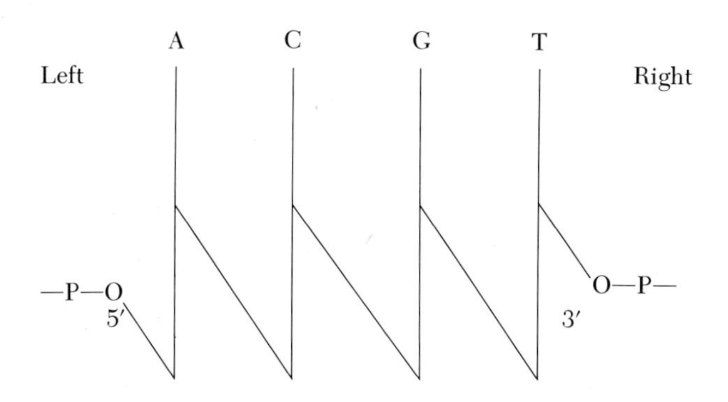

FIGURE 16-1 *a*, James Watson (1928–) Professor of Biology, Harvard University. Shared the Nobel Prize for Medicine and Physiology in 1962 with Crick and Wilkins. Double-helical structure of DNA and role of RNA in protein synthesis. *b*, Francis Crick (1916–) member British Medical Research Council, Laboratory on Molecular Biology. Shared the Nobel Prize for Medicine and Physiology in 1962 with Watson and Wilkins. Proposed model for double-helical structure of DNA.

By physical-chemical methods, such as x-ray diffraction, the DNA molecule has been shown to consist of two chains of nucleotides coiled in a double helical structure, forming a molecule that is very long by contrast to its diameter. Watson and Crick (Fig. 16–1) suggested that the pair of nucleotide chains run in opposite directions around the helix in such a way that an adenine of one chain is bonded to a thymine of the other by hydrogen bonds, and a guanine of one chain is joined by hydrogen bonds to a cytosine of the other as shown in Figure 16–2. Thymine of the first chain is connected by two hydrogen bonds to adenine of the second, and cytosine of the first chain is joined to guanine of the second chain by three hydrogen bonds. The chains consist of deoxyribose nucleotides joined together by phosphate groups with the bases projecting perpendicularly from the chain into the center of the helix. For every adenine or guanine projecting into the central portion, a thymine or cytosine molecule must project toward it from a second chain. Because of the base pairing, the two chains are not identical, and they do not run in the same direction with respect to the linkages between deoxyribose and the base. The chains in the DNA structure are therefore considered **antiparallel.** A portion of the helix emphasizing the hydrogen bonding and the antiparallel chain structure is shown in Figure 16–3.

The Structure of RNA

Information concerning the structure of RNA is not as extensive or as definite as that describing DNA. In general, RNA has been shown to have a structure similar to that of DNA, although the smaller molecules of viral RNA, and also messenger RNA, have single strands of nucleotides. The RNA molecules in the cytoplasm of the cell are of three major types. The largest molecules are **ribosomal RNA** (r-RNA), with molecular weights of a few million. They are associated with the structure of the ribosomes and serve as a template for protein synthesis within the cytoplasm. **Messenger RNA** (m-RNA) molecules are varied in size, with molecular weights from 300,000 to two million. The m-RNA molecules carry the genetic message from the DNA in the nucleus to the protein-synthesizing sites. The smallest RNA molecules are called **transfer RNA** (t-RNA) or

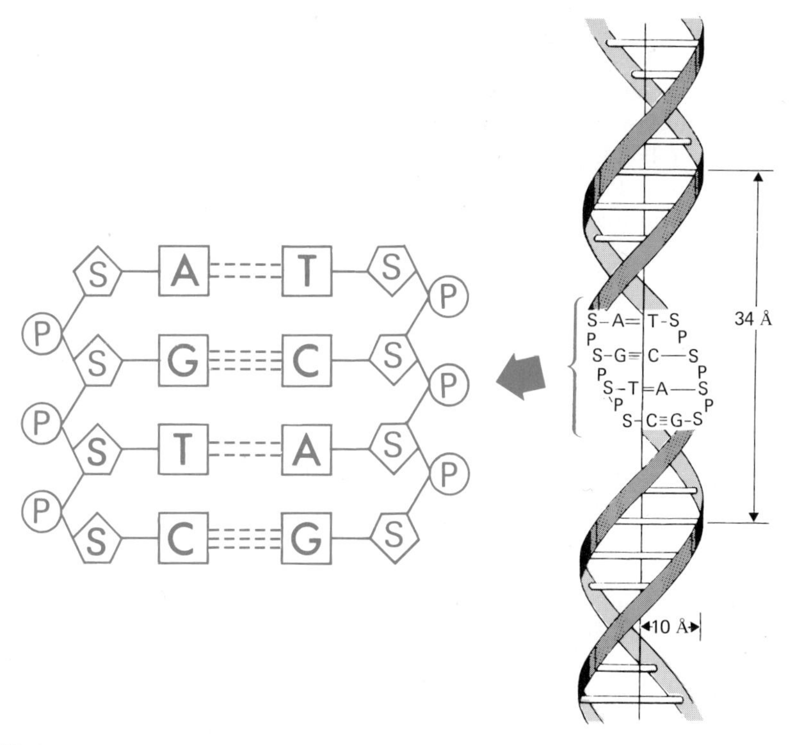

FIGURE 16-2 Double helix of DNA. Here P means phosphate diester, S means deoxyribose, A=T is the adenine-thymine pairing, and G≡C is the guanosine-cytosine pairing. (After Conn and Stumpf: *Outlines of Biochemistry*, 2nd Ed. New York, Wiley, 1963, p. 108.)

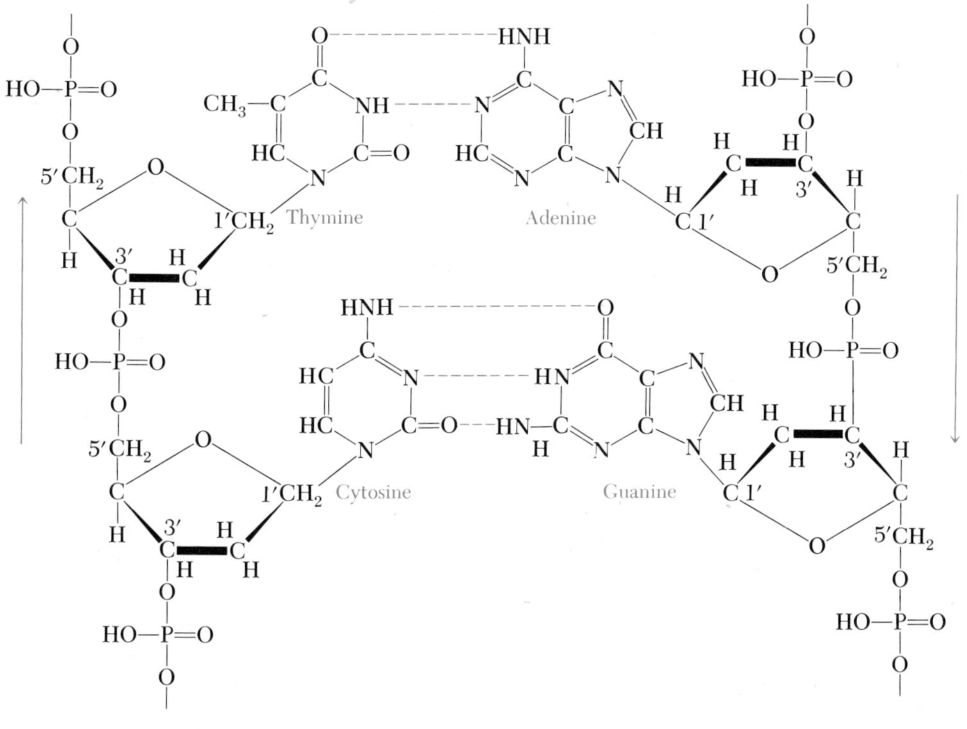

FIGURE 16-3 Hydrogen bonding with antiparallel chains. (Adapted from Conn and Stumpf: *Outlines of Biochemistry*, 2nd Ed. New York, Wiley, 1963.)

soluble RNA (s-RNA), and have molecular weights from 25,000 to 40,000. Their function is to transport specific amino acids to their specific sites on the protein-synthesizing template. The basic composition of three transfer RNAs is known. They are composed of single strands of nucleotides bent into cloverleaf type structures to give the maximum number of hydrogen-bonded pairs. The closed loop of the chain at one end of the cloverleaf contains a sequence of three bases that serves as an anticodon for a specific amino acid as Alanyl t-RNA. The role of these RNA molecules will be discussed.

THE BIOLOGIC IMPORTANCE OF THE NUCLEIC ACIDS

Originally RNA was associated only with yeast and was thought to be restricted to plant sources. DNA from thymus tissue represented the nucleic acids of animal tissues. As methods for their determination have been developed, both RNA and DNA have been found in practically all types of cells. DNA appears to be restricted to the nucleus, most specifically to the chromosomes, whereas RNA occurs both in the cytoplasm and nucleus of a cell.

From the standpoint of genetics it is highly significant that the amount of DNA in each cell nucleus is constant for a given species and the DNA is confined to the chromosomes. It has also been found that the number of genes in a cell nucleus is proportional to the number of DNA molecules. The DNA content of each somatic cell is constant for a given species, and the DNA in the sperm cell, which has only one-half the number of chromosomes, is one-half of that in the somatic cell. In cells that contain multiple sets of chromosomes there is a corresponding increase in the DNA content. This constancy of DNA content would be expected if DNA is an integral part of the chromosomes primarily concerned with the genetic process, carrying the code of genetic information for metabolic function and control within each cell. Evidence for the role of DNA in the genetic process evolved from studies of the synthesis of cellular proteins. As a result of the information stored in DNA in the nucleus, specific proteins are formed in accordance with the genetic code. The code consists of a sequence of three nucleotides which direct the incorporation of a specific amino acid in the particular sequence of amino acids that is characteristic of the primary structure of a protein (see Chapter 15). The overall function of DNA and its relation to RNA and protein synthesis are outlined in Figure 16–4.

The method of DNA replication was proposed by Watson and Crick based on their model of DNA. They suggested that the paired molecules of a DNA helix first separate from each other, as shown in Figure 16–5. Deoxyribonucleotides that were in the uncombined form in the cell then become attached by hydrogen bonding to the nucleotide bases of the separated strands of DNA. The base pairing follows the patterns of DNA with adenine and thymine, and also guanine and cytosine deoxyribonucleotides being joined by hydrogen bonds. The nucleotide chains are then linked together by a polymerizing catalyst or enzyme, and a new DNA helix is formed that is identical to the parent molecule.

In reduplication of cell nuclei, which is necessary in cell division, the double helix may unravel and each of the original chains may serve as a template for the synthesis of another chain. By adding labeled purine and pyrimidine intermediates to a synthesizing system it has been shown experimentally that one chain of newly synthesized DNA contains labeled intermediates from the system, whereas the original chain is unlabeled. Since the base-pairing pattern of DNA, adenine combining with thymine and guanine with cytosine deoxyribonucleotides, is followed, the newly synthesized chain will possess

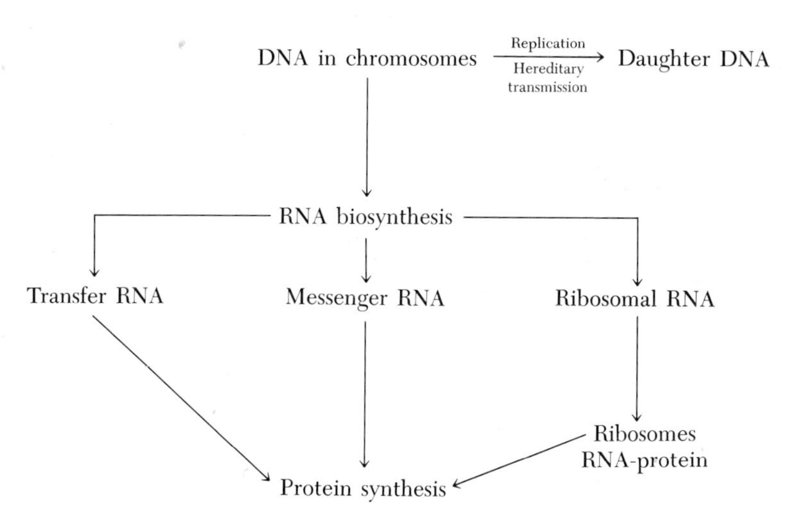

FIGURE 16-4 The relation of DNA to RNA in protein synthesis.

the exact nucleotide sequence of the original parent chain. The result is the synthesis of two pairs of DNA chains (Fig. 16–5) in which each pair is identical in nucleotide sequence and genetic coding information to the original pair of parent chains. In the laboratory it has been demonstrated that pure DNA preparations from a particular species of bacteria or bacteriophage, when added to another species of bacteria, will serve as a template to direct the recipient cells to develop the characteristics of the donor bacteria. It is quite possible that the multiplication of viruses within cells may occur by the same process. For example, type 1 poliomyelitis virus, which has been crystallized, contains an RNA that serves as a template to infect cells with this type of virus. This knowledge suggests that RNA functions mainly in the cytoplasm of a cell as a template for the synthesis of specific cellular proteins. A close relation exists between DNA of the nucleus

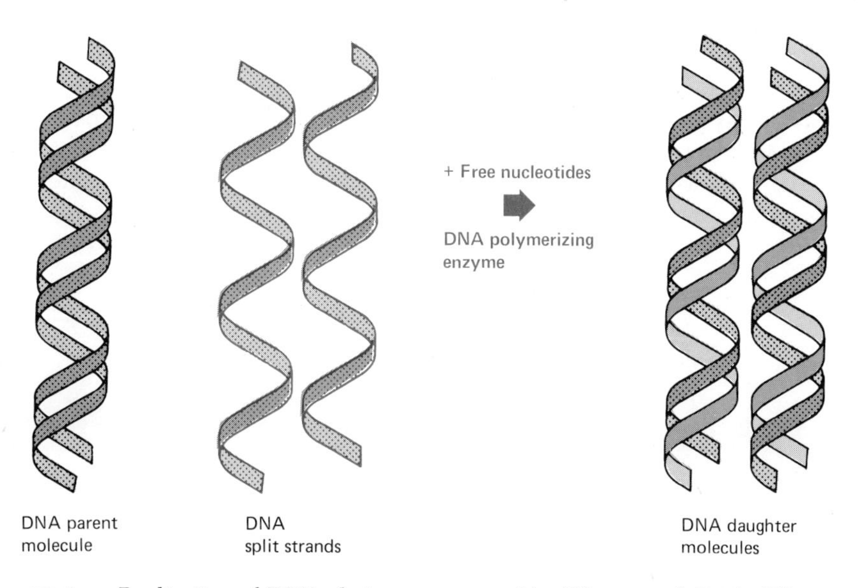

FIGURE 16-5 Replication of DNA chains as suggested by Watson and Crick. White strands are newly synthesized DNA.

and RNA of the cytoplasm, since one chain of DNA and one chain of RNA could twist around each other to form a double helix and thus influence the RNA template.

Multiplication of viruses within cells can occur by a similar process in which the viral genes consist of DNA that transmits information to the RNA of the cell and then into the cell proteins. Several common viruses which cause poliomyelitis, the common cold, and influenza are called RNA viruses, since the RNA replicates directly into new copies of RNA and translates information directly to proteins of the cell without DNA involvement in their replication. Recently evidence for a reverse flow of genetic information from RNA to DNA has been obtained in experiments using the Rous sarcoma virus. It has been suggested that in normal cells there are regions of DNA that serve as templates for the synthesis of RNA, and that this RNA serves in turn as a template for the synthesis of DNA which then becomes integrated with the cellular DNA. These experiments may be valuable in an explanation of the origin of cancer in man. Since it was shown that cancer-causing RNA viruses can produce a DNA transcript of the viral RNA, it is possible that the viral RNA may transmit genetic information to the genes of cells that will eventually surface as spontaneous cancer.

Since about 1955, a major objective in biochemistry has been to document and clarify the details of what is called the **central dogma of genetics:** that DNA is the hereditary material; that its information is encoded in the sequences of its subunits, the genes; and that this information is transcribed onto RNA and then translated into protein. At Oak Ridge National Laboratory, starting in 1967, attempts were made to produce electron micrographs of individual genes in action. Recently, in studies of the bacterial genes of *E. coli* pictures were obtained that show both an inactive chromosome segment and an active chromosome segment. In the active segment the DNA is seen being transcribed onto messenger RNA and the RNA being translated into protein. The electron micrographs obtained in these studies bear a striking resemblance to diagrams of transcription and translation that have been proposed by recognized research workers in this area.

In another recent study of how cells proceed through division in the cell cycle, an electron micrograph clearly depicting chromosome replication was obtained. The DNA double helix does not unwind as neatly as shown in Figure 16–5, but appears as tangled strands with segments of replicating DNA molecules joining the original DNA strand at several junctures in the micrograph.

From these and other studies a science of **genetic engineering** may result, which would have as its first objective the reversal of cancer growth and the repair of inborn deficiencies in genetic diseases. Just as viral RNA may transmit cancer in man, as mentioned earlier, scientists may be able to introduce chemotherapeutic RNA molecules into the body tissues for repair or curative purposes.

IMPORTANT TERMS AND CONCEPTS

adenine
cytosine
DNA
guanine
hydrogen bonding
nucleoside
nucleotide
purine
pyrimidine
replication
RNA
thymine
uracil
viruses

QUESTIONS

1. What type of protein is found in nucleoproteins? What products result from complete hydrolysis of nucleic acids?
2. Write the formula and name for a nucleoside containing β-2-deoxyribose.
3. What is the difference between ATP and ADP; ADP and AMP? Illustrate these differences with a composite formula.
4. What are the two major types of nucleic acids? List the composition of each type.
5. How are the nucleotides linked together chemically when they form nucleic acids?
6. Prepare a sketch of the DNA molecule showing the double helix and the hydrogen bonding between the bases.
7. What is meant by the antiparallel chain structure of DNA?
8. How does the structure of RNA differ from that of DNA? What are the three major types of RNA in the cell?
9. Illustrate a tetranucleotide portion of a DNA molecule in a simplified fashion and explain the notations used.
10. Briefly describe the function and importance of DNA in the cell.
11. Illustrate the process of replication of DNA.

SUGGESTED READINGS

Britten and Kohne: Repeated Segments of DNA. Scientific American, Vol. 222, No. 4, p. 24, 1970.
Crick: The Genetic Code. Scientific American, Vol. 215, No. 4, p. 55, 1966.
Fairley: Nucleic Acids, Genes, and Viruses. Journal of Chemical Education, Vol. 36, p. 544, 1959.
Fraenkel-Conrat and Stanley: The Chemistry of Life 2. Implications of Recent Studies of a Simple Virus. Chemical & Engineering News, May 15, 1961, p. 136.
Hanawalt and Haynes: The Repair of DNA. Scientific American, Vol. 216, No. 2, p. 36, 1967.
Holley: The Nucleotide Sequence of a Nucleic Acid. Scientific American, Vol. 214, No. 2, p. 30, 1966.
Lesk: Progress in Our Understanding of the Optical Properties of Nucleic Acids. Journal of Chemical Education, Vol. 46, p. 821, 1969.
Luria: Bacteriophage Genes and Bacterial Functions. Science, Vol. 136, p. 685, 1962.
Mazia: The Cell Cycle. Scientific American, Vol. 230, No. 1, p. 55, 1974.
Miller: The Visualization of Genes in Action. Scientific American, Vol. 228, No. 3, p. 34, 1973.
Research Reporter: Portrait of a Gene. Chemistry, Vol. 42, No. 8, p. 20, 1969.
Research Reporter: Trouble on the DNA Front. Chemistry, Vol. 43, No. 9, p. 24, 1970.
Temin: RNA-Directed DNA Synthesis. Scientific American, Vol. 226, No. 1, p. 24, 1972.
Watson: Double Helix. Atheneum, New York, 1968.
Yanofsky: Gene Structure and Protein Structure. Scientific American, Vol. 216, No. 5, p. 80, 1967.

BIOCHEMISTRY OF THE CELL

CHAPTER 17

The *objectives* of this chapter are to enable the student to:

1. Explain how the electron microscope has affected the study of the biochemistry of the cell.
2. Describe the nature of the major subcellular components.
3. Recognize the similarities and differences between plant and animal cells.
4. Describe the method of separation of cellular components.
5. State the functions of the major subcellular components.

Within a span of 25 years biochemistry has progressed from nutritional studies on the whole animal, to tissue slices of important organs, to tissue homogenates, and currently to a study of the cell. Considerable research is directed toward an understanding of the chemical reactions that occur in the cell and their relation to cellular function and structure. The use of labeled compounds and cytochemical techniques have assisted the biochemist in the location of the cellular site for specific reactions, especially those involving enzymes. The **light microscope** has been invaluable in the study of staining reactions and the rough morphology of the cell. With the advent of the **electron microscope** the fine structure of the cell was revealed, and a whole new area of biochemical research was made available. A more complete understanding of the relation of structure to function is now within the grasp of biochemists.

When faced with a large complex molecule to study, a chemist first resorts to hydrolysis to reduce the unknown molecule to smaller identifiable units. In like manner, to understand the function of a plant, animal, or bacterial cell we first study the subcellular components. Cells differ in size, appearance, and structure, depending on their function, but a typical animal cell has the features illustrated in Figure 17–1. Structures common to all cells include the cell membrane, nucleus, mitochondria, endoplasmic reticulum, ribosomes, and Golgi apparatus. A typical plant cell, Figure 17–2, also includes the same structures. Fortunately for cytological studies, the electron microscope can be focused on each subcellular component to reveal its structural details.

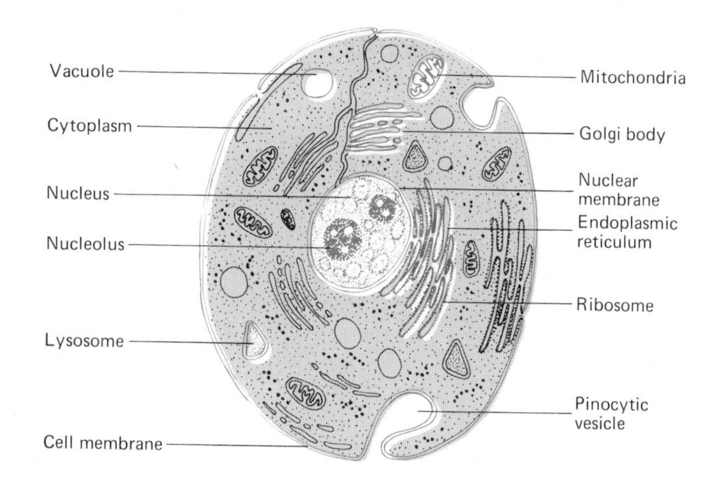

FIGURE 17-1 A diagram of a typical cell based on an electron micrograph. (After the Living Cell, by Jean Brachet. Copyright © 1961 by Scientific American, Inc. All rights reserved.)

SUBCELLULAR COMPONENTS

Cell Membrane

All the subcellular components of a cell are contained within a definite cell wall or membrane. This membrane plays a vital role in the passage of nutrient and waste material into and out of the cell. In addition to the cell membrane, plant cells have rigid walls that surround and protect the membrane. The cell walls consist of cellulose and other polysaccharides. The **cell membrane** is composed of lipids and protein arranged in such a fashion that water-soluble and lipid-soluble substances can pass through the membrane. The permeability of living membranes has never been adequately explained. Although many cells are bathed in a fluid rich in sodium and chloride ions and low in potassium ions, the cell contents are rich in potassium ions and low in sodium and chloride ions. The different rates of absorption of monosaccharides and amino acids from the small intestine emphasize the importance of the membrane in selective permeability toward small ions and molecules.

In general, large molecules do not pass directly through the cell membrane; however, they may be taken into the cytoplasm by pinocytosis. Pinocytosis starts with the indentation of the cell membrane to form a pinocytic vesicle (Fig. 17-1) that engulfs the large molecules and then closes with the formation of a vacuole or lysosome that moves into the cytoplasm of the cell. Membranes are also involved in a process of "active transport," or the movement of ions or molecules from a region of low concentration across the

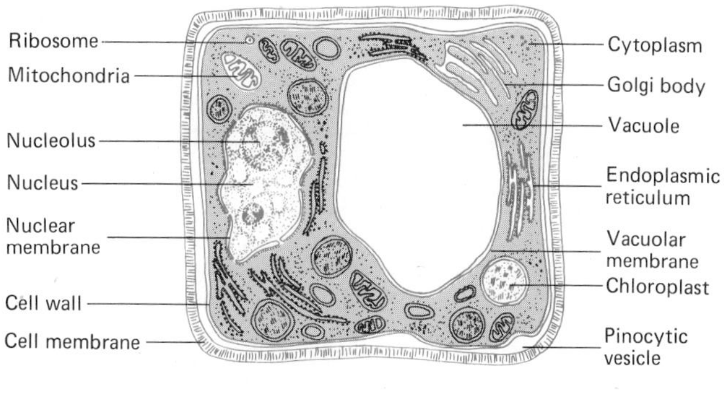

Ribosome
Mitochondria
Nucleolus
Nucleus
Nuclear membrane
Cell wall
Cell membrane
Cytoplasm
Golgi body
Vacuole
Endoplasmic reticulum
Vacuolar membrane
Chloroplast
Pinocytic vesicle

FIGURE 17-2 A diagram of a typical plant cell.

membrane into a region of higher concentration. This movement against a concentration gradient probably involves enzymes, coenzymes, and energy in the form of ATP.

Cytoplasm

The **cytoplasm** is the general protoplasmic mass in which the definite subcellular components described above are embedded. At present all of the essential compounds and macromolecules in the cell not associated with definite particles are thought to exist in the cytoplasm. Many soluble enzymes are found in the cytoplasm, particularly those associated with the conversion of glucose to pyruvic or lactic acids. Considerable research remains to be done on the components of the cell and the cytoplasm with respect to enzyme and coenzyme distribution and their role in various metabolic reactions.

Nucleus

The nucleus is roughly spherical in shape and is surrounded by a double layered membrane that is more porous than the cell membrane. In many cells the outer membrane is connected with the nuclear membrane by one or more channels through the cytoplasm. In addition there is usually a connection between the endoplasmic reticulum and the double-layered nuclear membrane. It has long been recognized that the **nucleus** serves as a site for the transmission and regulation of hereditary characteristics of the cell. This control is an essential feature of the **chromosomes** that are composed of **deoxyribose nucleic acid** (DNA), and basic protein (see p. 267). The nucleus also contains one or more small, dense, round bodies called nucleoli. These bodies contain DNA and **ribose nucleic acid** (RNA), and appear to be involved in the synthesis of RNA and proteins.

Mitochondria

These subcellular particles are shaped like an elongated oval 2 to 7μ in length and 1 to 3μ in diameter (Fig. 17–3). It is possible to stain mitochondria and observe them under a light microscope; however, the electron microscope reveals their fine structure. The walls of the mitochondria are double-layered membranes; the outer membrane is smooth and bag-shaped, the inner one also in the form of a closed sack that is folded to form ridges or tubes protruding into the center of the mitochondrion. These projections inside the mitochondria are called **crista.** The liquid in the matrix of the particle contains protein, neutral fat, phospholipids, and nucleic acids. In contrast to the nucleus, the nucleic acids in the mitochondria are mostly RNA with only small amounts of DNA. The **mitochondria** have been called the *powerhouses of the cell,* since they are the site of major oxidative processes and oxidative phosphorylation which result in the formation of ATP.

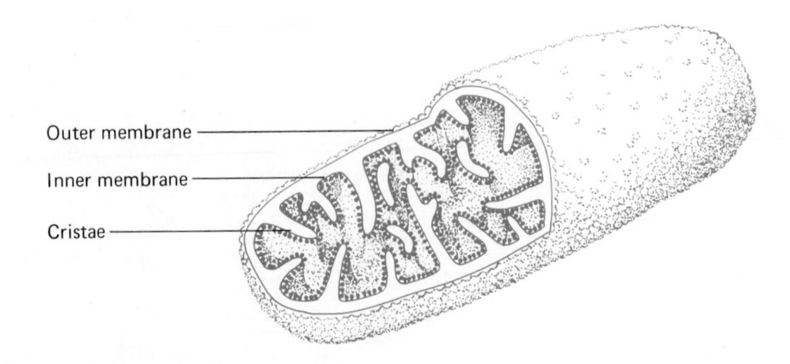

FIGURE 17–3 Tridimensional diagram of mitochondria showing outer and inner membranes with the cristae.

Chloroplasts

Plant cells contain highly pigmented particles 3 to 6μ in diameter called **chloroplasts** (Fig. 17–2). These particles contain the green pigment chlorophyll and play a major role in photosynthesis. Inside the chloroplast membrane is a series of laminated membrane structures called **grana.** Chlorophyll and lipids are concentrated in the grana which are active in the photosynthetic process. The structure and function of chloroplasts parallel those of mitochondria.

Endoplasmic Reticulum and Ribosomes

The **endoplasmic reticulum** is composed of a network of interconnected, thin, membrane-like tubules or vesicles. In some areas of the cytoplasm the membranes are covered with dark round bodies about 0.015μ in diameter which contain 80 per cent of the RNA in the cell and have therefore been termed ribosomes. These areas are known as rough endoplasmic reticulum contrasted to the smooth reticulum which does not have ribosomes adsorbed on its surface. The endoplasmic reticulum and accompanying ribosomes are also called microsomes or the microsomal region of a cell. The specific particles, the **ribosomes,** are the site for the synthesis of proteins within the cell.

The Golgi Apparatus

This is also called the Golgi body or complex and consists of an orderly array of flattened sacs with smooth membranes associated with small vacuoles of varying size. Although it is similar to the smooth endoplasmic reticulum, it exhibits different staining properties and its membrane has a different composition. The **Golgi apparatus** is often connected to the cell membrane by a channel and serves as a way station in the transport of substances produced in other subcellular particles. In liver cells, for example, the Golgi apparatus is usually located close to the small bile canals and is involved in the transport and excretion of substances such as bilirubin glucuronide into the bile.

Lysosomes

These particles are spherical in shape with an average diameter of 0.4μ. They contain several soluble hydrolytic enzymes (hydrolases) that exhibit an optimum pH in the acid range. The lysosomal membrane is lipoprotein in nature and prevents the enzymes from escaping into the cellular cytoplasm. The membrane also prevents the substrates for the enzymes from entering the cell. When the cell is injured and the membrane is broken, the released enzymes cause cellular breakdown. In autolysis of tissue, whether normal (as involution of the thymus gland at puberty), pathological, or post-mortem, the lysosomal enzymes destroy cellular tissue. In fact, one of the main functions suggested for these particles is to help clear tissues of dead cells. The processes of phagocytosis and pinocytosis involve the engulfment of foreign material into vesicles or vacuoles and the digestion of this material. These particles may be converted into lysosomes to assist in the hydrolysis of phagocytosed material.

Vacuoles and Vesicles

These particles are roughly spherical in shape and vary in size from 0.1 to 0.7μ in diameter. They are often found close to the Golgi apparatus and to channels involved in the entrance and excretion of material to and from the cell. Vacuoles may serve as temporary storage sacs, or as bodies involved in the removal of foreign material from the cell.

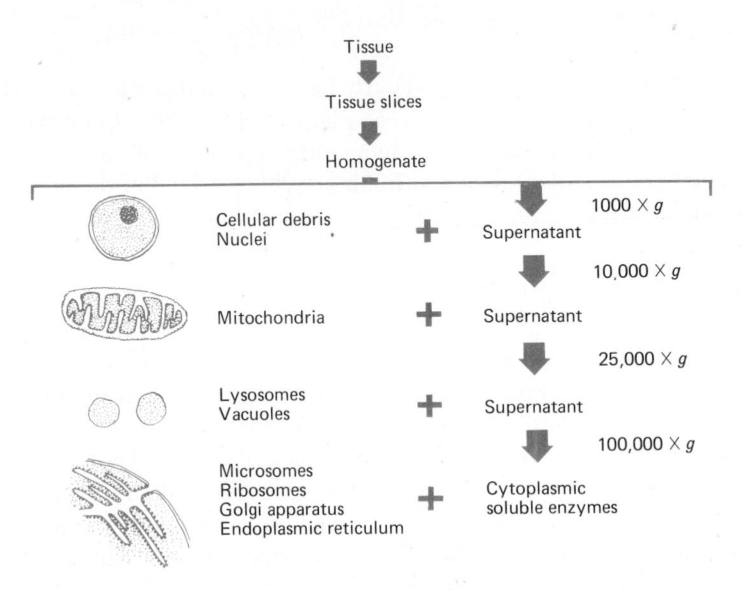

FIGURE 17-4 The separation of cell components by differential centrifugation. (Modified from Bennett and Frieden: Modern Topics in Biochemistry, The Macmillan Co., 1966.)

SEPARATION OF CELLULAR COMPONENTS

Since in most instances it is impossible to study the reactions in a single cell or in a particular particle within that cell, several schemes for the separation of cellular components have been proposed. To study enzyme systems in the mitochondria, for instance, it is desirable to have a large number of functioning mitochondria free from cytoplasm and other cell particles. The isolation procedure begins with thin slices of the proper tissue or organ which are homogenized by grinding the tissue in a glass or teflon homogenizer in cold isotonic sucrose solution. The homogenate, which consists mostly of disrupted cells, is then subjected to **differential centrifugation** to separate the subcellular particles. An example of a general scheme for the separation of the essential components of a cell is shown in the accompanying diagram (Fig. 17–4). The gravitational force of centrifugation in the scheme is represented as g (for example, 10,000 × g); the time of centrifugation is approximately 10 to 15 minutes for all steps but the final centrifugation, which requires one hour.

BIOCHEMICAL FUNCTION OF CELLULAR COMPONENTS

Many investigators have concentrated their research activities on the biochemical reactions that occur in a specific subcellular particle. The mitochondria and ribosomes particularly have been the subject of several research studies. It is obviously not possible at present to reconstruct the exact biochemical functions of the intact cell, but a combination of cytochemical techniques and research on reactions within the separated particles provide a greater understanding of the overall process. Some of the biochemical functions that have been associated with cellular components are listed in Table 17–1.

TABLE 17-1 BIOCHEMICAL FUNCTIONS OF CELLULAR COMPONENTS

CELLULAR COMPONENTS	BIOCHEMICAL COMPOSITION	BIOCHEMICAL FUNCTION
Cell membrane	Phospholipid, protein, lipoprotein	Structural support, ion transport, "active" transport
Nuclei	DNA, basic protein, RNA	Chromosomal genetic control, DNA-regulated metabolism
Ribosomes	RNA, proteins	Protein synthesis
Lysosomes	Protein, enzymes	Protective membrane, hydrolysis of engulfed material
Cytoplasm	Protein, enzymes	Soluble enzymes used in glycolytic scheme, and so forth
Golgi apparatus	Protein	Intracellular transfer and cellular excretion

IMPORTANT TERMS AND CONCEPTS

chloroplasts
cytoplasm
differential centrifugation
electron microscope
Golgi apparatus
lysosomes
membrane
mitochondria
nucleus
vacuoles

QUESTIONS

1. What major instrumental development enabled the biochemist to study subcellular components?

2. What is the nature of the cell membrane? What evidence is there for selective permeability of the cell membrane?

3. The membrane surrounding the nucleus not only has pores but has a direct connection to the cytoplasm through the endoplasmic reticulum. Is this arrangement an advantage to the cell? Why?

4. Biochemists have intensively studied the mitochondria for several years. Why should they be so interested in this subcellular particle?

5. In an electron micrograph of the cell, how can rough endoplasmic reticulum be differentiated from smooth? What is one of the major functions of the rough form?

6. Discuss the nature and function of the Golgi apparatus.

7. What is the nature and function of the lysosomes?

8. How do the enzymes in the lysosomes differ from those in the cytoplasm?

9. Based on the scheme of separation of cellular components, how would you rate the relative size and the density of microsomes *versus* nuclei *versus* mitochondria?

10. Given 10g of liver tissue, how would you proceed to demonstrate that liver cell lysosomes hydrolyze dead protein tissue?

SUGGESTED READING

Baserga and Kisieleski: Autobiographies of Cells. Scientific American, Vol. 209, No. 2, p. 103, 1963.

Everhart and Hayes: The Scanning Electron Microscope. Scientific American, Vol. 226, No. 1, p. 54, 1972.

Green: The Mitochondrion. Scientific American, Vol. 210, No. 1, p. 63, 1964.

Racker: The Membrane of the Mitochondrion. Scientific American, Vol. 218, No. 2, p. 32, 1968.

Rich: Polyribosomes. Scientific American, Vol. 209, No. 6, p. 44, 1963.

BIOCHEMICAL ENERGETICS

CHAPTER 18

The *objectives* of this chapter are to enable the student to:

1. Recognize the relation between ΔG and $\Delta G°$.
2. Calculate the $\Delta G°$ for a coupled reaction that utilizes ATP as the driving force.
3. Recognize ATP as a high energy compound and its relation to ADP and AMP.
4. Describe the formation of ATP under anaerobic conditions in the cell.
5. Explain the process of oxidative phosphorylation and its site of action in the mitochondria.
6. Recognize the difference in moles of ATP formed by $NADH_2$ and $FADH_2$ in the electron transport system.

To understand more fully the reactions catalyzed by enzymes in the metabolic processes of the body, we must consider the energy relationships that are involved. The energy released from one reaction within a cell may be used almost simultaneously in another reaction that is essential in cellular economy. Energy produced in the cell may also be used as heat, for mechanical work as in muscular contraction, or as an electric impulse in nerve transmission. Many of the reactions in metabolism produce chemical energy which is stored in **high energy compounds.** These high-energy compounds are used to drive essential reactions in the metabolic cycles of carbohydrate, lipid, and protein metabolism.

FREE ENERGY CHANGE IN REACTIONS

The basic thermodynamics of a chemical reaction was considered in general chemistry. The change in free energy of a biochemical reaction is most important in biological systems. In a solution containing biological products B and reactant A, the change in free energy, ΔG, is a measure of the useful chemical energy that can be derived from the reaction. An equation that expresses the relation between ΔG and a change in concentration of products and reactants is as follows:

$$\Delta G = \Delta G° + RT\ 2.3 \log_{10} \frac{[B]}{[A]}$$

where $\Delta G°$ is the **standard free-energy change** of the reaction, R is the gas constant, T is the absolute temperature, and the activities of B and A are expressed as concentrations in moles per liter. At equilibrium, $\Delta G = 0$ and the ratio of [B] to [A] equals K. By substituting in the previous equation

$$0 = \Delta G° + RT\ 2.3 \log_{10} K, \text{ or}$$
$$\Delta G° = -RT\ 2.3 \log_{10} K$$

From these equations we could calculate $\Delta G°$ knowing the equilibrium constant K or the concentration of the products and of the reactants at equilibrium. Once $\Delta G°$ is known, we can calculate ΔG or the free-energy change of any reaction. The ΔG of a reaction depends on the standard free-energy change and on the concentrations of reactants and products. For example, if, when equilibrium is reached in the enzymatic hydrolysis of adenosine monophosphate at 25°C to form adenosine plus phosphoric acid (Ⓟ), the concentration of AMP is 0.002M and the concentration of the products is 0.120M, we may calculate $\Delta G°$ as follows:

$$K = \frac{B}{A} = \frac{A + Ⓟ}{AMP} = \frac{0.120}{0.002} = 60$$

$$\Delta G° = -RT\ 2.303 \log_{10} K = -1.987 \times 298 \times 2.303 \log_{10} 60$$

$$\Delta G° = -1363 \times 1.78 = -2.20 \text{ kcal}$$

COUPLED REACTIONS

There are many chemical reactions in a cell that have a $+\Delta G°$ and will not proceed in a left to right direction without assistance. Removal of one or more of the products of the reaction may force it to the right, or it may be coupled to a highly exergonic reaction ($-\Delta G$). In general, an endergonic reaction ($+\Delta G$) may be coupled with an exergonic reaction so that energy is delivered to the endergonic reaction. In such coupled reactions the algebraic sum of the free-energy changes in the two reactions must be negative in sign (a net decline in free energy) for the coupled reaction to occur. The energy in ATP (adenosine triphosphate), as represented by a $-\Delta G°$ of 8.0 kcal when ATP is converted to ADP (adenosine diphosphate), is often used to drive endergonic reactions. In the formation of glucose-6-phosphate from glucose and phosphate, about 3.0 kcal are required. If this reaction is coupled with the ATP → ADP reaction in the presence of the enzyme hexokinase, the following $\Delta G°$ results:

$$\text{Glucose} + \text{ATP} \xrightarrow{\text{hexokinase}} \text{Glucose-6-}PO_4 + \text{ADP}$$

$$\Delta G° = -8.0 + 3.0 = -5.0 \text{ kcal (approximately)}$$

Many **coupled reactions** in the cell involve the formation of a common intermediate with the assistance of ATP. The formation of sucrose from glucose and fructose, for example, has a $\Delta G°$ of +5.5 kcal and requires the conversion of ATP to ADP to drive the reaction to completion. The coupled reaction and formation of a common intermediate, glucose-1-PO_4, can be represented as follows:

$$\text{Glucose} + \text{ATP} \rightarrow \text{Glucose-1-}PO_4 + \text{ADP}$$
$$\text{Glucose-1-}PO_4 + \text{Fructose} \rightarrow \text{Sucrose} + P_i$$
$$\Delta G° = +5.5 - 8.0 = -2.5 \text{ kcal (approximate)}$$

In the first reaction, the terminal PO_4 group of ATP was transferred to glucose and with it some of the energy of the ATP. In the second reaction, the energy-enriched glucose-1-PO_4 reacts with fructose to form sucrose.

HIGH-ENERGY COMPOUNDS

In the preceding section we learned that many endergonic reactions in a cell can be coupled to an exergonic reaction to obtain the energy to drive the cellular reaction to the right. Early investigations on the nature of muscular contraction revealed that the presence of the high-energy compound creatine phosphate has a driving force in muscle reactions. Studies on the oxidation of glucose and especially the metabolic cycles of carbohydrate oxidation emphasized the role of **adenosine triphosphate, ATP,** and this energy-rich compound has become the key in linking endergonic processes to those that are exergonic.

High-energy compounds are often complex phosphate esters that yield large amounts of free energy on hydrolysis. A more detailed consideration of the energy released by the stepwise hydrolysis of ATP will illustrate the high energy concept.

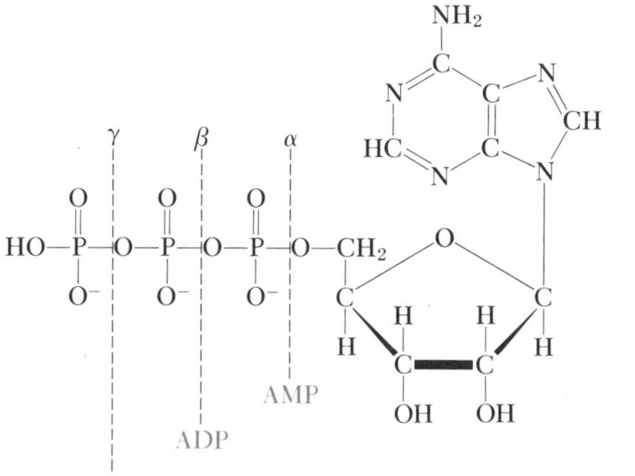

Adenosine triphosphate, ATP

$$\text{ATP} \xrightarrow{\text{hydrolysis}} \text{ADP} + H_3PO_4 \qquad \Delta G^\circ = -8.0 \text{ kcal}$$

$$\text{ADP} \xrightarrow{\text{hydrolysis}} \text{AMP} + H_3PO_4 \qquad \Delta G^\circ = -6.5 \text{ kcal}$$

$$\text{AMP} \xrightarrow{\text{hydrolysis}} \text{Adenosine} + H_3PO_4 \qquad \Delta G^\circ = -2.2 \text{ kcal}$$

Several explanations have been proposed for the release of energy on the hydrolysis of high-energy compounds. These include the fact that these compounds are unstable in acid and alkaline solutions and are readily hydrolyzed. Also, the hydrolysis products, inorganic phosphate, ADP, and AMP, have many more resonance possibilities than the parent ATP. A major reason for the release of energy involves the *type of bond structure* in these compounds. The β and γ bonds in ATP are anhydride linkages that involve a large amount of repulsion energy between the phosphates, which is released on hydrolysis.

Other phosphorus-containing, high-energy compounds include:

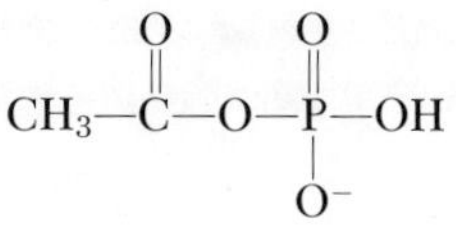

```
           O      O
           ‖      ‖
     CH3—C—O—P—OH
                  |
                  O⁻
```

Acetyl phosphate
$\Delta G° = -10.0$ kcal

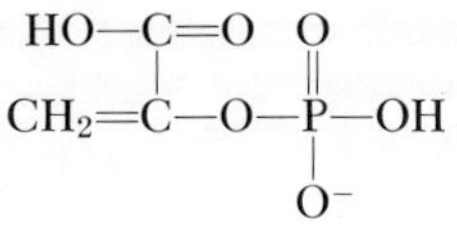

```
        HO—C=O  O
            |   ‖
     CH2=C—O—P—OH
                |
                O⁻
```

Phosphoenolpyruvic acid
$\Delta G° = -12.0$ kcal

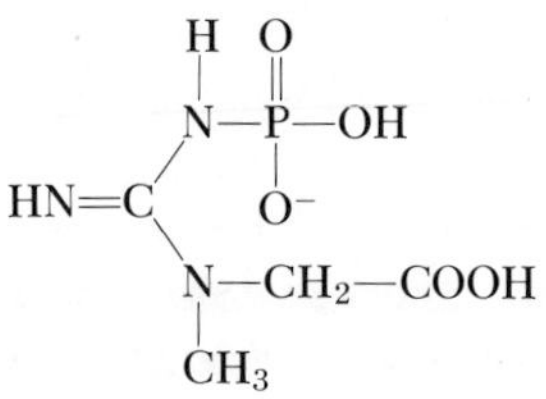

```
            H  O
            |  ‖
            N—P—OH
           /   |
     HN=C      O⁻
           \
            N—CH2—COOH
            |
            CH3
```

Creatine phosphate
$\Delta G° = -10.0$ kcal

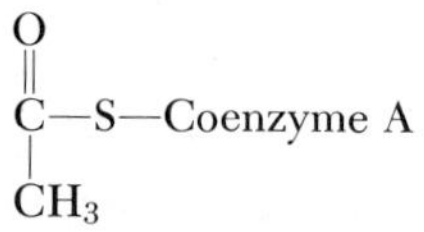

```
     O
     ‖
     C—S—Coenzyme A
     |
     CH3
```

Acetyl coenzyme A
$\Delta G° = -8.2$ kcal

The top two compounds in the previous illustrations have anhydride linkages between a phosphate and either a carbonyl or acid enol group. Creatine phosphate, the high-energy compound in muscle, has a direct linkage between phosphate and nitrogen, whereas the acyl mercaptide linkage in acetyl coenzyme A is also characteristic of an energy-rich compound. In every instance the high-energy compound is readily hydrolyzed to products that undergo spontaneous reactions. These reactions result in forms that are thermodynamically more stable.

Simple phosphate esters, such as AMP, glucose-6-phosphate, and 3-phosphoglyceric acid, are not considered as high-energy compounds, and yield less energy on hydrolysis.

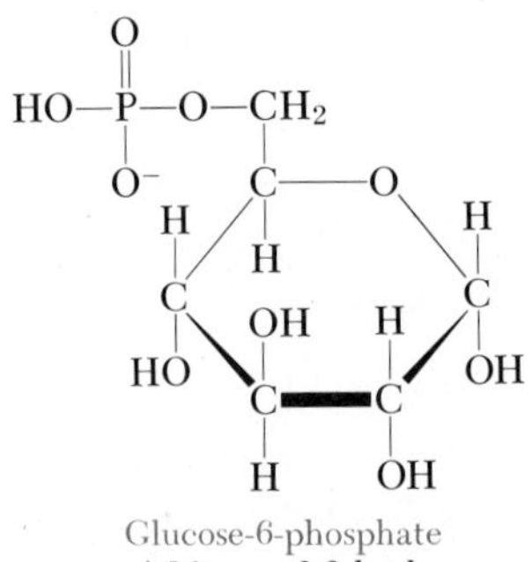

```
          O
          ‖
     HO—P—O—CH2
          |       |
          O⁻      C———O
            H   / |      \   H
            |  /  H       \  |
            C                C
            | \  OH   H   /  |
           HO  \ |    |  /   OH
                 C━━━━C
                 |    |
                 H    OH
```

Glucose-6-phosphate
$\Delta G° = -3.3$ kcal

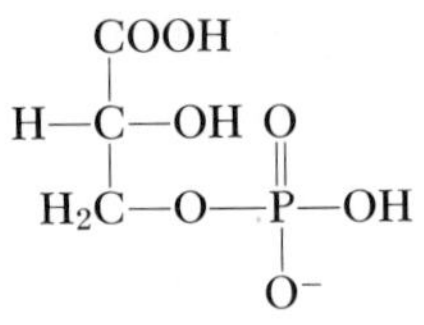

```
          COOH
           |
        H—C—OH  O
           |     ‖
         H2C—O—P—OH
                 |
                 O⁻
```

3-Phosphoglyceric acid
$\Delta G° = -3.3$ kcal

The distinction between high-energy compounds and low-energy phosphate esters is arbitrary, but a dividing line of 5.0 kcal per mole is usually set by biochemists.

In the discussion of metabolism that follows, there will be many examples of the use of high-energy compounds in the storage of energy, the transmission of energy, and the coupling of energy obtained from foodstuffs to the utilization of that energy for cellular reactions.

THE FORMATION OF ATP

Since adenosine triphosphate has been marked as a key compound in the storage of chemical energy and in the coupling of exergonic reactions to endergonic reactions in the cell, it is a major driving force in the metabolic reactions in the tissue. Although ATP can be formed by light energy in the process of photosynthesis, which will be discussed in Chapter 20, the present discussion will consider its formation in the cytoplasm in the absence of oxygen (substrate level phosphorylation) and in the **mitochondria** by the process of oxidative phosphorylation.

Substrate Level ATP

In the anaerobic scheme of carbohydrate metabolism (Embden-Meyerhof pathway) glucose is phosphorylated and is eventually broken down to the 3-carbon phosphorylated derivative. In the following two reactions ADP is converted into the energy-rich compound ATP with the assistance of catalysts, called **enzymes.** These reactions can take place in the absence of O_2 and in the cytoplasm, and are termed **substrate level phosphorylations.**

$$\begin{array}{c}\mathrm{O{=}C{-}O{-}P({=}O)(O^-){-}OH}\\ \mathrm{H{-}C{-}OH}\\ \mathrm{H_2C{-}O{-}P({=}O)(O^-){-}OH}\end{array} + \mathrm{ADP} \xrightarrow[\text{kinase}]{\text{phosphoglycero-}} \begin{array}{c}\mathrm{COOH}\\ \mathrm{H{-}C{-}OH}\\ \mathrm{H_2C{-}O{-}P({=}O)(O^-){-}OH}\end{array} + \mathrm{ATP}$$

1,3-Diphosphoglyceric acid → 3-Phosphoglyceric acid

$$\begin{array}{c}\mathrm{COOH}\\ \mathrm{C{-}O{-}P({=}O)(O^-){-}OH}\\ \|\\ \mathrm{CH_2}\end{array} + \mathrm{ADP} \xrightarrow[\text{kinase}]{\text{pyruvic}} \begin{array}{c}\mathrm{COOH}\\ \mathrm{C{=}O}\\ \mathrm{CH_3}\end{array} + \mathrm{ATP}$$

Phosphoenolpyruvic acid → Pyruvic acid

Oxidative Phosphorylation ATP

One of the major aerobic or oxidative schemes of carbohydrate metabolism (Krebs cycle) involves the reaction of intermediate compounds with the production of several moles of ATP. An **electron transport system** in the mitochondria of the cell actively transports electrons from a reduced metabolite to oxygen with the assistance of enzymes and coenzymes, as shown in Figure 18–1. P_i is inorganic phosphate. NAD is nicotinamide adenine dinucleotide, FMN is flavin mononucleotide, and FAD is flavin adenine dinucleotide; these three coenzymes will be discussed in the following chapter. Other intermediate compounds in Figure 18–1 between the reduced metabolite and oxygen, besides NAD and FAD, are coenzyme Q and the cytochromes. The overall reactions involve first

$$\mathrm{NAD + Metabolite \cdot H_2 \rightarrow Metabolite + NADH_2}$$

then

$$\mathrm{NADH_2 + 3\,ADP + 3\,P_i + \tfrac{1}{2}O_2 \rightarrow NAD + 3\,ATP + H_2O}$$

As may be seen from Fig. 18–1, ATP is generated at three sites on the electron transport chain with the following electron transfers: from $NADH_2$ to coenzyme Q through flavoprotein FMN, from coenzyme Q to cytochrome c through cytochrome b, and from cytochrome c to O_2 through cytochrome a_3. Three moles of ATP may therefore be generated by the passage of electrons from a mole of substrate through NAD to molecular oxygen, but only two moles may be generated if electrons are transferred directly from the substrate to coenzyme Q through FAD because this transfer bypasses the first phosphorylation site.

The enzymes of electron transport are located on the inner membrane of the mitochondria. This membrane is thought to consist of repeating units, each composed of a headpiece which projects into the matrix, joined by a stalk to a basepiece (Fig. 18–2). The basic electron transport chain is located in the basepieces. Each basepiece is considered a complex containing a portion of the electron transport enzymes. Four such complexes plus $NADH_2$, coenzyme Q, and cytochrome c constitute a complete electron transport system. The energy generated by electron transport in the basepiece complex is transmitted through the stalk protein to the headpiece, where it is converted into the high-energy bond of ATP.

NAD, nicotinamide adenine dinucleotide (structure shown in the next chapter), is a dinucleotide composed of AMP linked to nicotinamide-ribose-phosphate. The nicotinamide portion of the molecule is involved in the oxidation and reduction reactions in oxidative phosphorylation. The remaining portion of the molecule is represented as Ribose-Ⓟ-O-Ⓟ-adenosine, where Ⓟ stands for phosphoric acid. The reduction of NAD to form $NADH_2$ may be represented as:

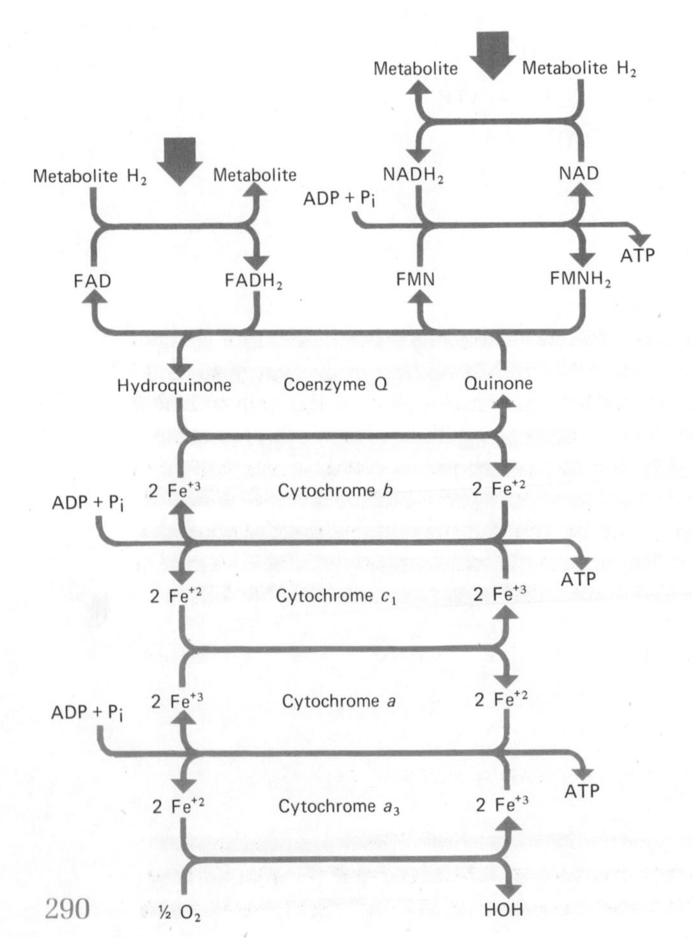

FIGURE 18–1 Oxidative phosphorylation and the electron transport system.

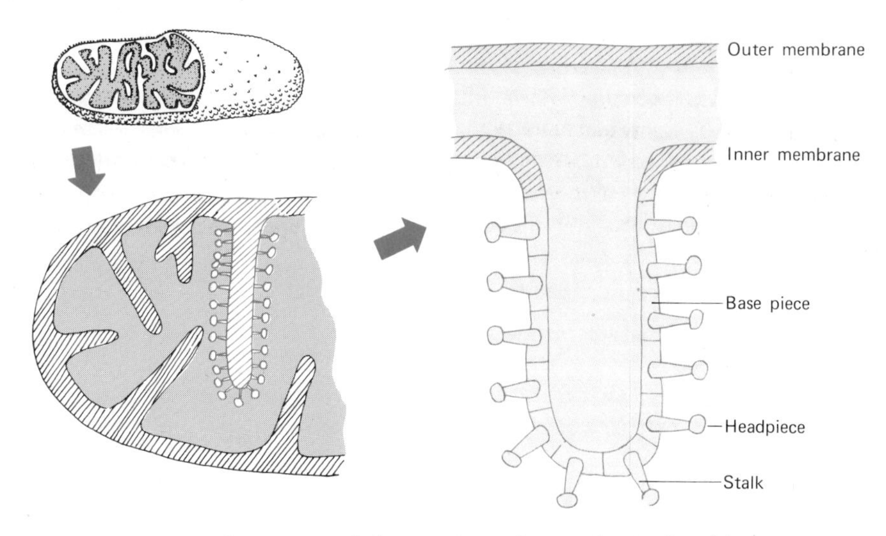

FIGURE 18-2 Proposed structure of the inner membrane of mitochondria.

$$\text{NAD} \underset{-2H^+}{\overset{+2H^+}{\rightleftharpoons}} \text{NADH}_2 + H^+$$

H, O, C—NH_2, N, Ribose-Ⓟ-O-Ⓟ-adenosine
NAD

H H, O, C—NH_2, N, Ribose-Ⓟ-O-Ⓟ-adenosine
$NADH_2$

FAD, **flavin adenine dinucleotide** (structure shown in next chapter), is a dinucleotide composed of a flavin-ribose-phosphate linked to AMP. The reduction of FAD in the electron transport system may be represented as follows:

$$\text{FAD} \underset{-2H^+}{\overset{+2H^+}{\rightleftharpoons}} \text{FADH}_2$$

CH_3, CH_3, N, N, N, C, O, NH, C=O, Ribose-Ⓟ-O-Ⓟ-adenosine
FAD

CH_3, CH_3, H, N, N, N, H, C, O, NH, C=O, Ribose-Ⓟ-O-Ⓟ-adenosine
$FADH_2$

Coenzyme Q is a lipid soluble quinone, sometimes called ubiquinone-10 for the ten isoprene units found in the side chain (the number may vary from 0 to 10). This coenzyme is readily reduced to the hydroquinone form during the transport of hydrogen, as shown:

O, CH_3O, C, CH_3, CH_3O, C, O, CH_3, $(CH_2CH{=}C{-}CH_2)_{10}H$
Coenzyme Q
(quinone form)

OH, CH_3O, C, CH_3, CH_3O, C, OH, CH_3, $(CH_2CH{=}C{-}CH_2)_{10}H$
Coenzyme Q
(hydroquinone form)

The **cytochromes** are oxidation-reduction pigments that consist of iron-porphyrin complexes known as **heme,** which is also an integral part of hemoglobin, the respiratory pigment of the red blood cells. The heme in cytochrome c, for example, is attached to a protein molecule by coordination with two basic amino acid residues, and by thioether linkages formed by the addition of a sulfhydryl group from each of two molecules of cystëine in the protein molecule. Cytochrome c is an electron carrier in the oxidative phosphorylation cycle, in which the iron atom of heme is changed from Fe^{+++} to Fe^{++} as shown:

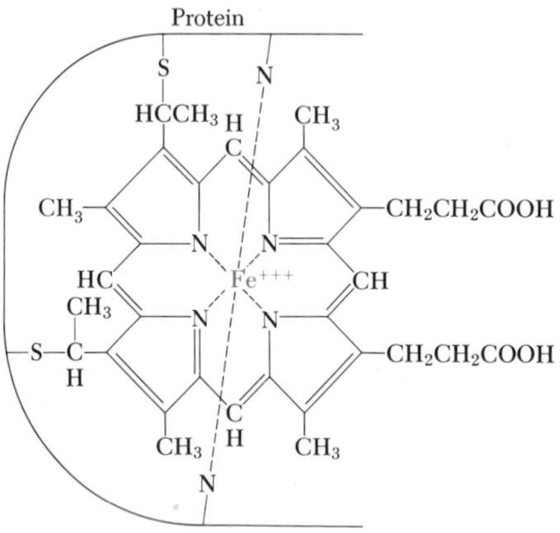

Cytochrome c (oxidized)

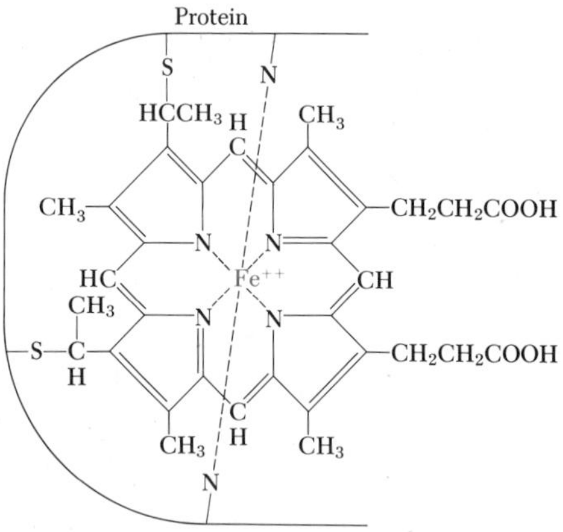

Cytochrome c (reduced)

The exact nature and detailed functional mechanism of the oxidative phosphorylation cycle is as yet not completely understood. Recently Green and his co-workers have isolated a large **electron transport particle** from the mitochondria of cells. They then separated the particle into four complexes, and, after a study of their composition and function, proposed the following relationship between the complexes and the electron transport system:

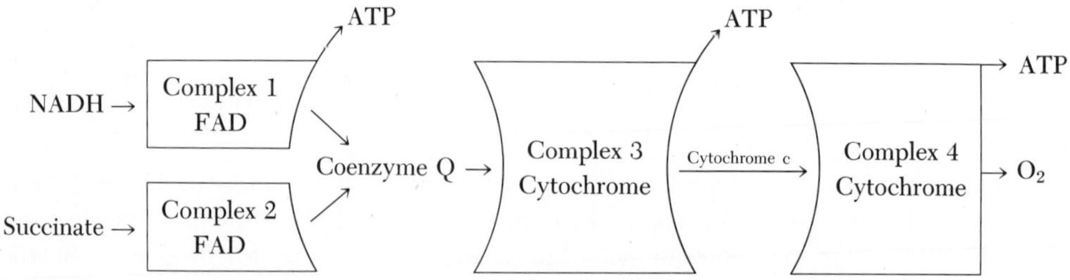

NADH → Complex 1 FAD → Coenzyme Q → Complex 3 Cytochrome → (Cytochrome c) → Complex 4 Cytochrome → O_2

Succinate → Complex 2 FAD → Coenzyme Q

ATP (Complex 1), ATP (Complex 3), ATP (Complex 4)

Biochemical energy in the form of ATP is an essential driving force in many metabolic reactions in the cells and tissues. As has been described, several complex reactions are involved in the synthesis of this vital compound, and it should be emphasized that three moles of ATP are formed when the electrons from $NADH_2$ are transported through the system to oxygen. Also two moles of ATP are formed when electrons from $FADH_2$ are transported to oxygen. These relationships will assist in the understanding of the energy balance in the metabolic cycles.

IMPORTANT TERMS AND CONCEPTS

ATP
coenzyme Q
coupled reactions
cytochromes
electron transport system
$FADH_2$
free-energy change
high-energy compounds
$NADH_2$
oxidative phosphorylation

QUESTIONS

1. What are the different forms of energy that may be produced in the living cell?

2. What is the relation between the standard free-energy change of a reaction, $\Delta G°$, and the free-energy change, ΔG?

3. In the conversion of glucose-1-phosphate to glucose-6-phosphate, an equilibrium concentration of 0.001 M glucose-1-PO_4 and 0.022 M glucose-6-PO_4 is obtained. Calculate the $\Delta G°$ of the reaction.

4. The formation of an ester from an acid and an alcohol resulted in a $\Delta G°$ of +2.0 kcal. ATP formed an intermediate with the acid to drive the reaction to completion. Represent the reaction and calculate the new $\Delta G°$.

5. Briefly explain why phosphoenolpyruvic acid is a high-energy compound.

6. If 3-phosphoglyceric acid were converted to 1,3-diphosphoglyceric acid, what would happen to the ATP and what energy change would you expect?

7. Why is the oxidative phosphorylation mechanism also called the electron transport system? Explain.

8. Trace the process of the electron transport system from the basepiece to the headpiece on the inner membrane of the mitochondria.

9. Explain why only two moles of ATP are formed in the electron transport system when electrons from $FADH_2$ are transported to oxygen.

SUGGESTED READING

Alberty: Thermodynamics of the Hydrolysis of Adenosine Triphosphate. Journal of Chemical Education, Vol. 46, p. 713, 1969.
Changeux: The Control of Biochemical Reactions. Scientific American, Vol. 212, No. 4, p. 36, 1965.
Heidt: The Path of Oxygen from Water to Molecular Oxygen. Journal of Chemical Education, Vol. 43, p. 623, 1966.
Kirschbaum: Biological Oxidations and Energy Conservation. Journal of Chemical Education, Vol. 45, p. 28, 1968.
Lehninger: Bioenergetics. New York, W. A. Benjamin, Inc., 1965.

ENZYMES AND COENZYMES

CHAPTER 19

The *objectives* of this chapter are to enable the student to:

1. Define and describe the chemical nature of enzymes.
2. Recognize the systematic code number of a common enzyme.
3. Recognize the relationship between the active site of an enzyme and its specificity of action.
4. Explain the value of the Michaelis constant, K_m, of an enzyme.
5. Describe the effect of pH, temperature, and end products on an enzyme reaction.
6. Explain the difference between a coenzyme, a prosthetic group, and an apoenzyme.
7. Discuss the function of a particular coenzyme and the reactions that are involved.
8. Recognize and illustrate the difference between a competitive and a noncompetitive inhibitor.

A rapidly developing field of biochemical research involves the study of enzyme-controlled reactions in the cells of animals, plants, and microorganisms. It has been estimated that over one fourth of the biochemists in this country are directly engaged in some form of enzyme research. Many of the reactions that occur in living organisms are not only accelerated by enzymes but would not occur to any appreciable extent at body temperature. The magnitude of the problem facing the enzyme chemist is indicated by the estimate of as many as 1000 separate enzymes in a single cell. These enzymes are not randomly scattered in the cell as in a sack of enzymes, but are located in specific areas or subcellular components concerned with various types of metabolic activity. The enzymes of the Embden-Meyerhof pathway for glycolysis (see p. 322) are found in the cytoplasm, whereas the enzymes involved in the Krebs cycle (see p. 324) are located in the mitochondria.

THE CHEMICAL NATURE OF ENZYMES

Enzymes have always been considered as catalysts and are often compared to inorganic catalytic agents such as platinum and nickel. These inorganic agents are often used in conjunction with high temperatures, high pressures, and favorable chemical conditions. Few of these conditions occur when an enzyme reacts in body tissue, at body temperature, and at the pH of body fluids. Enzymes were originally defined as catalysts, organic in nature, formed by living cells, but independent of the presence of the cells in their action. A more current definition would state that *enzymes are proteins, formed by a living cell, which catalyze a thermodynamically possible reaction by lowering the activation energy so the rate of reaction is compatible with the conditions in the cell.* An enzyme does not change the $\Delta G°$ or equilibrium constant of a reaction.

By now it is generally accepted that all enzymes are protein in nature. The purification, crystallization, and inactivation procedures exactly parallel those for pure proteins. For example, excessive heat, alcohol, salts of heavy metals, and concentrated inorganic acids will cause coagulation or precipitation of the protein material and thus inactivate an enzyme.

The naming of enzymes has become increasingly complex as many new specific enzymes have been described. Originally they were named according to their source or according to the method of separation when they were discovered. As the family of enzymes grew, they were named in a more orderly fashion by adding the ending **-ase** to the root of the name of the substrate. An enzyme's **substrate** is the compound or type of substance upon which it acts. For example, sucrase catalyzes the hydrolysis of sucrose, lipase is an enzyme that hydrolyzes lipids, and urease is the enzyme that splits urea. This system also was used to name types of enzymes such as proteases, oxidases, and hydrolases. The discovery of so many enzyme mechanisms in the past few years has resulted in a mass of complex substrates and enzyme nomenclature. The problem was assigned to a Commission on Enzymes of the International Union of Biochemistry, whose members studied the system of nomenclature for six years before publishing a report in 1961. They were not in favor of eliminating all the names in common usage, but recommended the use of two names for an enzyme. One was the trivial name, either the one in common use or a simple name describing the activity of the enzyme. The other was constructed by the addition of the ending **-ase** to an accurate chemical name for the substrate.

They also devised a systematic code number (E.C.) for each enzyme. This number characterizes the type of reaction catalyzed by the enzyme as to class (first number), subclass (second number), and sub-subclass (third number). The fourth number is specific for the particular enzyme named. As an example, pancreatic lipase is assigned the number 3.1.1.3, which specifies a hydrolase (3) that acts on ester bonds (3.1) that are carboxylic esters (3.1.1.), with the specific number for the lipase (3.1.1.3).

CLASSIFICATION OF ENZYMES

The current classification of enzymes is based on both the type of chemical reaction catalyzed and the numbering system devised by the Commission. The six major classes of enzymes, with examples of their subclasses and their number codes, are shown in the following classification.

1. ***Oxidoreductases.*** Enzymes in this group are involved in physiological oxidation processes. An example would be alcohol:NAD oxidoreductase or alcohol dehydrogenase, designated 1.1.1.1, acting on a CH · OH group (subclass), with NAD as the coenzyme (sub-subclass).
2. ***Transferases.*** These enzymes catalyze the transfer of a chemical group from one substrate to another. The transfer of amino, methyl, alkyl, and acyl groups and groups containing phosphorus or sulfur is catalyzed by these enzymes. Enzyme number 2.1.1.1 is a one-carbon transferase (2.1) that transfers methyl groups (2.1.1), and is specifically named adenosylmethionine:nicotinamide methyltransferase (2.1.1.1).

3. ***Hydrolases.*** This large group of enzymes catalyzes hydrolytic reactions and includes digestive enzymes such as amylase, sucrase, lipase, and the proteases. A specific example may be lipase 3.1.1.3, described in the preceding section.
4. ***Lyases.*** These enzymes catalyze the removal of chemical groups without hydrolysis. These enzymes act on C—C, C—O, C—N, and C—S bonds, and include 2-oxoacid carboxy-lyase or pyruvate decarboxylase (4.1.1.1), which acts on a C—C bond (4.1) removing a carboxyl group (4.1.1).
5. ***Isomerases.*** This group of enzymes catalyzes isomerization reactions. Examples are *cis-trans* isomerases, racemases, and epimerases. An example related to the process of vision (see p. 239) is retinine isomerase (5.2.1.3), in which 5.2 designates a *cis-trans* isomerase and 5.2.1 denotes a *cis-trans* isomerase acting on polyunsaturated hydrocarbons.
6. ***Ligases.*** These enzymes catalyze the linking together of two molecules with the breaking of a pyrophosphate bond of ATP or a similar triphosphate. An example of a ligase involved in the synthesis of protein would be tyrosine:t-RNA ligase or tyrosyl-t-RNA synthetase, 6.1.1.1, which links two molecules together forming C—O bonds (6.1) or, more specifically, an amino acid-RNA ligase forming C—O bonds (6.1.1).

PURIFICATION OF ENZYMES

Crude extracts of enzymes from tissues or cells may be obtained by grinding the tissue in metal grinders or between ground glass surfaces, by alternate freezing and thawing, or by exposure to ultra-sound. The enzymes may be separated from the extracts by protein precipitation, adsorption on ion exchange resins, electrophoresis (adsorption assisted by the passage of an electrical current), and extraction with various solvents. The first enzyme to be obtained in crystalline form was urease, which was crystallized by Sumner in 1926. Since then about 75 other enzymes have been crystallized.

As enzyme preparations are purified and crystallized they naturally increase their potency of action. The activity of an enzyme, or more specifically one unit of an enzyme, is defined as that amount which will catalyze the transformation of 1 micromole of substrate per minute. A **micromole** is 1 millionth of a gram molecular weight. The potency, or **specific activity,** is expressed as units of enzyme per milligram of protein; the **molecular activity** is defined as units per micromole of enzyme at optimal substrate concentration. To compare the relative activity of different enzymes the **turnover number** is used. This is defined as the number of moles of substrate transformed per mole of enzyme per minute at a definite temperature. Turnover numbers vary from about 10,000 to 5,000,000, with the enzyme catalase exhibiting the highest activity.

PROPERTIES OF ENZYMES

Specificity of Action

Perhaps the major difference between the classical inorganic catalysts, such as platinum and nickel, and enzymes is the specificity of action of the latter. Platinum, for example, will act as a catalyst for several reactions. Enzymes may exhibit different types of specificities, as listed below:

1. Absolute specificity
2. Stereochemical specificity
3. Reaction, or linkage, specificity
4. Group specificity

Urease exhibits **absolute specificity** in action in that it catalyzes the splitting of a single compound, urea. Other enzymes exhibit **stereochemical specificity;** D-amino acid oxidase is specific for D-amino acids and will not affect the natural L-amino acids. Arginase catalyzes the hydrolysis of L-arginine to urea and ornithine, but will not act on the D-isomer. The comparison between the specificity of enzymes and inorganic catalysts may need reconsideration as more information concerning organometallic catalysts becomes available. For example, optically active rhodium catalysts have been developed that synthesize optically active products.

In Chapter 15 enzymes that split particular bonds between amino acids in peptide fragments were used to help determine the amino acid sequence in proteins. These enzymes are **linkage specific;** for example, trypsin splits peptide bonds adjacent to lysine or arginine residues, whereas chymotrypsin splits bonds next to aromatic amino acids, such as tyrosine in a polypeptide chain. Enzymes that exhibit **group specificity** are also valuable in sequence studies. Carboxypeptidase is specific for terminal amino acids containing a free carboxyl group, and aminopeptidase splits terminal amino acids with a free amino group off the end of a peptide chain. In general, the specificity of action accounts for the large numbers of enzymes found in cells and tissues, and for the fact that enzymes are involved in all the metabolic reactions that occur in the cell.

Energy of Activation

Cellular processes in biochemistry are designed to obtain energy from food and to transform it to forms of energy useful in body growth and maintenance. As discussed in the last chapter on page 286, chemical reactions may either release energy, or require energy to proceed. The first type of reaction is characterized by a negative ΔG, with energy content of the reactants greater than that of the products, and a tendency toward spontaneous reaction. The latter type of reaction has a positive ΔG, with less energy content in the reactants and no tendency toward spontaneous reaction. Since all chemical reactions tend to proceed in the direction of forming products of a lower energy content than the reactants, the problem is how to initiate the reaction and make it proceed spontaneously. In the reaction

$$A + B \rightarrow \text{Activated state} \rightarrow C + D$$

energy is required to produce the activated state before the reaction can proceed to products C + D. The energy of activation, E_a, required to change A + B to the activated state is affected by a catalyst. Figure 19–1 illustrates the effect of an enzyme on the energy of activation compared to the effect of the common catalyst, platinum, on the decomposition of hydrogen peroxide to water and oxygen.

To decompose hydrogen peroxide to water and oxygen without a catalyst requires 18.0 kcal/mole. An effective metallic catalyst such as platinum lowers the activation energy of the reaction to 12.0 kcal/mole, compared to the low activation energy required by the enzyme catalase of about 3.0 kcal/mole. In general, enzymes lower activation energies of reactions to the point where they can be readily carried out at body temperature under the conditions of living tissue.

ENZYME ACTIVITY

The activity of an enzyme is affected by many factors. Most important are the concentrations of the substrate and the enzyme, and the temperature and the pH of the reaction. In addition, the rate of enzyme reaction is affected by the nature of the end

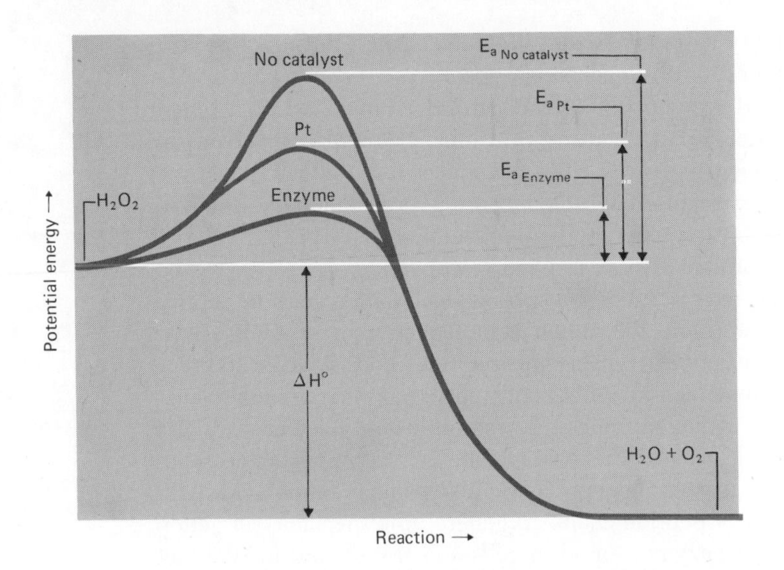

FIGURE 19-1 Effect of catalysts on the energy of activation in the decomposition of hydrogen peroxide.

products, the presence of inhibitors, and light. The activity may be measured by following the chemical change that is catalyzed by the enzyme. The substrate is incubated with the enzyme under favorable conditions, and samples are withdrawn at short intervals for analysis of the end products or analysis of the decrease of substrate concentration. The enzyme lipase, for example, catalyzes the hydrolysis of fat molecules to fatty acids and glycerol. A simple method of measuring the activity of lipase would involve a determination of the rate of appearance of fatty acid molecules.

Effect of Substrate

Michaelis and Menten in 1913 first expressed the concept of the **enzyme substrate complex** as a transition state in enzyme reactions.

$$\underset{\text{Enzyme}}{E} + \underset{\text{Substrate}}{S} \underset{k_{-1}}{\overset{k_1}{\rightleftharpoons}} \underset{\text{Enzyme substrate complex}}{ES} \xrightarrow{k_2} \underset{\text{Products}}{E + P}$$

The formation of the ES complex permits the overall reaction to proceed at a lower energy of activation. A constant, K_m, known as the **Michaelis constant,** is related to the three rate constants as follows:

$$K_m = \frac{k_2 + k_{-1}}{k_1}$$

Since the reaction would readily proceed to E + P once the ES complex is formed, the rate constant k_2 would be lower than k_1 and definitely lower than k_{-1}, which would tend to dissociate ES into E + S. When k_{-1} is much greater than k_2 we could neglect k_2, and K_m would equal k_{-1}/k_1, in which case K_m would be the thermodynamic equilibrium constant for the reversible formation of the ES complex. Under proper conditions of temperature and pH, the Michaelis constant, K_m, is approximately equal to the dissociation constant of the ES complex. Conversely, the reciprocal of K_m, or $1/K_m$, is a measure of the *affinity of an enzyme for its substrate.* The substrate concentration required to yield half the $v_{max.}$ can be readily determined from Figure 19-2, and is equal to K_m. A K_m of 0.001M would indicate that the active site of the enzyme is half saturated when

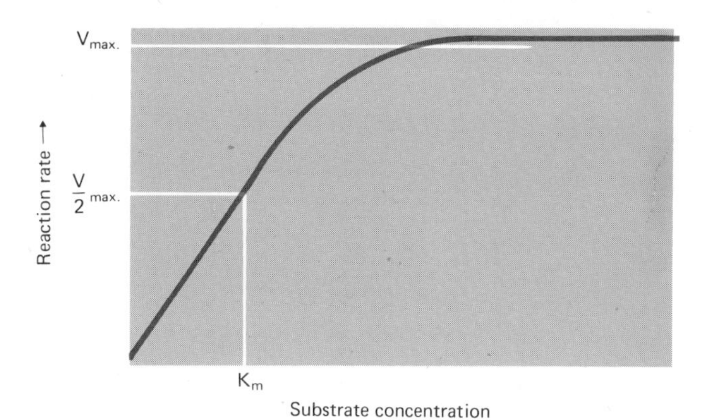

FIGURE 19–2 Effect of substrate concentration on the reaction rate when the enzyme concentration is held constant.

its substrate is present at that concentration. The enzyme, therefore, has a high affinity for its substrate.

As seen from Figure 19–2, K_m by definition equals the substrate concentration [S] at one half the maximal rate, $v_{max.}$ where $v_{max.}$ equals the maximum rate at the saturation concentration of the substrate. The rate of the reaction, v, at a given substrate concentration can be expressed as:

$$v = \frac{v_{max.}[S]}{K_m + [S]}$$

This relationship between the rate of reaction, $v_{max.}$, K_m, and [S] is known as the Michaelis-Menten equation. If a saturation concentration of substrate is used, $v = v_{max.}$; however, if we choose a rate where $v = \frac{v_{max.}}{2}$ (Fig. 19–2), then in the equation

$$\frac{v_{max.}}{2} = \frac{v_{max.}[S]}{K_m + [S]}$$

Cancelling $v_{max.}$ on each side of the equation

$$K_m + [S] = 2[S] \quad \text{or} \quad K_m = 2[S] - [S]$$
$$\text{or} \quad K_m = [S]$$

and K_m is the substrate concentration in moles per liter that will yield the half-maximum rate.

With reference to Figure 19–2 and the relation of K_m to ES, [S], and v, several conclusions can be drawn concerning enzyme activity. At a very low substrate concentration [S], most of the enzyme molecules would exist in the free state, not combined in the complex ES. In this region the amount of ES formed is proportional to the amount of substrate, and therefore the rate (v) of product formation would be directly proportional to the concentration of the substrate, as seen by the linear portion of the curve in Figure 19–2. At high substrate concentrations, the enzyme exists in the combined state as ES. At this point the rate of reaction, v, is proportional to the concentration of ES, but independent of the concentration of the substrate [S]. Since the enzyme E must be combined with the substrate S to form the products of the reaction, the K_m for any enzyme-substrate system is independent of the enzyme concentration.

The curve shown in Figure 19–2 represents the ideal relationship between the substrate concentration [S] and the maximum velocity $v_{max.}$. Experimentally it is difficult

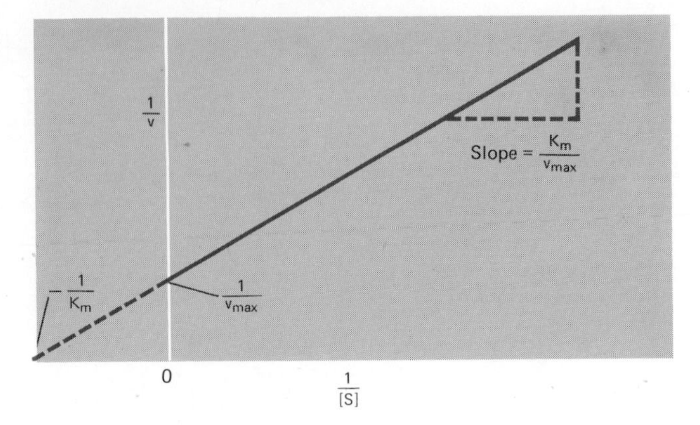

FIGURE 19-3 The Lineweaver-Burk plot of an enzyme reaction.

to reproduce this curve for a particular enzyme to obtain an accurate measure of $v_{max.}$ or K_m. By inverting the Michaelis-Menten equation and expressing it as an equation for a straight line, the following relationship may be obtained:

$$\frac{1}{v} = \frac{K_m}{v_{max.}} \times \frac{1}{[S]} \times \frac{1}{v_{max.}}$$

This equation may then be used to construct a double-reciprocal plot of $\frac{1}{v}$ versus $\frac{1}{[S]}$ to obtain a graphic evaluation of K_m and $v_{max.}$. This is called a Lineweaver-Burk plot, after the investigators who proposed its use (Fig. 19–3). Experimentally, it requires only a few points on the curve to determine K_m; therefore, the Lineweaver-Burk plot method is most often used for this purpose in the laboratory. In addition to the K_m representing a measure of the affinity of an enzyme for its substrate, it also is of practical value in the assay of enzymes. At a substrate concentration of 100 times the K_m value, the enzyme will exhibit a maximum rate of activity or $v_{max.}$. The K_m value, therefore, determines the amount of substrate to use in an enzyme assay.

The Active Site

In view of the specificity of action of many enzymes, it is reasonable to assume that only a small portion of the enzyme protein is involved in its catalytic activity. The portion

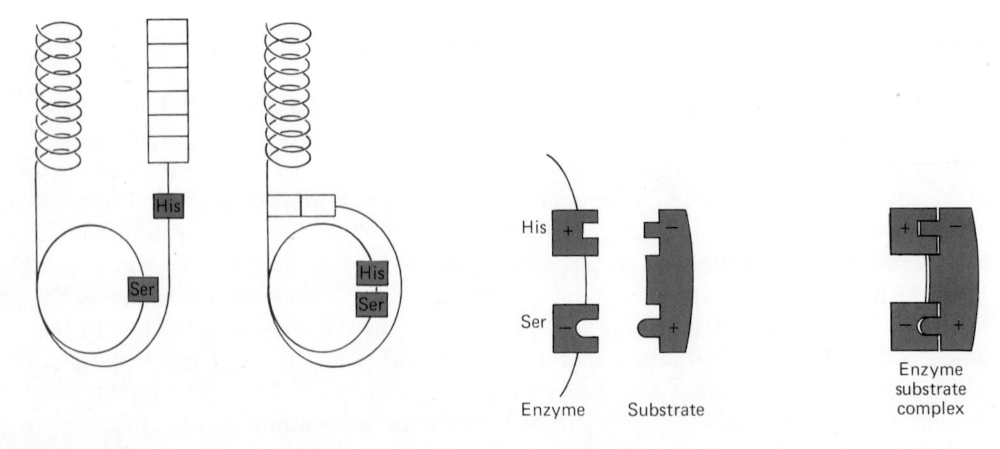

FIGURE 19-4 The conversion of trypsinogen to trypsin, showing the amino acids at the active site.

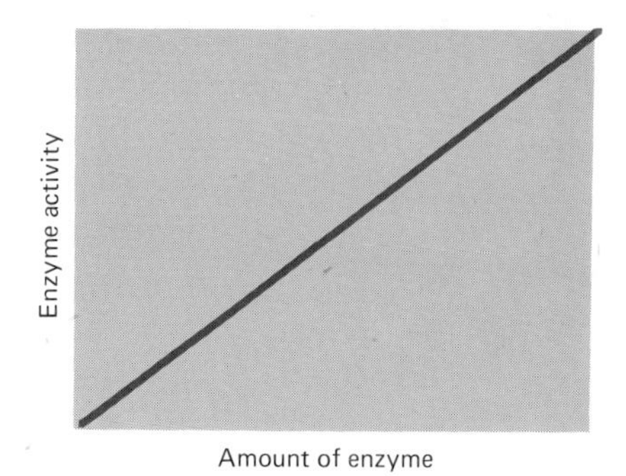

FIGURE 19-5 The effect of increasing amounts of enzyme on the activity of the enzyme.

of the enzyme molecule to which the substrate binds is called the **active site.** This site often consists of specific amino acids or specific charges in a sidechain. Trypsinogen, for example, is synthesized in the pancreas and is converted to the active enzyme trypsin by the enzyme enterokinase or even by trypsin itself. The primary structure of trypsinogen is a single polypeptide chain, and the conversion to trypsin is thought to involve the splitting off of a hexapeptide followed by a change in the conformation of the polypeptide to expose the catalytically active sites containing serine and histidine residues (Fig. 19–4).

Effect of Enzyme

When a purified enzyme is used, the rate of reaction is proportional to the concentration of the enzyme over a fairly wide range (Fig. 19–5). The substrate concentration must be kept constant and remain in excess of that required to combine with the enzyme. This relationship may also be used to measure the amount of an enzyme in a tissue extract or a biologic fluid. At the proper substrate concentration ($100 \times K_m$), temperature, and pH, the measured rate of activity is proportional to the quantity of enzyme present.

Effect of pH

The hydrogen ion concentration, or pH, of the reaction mixture exerts a definite influence on the rate of enzyme activity. If a curve is plotted comparing changes in pH with the rate of enzyme activity, it takes the form of an inverted U or V (Fig. 19–6). The maximum rate occurs at the **optimum pH,** with a rapid decrease of activity on either side of this pH value. The optimum pH of an enzyme may be related to a certain electric charge on the surface or to optimum conditions for the binding of the enzyme to its substrate. Most enzymes exhibit an optimum pH value close to 7, although pepsin is most

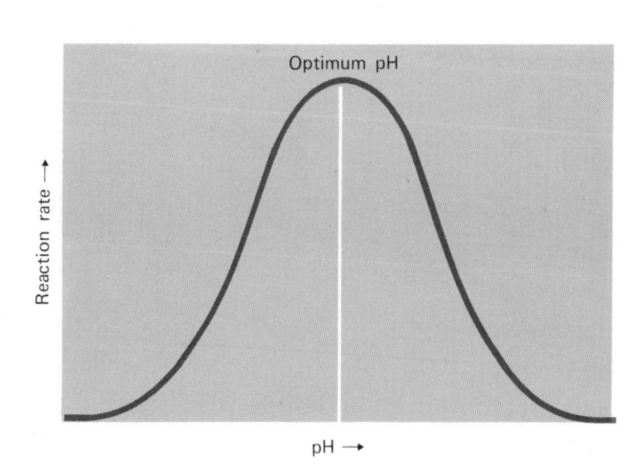

FIGURE 19-6 The effect of pH on enzyme activity.

active at pH 1.6 and trypsin at pH 8.2. Pepsin has no activity in an alkaline solution, whereas trypsin is inactive in an acid solution.

Effect of Temperature

The speed of most chemical reactions is increased two or three times for each 10°C rise in temperature. This is also true for reactions in which an enzyme is the catalyst, although the temperature range is fairly narrow. The activity range for most enzymes occurs between 10° and 50°C; the **optimum temperature** for enzymes in the body is around 37°C. The increased rate of activity observed at 50°C or above is short-lived, because the increased temperature first denatures and then coagulates the enzyme protein, thereby destroying its activity. The optimum temperature of an enzyme is therefore dependent on a balance between the rise in activity with increased temperature and the denaturation or inactivation by heat (Fig. 19–7). For any 10° rise in temperature the change in rate of enzyme activity is known as the Q_{10} value, or temperature coefficient. The Q_{10} value for most enzymes varies from 1.5 to 3.0.

Effect of End Products

The end products of an enzyme reaction have a definite effect on the rate of activity of the enzyme. If they are allowed to increase in concentration without removal, they will slow the reaction. Some end products, when acid or alkaline in nature, may affect the pH of the mixture and thus decrease the rate of reaction. The effect of the end products on the activity of the enzyme is sometimes expressed as a chemical feedback system, with inhibition or decrease in rate called **negative feedback.** The activity of enzymes in the cell may be controlled to some extent by this chemical feedback system as in a sequence of cellular reactions in which the product D inhibits the enzyme reaction A → B.

$$A \nrightarrow B \rightarrow C \rightarrow D$$

Several enzymes contain sulfhydryl groups (SH) which are associated with their active centers. Oxidizing agents change these groups to disulfide linkages and cause inactivation of the enzyme, whereas reducing agents restore the SH groups and activate the enzymes.

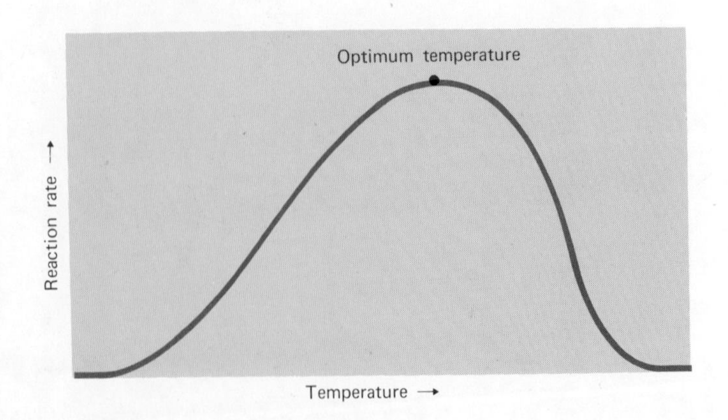

FIGURE 19–7 The effect of temperature on enzyme activity.

ACTIVATION OF ENZYMES

In the body many enzymes are secreted in an inactive form to prevent their action on the very glands and tissues that produce them. Also, during the process of purification an enzyme may become inactive. They may be activated by several agents: a change of pH, the addition of inorganic ions, or the addition of organic compounds.

The requirements for enzyme activity are further complicated by the fact that several enzymes require the presence of a metal ion for their activity. Enzymes have been characterized that require zinc, magnesium, iron, cobalt, and copper. Carbonic anhydrase, an enzyme that catalyzes the formation of carbonic acid from CO_2 and H_2O, requires zinc and is inactivated when this metal is removed.

Proenzymes

A proenzyme is the precursor of the active enzyme in the body. For example, pepsinogen is the proenzyme of pepsin and trypsinogen is the inactive form of trypsin. When pepsinogen is secreted into the stomach, it is converted into pepsin by hydrogen ions of the hydrochloric acid. The pepsin then activates more pepsinogen to form more of the active enzyme. Trypsinogen is secreted by the pancreas and is activated in the intestine by enterokinase (see p. 301 and Fig. 19–4).

COENZYMES

In early studies on the enzymes of yeast it was observed that the dialysis of a solution of the yeast material inactivated the enzymes. When the dialyzed material was added to the enzymes, they again exhibited activity. The cofactor in the dialysate was called a **coenzyme.** Since that time several coenzymes have been discovered, and they have been found to consist of small organic molecules. If the organic molecule, or non-protein portion, is readily separated from the enzyme, it is called a coenzyme. If it is firmly attached to the protein portion of the enzyme, it is called a **prosthetic group.** Most enzymes may therefore be considered as conjugated proteins composed of an inactive protein molecule called the **apoenzyme** combined with the prosthetic group or coenzyme. The complete, conjugated, active molecule is called a **holoenzyme.**

As our knowledge of intermediary metabolism increases, it becomes apparent that *vitamins or derivatives of vitamins serve as coenzymes or prosthetic groups* in enzymatic reactions involving oxidation, reduction, and decarboxylation (removal of carboxyl groups). The water-soluble vitamin B complex contains several vitamins that exhibit the properties of coenzymes.

Vitamin B_1

Vitamin B_1, or thiamine, contains a pyrimidine ring and a sulfur-containing thiazole ring.

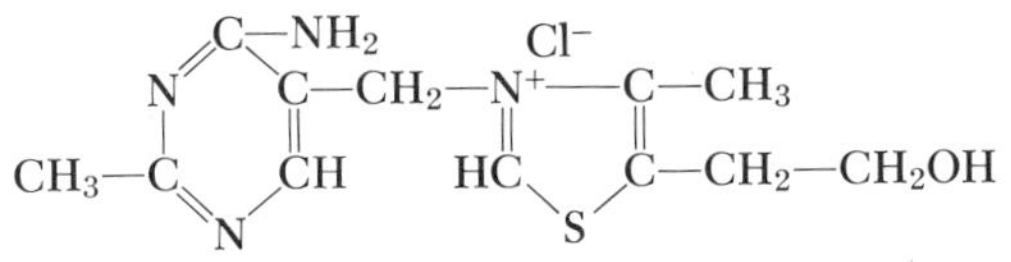

```
         C—NH2          Cl⁻
       //     \
      N        C—CH2—N⁺——C—CH3
      |        ||     ||   ||
 CH3—C         CH    HC    C—CH2—CH2OH
       \\     /        \  /
         N              S
```

Thiamine chloride

A deficiency of the vitamin in the diet results in a disease called **polyneuritis** in animals and **beriberi** in man. The peripheral nerves of the body are involved, with muscle cramps, numbness of the extremities, pain along the nerves, and eventually atrophy of muscles, edema, and circulatory disturbances occurring in the body.

Yeast, whole-grain cereals, eggs, and pork are good sources of the vitamin. Thiamine occurs free in cereal grains, but occurs as the coenzyme, thiamine pyrophosphate, in yeast and meat.

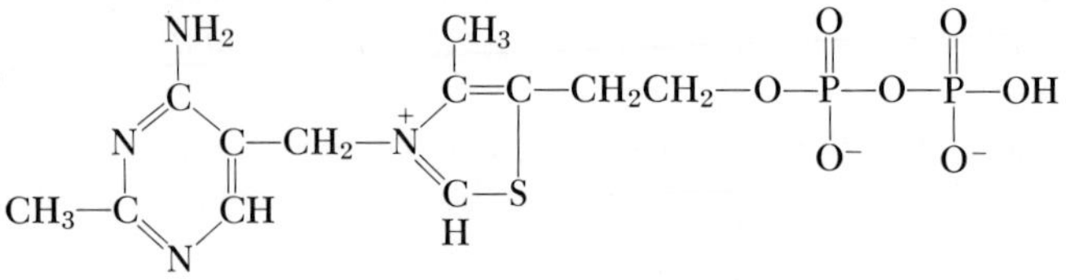

Thiamine pyrophosphate (cocarboxylase) (TPP)

Cocarboxylase functions in the oxidative decarboxylation of pyruvic acid to form acetaldehyde and carbon dioxide (p. 324). The thiazole ring is the active site of this function, with the hydrogen atom dissociating as a proton from carbon-2 and the formation of a carbanion. The carbanion structure then reacts with pyruvic acid to form CO_2 and acetaldehyde. The essential reactions may be outlined as follows:

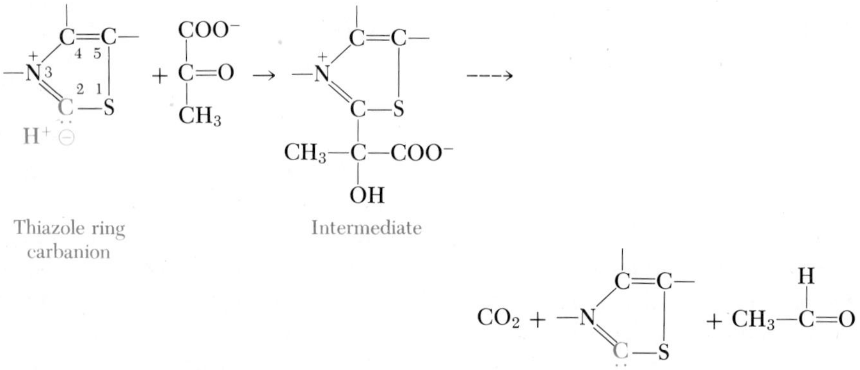

Thiazole ring carbanion

Intermediate

Acetaldehyde

Cocarboxylase is therefore essential in the conversion of pyruvic acid to acetaldehyde. If this reaction does not occur at a normal rate, pyruvic acid may accumulate in the blood and tissues and give rise to the neuritis that is common in thiamine deficiency. **Thiamine pyrophosphate, TPP,** also serves as a coenzyme for enzymes such as α-keto acid oxidase, phosphoketolase, and transketolase.

Riboflavin

Riboflavin, or vitamin B_2, is composed of a pentose alcohol, ribitol, and a pigment, flavin.

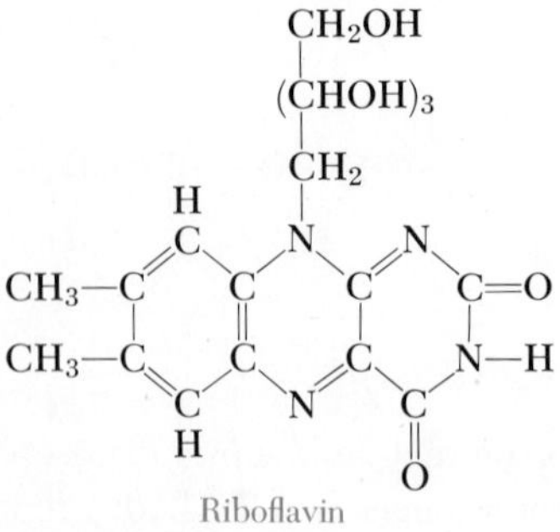

Riboflavin

A deficiency of vitamin B_2 in the diet of animals such as the rat, dog, and chicken causes lack of growth, loss of hair, and cataracts of the eyes. Lack of the vitamin in the human affects vision and causes inflammation of the cornea, and sores and cracks in the corners of the mouth.

Foods rich in riboflavin are yeast, liver, eggs, and leafy vegetables. Milled cereal products lose both their thiamine and vitamin B_2, and at present there is a trend toward the fortification of white flour with these two vitamins.

The vitamin functions as a coenzyme; in fact, it occurs in foods as a component of two flavin coenzymes, FMN and FAD. The structures of **flavin mononucleotide, FMN,** and **flavin adenine dinucleotide, FAD,** are represented as follows:

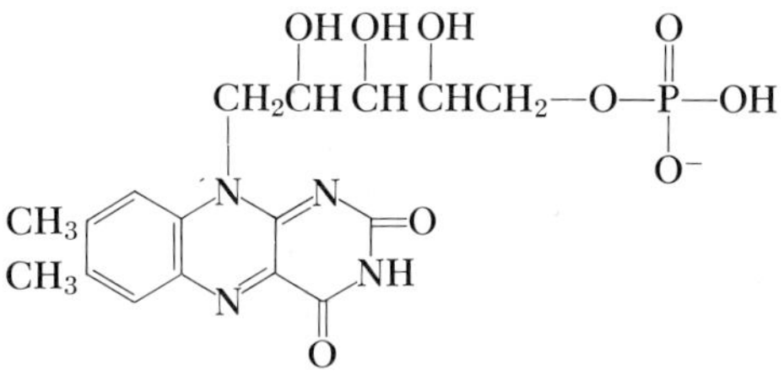

Flavin mononucleotide (FMN)

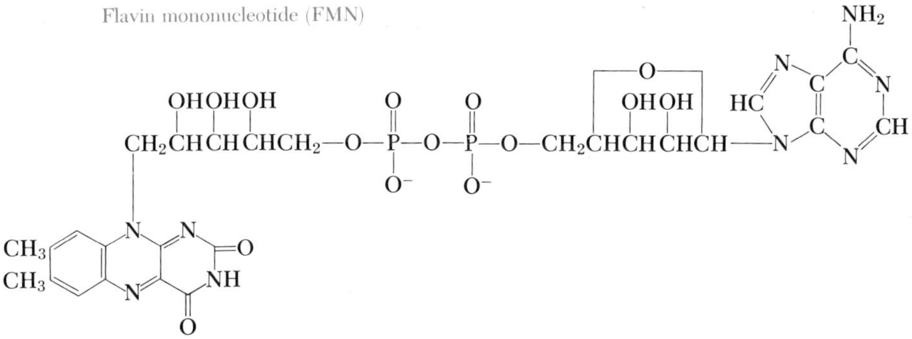

Flavin adenine dinucleotide (FAD)

Both FMN and FAD serve as coenzymes for a group of enzymes which catalyze oxidation-reduction reactions. Glutathione reductase, succinic dehydrogenase, and D-amino acid oxidase are examples of these enzymes. In the preceding chapter the mechanism for the reduction of FAD to $FADH_2$ by $NADH_2$ in the oxidative phosphorylation process was described. The flavin portion of the molecule is the active site for the oxidation-reduction reactions.

Nicotinic Acid and Nicotinamide

These two compounds have comparatively simple structures and as vitamins are called **niacin.**

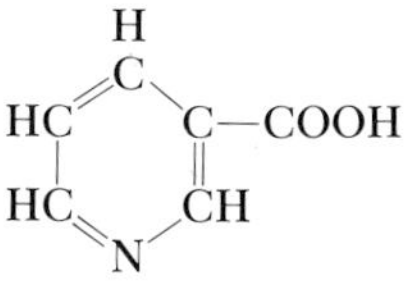

Nicotinic acid

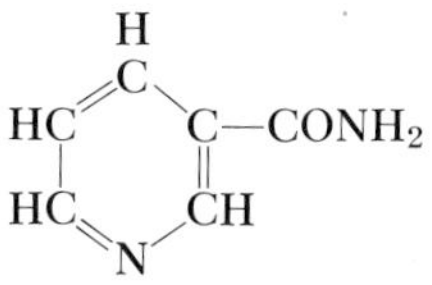

Nicotinamide

A deficiency of niacin in the diet results in **pellagra** in man and blacktongue in dogs. Pellagra is a disease characterized by skin lesions that develop on parts of the body that are exposed to sunlight. A sore and swollen tongue, loss of appetite, diarrhea, and nervous and mental disorders are typical symptoms of the disease.

Liver, lean meat, and yeast are good sources of niacin, whereas corn, molasses, and fat meat are very poor sources. Pellagra is more prevalent in the South, where the latter three foods are major constituents of the diet.

Niacin is an essential component of two important coenzymes, **nicotinamide-adenine dinucleotide, NAD,** and **nicotinamide-adenine dinucleotide phosphate, NADP.**

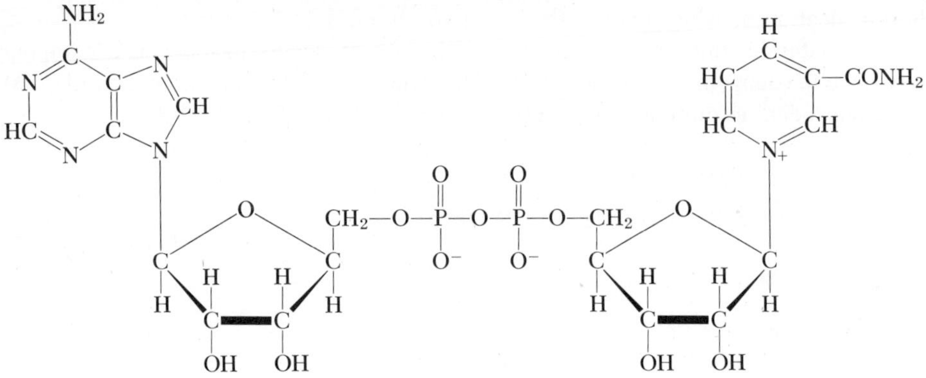

Nicotinamide-adenine dinucleotide (NAD)

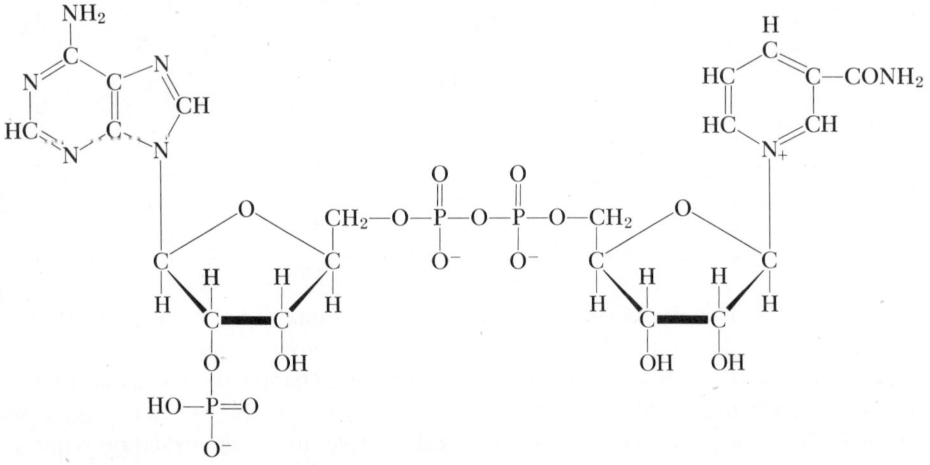

Nicotinamide-adenine dinucleotide phosphate (NADP)

The nicotinamide portion of NAD and NADP is involved in the mechanism of the oxidation-reduction reactions with which these coenzymes are involved. This mechanism was outlined in Chapter 18.

Both NAD and NADP are coenzymes for dehydrogenases, which are enzymes that catalyze oxidation-reduction reactions. Lactic dehydrogenase, for example, catalyzes the oxidation of lactic acid to form pyruvic acid, with NAD serving as a coenzyme and being reduced to $NADH_2$ in the reaction. Alcohol dehydrogenase and glucose-6-phosphate dehydrogenase also require NAD as a coenzyme. Many enzymes of clinical diagnostic significance, such as lactic dehydrogenase, may be determined quantitatively in body fluids by the change in form of the coenzyme that occurs in the reaction.

$$\text{NAD} \rightleftharpoons \text{NADH}_2$$

Zero absorption of light at 340 nm — Strongly absorbs light at 340 nm

By measuring the change of absorbance of the solution at a wavelength of 340 nm in a spectrophotometer, the concentration of **dehydrogenase enzyme** responsible for the change can be determined.

Pyridoxine

The original name for this vitamin was **vitamin B_6**, which is a general name for **pyridoxine** and two closely related compounds, **pyridoxal** and **pyridoxamine.** These compounds, like nicotinic acid, are pyridine derivatives.

```
       CH2OH                      H                          CH2NH2
        |                        C=O                          |
        C                         |                           C
      /   \\                      C                          /   \\
 HO—C       C—CH2OH          /   \\                     HO—C       C—CH2OH
    ||      |           HO—C       C—CH2OH                ||      |
CH3—C       CH              ||      |                CH3—C       CH
      \   //            CH3—C       CH                     \   //
        N                     \   //                         N
                                N
   Pyridoxine                 Pyridoxal                  Pyridoxamine
```

A deficiency of vitamin B_6 in the diet of young rats results in a dermatitis called **acrodynia,** which is characterized by swelling and edema of the ears, nose, and paws. Pigs, cows, dogs, and monkeys exhibit central nervous system disturbances on a pyridoxine-deficient diet.

Vitamin B_6 is widely distributed in nature, with yeast, eggs, liver, cereals, legumes, and milk serving as good sources. The phosphate derivatives of pyridoxal and pyridoxamine occur in vitamin B_6 sources and serve as the coenzyme forms of the vitamin.

```
        H
       C=O                                       CH2NH2
        |                  O                       |                   O
        C                  ||                      C                   ||
      /   \\                                     /   \\
 HO—C       C—CH2—O—P—OH                    HO—C       C—CH2—O—P—OH
    ||      |          |                        ||      |          |
CH3—C       CH         O⁻                   CH3—C       CH         O⁻
      \   //                                      \   //
        N                                           N
     Pyridoxal phosphate                       Pyridoxamine phosphate
```

Pyridoxal phosphate is the major coenzyme for several enzymes involved in amino acid metabolism. Processes such as transamination, decarboxylation, and racemization of amino acids require pyridoxal phosphate as a cofactor. The active site in pyridoxal phosphate is the aldehyde group and the adjacent hydroxyl group. The functional mechanism of pyridoxal and pyridoxamine phosphates in transamination is described in Chapter 22.

Pantothenic Acid

Pantothenic acid is an amide of dihydroxydimethylbutyric acid and alanine.

```
           CH3 OH   O  H
            |   |   || |
HO—CH2—C——CH—C—N—CH2—CH2—COOH
            |
           CH3
```

Pantothenic acid

Many animals show deficiency symptoms on diets lacking pantothenic acid; for example, the rat fails to grow, and exhibits a dermatitis, graying of hair, and adrenal cortical failure.

In recent dietary research on pantothenic acid deficiency in man, such symptoms as emotional instability, gastrointestinal tract discomfort, and a burning sensation in the hands and feet have been observed.

Pantothenic acid is so widespread in nature that it was named from the Greek word *pantos,* meaning everywhere. Yeast, eggs, liver, kidney, and milk are good sources. The coenzyme form of this vitamin is known as **coenzyme A.**

Coenzyme A

The functional group of the coenzyme is the —SH group, resulting in the abbreviated form CoASH. In biological systems it functions mainly as **acetyl CoA,** and is involved in acetylation reactions, synthesis of fats, synthesis of steroids, and the metabolic reactions that will be discussed in subsequent chapters. The formation of acetyl CoA involves a reaction of the functional —SH group with a lipoic acid complex. Lipoic acid is a S-containing vitamin which is a derivative of valeric acid. Acetyl CoA is also important as a source of acetate for the Krebs cycle.

Folic Acid

Folic acid is a complex molecule consisting of three major parts: a yellow pigment called a pteridine, *p*-aminobenzoic acid, and glutamic acid. Its composition led to the name **pteroylglutamic acid** and its structure may be represented as follows:

Folic acid

A lack of this vitamin in the diets of young chickens and monkeys causes anemia and other blood disorders. Recently favorable clinical results have been reported in man following the use of folic acid in **macrocytic anemias,** which are characterized by the presence of giant red corpuscles in the blood. This type of anemia can occur in sprue, pellagra, pregnancy, and in gastric and intestinal disorders.

Folic acid occurs in many plant and animal tissues, especially in the foliage of plants, from which it was named. Yeast, soybeans, wheat, liver, kidney, and eggs are good sources of this vitamin.

To function as a coenzyme, folic acid must be reduced to either **dihydrofolic acid** or **tetrahydrofolic acid.** The enzymes folic reductase and dihydrofolic reductase convert the vitamin to the active coenzyme tetrahydrofolic acid. The major role of tetrahydrofolic acid is as a carrier of one-carbon or formate units in the biosynthesis of purines, serine, glycine, and methyl groups.

Vitamin B_{12}

Vitamin B_{12} has a complex chemical structure that is centered about an atom of cobalt bound to the four nitrogen atoms of a tetrapyrrole, to a nucleotide, and to a cyanide group. It is called **cyanocobalamin,** and is represented as follows:

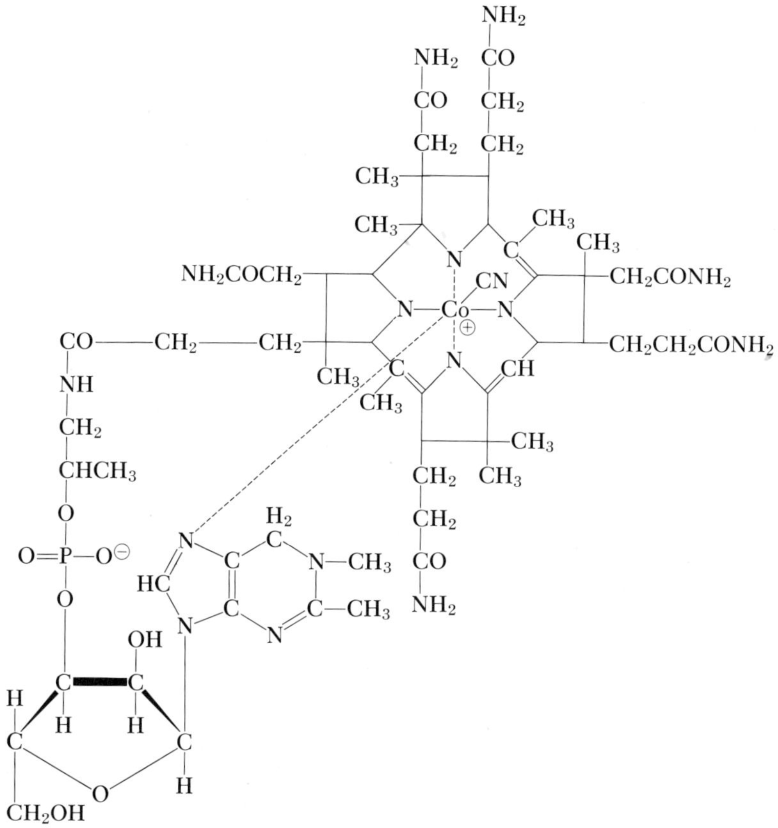

Vitamin B_{12} (cyanocobalamin)

Vitamin B_{12}, like folic acid, is useful in the treatment of the anemias that develop in humans and animals. **Pernicious anemia** in particular responds most readily to treatment with the vitamin. In addition to increasing the hemoglobin and the red cell count, vitamin B_{12} administration also produces a remission of the neurological symptoms of anemia.

The best source of vitamin B_{12} is liver. Other sources include milk, beef extract, and culture media of microorganisms. Liver extracts also contain a hydroxycobalamin on which the cyanide group is replaced by a hydroxyl group. The coenzyme form of the vitamin occurs in nature and is known as coenzyme B_{12}. It is an unstable compound in which the CN or OH group attached to the cobalt atom in vitamin B_{12} is replaced by the nucleoside, adenosine, as shown:

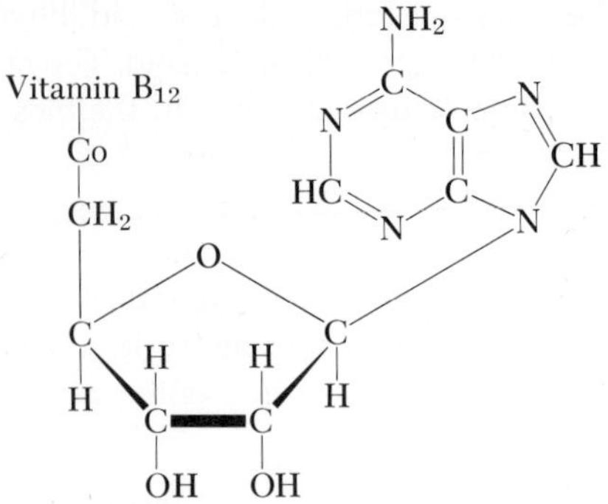

Coenzyme B_{12}

The coenzyme is readily converted into either cyano- or hydroxycobalamin in the presence of cyanide or light. Coenzyme B_{12} functions in several important reactions in metabolism. It is involved in the isomerization of dicarboxylic acids; for example, it catalyzes the conversion of glutamic acid to methyl aspartic acid. The coenzyme also assists in the conversion of glycols and glycerol to aldehydes, in the biosynthesis of methyl groups, and in the synthesis of nucleosides.

ENZYME INHIBITORS

The activity of an enzyme may be inhibited by an increase in temperature, a change in pH, and the addition of a variety of protein precipitants. More specific inhibition can be achieved by the addition of an oxidizing agent to attack SH groups, or inhibitors such as iodoacetamide and *p*-chloromercuribenzoate that react with SH groups. Cyanide forms compounds with metals essential for enzyme action, whereas fluoride combines with magnesium and inhibits enzymes that require magnesium. Cyanide, for example, may remove a metal such as copper that is essential for the activity of the enzyme.

Sodium azide and monoiodoacetate are also potent inhibitors. This type of compound usually combines with a group at the active site of the enzyme and cannot be displaced by additional substrate. These inhibitors are called **noncompetitive inhibitors,** since their degree of inhibition is not related to the concentration of the substrate.

Compounds that directly compete with the substrate for the active site on the enzyme surface in the formation of the enzyme-substrate complex are called **competitive inhibitors.** An example of competitive inhibition would be the action of sulfanilamide on the utilization of *p*-aminobenzoic acid in the body. The similarity of these two compounds may readily confuse the enzyme involved in the utilization of this B vitamin in the synthesis of tetrahydrofolic acid, the active coenzyme of folic acid.

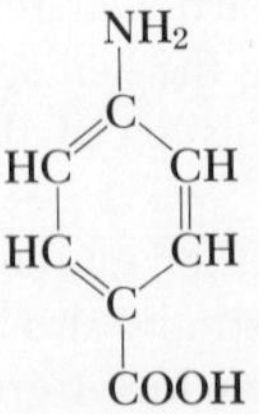

p-Aminobenzoic acid 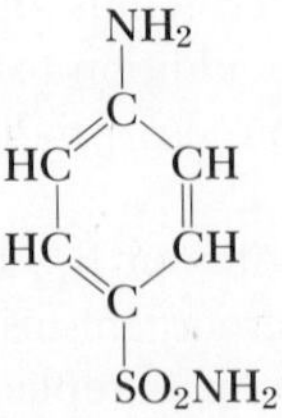 Sulfanilamide

The enzyme succinic dehydrogenase catalyzes the oxidation of succinic acid to fumaric acid. Malonic acid inhibits this reaction, but the degree of inhibition can be reduced by the addition of more substrate, succinic acid. Malonic acid is similar to but has one less carbon atom than succinic acid as shown:

Malonic acid	Succinic acid
COOH	COOH
|	|
CH_2	CH_2
|	|
COOH	CH_2
	|
	COOH

Malonic acid probably fits into the active site of succinic dehydrogenase, and thus it classifies as a competitive inhibitor. The concentration of the substrate and the concentration of the inhibitor both govern the action of malonic acid.

The Lineweaver-Burk plot (see p. 300) is frequently used to illustrate competitive or noncompetitive inhibition. In the example of malonic acid as an inhibitor of succinic dehydrogenase, the malonic acid combines with the dehydrogenase forming an inactive enzyme inhibitor (EI) complex. The rate of formation of the product (fumaric acid) depends solely on the concentration of ES.

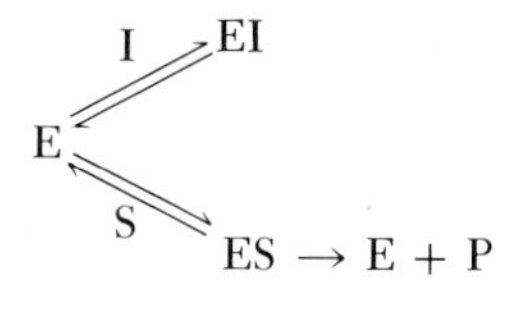

As long as there is sufficient substrate present in the reaction, the same $v_{max.}$ will be obtained, but the presence of EI alters the K_m for the substrate. In contrast to this behavior, noncompetitive inhibitors combine irreversibly with the enzyme molecule:

$$E + I \rightarrow EI$$

As the inhibitor concentration is increased the amount of enzyme, E, is decreased and the $v_{max.}$ is proportionately decreased. The K_m value for the substrate remains constant, since $\frac{v_{max.}}{2}$ is proportional to $v_{max.}$. Competitive and noncompetitive inhibition are represented graphically in the Lineweaver-Burk plots shown in Figure 19–8.

The action of drugs in the body may depend on specific inhibitory effects on a particular enzyme in the tissues. The highly toxic nerve poison diisopropylfluorophosphate inhibits acetylcholine esterase, an enzyme essential for normal nerve function (p. 376)

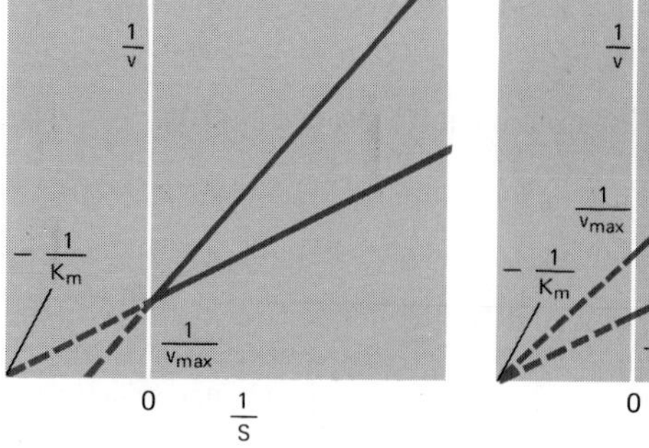

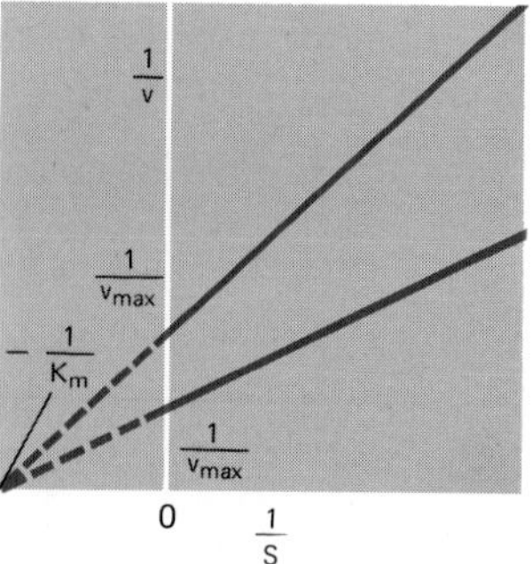

FIGURE 19–8 Lineweaver-Burk plot of competitive inhibition on the left compared to noncompetitive inhibition on the right.

by forming an enzyme inhibitor compound by attachment to a hydroxyl group on a serine residue in the enzyme.

$$\text{Enzyme—OH} + \text{F—}\overset{\displaystyle O}{\overset{\|}{P}}\text{—}\left(\text{O—}\underset{H}{C}\begin{matrix}CH_3\\CH_3\end{matrix}\right)_2 \rightarrow \text{Enzyme—O—}\overset{\displaystyle O}{\overset{\|}{P}}\text{—}\left(\text{O—}\underset{H}{C}\begin{matrix}CH_3\\CH_3\end{matrix}\right)_2$$

Acetylcholine esterase — Diisopropyl-fluorophosphate (DFP) — Enzyme inhibitor compound

Antibiotic drugs may act by inhibiting enzyme and coenzyme reactions in microorganisms. Penicillin, for example, adversely affects cell wall construction in bacteria. A similar mechanism may be involved in the action of insecticides and herbicides.

Inhibitors of enzyme action in the body are called **antienzymes.** The tapeworm is a classic example of a protein-rich organism that is not digested in the intestine of the host. Substances that inhibit the activity of pepsin and trypsin have been isolated from the tapeworm. A trypsin inhibitor has been found in the secretion of the pancreas and in milk made from fresh soya beans. This substance exhibits properties similar to those of enzymes, and its activity is destroyed by heat. It may be formed by the pancreas to control the production of trypsin.

IMPORTANT TERMS AND CONCEPTS

active site
code number
coenzyme
coenzyme A
competitive inhibitors
cyanocobalamin
energy of activation
enzyme
flavin adenine dinucleotide
Lineweaver-Burk plots
Michaelis constant
nicotinamide-adenine dinucleotide
noncompetitive inhibitors
optimum pH
optimum temperature
pyridoxal phosphate
specificity
substrate
tetrahydrofolic acid
thiamine pyrophosphate

QUESTIONS

1. What is the nature of an enzyme and how may it be defined?
2. Briefly explain how enzymes are classified.
3. How would you describe the activity of an enzyme with a systematic code number of 3.1.1.1?
4. What is meant by the energy of activation of a reaction? How is it affected by an enzyme?
5. What is a substrate? Give an example of a substrate and enzyme using the trivial name for the enzyme. Give an example of an enzyme and its substrate based on modern nomenclature.
6. Give examples of three types of specificity of action of enzymes.
7. What role does the formation of an enzyme-substrate complex play in the action of an enzyme? Explain.
8. What is meant by the active site of an enzyme? Explain.
9. Draw a graph representing the change in activity of an enzyme as the amount of its substrate is increased from zero to a maximum concentration. Explain the shape of the curve obtained.

10. What are the advantages of expressing an enzyme reaction as a Lineweaver-Burk plot?

11. How would you define the Michaelis constant, K_m, of an enzyme? Why is the K_m value important in enzyme reactions?

12. What is meant by (1) optimum pH? (2) optimum temperature of enzyme reactions?

13. How is the phenomenon of negative feedback related to the end products of an enzyme reaction? Could negative feedback control be of value in the body? Explain.

14. What is the difference between a coenzyme and a prosthetic group? Why is an apoenzyme inactive?

15. Name the coenzymes that are involved in oxidation-reduction reactions in the body.

16. Discuss the mechanism of function of any one coenzyme, including chemical structures in your discussion.

17. Cyanide is a very potent poison. Explain how cyanide may exert its toxic properties.

18. Explain the difference between competitive and noncompetitive inhibitors, and give an example of each.

19. Illustrate competitive and noncompetitive inhibition using Lineweaver-Burk plots.

SUGGESTED READING

Classification and Nomenclature of Enzymes. Science, Vol. 137, p. 405, 1962.
Green: Enzymes in Teams. Scientific American, Vol. 181, No. 3, p. 48, 1949.
Enzyme Nomenclature. Recommendations 1964 of the International Union of Biochemistry. New York, Elsevier Publishing Co., 1965.
Miller and Cory: Activation Energies for a Base-Catalyzed and Enzyme-Catalyzed Reaction. Journal of Chemical Education, Vol. 48, p. 475, 1971.
Neurath: Protein-Digesting Enzymes. Scientific American, Vol. 211, No. 6, p. 68 1964.
Phillips: Three-Dimensional Structure of an Enzyme Molecule. Scientific American, Vol. 215, No. 5, p. 78, 1966.
Research Reporter: First Synthesis of an Enzyme, Ribonuclease. Chemistry, Vol. 42, No. 4, p. 21, 1969.
Shaw: The Kinetics of Enzyme Catalyzed Reactions. Journal of Chemical Education, Vol. 34, p. 22, 1957.
Sumner: The Story of Urease. Journal of Chemical Education, Vol. 14, p. 255, 1937.
Sumner: Enzymes, the Basis of Life. Journal of Chemical Education, Vol. 29, p. 114, 1952.
Thayer: Some Biological Aspects of Organometallic Chemistry. Journal of Chemical Education, Vol. 48, p. 807, 1971.
Wroblewski: Enzymes in Medical Diagnosis, Scientific American, Vol. 205, No. 2, p. 99, 1961.

CARBOHYDRATE METABOLISM

CHAPTER 20

The *objectives* of this chapter are to enable the student to:

1. Describe the processes of digestion and absorption of the carbohydrates.
2. Recognize the factors involved in the control of the normal blood sugar level.
3. Discuss the role of hormones in the control of the blood sugar level.
4. Describe the process of glycogenesis.
5. Recognize the essential reactions in the Embden-Meyerhof pathway.
6. Outline the essential reactions in the Krebs cycle.
7. Account for the total number of moles of ATP formed in the Embden-Meyerhof and Krebs pathways.
8. Recognize the relationship between the hexose monophosphate and the Embden-Meyerhof pathways.
9. Recognize the essential nature of chlorophyll in the light reaction of photosynthesis.
10. Outline the essential reactions in the dark reaction of photosynthesis.

In the preceding chapters on biochemistry we have considered the chemistry of carbohydrates, lipids, proteins, and nucleic acids. These substances are taken into the body in food and ordinarily cannot be utilized directly in the form in which they are ingested. Before the food can be absorbed and utilized by the body, it must be broken down into small, relatively simple molecules. The process by which complex food material is changed into simple molecules is called **digestion.** Digestion involves the hydrolysis of carbohydrates into monosaccharides, fats into glycerol and fatty acids, and proteins into amino acids, by the action of the hydrolases or hydrolytic enzymes.

DIGESTION

Salivary Digestion

Food taken into the mouth is broken into smaller pieces by chewing and is mixed with saliva, which is the first of the digestive fluids. Saliva contains **mucin,** a glycoprotein that makes the saliva slippery, and **ptyalin,** an enzyme that catalyzes the hydrolysis of starch to maltose. Since this enzyme has little time to act on starches in the mouth, its main activity takes place in the stomach before it is inactivated by the acid gastric contents. The most important functions of saliva are to moisten and lubricate the food for swallowing and to initiate the digestion of starch to dextrins and maltose.

Gastric Digestion

When food is swallowed it passes through the esophagus into the stomach. During the process of digestion the food is mixed with **gastric juice,** which is secreted by many small tubular glands located in the walls of the stomach. Gastric juice is a pale yellow, strongly acid solution containing the enzymes **pepsin** and **rennin.** Pepsin initiates the hydrolysis of large protein molecules, whereas rennin converts casein of milk into a soluble protein. The mixing action of the stomach musculature and the process of digestion produce a liquid mixture called chyme which passes through the pyloric opening into the intestine.

Intestinal Digestion

The acid chyme is neutralized by the alkalinity of the three digestive fluids, **pancreatic juice, intestinal juice,** and **bile,** in the first part of the small intestine known as the duodenum. There are enzymes in the pancreatic juice that are capable of digesting proteins, fats, and carbohydrates. The pancreatic proteases are **trypsin, chymotrypsin,** and **carboxypolypeptidase,** whereas the pancreatic lipase is called **steapsin.** The enzyme **amylopsin** in pancreatic juice is an amylase similar to ptyalin in the saliva. This enzyme splits starch into maltose. The most important enzymes present in the intestinal juice are **aminopolypeptidase** and **dipeptidase,** and the three disaccharide-splitting enzymes, **sucrase, lactase,** and **maltase.** Cane sugar is the main source of dietary sucrose; milk contains lactose, and maltose comes from the partial digestion of starch by ptyalin and amylopsin. Sucrase, lactase, and maltase split these disaccharides into their constituent monosaccharides, thus completing the digestion of carbohydrates.

ABSORPTION

The monosaccharides glucose, fructose, and galactose are absorbed directly into the bloodstream through the capillary blood vessels of the **villi.** The villi are finger-like projections on the inner surface of the small intestine that greatly increase the effective absorbing surface. There are approximately five million villi in the human small intestine, and each villus is richly supplied with both lymph and blood vessels. Considerable evidence exists to indicate that the intestinal mucosa possesses the property of **selective absorption,** which is not possessed by a nonliving membrane. The exact mechanism by which hexose sugars are absorbed by the intestine against a concentration gradient has not been established, although it probably involves a phosphorylation process. The three monosaccharides are absorbed at different rates. Galactose is absorbed more rapidly than glucose, which is absorbed more rapidly than fructose.

THE BLOOD SUGAR

After the monosaccharides are absorbed into the blood stream, they are carried by the portal circulation to the liver. Fructose and galactose are phosphorylated by liver enzymes and are either converted into glucose or follow similar metabolic pathways. The metabolism of carbohydrates, therefore, is essentially the metabolism of glucose.

The concentration of glucose in the general circulation is normally 70 to 90 mg per

100 ml of blood. This is known as the **normal fasting level** of blood sugar. After a meal containing carbohydrates, the glucose content of the blood increases, causing a temporary condition of **hyperglycemia.** In cases of severe exercise or prolonged starvation, the blood sugar value may fall below the normal fasting level, resulting in the state of **hypoglycemia.** After an ordinary meal the glucose in the blood reaches hyperglycemic levels; this may be returned to normal by the following processes:

1. Storage
   (a) as glycogen
   (b) as fat
2. Oxidation to produce energy
3. Excretion by the kidneys

The operation of these factors in counteracting hyperglycemia is illustrated in Figure 20–1. The space between the vertical lines may be compared to a thermometer, with values expressed as milligrams of glucose per 100 ml of blood. During active absorption of carbohydrates from the intestine the blood sugar level rises, causing a temporary hyperglycemia. In an effort to bring the glucose concentration back to normal, the liver may remove glucose from the blood stream, converting it into glycogen for storage. The muscles will also take glucose from the circulation to convert it to glycogen or to oxidize it to produce energy. If the blood sugar level continues to rise, the glucose may be converted into fat and stored in the fat depots. These four processes usually control the hyperglycemia; but if large amounts of carbohydrates are eaten and the blood sugar level exceeds an average of 160 mg of glucose per 100 ml, the excess is excreted by the kidneys. The blood sugar level at which the kidney starts excreting glucose is known as the **renal threshold** and has a value from 150 to 170 mg per 100 ml.

In addition to the above factors, there are more specific reactions of the liver and the hormones to bring about regulation of the level of the glucose in the blood. The liver, for example, functions both in the removal of sugar from the blood and in the addition of sugar to the blood. During periods of hyperglycemia the liver stops pouring sugar into the blood stream and starts to store it as liver glycogen. During fasting the liver supplies glucose to the blood by breaking down its glycogen and by forming glucose from other food material such as amino acids or glycerol. The liver is assisted in this control process by several hormones.

HORMONES AND THE BLOOD SUGAR LEVEL

The properties and action of enzymes have already been discussed. In metabolism there are many related chemical reactions under the influence of enzymes. Another

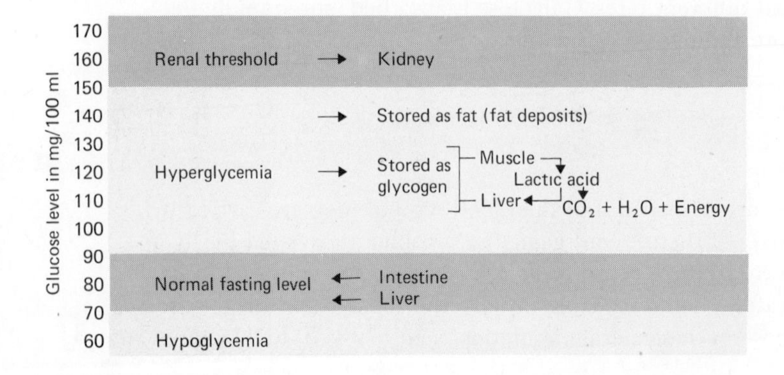

FIGURE 20–1 Factors involved in the regulation of the glucose level of the blood.

important group of regulating agents is the **hormones.** The hormones are formed mainly in the endocrine glands, which are also called ductless glands since their secretions diffuse directly into the blood stream. Enzyme action is more specific than hormone action, and the factors involved in the action of a hormone appear to be related to the action of other hormones. In a normal individual, major cellular processes depend on an endocrine or **hormone balance,** and a disturbance in this balance results in metabolic abnormalities. In the regulation of a body process, a hormone probably has control over several specific enzyme-catalyzed reactions.

Insulin

Although it was demonstrated as early as 1889 that removal of the pancreas of an animal would result in diabetes mellitus, it was not until 1922 that Banting and Best developed a method for obtaining active extracts of the pancreas. Within a short time insulin became available in sufficient quantities for the treatment of diabetes. It was first crystallized in 1926. More recently, as a result of the brilliant work of Sanger and his co-workers, a molecule consisting of two chains of amino acids with a molecular weight of 6000 has been described. The native molecule is thought to consist of four chains with a molecular weight of 12,000. Since it is a protein, it is not effective when taken by mouth, because the proteolytic enzymes of the gastrointestinal tract cause its hydrolysis and destroy its activity. Insulin is usually injected subcutaneously when administered to a diabetic.

Insulin lowers the blood sugar level by increasing the conversion of glucose into liver and muscle glycogen, by regulating the proper oxidation of glucose by the tissues, and by preventing the breakdown of liver glycogen to yield glucose. In muscle and adipose tissue, insulin acts by increasing the rate of transport of glucose across membranes into the cells. Also, there is considerable evidence that in liver tissue insulin acts by controlling the phosphorylation of glucose to form glucose-6-phosphate, which is the first step in the formation of glycogen. In the absence of an adequate supply of insulin the transformation of extracellular glucose to intracellular glucose-6-phosphate is retarded.

Diabetes Mellitus. If the pancreas fails to produce sufficient insulin, the condition of **diabetes mellitus** results. The failure of the storage mechanisms in the absence of insulin causes a marked increase in the blood sugar level. Glucose is ordinarily excreted in the urine because the renal threshold is exceeded. The impairment of carbohydrate oxidation causes the formation of an excess of ketone bodies. Some of these ketone bodies are acid in nature, and the severe acidosis that results from the lack of insulin causes **diabetic coma,** which is sometimes fatal to a diabetic patient. When the correct dosage of insulin is injected, carbohydrate metabolism is properly regulated and the above symptoms do not appear.

Glucagon

Glucagon is a hormone that is produced by the α-cells of the pancreas. It is a polypeptide of known amino acid sequence with a molecular weight of about 3500. Glucagon causes a rise in the blood sugar level by increasing the activity of the enzyme liver phosphorylase, which is involved in the conversion of liver glycogen to free glucose. The activation of phosphorylase depends on the presence of the compound cyclic-3′,5′-adenosine monophosphate (AMP), whose formation is increased by the action of glucagon.

Epinephrine

This hormone is produced by the central portion, or medulla, of the adrenal glands. Epinephrine is antagonistic to the action of insulin in that it causes glycogenolysis in the liver with the liberation of glucose. It stimulates an enzyme to produce cyclic-3′,5′-

AMP from ATP and is also involved in the activation of phosphorylase. In addition to hyperglycemia, it also increases blood lactic acid by converting muscle glycogen to lactic acid. Continued secretion of epinephrine occurs under the influence of strong emotions such as fear or anger. This mechanism is often used as an emergency function to provide instant glucose for muscular work. The hyperglycemia that results often exceeds the renal threshold, and glucose is excreted in the urine.

Adrenal Cortical Hormones

Hormones such as **cortisone** and **cortisol** are produced by the outer layer or cortex of the adrenal gland. These hormones, especially those with an oxygen on position 11, have an effect on carbohydrate metabolism. In general they stimulate the production of glucose in the liver by increasing gluconeogenesis from amino acids. The cortical hormones are therefore antagonistic to insulin.

Anterior Pituitary Hormones

Of the many hormones secreted by the anterior lobe of the pituitary gland, the growth hormone, ACTH and the diabetogenic hormone affect the blood sugar level. The **growth hormone** causes the liver to increase its formation of glucose, but at the same time it stimulates the formation of insulin by the pancreas. Its action is complex and not completely understood. **ACTH,** the adrenocorticotropic hormone, stimulates the function of the hormones of the adrenal cortex and their action on the blood sugar level. The **diabetogenic hormone,** when injected into an animal, causes permanent diabetes and exhaustion of the islet tissue of the pancreas.

Although the overall control of the blood sugar level depends on the action of the liver and a balanced action of several hormones, it is readily apparent that insulin plays a major role in the normal process and is an important factor in the control of diabetes mellitus.

GLYCOGEN

As may be recalled from Chapter 13, glycogen is a polysaccharide with a branched structure composed of linear chains of glucose units joined by α-1,4 linkages and with α-1,6 linkages at the branch points. During absorption of the carbohydrates, the excess glucose is stored as glycogen in the liver. Normally this organ contains about 100g of glycogen, but it may store as much as 400g. The glycogen in the liver is readily converted into glucose and serves as a reservoir from which glucose may be drawn if the blood sugar level falls below normal. The formation of glycogen from glucose is called **glycogenesis,** whereas the conversion of glycogen to glucose is known as **glycogenolysis.** The muscles also store glucose as glycogen, but muscle glycogen is not as readily converted into glucose as is liver glycogen.

Glycogenesis

The process of glycogenesis is not just a simple conversion of glucose to glycogen. As we have learned previously, insulin is involved in the action of glucokinase in the phosphorylation of glucose to glucose-6-phosphate. The glucose-6-phosphate is then converted to glucose-1-phosphate with the aid of the enzyme phosphoglucomutase. The glucose-1-phosphate then reacts with uridine triphosphate (UTP) to form an active nucleotide, uridine diphosphate glucose (UDPG). In the presence of a branching enzyme and the enzyme UDPG-glycogen-transglucosylase, the activated glucose molecules of UDPG are joined in glucosidic linkages to form glycogen. These reactions may be represented as follows:

$$\text{Glucose} \xrightarrow[\text{ATP} \;\to\; \text{ADP}]{\text{(insulin)}\ \text{glucokinase}} \text{Glucose-6-phosphate}$$

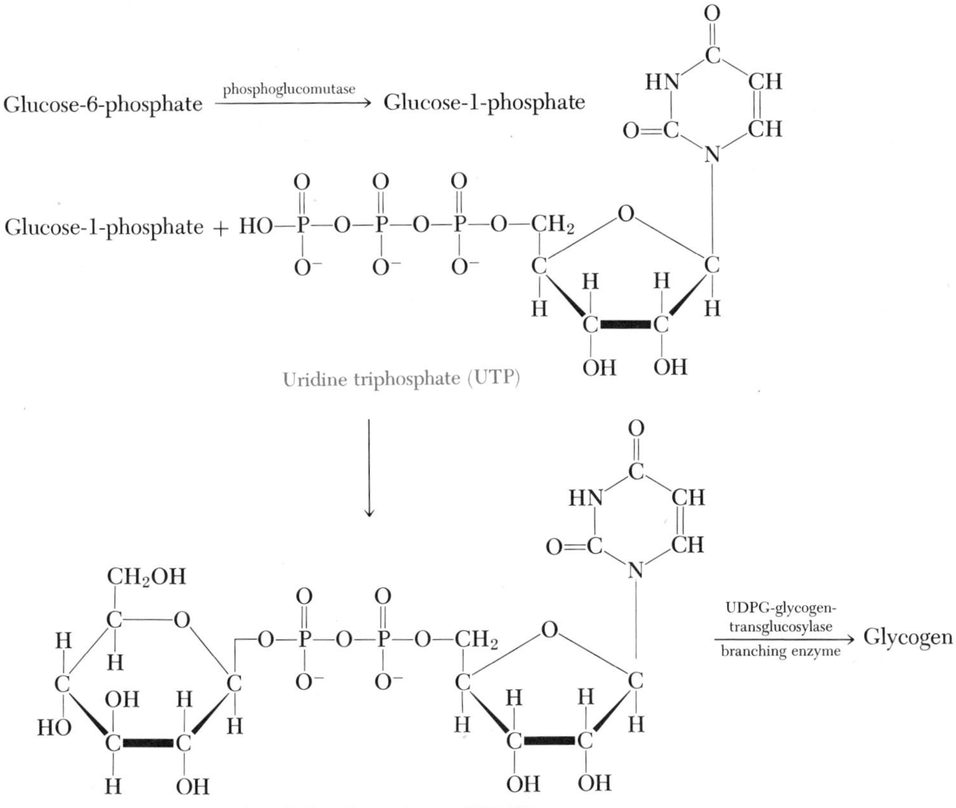

$$\text{Glucose-6-phosphate} \xrightarrow{\text{phosphoglucomutase}} \text{Glucose-1-phosphate}$$

Glucose-1-phosphate + Uridine triphosphate (UTP) → Uridine diphosphate glucose (UDPG)

Uridine triphosphate (UTP)

Uridine diphosphate glucose (UDPG)

$$\text{UDPG} \xrightarrow[\text{branching enzyme}]{\text{UDPG-glycogen-transglucosylase}} \text{Glycogen}$$

Glycogenolysis

The process of glycogenolysis liberates glucose into the blood stream to maintain the blood sugar level during fasting and to supply energy for muscular contraction. In the liver the reaction is initiated by the action of the enzyme phosphorylase, which splits the 1,4 glucosidic linkages in glycogen. The enzyme phosphorylase exists in two forms: an active form, **phosphorylase a,** and an inactive form, **phosphorylase b.** Phosphorylase b is converted to the active form of the enzyme by ATP in the presence of Mg^{+2} and phosphorylase b kinase, as shown:

$$\text{2 Phosphorylase b} + \text{4 ATP} \xrightarrow[\text{cyclic-3',5'-AMP}]{\text{kinase, } Mg^{+2}} \text{Phosphorylase a} + \text{4 ADP}$$

The phosphorylase b kinase is activated by cyclic-3′,5′-AMP, a derivative of adenylic acid.

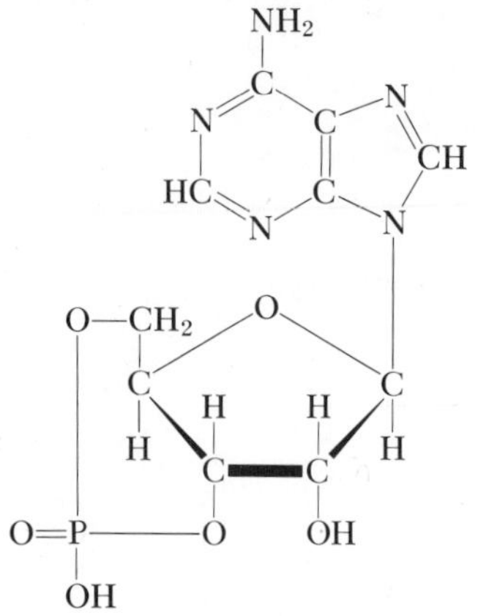

Cyclic-3′,5′-AMP

Cyclic-3′,5′-AMP is formed from ATP by the action of an enzyme called *cyclase enzyme*. This enzyme is activated by epinephrine and glucagon which therefore indirectly activate the phosphorylase responsible for the initiation of glycogenolysis. Other enzymes assist the breakdown to glucose-1-phosphate, which is subjected to the reversed action of phosphoglucomutase to yield glucose-6-phosphate. A specific enzyme in the liver, glucose-6-phosphatase, acts on glucose-6-phosphate to produce glucose. This enzyme is not present in muscle; therefore, muscle glycogen cannot serve as a source of blood glucose. These reactions may be illustrated as follows:

$$\text{Glycogen} \xrightarrow[\text{debranching enzyme}]{\text{phosphorylase}} \text{Glucose-1-phosphate}$$

$$\text{Glucose-1-phosphate} \xrightarrow{\text{phosphoglucomutase}} \text{Glucose-6-phosphate}$$

$$\text{Glucose-6-phosphate} \xrightarrow[\text{in liver}]{\text{glucose-6-phosphatase}} \text{Glucose}$$

OXIDATION OF CARBOHYDRATES

Glucose is ultimately oxidized in the body to form CO_2 and H_2O with the liberation of energy. **Glucose-6-phosphate** is a principal compound in the metabolism of glucose. As discussed earlier, it may be formed by the phosphorylation of glucose under the control of insulin. Once it is formed, it may be converted to glycogen or to free glucose, or it may be metabolized by several mechanisms or pathways. The two major pathways of glucose-6-phosphate metabolism are the **anaerobic,** or **Embden-Meyerhof, pathway** followed by the **aerobic,** or **Krebs, cycle.** The largest proportion of energy available from the oxidation of the glucose molecule is liberated from the Krebs cycle, but the Embden-Meyerhof pathway is essential in the formation of pyruvic acid used in the Krebs cycle.

GLYCOLYSIS

The ready availability of muscle preparations and their use in the development of physiology led to an early study of the biochemical changes associated with muscular contraction. It was observed that when a muscle contracts in an anaerobic medium, glycogen disappears and pyruvic and lactic acids are formed. In the presence of oxygen, or under aerobic conditions, the glycogen is re-formed, and the pyruvic and lactic acids disappear. Further studies demonstrated that one fifth of the lactic acid formed during glycolysis is oxidized to CO_2 and water, whereas the remaining four fifths is converted to glycogen.

Substances other than the carbohydrates in food and the lactic acid from muscular contraction may be converted into glycogen. These glycogenic compounds are formed by the process of **gluconeogenesis,** which is the conversion of non-carbohydrate precursors into glucose. Examples of these precursors are the glycogenic amino acids, the glycerol portion of fat, and any of the metabolic breakdown products of glucose, such as pyruvic acid, which may form glucose by reversible reactions in metabolism. The reactions discussed in the above section can be summarized in the **lactic acid cycle.**

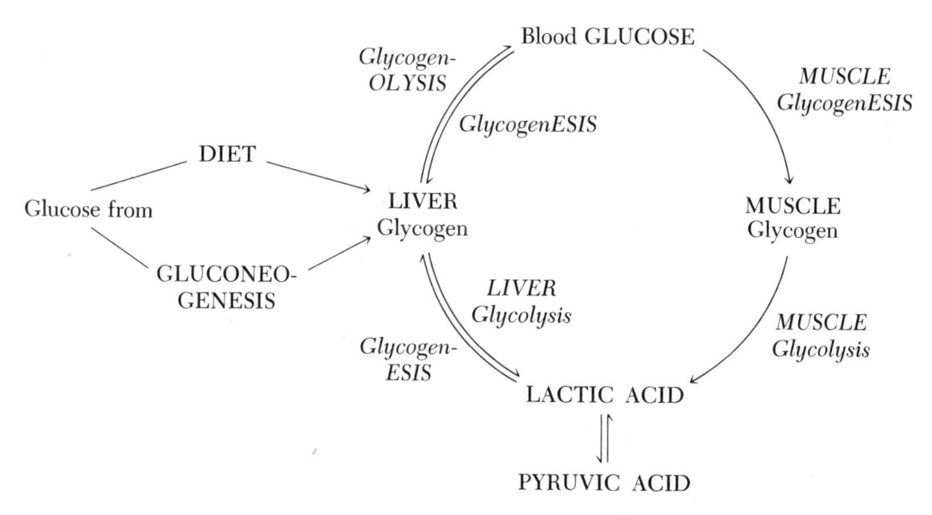

The Lactic Acid Cycle

Anaerobic, or Embden-Meyerhof, Pathway of Glycolysis

The chemical reactions in metabolic pathways like the Embden-Meyerhof pathway are detailed and complex, and may lead to confusion at first examination. An understanding of this and other metabolic pathways may be expedited by consideration of a preliminary outline of the essential reactions. Glucose from any source is converted to glucose-6-phosphate, which in turn is converted into fructose-6-phosphate, which is further phosphorylated to fructose-1,6-diphosphate. This set of reactions has converted the hexose glucose into a hexose diphosphate. The pathway diverges at this point with the splitting of fructose diphosphate into two triose monophosphates (see p. 161). One of these, glyceraldehyde-3-phosphate, is first transformed into 1,3-diphosphoglyceric acid, then successively into 3-phospho- and 2-phosphoglyceric acid. The last compound then forms phosphoenolpyruvic acid which is converted into pyruvic acid. The pyruvic acid is a key compound which may be reduced to lactic acid (pp. 155 and 207) or further oxidized in the Krebs cycle.

The detailed chemical compounds and enzymes involved in the Embden-Meyerhof pathway are shown in the following scheme.

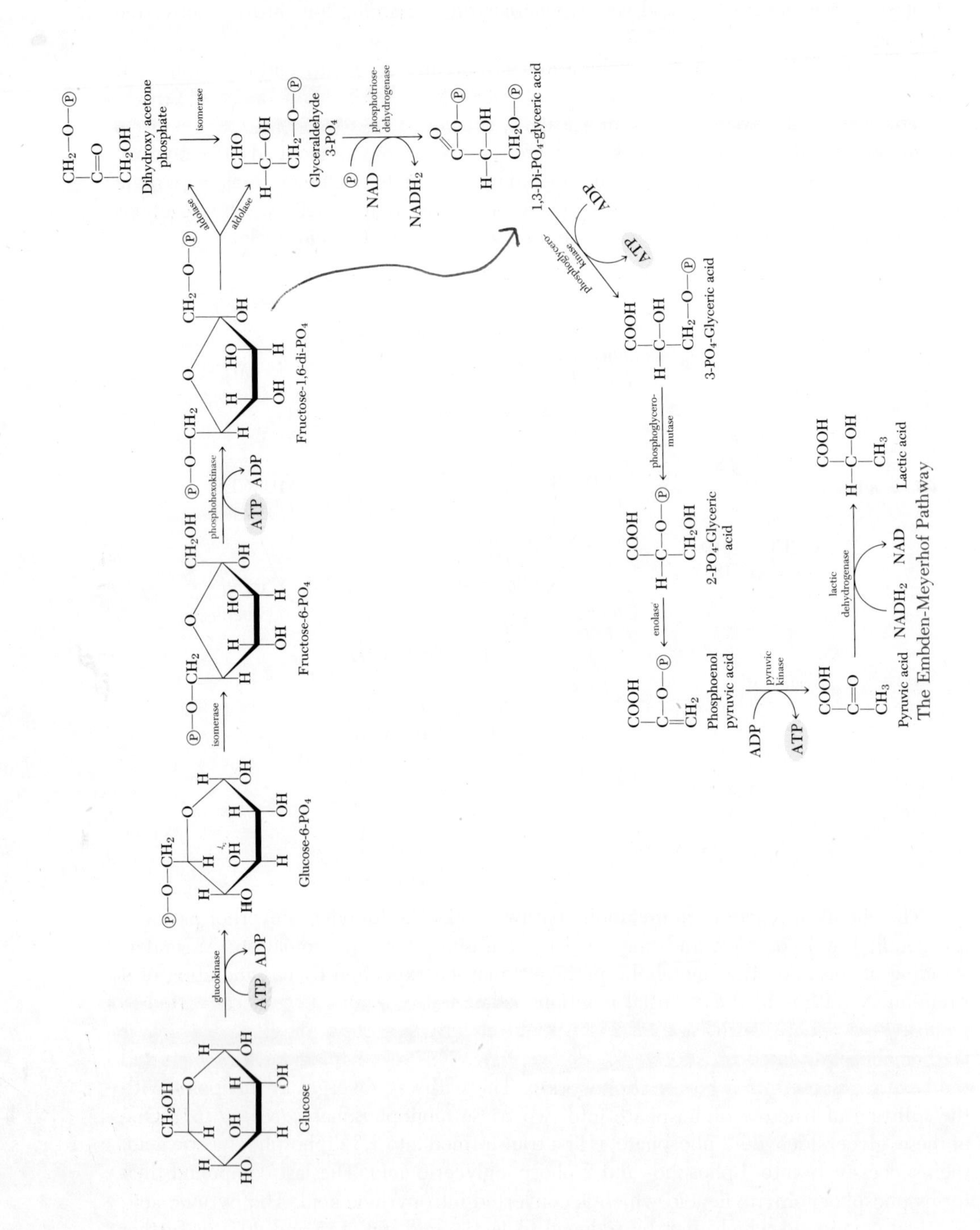

The Embden-Meyerhof Pathway

The requirement for and liberation of ATP in this anaerobic pathway is emphasized by the shaded areas. One mole of ATP is required for the phosphorylation of glucose and one more for the conversion of fructose-6-phosphate to fructose-1,6-diphosphate. The reaction of 1,3-diphosphoglyceric acid to form 3-phosphoglyceric acid liberates 1 mole of ATP per triose molecule or 2 moles of ATP per glucose molecule. The conversion of phosphoenolpyruvic acid to pyruvic acid also yields 2 moles of ATP per glucose molecule. For each mole of glucose broken down in the Embden-Meyerhof pathway, therefore, *2 moles of ATP are consumed and 4 moles are liberated, with a net gain of 2 moles of ATP.*

Aerobic, or Krebs, Cycle

The pyruvic acid formed in the Embden-Meyerhof pathway and the lactic acid from the lactic acid cycle or from the reduction of pyruvic acid are eventually oxidized with the formation of CO_2 and energy in the form of ATP. These reactions are carried out in the Krebs cycle, which may be outlined as follows:

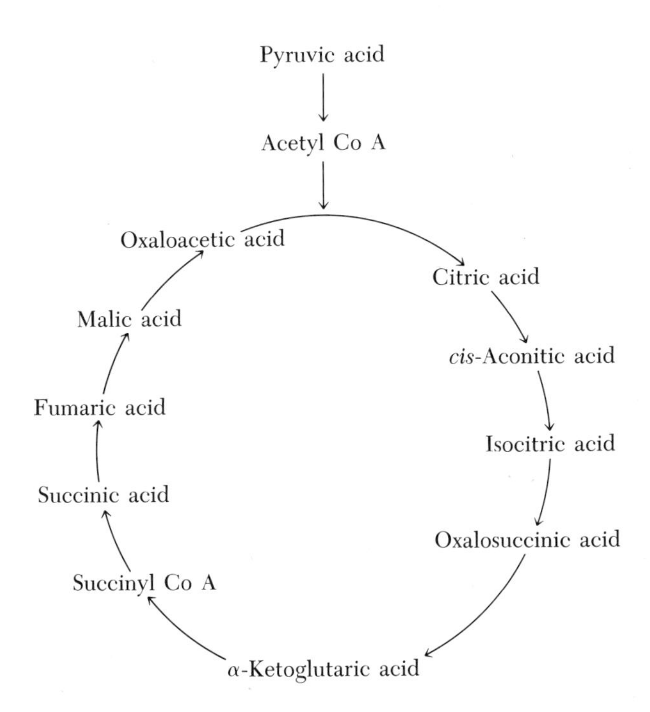

Pyruvic acid from any source forms acetyl Co A, which transfers the acetyl group to oxaloacetic acid to make the tricarboxylic acid, citric acid. Citric acid is successively transformed into *cis*-aconitic acid, then to isocitric and to oxalosuccinic, all tricarboxylic acids. The hydration of *cis*-aconitic to isocitric acid as well as the later conversion of fumaric to malic acid involves the electrophilic addition of H^+ as discussed on page 78. The oxalosuccinic acid then loses CO_2 to form α-ketoglutaric acid, which is converted to succinyl Co A and then to succinic acid and a series of dicarboxylic acids, including fumaric acid and malic acid, back to oxaloacetic acid, and the cycle is completed. The chemical structures, enzymes, and coenzymes of the Krebs cycle are shown in the following diagram.

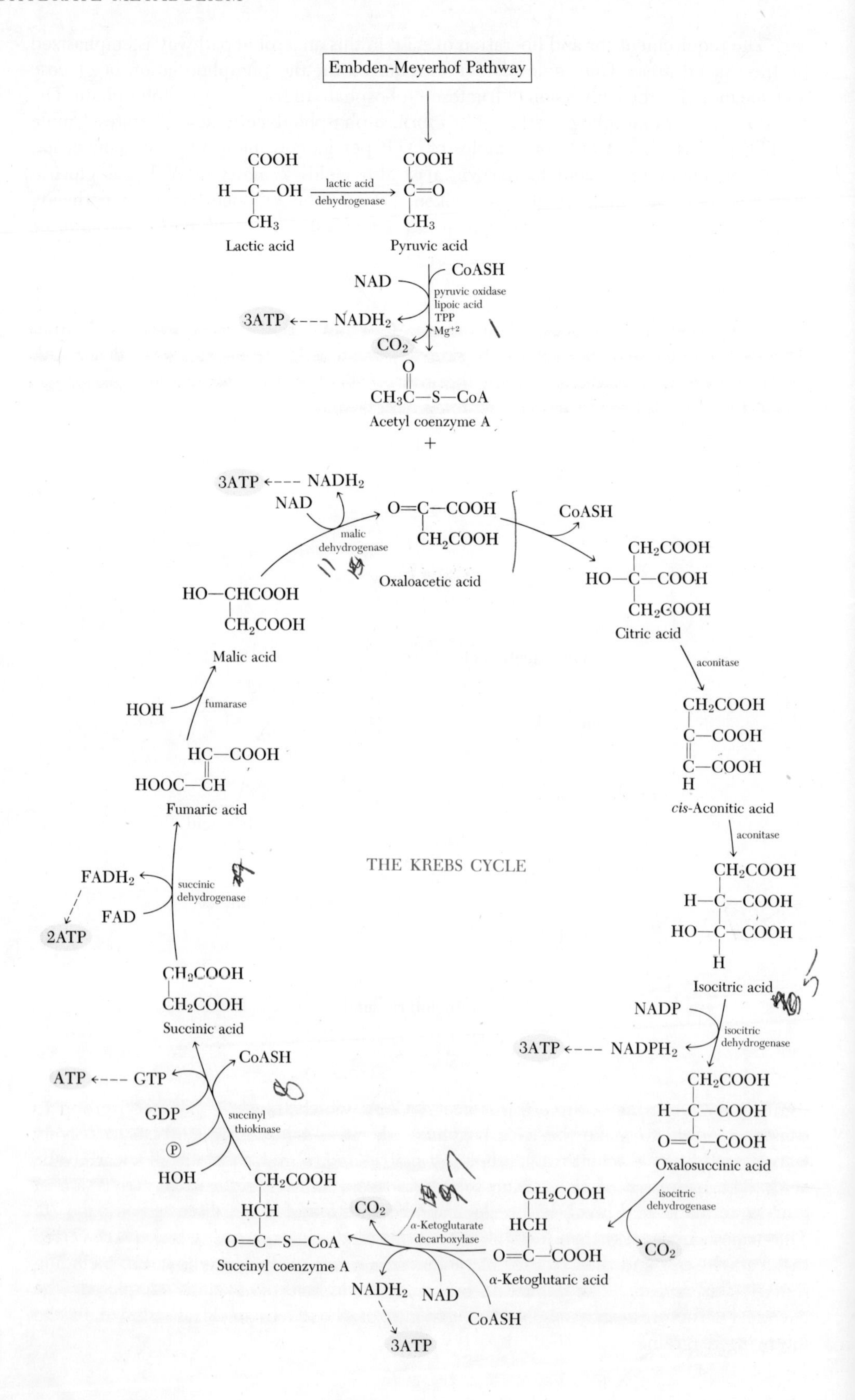

Embden-Meyerhof Pathway
COOH
H—C—OH
CH3
Lactic acid
lactic acid dehydrogenase
COOH
C=O
CH3
Pyruvic acid
NAD
CoASH
pyruvic oxidase
lipoic acid
TPP
Mg+2
3ATP
NADH2
CO2
O
CH3C—S—CoA
Acetyl coenzyme A
+
3ATP
NADH2
NAD
malic dehydrogenase
O=C—COOH
CH2COOH
Oxaloacetic acid
CoASH
CH2COOH
HO—C—COOH
CH2COOH
Citric acid
aconitase
CH2COOH
C—COOH
C—COOH
H
cis-Aconitic acid
aconitase
CH2COOH
H—C—COOH
HO—C—COOH
H
Isocitric acid
NADP
isocitric dehydrogenase
3ATP
NADPH2
CH2COOH
H—C—COOH
O=C—COOH
Oxalosuccinic acid
isocitric dehydrogenase
CO2
CH2COOH
HCH
O=C—COOH
α-Ketoglutaric acid
CO2
α-Ketoglutarate decarboxylase
NADH2
NAD
CoASH
3ATP
CH2COOH
HCH
O=C—S—CoA
Succinyl coenzyme A
HOH
P
GDP
ATP
GTP
CoASH
succinyl thiokinase
CH2COOH
CH2COOH
Succinic acid
FADH2
FAD
2ATP
succinic dehydrogenase
HC—COOH
HOOC—CH
Fumaric acid
HOH
fumarase
HO—CHCOOH
CH2COOH
Malic acid
THE KREBS CYCLE

The overall reaction for the conversion of pyruvic acid to carbon dioxide and water may be written as:

$$C_3H_4O_3 + {}^5\!/\!_2O_2 + 15ADP + 15P_i \rightarrow 3CO_2 + 2H_2O + 15ATP$$

The moles of ATP formed and CO_2 liberated in one turn of the Krebs cycle are shown in shaded areas. It may be recalled from a consideration of the electron transport mechanism in oxidative phosphorylation, page 289, that the oxidation of $NADH_2$ or $NADPH_2$ (through $NADH_2$) by way of the cytochrome system yields 3 moles of ATP per mole of $NADH_2$. Starting with $FADH_2$, the system yields 2 moles of ATP per mole of $FADH_2$.

When one molecule of glucose is completely oxidized it liberates 686.0 kilocalories. Each molecule of glucose subjected to the Embden-Meyerhof pathway liberates 8 moles of ATP (6 moles from the $NADH_2$ formed in the conversion of glyceraldehyde-3-phosphate to 1,3-diphosphoglyceric acid, and 2 moles net yield of ATP formed directly). Since each mole of glucose forms 2 moles of pyruvic acid, the Krebs cycle will yield 2 × 15 or 30 moles of ATP per molecule of glucose. *A total of 38 moles of ATP are therefore formed by the oxidation of a molecule of glucose in the Embden-Meyerhof and Krebs cycles.* Since each mole of ATP will yield approximately 8.0 kcal on hydrolysis, the 38 moles are equivalent to 304.0 kcal. This series of reactions is therefore capable of storing about 44 per cent of the available calories in the form of the high energy compound ATP, to be used in muscular work and for other energy requirements.

ALTERNATE PATHWAYS OF CARBOHYDRATE OXIDATION

Pathways other than the Embden-Meyerhof and Krebs cycles are involved in the oxidation of carbohydrates. The most generally accepted alternate pathway is called the **hexose monophosphate shunt,** or the **pentose phosphate pathway.** The key metabolic compound, glucose-6-phosphate, is oxidized in this alternate pathway as outlined.

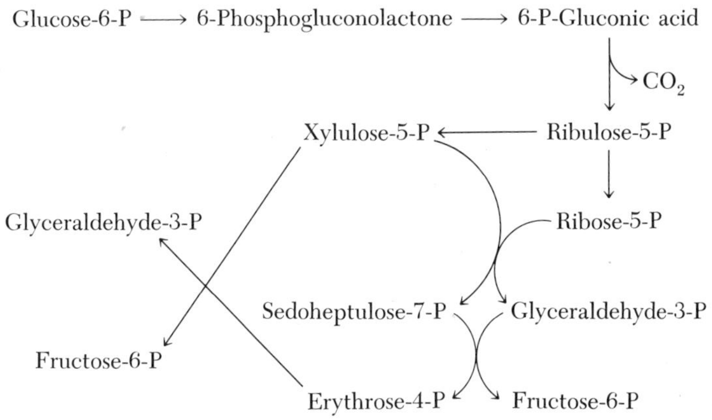

Glucose-6-P ⟶ 6-Phosphogluconolactone ⟶ 6-P-Gluconic acid
CO_2
Xylulose-5-P ⟵ Ribulose-5-P
Ribose-5-P
Glyceraldehyde-3-P
Sedoheptulose-7-P
Glyceraldehyde-3-P
Fructose-6-P
Erythrose-4-P
Fructose-6-P

Glucose-6-phosphate from any source is oxidized to 6-phosphogluconolactone and then to 6-phosphogluconic acid. The acid is converted to a pentose, ribulose-5-phosphate, by the loss of CO_2, and the pentose is transformed into two other pentoses, xylulose-5-phosphate and ribose-5-phosphate. The two latter compounds are converted into a seven-carbon, sedoheptulose-7-phosphate, and the three-carbon glyceraldehyde-3-phosphate. These two compounds are then transformed into a four-carbon, erythrose-4-phosphate, and the hexose fructose-6-phosphate. The final reaction involves xylulose-5-phosphate and erythrose-4-phosphate, forming glyceraldehyde-3-phosphate and more fructose-6-phosphate.

The Phosphogluconate Oxidative Pathway

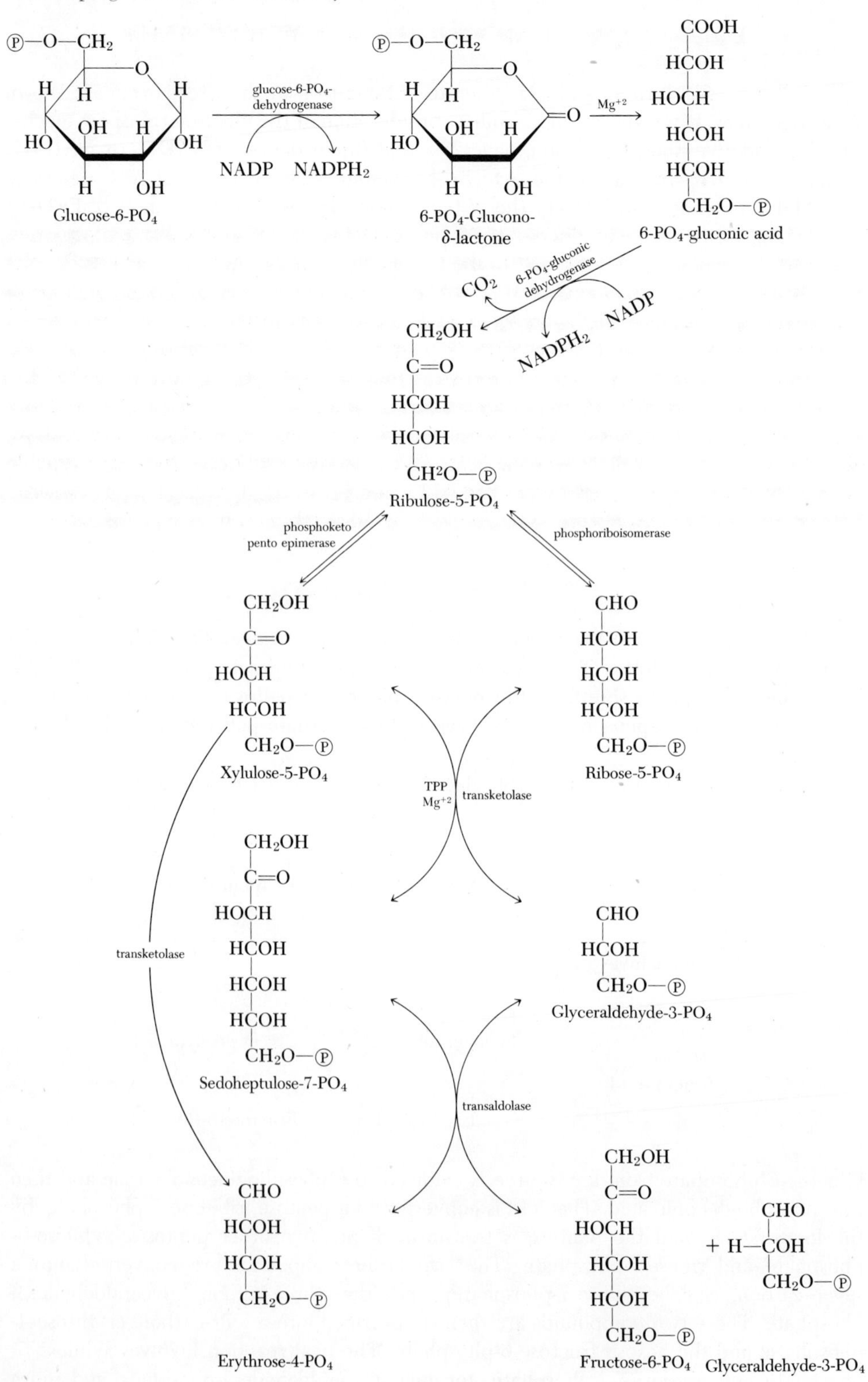

PHOTOSYNTHESIS

Carbohydrates are formed in the cells of plants from carbon dioxide and water. In the presence of sunlight and chlorophyll, the green pigment of leaves, these two compounds react to form pentoses, trioses, fructose, and more complex sugars. **Chlorophyll** is a protoporphyrin derivative containing magnesium that is located in the chloroplasts of green leaves.

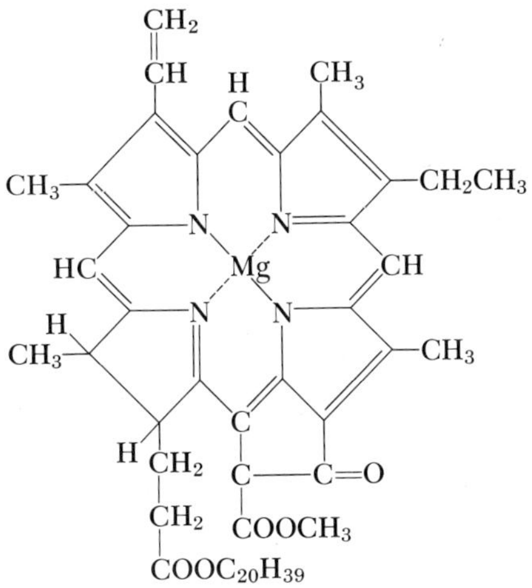

Chlorophyll a

Originally the reaction between carbon dioxide and water to form carbohydrates was represented as follows:

$$CO_2 + H_2O \xrightarrow[\text{chlorophyll}]{\text{sunlight}} \underset{\text{Simple sugar}}{C_6H_{12}O_6} + 6\ O_2$$

This process by which plants convert the energy of sunlight to form food material is called **photosynthesis.** Although photosynthesis is represented as a simple chemical reaction, it is more complex and includes several intermediates of the phosphogluconate and Embden-Meyerhof pathways.

The use of isotopes and radioactive tracers has greatly assisted the research workers in this field. For example, using isotopic oxygen, ^{18}O, as a tracer, it was shown that the oxygen liberated during photosynthesis is derived from the water molecules and not from carbon dioxide. When a green leaf is grown in an atmosphere of $^{14}CO_2$, the radioactive carbon appears very rapidly in a three-carbon atom and a five-carbon atom intermediate, and later in glucose and starch.

The Light Reaction

The reaction involving the conversion of light energy into chemical energy is called the **light reaction.** This transformation of energy occurs during photosynthetic phosphorylation and takes place in the chloroplasts of plants (Fig. 17–2). The essential reaction that occurs in **photophosphorylation** can be represented as follows:

$$ADP + P_i \xrightarrow{\text{light energy}} ATP$$

The photochemical process is initiated by the absorption of light by chlorophyll, which produces an excited-state molecule in which several electrons are raised from their normal energy level to a higher level in the double bond structure of chlorophyll. These excited electrons flow from chlorophyll to an iron-containing protein, **ferredoxin,** and bring about the reduction of NADP to form $NADPH_2$, which is used in the CO_2 fixation reactions of photosynthesis. Some of the excited electrons flow from ferredoxin through flavin pigments to a quinone structure called **plastoquinone,** then to **cytochrome pigments,** and then back to chlorophyll and their normal energy level. During this cycle some of the energy is given up by coupling in the reaction of ADP with P_i to form ATP. The ATP is also used in the CO_2 fixation reactions of photosynthesis. The electrons that are used in the formation of $NADPH_2$ and ATP are replenished by a reaction in which the OH ions of water form molecular oxygen and donate electrons to chlorophyll through a cytochrome chain. The process of **photosynthetic phosphorylation** may be represented as follows:

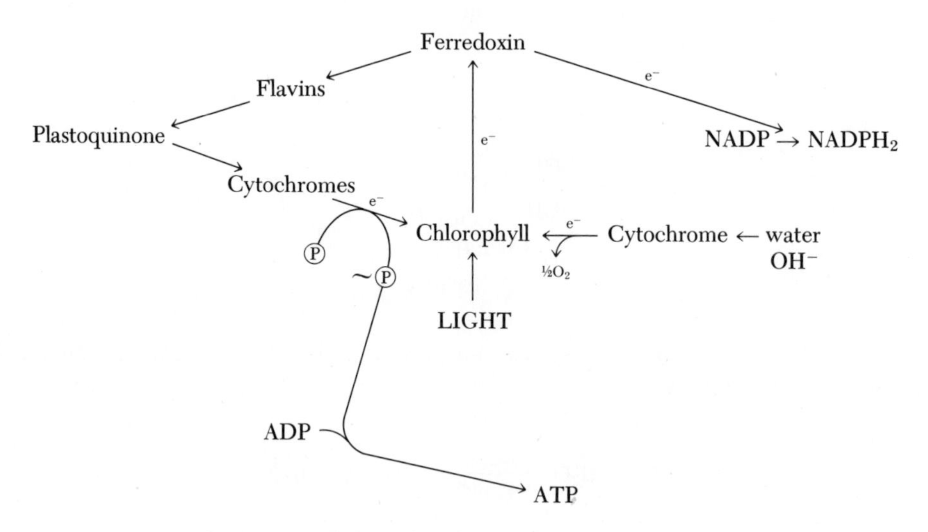

The Process of Photophosphorylation in Photosynthesis

The Dark Reaction

The reactions of photosynthesis that are not dependent on light energy have been termed the **dark reaction.** This process involves the incorporation of carbon into carbohydrates, **carbon fixation,** and requires the energy from ATP and a quantity of $NADPH_2$ formed in the light reaction. The dark reaction may be outlined as follows:

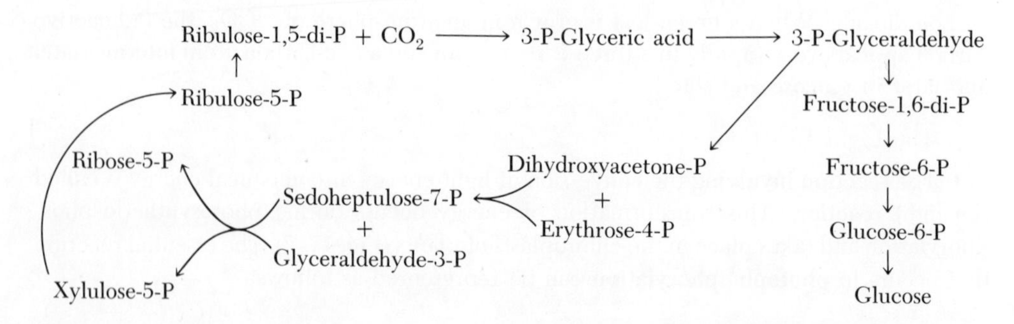

Ribulose-1,5-diphosphate plus carbon dioxide forms a complex that dissociates into two molecules of 3-phosphoglyceric acid, which is converted to 3-phosphoglyceraldehyde.

This latter compound can be converted to glucose-6-phosphate through fructose-1,6-diphosphate and fructose-6-phosphate, and eventually to glucose and polysaccharides. The 3-phosphoglyceraldehyde may also be converted to dihydroxyacetone phosphate, which reacts with erythrose-4-phosphate to form sedoheptulose-7-phosphate, which reacts with glyceraldehyde-3-phosphate to form two molecules of pentose, ribose-5-phosphate and xylulose-5-phosphate. The pentoses can be converted into ribulose-1,5-diphosphate. The structures of the compounds and the enzymes involved in the dark reaction are shown in the following diagram.

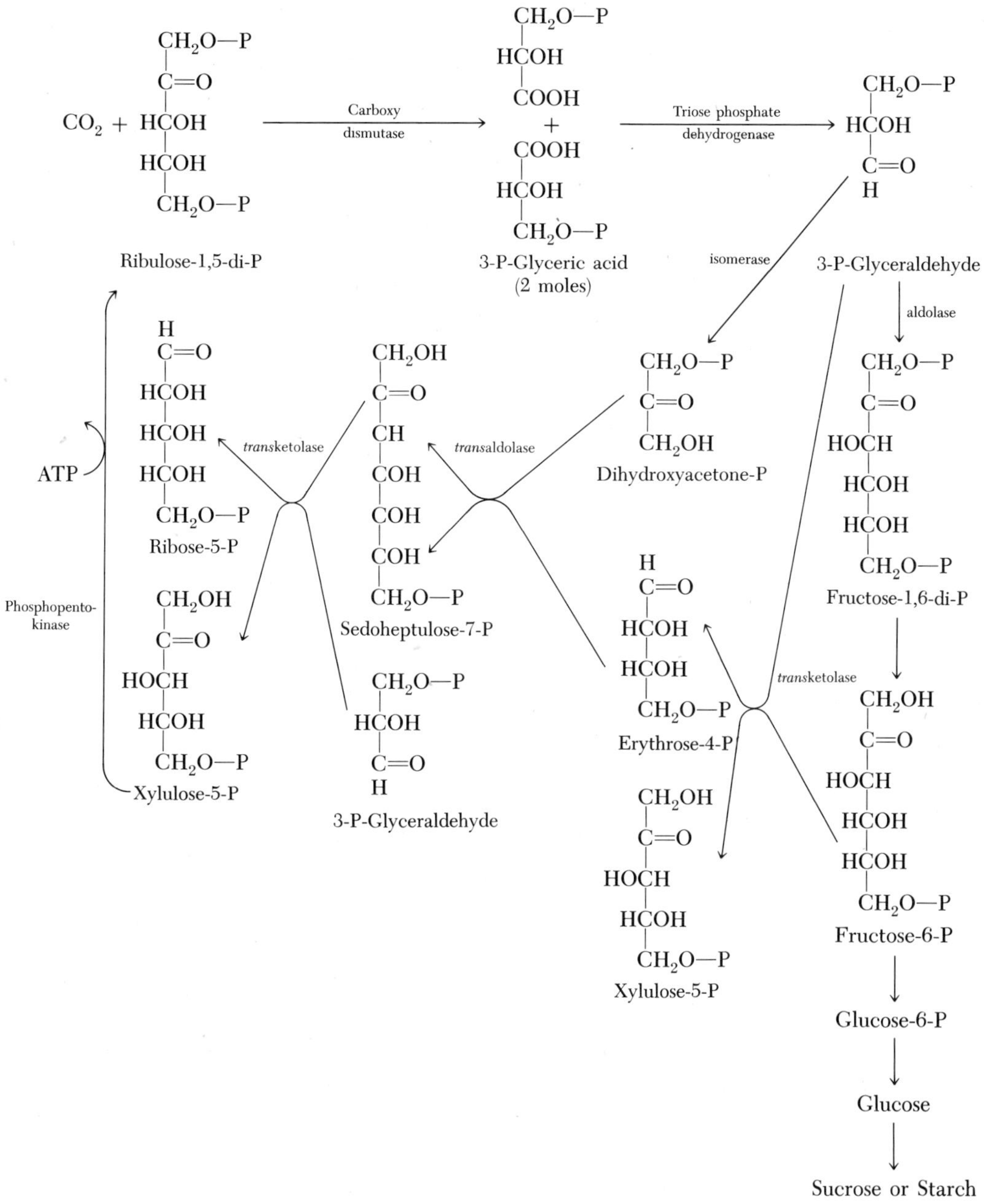

CO_2 + Ribulose-1,5-di-P ($CH_2O—P$, C=O, HCOH, HCOH, $CH_2O—P$) —Carboxy dismutase→ 3-P-Glyceric acid (2 moles) ($CH_2O—P$, HCOH, COOH + COOH, HCOH, $CH_2O—P$) —Triose phosphate dehydrogenase→ 3-P-Glyceraldehyde ($CH_2O—P$, HCOH, C=O, H)

3-P-Glyceraldehyde —isomerase→ Dihydroxyacetone-P ($CH_2O—P$, C=O, CH_2OH)

3-P-Glyceraldehyde —aldolase→ Fructose-1,6-di-P ($CH_2O—P$, C=O, HOCH, HCOH, HCOH, $CH_2O—P$) → Fructose-6-P (CH_2OH, C=O, HOCH, HCOH, HCOH, $CH_2O—P$) → Glucose-6-P → Glucose → Sucrose or Starch

Fructose-6-P + 3-P-Glyceraldehyde —*trans*ketolase→ Erythrose-4-P (H, C=O, HCOH, HCOH, $CH_2O—P$) + Xylulose-5-P (CH_2OH, C=O, HOCH, HCOH, $CH_2O—P$)

Erythrose-4-P + Dihydroxyacetone-P —*trans*aldolase→ Sedoheptulose-7-P (CH_2OH, C=O, CH, COH, COH, COH, $CH_2O—P$)

Sedoheptulose-7-P + 3-P-Glyceraldehyde ($CH_2O—P$, HCOH, C=O, H) —*trans*ketolase→ Ribose-5-P (H, C=O, HCOH, HCOH, HCOH, $CH_2O—P$) + Xylulose-5-P (CH_2OH, C=O, HOCH, HCOH, $CH_2O—P$)

Ribose-5-P, Xylulose-5-P —Phosphopento-kinase, ATP→ Ribulose-1,5-di-P

The relationship between the hexose monophosphate shunt and the dark reaction is apparent from the common compounds, enzymes and coenzymes, especially $NADPH_2$.

Some of the trioses and hexoses are also involved in the Embden-Meyerhof pathway, which illustrates the interlocking relationships that exist between the carbohydrate metabolic cycles.

MUSCLE CONTRACTION

Contraction of muscle fibers under anaerobic conditions leads to the formation of lactic acid and eventually to muscle fatigue. Muscle glycogen disappears during this process. In the presence of oxygen, the muscle regains its glycogen, loses lactic acid, and recovers its ability to contract. The high-energy compound **creatine phosphate** also changes form during contraction and subsequent recovery. The process of glycolysis supplies ATP, and the creatine phosphate and ATP join forces in muscular contraction.

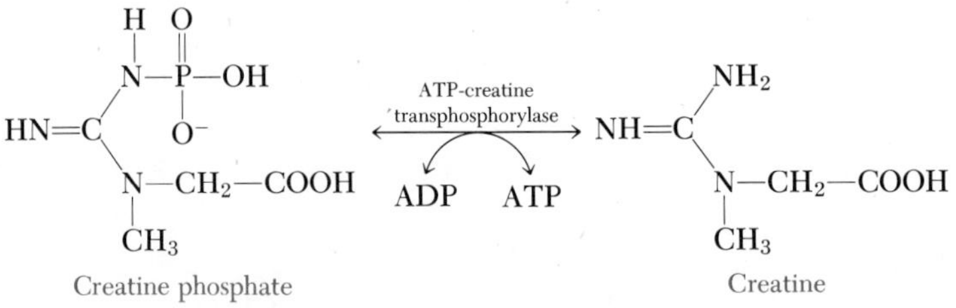

$$HN{=}C(-NH-PO(OH)O^-)(-N(CH_3)-CH_2-COOH) \xrightleftharpoons[ADP \;\rightleftharpoons\; ATP]{\text{ATP-creatine transphosphorylase}} NH{=}C(-NH_2)(-N(CH_3)-CH_2-COOH)$$

Creatine phosphate — Creatine

The straight and curved arrows with heads on each end denote reversible reactions, and you may recall that a direct linkage between nitrogen and phosphorus, as in creatine phosphate, denotes a high-energy compound. The reaction is readily reversible, and the muscle continues to contract as long as creatine phosphate is present.

During recovery, when more ATP is formed from glycolysis, the creatine phosphate is regenerated. The ATP is the direct source of energy for muscular work, and the function of creatine phosphate and glycolysis is to supply the ATP. Since there is only a small amount of ATP in the muscle at any instant, the supply of ATP needed for muscular work is obtained from creatine phosphate, the process of glycolysis, and the recovery of muscle glycogen. Thus muscular contraction is dependent on the cooperative action of several systems in carbohydrate metabolism.

IMPORTANT TERMS AND CONCEPTS

absorption
ATP formation
blood sugar level
chlorophyll
cyclic-3′,5′-AMP
dark reaction
digestion
Embden-Meyerhof pathway
epinephrine
glucose-6-phosphate
glycogenesis
glycogenolysis
glycolysis
hexose monophosphate shunt
insulin
Krebs cycle
lactic acid cycle
light reaction
phosphorylase
photosynthesis

QUESTIONS

1. Outline the process of digestion and absorption of carbohydrates in the gastrointestinal tract.

2. What is the normal fasting level of blood glucose? What values of blood glucose would be considered hypoglycemic? hyperglycemic?

3. Discuss the factors involved in counteracting the normal hyperglycemia that occurs after a meal.

4. Discuss the role of insulin in the control of the normal blood sugar level.

5. List the hormones other than insulin involved in the control of the blood sugar level. Discuss the function of one of these hormones.

6. Describe the process of glycogenesis.

7. What is cyclic-3′,5′-AMP? What role does it play in glycogenesis?

8. How do the reactions of the lactic acid cycle explain the fate of the lactic acid formed by the process of glycolysis? Explain.

9. Outline the essential reactions in the Embden-Meyerhof pathway of glycolysis.

10. How many moles of ATP per glucose molecule are required to run the Embden-Meyerhof pathway, and how many moles are produced?

11. Outline the essential reactions in the aerobic Krebs cycle.

12. One turn of the Krebs cycle will yield how many moles of ATP? Why is this number so large compared to the Embden-Meyerhof pathway?

13. Outline the essential reactions in the hexose monophosphate shunt or pentose phosphate pathway.

14. What are the major products of the hexose monophosphate shunt? How are they used in other pathways and cycles?

15. What is chlorophyll and what is its function in the process of photophosphorylation?

16. Explain how any three compounds formed in the CO_2 fixation scheme in photosynthesis could enter the reactions of the Embden-Meyerhof or Krebs cycle.

17. Outline the essential reactions in the light reaction of photosynthesis.

18. Outline the essential reactions in the dark reaction of photosynthesis.

19. What is the relation between ATP and creatine phosphate in muscular contraction?

SUGGESTED READING

Arnon: The Role of Light in Photosynthesis. Scientific American, Vol. 203, No. 5, p. 104, 1960.
Bergen: Tracer Isotopes in Biochemistry. Journal of Chemical Education, Vol. 29, p. 84, 1952.
Calvin and Bassham: The Photosynthesis of Carbon Compounds. New York, W. A. Benjamin Inc., 1962.
Heidt: The Path of Oxygen from Water to Molecular Oxygen. Journal of Chemical Education, Vol. 43, p. 623, 1966.
Horecker: Pathways of Carbohydrate Metabolism and Their Physiological Significance. Journal of Chemical Education, Vol. 42, p. 244, 1965.
Lehninger: Energy Transformation in the Cell. Scientific American, Vol. 202, No. 5, p. 102, 1960.
Park: Advances in Photosynthesis. Journal of Chemical Education, Vol. 39, p. 424, 1962.
Smith and York: Stereochemistry of the Citric Acid Cycle. Journal of Chemical Education, Vol. 47, p. 588, 1970.

LIPID METABOLISM

CHAPTER 21

The *objectives* of this chapter are to enable the student to:

1. Describe the process of digestion and absorption of dietary fat.
2. Recognize the normal blood lipids and their role in metabolism and storage of fat.
3. Describe the reactions involved in the oxidation of fatty acids.
4. Account for the total number of moles of ATP formed in the oxidation of a fatty acid.
5. Outline the reactions involved in the synthesis of fatty acids.
6. Describe the reactions that take place when ketone bodies are formed in the liver.
7. Discuss the synthesis of cholesterol and its relationship to atherosclerosis.
8. Describe the correlation between carbohydrate and lipid metabolism.

In the preceding chapter it was pointed out that the energy for many of the body activities is derived from the metabolism of carbohydrates. Fat stored in the fat depots of the body also represents a rich source of energy, especially since the caloric value of fats is more than twice that of carbohydrate or protein. The major stores of foodstuff in the body are therefore carbohydrate or fat in nature, and the body obtains most of its required energy from the oxidation of carbohydrates or fats.

Although the major energy source is fat, the metabolism of the lipids also involves phospholipids, glycolipids, and sterols. These latter substances are not stored in the fat depots but are essential constituents of tissues that play a role in fat transport and in many cellular metabolic reactions. These lipid derivatives also function as components of cell membranes, nerve tissue, membranes of subcellular particles such as microsomes and mitochondria, and chloroplasts in green leaves.

Unsaturated fatty acids such as **linoleic** and **linolenic** are essential components of cellular lipids that must be obtained in the diet, since they cannot be synthesized by the body. Cholesterol can be readily synthesized by the tissues and is currently a topic of considerable controversy, since there may be a relation between dietary cholesterol, blood cholesterol levels, and atherosclerosis.

DIGESTION

In the normal adult, little or no digestion of fat occurs as the food passes through the mouth and the stomach. When fat enters the duodenum, the gastrointestinal tract hormone **cholecystokinin** is secreted and is carried by the blood to the gallbladder, where it stimulates that organ to empty its bile into the small intestine. Bile acids and bile salts are good detergents and emulsify fats for digestion by **pancreatic lipase.** Another hormone that is secreted when the chyme enters the duodenum is **secretin.** This hormone enters the circulation and stimulates the pancreas to release pancreatic juice into the intestine. The enzyme pancreatic lipase is activated by the bile salts and splits fat into fatty acids, glycerol, soaps, mono-, and diglycerides.

ABSORPTION

In the absorption process, as the end products of fat digestion pass through the intestinal mucosa they are reconverted into triglycerides, which then enter the lymph circulation. Bile salts are essential in absorption, both because of their effect on the solubility of fatty acids and because of direct involvement in the absorption process. Phospholipids are split into their component structures by digestive enzymes and are also resynthesized in the intestinal mucosa during absorption. Short-chain fatty acids may be directly absorbed into the blood and carried to the liver.

BLOOD LIPIDS

The blood lipids to a certain extent parallel the behavior of the blood sugar. Their concentration in the blood increases after a meal and the level is returned to normal by processes of storage, oxidation, and excretion.

The lipids of the blood are constantly changing in concentration as lipids are added by absorption from the intestine, by synthesis, and by removal from the fat depots; they are removed by storage in the fat depots, oxidation to produce energy, synthesis to produce tissue components, and excretion into the intestine. The **normal fasting level** of blood lipids is usually measured in the plasma. Average values for young adults are as follows:

	mg/100 ml
Total lipids	510
Triglycerides	150
Phospholipids	200
Total cholesterol	160

The triglycerides, phospholipids, and cholesterol in the plasma are combined with protein as lipoprotein complexes. These **lipoproteins** are bound to the α- and β-globulin fractions of the plasma proteins and are transported in this form. A small amount of **nonesterified fatty acids (NEFA)** is always present in the blood and is bound to the albumin fraction of the plasma for transportation. These free fatty acids are thought to be the most active form of the lipids involved in metabolism. Their concentration is affected by the mobilization of fat from fat depots and by the action of several hormones.

FAT STORAGE

Fats may be removed from the blood stream by storage in the various fat depots. When fat is stored under the skin, it is usually called **adipose tissue.** However, considerable quantities of fat may be stored around such organs as the kidneys, heart, lungs, and spleen. This type of depot fat acts as a support for these organs and helps to protect them from injury. Recent studies employing the electron microscope reveal two major types of storage fat. One type is composed almost entirely of fat globules and has the characteristics of a storage depot. The second type contains many cells and a more extensive blood circulation, and is metabolically active, converting glycogen to fat and releasing fatty acids to other tissues as energy sources.

Obesity. Obesity is the condition in which excessive amounts of fat are stored in the fat depots. In a small percentage of cases, obesity is due to a disorder of certain endocrine glands, but as a general rule it results from eating more food than the body requires. Most of the food consumed by an adult is used to produce energy, and food in excess of that necessary to fulfill the energy requirements of the body is stored as fat. Thin people generally are more active than fat people and are able to eat larger amounts of food without putting on weight.

Many people apparently eat all they want and yet maintain a fairly constant weight over long periods of time. This weight control may be due to the appetite, which is abnormally increased in people that are gaining weight and decreased in those who are losing weight. Recent investigations, however, have cast doubt on the simple explanation of overeating as the only factor responsible for obesity. Apparently there are some individuals who can maintain obesity or increase their weight on a low calorie diet. If they attempt to lose weight by decreasing the intake of food, their rate of metabolism decreases and they require fewer calories to maintain their activity. This combination of *endocrine balance, rate or metabolism,* and *difference in requirement for calories* is as yet not completely understood.

THE SYNTHESIS OF TISSUE LIPIDS

Lipids such as **phospholipids, glycolipids,** and **sterols** are essential constituents of cells, protoplasm, and tissues in various parts of the body. They are also involved in specialized functions, i.e., blood clotting mechanisms and in transportation of lipids in the blood. The adipose tissue that is stored around organs of the body does not contain the same proportion of saturated or unsaturated fatty acids as the food fat and therefore must also be synthesized. The most important organ in the body concerned with lipid synthesis is the liver. It is able to synthesize phospholipids and cholesterol and to modify all blood fats by lengthening or shortening, and saturating or unsaturating, the fatty acid chains.

Lecithin is used in transporting fats to the various tissues and may be involved in the oxidation of fats. Another essential phospholipid is **cephalin,** which is a vital factor in the clotting of blood. Special fats and oils in the body such as milk fat, various sterols, the natural oil of the scalp, and the wax of the ear are examples of lipids synthesized from the fats of the food.

LIPOLYSIS

The role of hormones and cyclic-3′,5′-AMP in glycogenolysis (p. 320) has already been described. **Lipolysis,** the hydrolysis of triglycerides to free fatty acids, is also under the control of hormones and cyclic-3′,5′-AMP.

$$\text{Nonactivated lipase} + \text{ATP} \xrightarrow[\text{Cyclic-3}',5'\text{-AMP}]{\text{Protein kinase}} \text{Activated lipase} + \text{ADP}$$

$$\text{Activated lipase} + \text{Triglycerides} \longrightarrow \text{Free fatty acids} + \text{Glycerol}$$

Hormones such as epinephrine and glucagon stimulate the production of cyclic-3′,5′-AMP and therefore the process of lipolysis, while prostaglandins (p. 227) depress the levels of cyclic-3′,5′-AMP and decrease the rate of lipolysis.

OXIDATION OF FATTY ACIDS

Fatty acids that arise from the breakdown of any lipid, but especially from fats, are oxidized completely to form CO_2, water, and energy. The glycerol portion of fats is phosphorylated in the liver to form glycerophosphate, which is then oxidized to dihydroxyacetone phosphate.

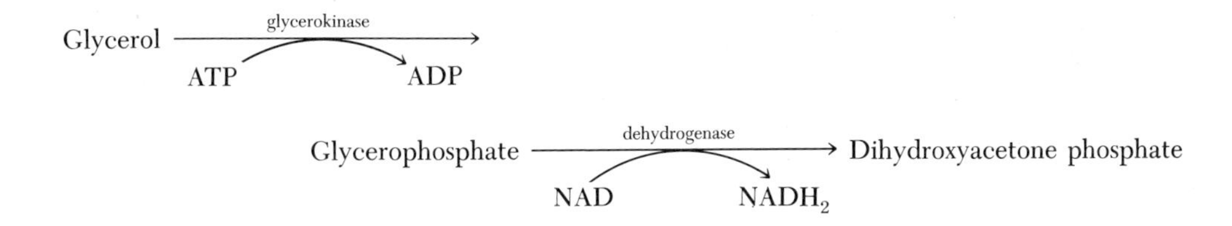

Both of these products can enter the Embden-Meyerhof pathway of carbohydrate metabolism.

The oxidation of fatty acids occurs in a series of reactions that require several enzymes and cofactors, with the production of acetyl coenzyme A. The acetyl CoA molecules then enter the Krebs cycle to form CO_2, H_2O, and energy. Early research by Knoop in 1904 established the fact that fatty acids were oxidized on the beta-carbon atom with the subsequent splitting off of two carbon fragments. In his **theory of beta-oxidation** he stated that acetic acid was split off in each stage of the process that reduced an 18-carbon fatty acid to a two-carbon acid.

In the past few years the detailed reactions, with their enzymes and cofactors, have been worked out, and Knoop's theory has been confirmed. Instead of acetic acid, the key compound in the reactions is acetyl CoA. Five reactions are involved in the conversion of a long chain fatty acid into a CoA derivative with two less carbon atoms and a molecule of acetyl CoA. These reactions are outlined in the scheme that is shown on the next page.

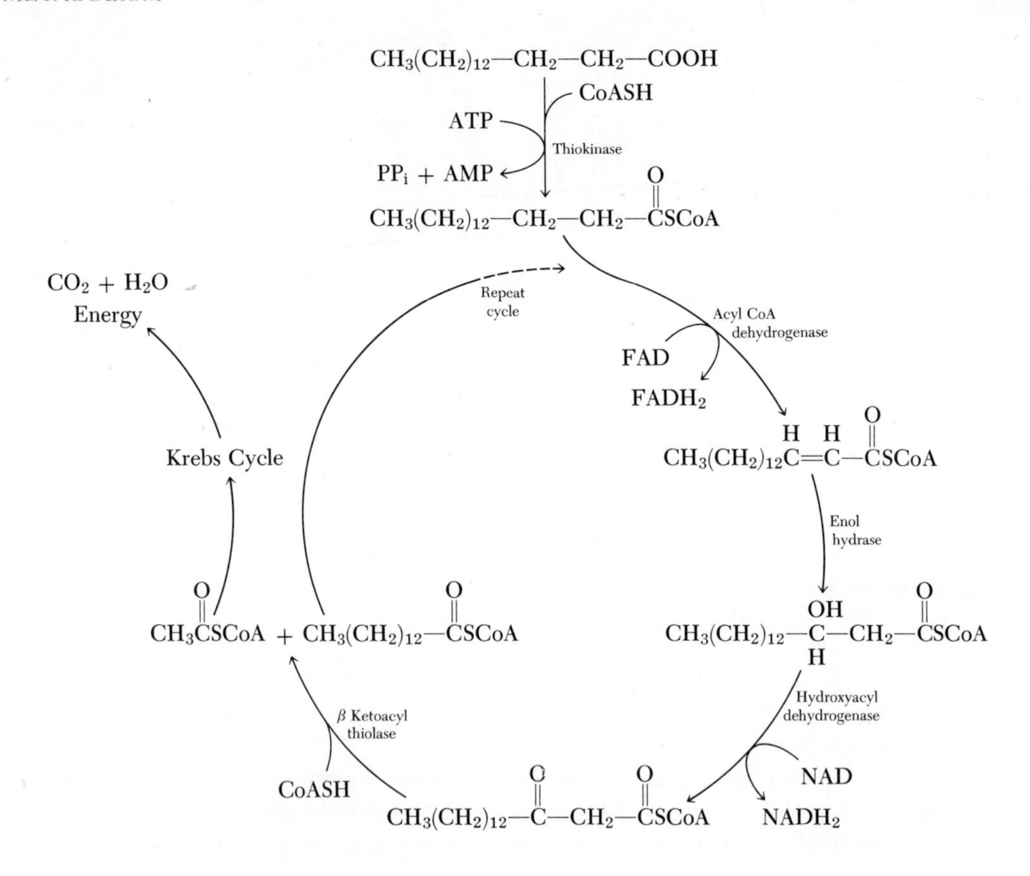

The first reaction initiates the series and involves the activation of a fatty acid molecule by conversion into a coenzyme A derivative. A dehydrogenase enzyme, with FAD as the coenzyme, desaturates the fatty acid; then a hydration is catalyzed by an enol hydrase. The hydroxyl group on the β-carbon atom is oxidized by a dehydrogenase with NAD as a coenzyme. The oxidized derivative plus coenzyme A is split into a fatty acid molecule with 2 less carbons, and acetyl CoA is formed. The acetyl CoA enters the Krebs cycle to form CO_2 and H_2O, plus energy. The new fatty-acid coenzyme A derivative does not have to be activated, but directly re-enters the cycle and again loses an acetyl CoA molecule. *Palmitic acid would require seven turns of the cycle to form 8 acetyl CoA moles.*

During the oxidation of palmitic acid, 7 $FADH_2$ and 7 $NADH_2$ moles would be formed. When these compounds enter the electron transport chain, they would form ATP as shown:

$$
\begin{array}{rl}
7\,FADH_2 \rightarrow & 14\ ATP \\
7\,NADH_2 \rightarrow & 21\ ATP \\
 & 35\ ATP \\
 & -\ \underline{1\ ATP}\ \text{used in first reaction} \\
 & 34\ \text{Net ATP}
\end{array}
$$

In the seven turns of the cycle, 8 acetyl CoA moles are formed. As may be recalled from the oxidation of this compound in the Krebs cycle (p. 324), each mole of acetyl CoA will give rise to 12 moles of ATP. The acetyl CoA formed from the oxidation of palmitic acid will therefore account for the formation of $8 \times 12 = 96$ moles of ATP.

The sum of $34 + 96 = 130$ ATP for the complete oxidation of palmitic acid in the above scheme. The total combustion of palmitic acid yields 2338.0 kcal, and when compared to cellular oxidation (p. 325),

$$\frac{130 \times 8.0 \text{ kcal} \times 100}{2338.0} = 48\%$$

This represents a very efficient conservation of energy in the form of ATP molecules when palmitic acid is completely oxidized by the tissues. The previous discussion emphasizes the statement that food fat is an effective source of available energy. Also, a contributing factor to this efficient utilization is the fact that all the enzymes utilized in the β-oxidation scheme, the Krebs cycle, oxidative phosphorylation, and electron transport are found in the **mitochondria** of the cell.

SYNTHESIS OF FATTY ACIDS

The β-oxidation pathway in the mitochondria can be reversed to form fatty acid molecules, but this accounts for only a small percentage of the fatty acids synthesized in the tissues. The **cytoplasm** of the cell is the major site, and acetyl coenzyme A is the starting material, for the synthesis. Acetyl coenzyme A is carboxylated to form **malonyl coenzyme A** under the influence of acetyl CoA carboxylase in the presence of ATP and the vitamin biotin. An **enzyme-biotin complex** adds CO_2 with the help of ATP. Acetyl CoA then reacts with this complex to form malonyl CoA, as follows:

```
                          O                                                   O
                          ‖     O                                             ‖
                          C     ‖                                             C
                      HN     N—C—O⁻           COO⁻                        HN     NH
                       |     |                 |                           |     |
    O                 HC————CH                 |   O                      HC————CH
    ‖                  |     |                 |   ‖                       |     |
CH3C—SCoA + Enzyme—C        C—R         → HC—C—SCoA + Enzyme—C        C—R
                    H    S   H                 H                         H   S   H

 Acetyl CoA          Biotin-enzyme-CO2          Malonyl CoA
                         complex
```

Malonyl CoA and acetyl CoA then form complexes with a multienzyme system called *fatty acid synthetase,* which includes an acyl carrier protein (ACP) that binds acyl intermediates during the formation of long-chain fatty acids. These two complexes then condense to form acetoacetyl-S-ACP, which is reduced to β-hydroxybutyryl-S-ACP with the assistance of NADPH, followed by the loss of a molecule of water to form an α,β-unsaturated-S-ACP. The unsaturated compound is reduced to butyryl-S-ACP, which combines with another molecule of malonyl-S-ACP to continue elongation of the chain.

SYNTHESIS OF TRIGLYCERIDES

Triglycerides are synthesized in the tissues from glycerol and fatty acids in activated forms. The active form of glycerol is **glycerophosphate,** which reacts with two fatty acid

CoA derivatives to form a **diglyceride,** which then reacts with another mole of fatty acid CoA to form a **triglyceride.** An outline of the process can be illustrated by the following scheme:

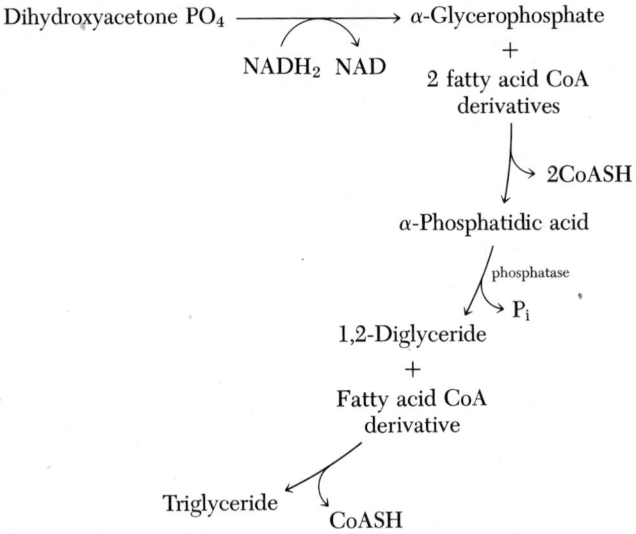

$$\text{Dihydroxyacetone } PO_4 \xrightarrow[NADH_2 \;\; NAD]{} \alpha\text{-Glycerophosphate} + \text{2 fatty acid CoA derivatives}$$

$$\xrightarrow{} 2CoASH$$

α-Phosphatidic acid

$$\xrightarrow{\text{phosphatase}} P_i$$

1,2-Diglyceride + Fatty acid CoA derivative

$$\longrightarrow \text{Triglyceride} + \text{CoASH}$$

FORMATION OF KETONE BODIES

The ketone, or acetone, bodies consist of **acetoacetic acid, β-hydroxybutyric acid,** and **acetone.** In a normal individual they are present in the blood in small amounts, averaging about 0.5 mg per 100 ml. Also, about 100 mg of ketone bodies is excreted per day in the urine. This low concentration in the blood and the small amount excreted in the urine are insignificant. But large amounts are present in the blood and urine during starvation and in the condition of diabetus mellitus. In general, any condition that results in a restriction of carbohydrate metabolism, with a subsequent increase in fat metabolism to supply the energy requirements of the body, will produce an increased formation of ketone bodies. This condition is called **ketosis.**

The precursor of the ketone bodies is acetoacetic acid which is formed in the liver from acetoacetyl CoA, a normal intermediate in the beta-oxidation of fatty acids. It may also be formed by the condensation of two molecules of acetyl CoA. Both methods of formation can be represented in the normal reversible reaction as follows:

$$\underset{\text{Acetyl CoA}}{2\ CH_3\overset{O}{\overset{\|}{C}}SCoA} \overset{\text{thiolase}}{\rightleftharpoons} \underset{\text{Acetoacetyl CoA}}{CH_3\overset{O}{\overset{\|}{C}}CH_2\overset{O}{\overset{\|}{C}}SCoA} + CoASH$$

The liver contains a deacylase enzyme which readily converts acetoacetyl CoA to the free acid.

$$\underset{\text{Acetoacetyl CoA}}{CH_3\overset{O}{\overset{\|}{C}}CH_2\overset{O}{\overset{\|}{C}}SCoA} + H_2O \xrightarrow{\text{deacylase}} \underset{\text{Acetoacetic acid}}{CH_3\overset{O}{\overset{\|}{C}}CH_2COOH} + CoASH$$

The other ketone bodies are formed from acetoacetic acid; acetone by decarboxylation and β-hydroxy-butyric acid by the action of a specific enzyme, as shown in the accompanying diagram.

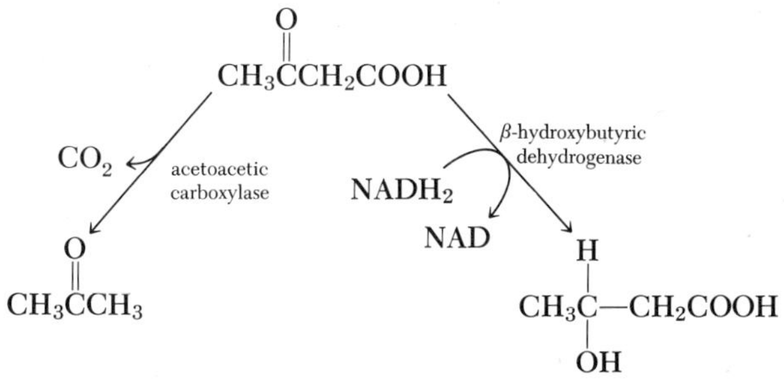

$$CH_3\overset{O}{\overset{\|}{C}}CH_2COOH$$

acetoacetic carboxylase ($\rightarrow CO_2$) · β-hydroxybutyric dehydrogenase ($NADH_2 \rightarrow NAD$)

$$CH_3\overset{O}{\overset{\|}{C}}CH_3 \qquad\qquad CH_3\overset{H}{\underset{OH}{C}}—CH_2COOH$$

Acetone β-Hydroxybutyric acid

PHOSPHOLIPID METABOLISM

Knowledge concerning the metabolism of the phospholipids is incomplete, although they are known to serve many important functions in the body. Because their molecules are more strongly dissociated than any of the other lipids, they tend to be more soluble in water, to lower surface tension at oil-water interfaces, and to be involved in the electron transport system in the tissues. They would have a tendency to concentrate at cell membranes, and are probably involved in the transport mechanisms for carrying fatty acids and lipids across the intestinal barrier and from the liver and fat depots to other body tissues. Further evidence for their function in transporting lipids is found in their presence in the lipoproteins of the plasma. Phospholipids are essential components of the blood clotting mechanism, and sphingomyelin is one of the principal components of the myelin sheath of nerves.

Dietary phospholipids are probably broken into their constituents by enzymes in the gastrointestinal tract. The synthesis of most of the phospholipids has been established in recent years by the use of isotopes to tag precursors and intermediate compounds. For example, the synthesis of lecithin involves the reaction of cytidine diphosphate choline and a 1,2-diglyceride. The phosphatidyl ethanolamines, or cephalins, are synthesized by similar reactions, and the sphingomyelins are synthesized by the reaction of N-acylsphingosine with **cytidine diphosphate choline.** Cytidine diphosphate choline (CDP-choline) has the structure:

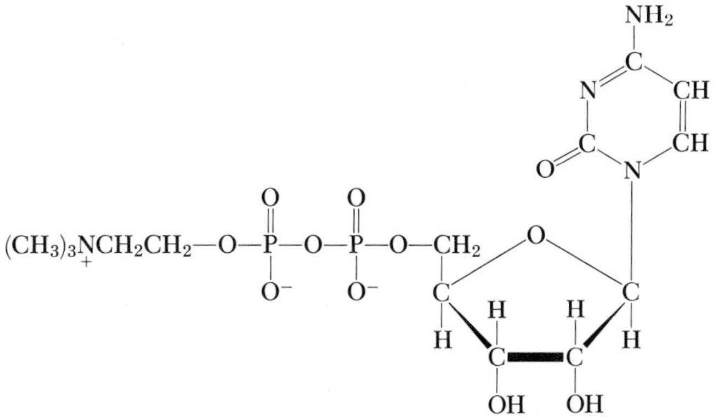

$$(CH_3)_3\overset{+}{N}CH_2CH_2—O—\underset{O^-}{\overset{O}{P}}—O—\underset{O^-}{\overset{O}{P}}—O—CH_2—\text{ribose}(\text{OH},\text{OH})—\text{cytosine}(NH_2)$$

Cytidine diphosphate choline (CDP-choline)

The synthesis of lecithin may be outlined as illustrated utilizing the 1,2-diglyceride as formed in triglyceride synthesis.

$$\text{1,2-Diglyceride} + \text{CDP-choline} \xrightarrow[\text{enzyme}]{\text{transferase}} \text{Lecithin} + \text{Cytidine monophosphate}$$

STEROL METABOLISM

The metabolism of sterols is mainly concerned with cholesterol and its derivatives. The **synthesis of cholesterol** and its relation to the other steroids of the body has been the subject of considerable research. Using either stable or radioactive isotopes, it has been shown that cholesterol can be synthesized from two-carbon compounds such as acetyl CoA. It can also be synthesized from acetoacetyl CoA and other intermediates. Two of the intermediate cholesterol precursors are β-hydroxy-β-methylglutaryl-SCoA and mevalonic acid, as shown in the following scheme:

Acetyl CoA + acetoacetyl CoA ⟶ β-hydroxy-β-methylglutaryl-SCoA

↓ → CoASH

$$HOOC{-}CH_2{-}\overset{\displaystyle H}{\underset{\displaystyle OH}{C}}{-}CH_2{-}CH_2OH$$

↓

Mevalonic acid

↓

Isopentyl pyrophosphate ⟵ 5-Phosphomevalonic acid

+

Dimethylallyl pyrophosphate

↓

Geranyl pyrophosphate

↓

Farnesyl pyrophosphate ⟶ Squalene

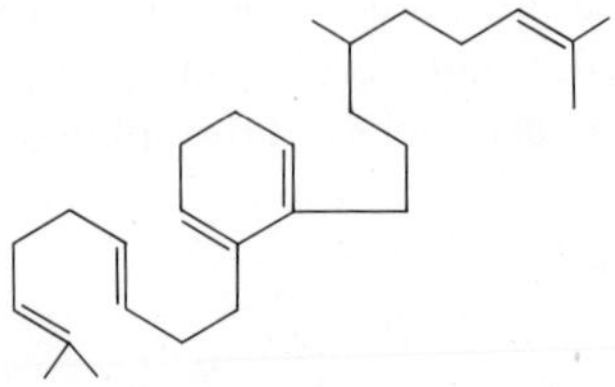

↓

Lanosterol

↓

Zymosterol

↓

Cholesterol ⟵ Demosterol

HO

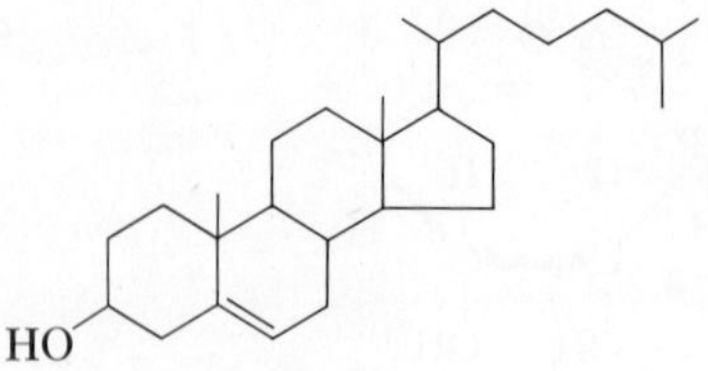

Cholesterol

Although the synthesis of cholesterol occurs in many tissues in the body, the liver is the main site of cholesterol formation.

Cholesterol is a key compound in the synthesis of essential steroids such as bile acids, sex hormones, adrenal cortical hormones, and vitamin D. Not only is cholesterol converted to bile acids by the liver, but it is also excreted as such in the bile. As mentioned previously, cholesterol in the bile can give rise to gall stones by accumulating on insoluble objects or particles. The concentration of cholesterol in the blood is apparently dependent on the dietary intake of sterols and neutral fats, and the synthesis of cholesterol by the liver. The normal level in the blood gradually increases with age and ranges from 150 to 200 mg/100 ml. Blood cholesterol levels are often determined in patients to assess their cholesterol status. Many methods have been devised for this determination, and many of them are modifications of the Liebermann-Burchard reaction described in Chapter 14. If the cholesterol level in the blood is maintained at an abnormally high concentration, such as 200 to 300 mg/100 ml, deposition of cholesterol plaques may occur in the aorta and lesser arteries. This condition, known as **atherosclerosis** or **arteriosclerosis,** is seen in older persons and often results in circulatory or heart failure. Considerable research effort is being directed at this problem in an attempt to reduce the cholesterol level in the blood of these patients and thus alleviate the symptoms of the disease.

CORRELATION OF CARBOHYDRATE AND FAT METABOLISM

From a nutritional standpoint it has long been apparent that carbohydrate can be converted into fat in the body. When glucose tagged with ^{14}C was fed to animals, the fatty acids of liver and other tissue fat were found to be labeled with ^{14}C. The conversion of fat to carbohydrate has long been open to question. The glycerol portion of fat is closely related to the three-carbon intermediates of carbohydrate metabolism, but it has been more difficult to demonstrate a direct relation between fatty acids and glucose. Since the role of acetyl CoA has been established in both carbohydrate and fat metabolism, it is apparent that the acetyl CoA from fatty acid oxidation can enter the Krebs cycle in the same fashion as this compound formed from pyruvic acid. More recently it has been shown that ^{14}C labeled fatty acids are converted to ^{14}C labeled glucose in a diabetic animal. The correlation between carbohydrate and fat metabolism in the body may be represented in the scheme shown on the next page.

IMPORTANT TERMS AND CONCEPTS

absorption
acetyl CoA
adipose tissue
ATP formation
blood lipids
cholesterol synthesis
digestion
fatty acid oxidation
fatty acid synthesis
ketone bodies

QUESTIONS

1. Briefly describe the digestion and absorption of dietary fat.

2. List the important lipids in a normal individual's blood and their approximate concentration.

3. What are chylomicrons? lipoproteins?

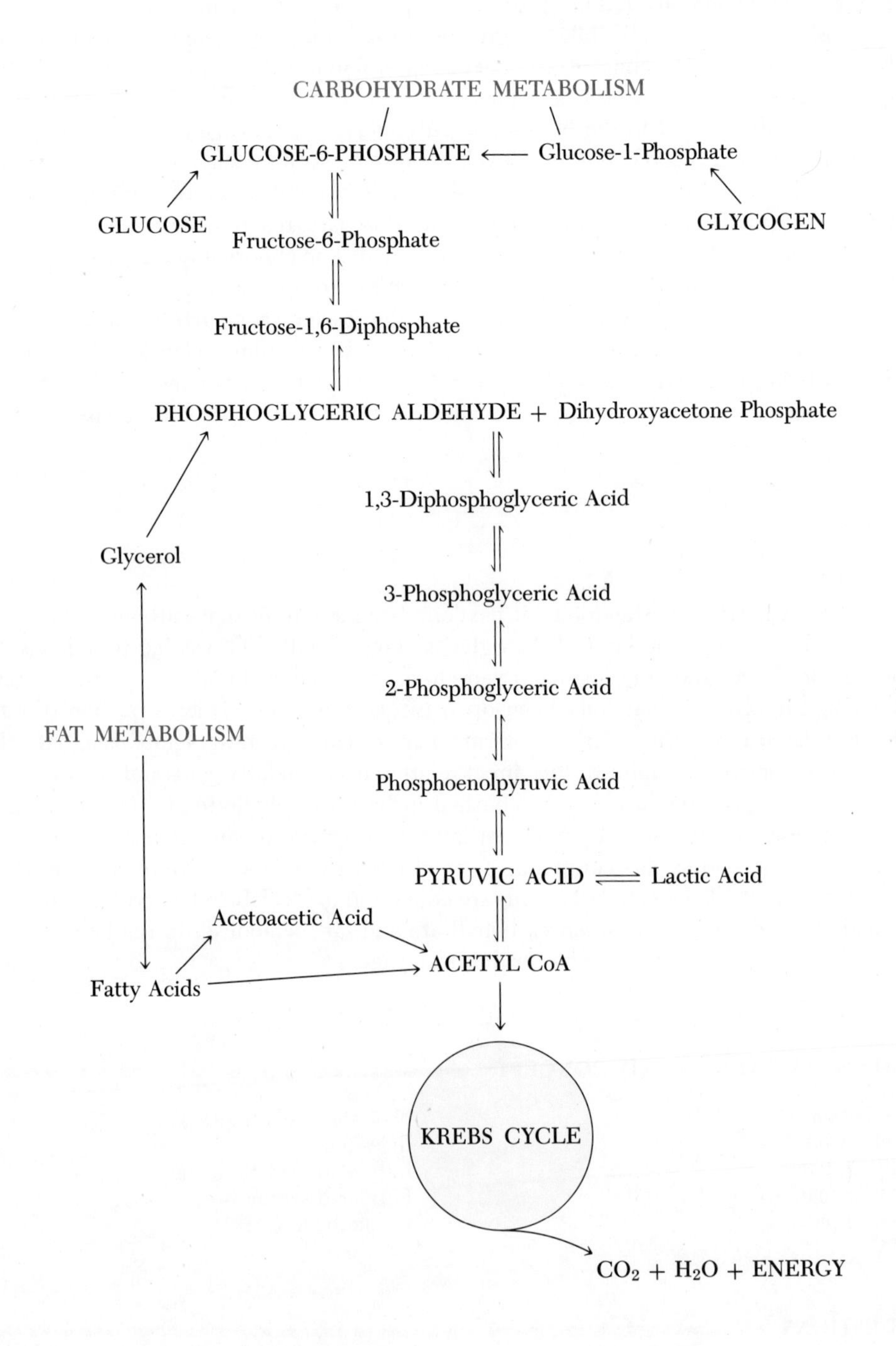
CARBOHYDRATE METABOLISM
GLUCOSE-6-PHOSPHATE
Glucose-1-Phosphate
GLUCOSE
GLYCOGEN
Fructose-6-Phosphate
Fructose-1,6-Diphosphate
PHOSPHOGLYCERIC ALDEHYDE + Dihydroxyacetone Phosphate
1,3-Diphosphoglyceric Acid
Glycerol
3-Phosphoglyceric Acid
2-Phosphoglyceric Acid
FAT METABOLISM
Phosphoenolpyruvic Acid
PYRUVIC ACID
Lactic Acid
Acetoacetic Acid
ACETYL CoA
Fatty Acids
KREBS CYCLE
$CO_2 + H_2O$ + ENERGY

4. Discuss some of the advantages and disadvantages of a generous supply of adipose tissue.

5. Outline the essential reactions in the scheme for oxidation of fatty acids.

6. How do you account for the large number of moles of ATP formed in the oxidation of fatty acids?

7. Outline the reactions involved in the synthesis of fatty acids.

8. Describe the process of synthesis of triglycerides.

9. Discuss two reactions that may take place in the liver for the formation of acetoacetic acid.

10. What is ketosis? Explain the cause of this condition in the body.

11. Show in outline form the compounds involved in the synthesis of cholesterol.

12. Name four important compounds in the body that are synthesized from cholesterol.

13. Why is atherosclerosis receiving so much attention in our society?

SUGGESTED READING

Bergen: Tracer Isotopes in Biochemistry. Journal of Chemical Education, Vol. 29, p. 84, 1952.
Gibson: The Biosynthesis of Fatty Acids. Journal of Chemical Education, Vol. 42, p. 236, 1965.
Green: Metabolism of Fats. Scientific American, Vol. 190, No. 1, p. 32, 1954.
Green: The Synthesis of Fat. Scientific American, Vol. 202, No. 2, p. 46, 1960.
Pike: Prostaglandins. Scientific American, Vol. 225, No. 5, p. 85, 1971.
Spain: Atherosclerosis. Scientific American, Vol. 215, No. 2, p. 48, 1966.

PROTEIN METABOLISM

CHAPTER 22

The *objectives* of this chapter are to enable the student to:

1. Describe the process of digestion and absorption of dietary protein.
2. Recognize the relationship between the amino acid pool and the dynamic state of tissue proteins.
3. Describe the activation of amino acids and their combination with t-RNA prior to protein synthesis.
4. Illustrate the processes of transcription and translation in protein synthesis.
5. Explain the relationship between codons and anticodons in protein synthesis.
6. Describe the function of the operon in the synthesis of enzyme proteins.
7. Recognize the difference between deamination and transamination.
8. Outline the essential reactions in the urea cycle.
9. Recognize the reactions involved in purine and pyrimidine metabolism.
10. Describe the correlation between carbohydrate, lipid, and protein metabolism.

The metabolism of proteins is concerned with a large variety of complex molecules. Tissue proteins of various species of animals, plants, and microorganisms all have specific structures and compositions. Protein enzymes and hormones, plasma proteins, and the protein of hemoglobin and of various nucleoproteins represent other types of proteins. **Anabolism,** or the synthesis of new proteins for growth and development, involves the building of different amino acids into the proper sequences and spatial arrangements to produce specific protein molecules. The process of **catabolism** of proteins to produce energy involves many general metabolic reactions and many that are specific for the metabolism of each of the twenty different amino acids.

DIGESTION

The digestion of the large, complex food protein molecules starts in the stomach. **Pepsin** is the principal enzyme in gastric juice, and it catalyzes the hydrolysis of large protein molecules into smaller, more soluble molecules of **proteoses** and **peptones.** The optimum pH of pepsin is 1.5 to 2; thus, it is ideally suited for the digestion of protein in normal stomach contents, of which the pH is 1.6 to 1.8. As the chyme passes into the small intestine, **trypsin, chymotrypsin, carboxypeptidase, aminopeptidase,** and **di-**

peptidase act on native proteins, proteoses and peptones, and polypeptides. The molecules are gradually split into **amino acids,** which are the end products of protein digestion.

ABSORPTION

Amino acids are absorbed through the intestinal mucosa directly into the blood stream by an active process that requires energy and enzymes. Each amino acid, like the monosaccharides, has a different rate of absorption, and there may be a different mechanism for types of amino acids such as acidic, basic, and neutral, L forms and D forms. After absorption, the amino acids are carried by the portal circulation to the liver and subsequently to all the tissues of the body. The liver removes some amino acids for its requirements and adds to the blood those it has synthesized. Other tissues add amino acids to the blood as their proteins undergo catabolism, or breakdown. In contrast to carbohydrate and fat metabolism, there are no storage depots for proteins or amino acids. The increased concentration of amino acids that occurs from the process of absorption, synthesis, or catabolism represents a temporary pool of amino acids which may be used for metabolic purposes. This pool of amino acids is available to all tissues and may be synthesized into new tissue proteins, blood proteins, hormones, enzymes, or nonprotein nitrogenous substances such as creatine and glutathione. The relationship that exists between the **amino acid pool** and protein metabolism in general may be represented as shown in the accompanying diagram.

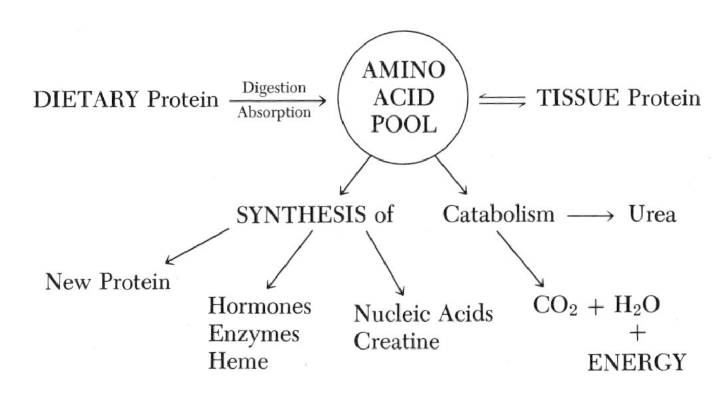

THE DYNAMIC STATE OF BODY PROTEIN

Until the late 1930's it was believed that the body proteins of the adult human were stable molecules and that the majority of the amino acids from the diet were catabolized to produce energy. A small proportion was thought to be used for maintenance and repair of the existing tissue proteins. When isotopes became available, Schoenheimer and his associates demonstrated that tissue proteins exist in a *dynamic state of equilibrium.* When the nitrogen of an amino acid was labeled with ^{15}N and incorporated in the diet of an animal, about 50 per cent of the ^{15}N was found in the tissues of the animal, and a greater percentage was found in the nitrogen of amino acids other than that specifically fed. This indicated that the amino acids of tissue proteins were constantly changing places with those in the amino acid pool, and that the body proteins were extremely labile molecules.

More recent research using isotopically labeled amino acids has shown that tissue proteins vary considerably in their rate of turnover of amino acid molecules. The **turnover rate** represents the amount of protein synthesized or degraded per unit time, and the

turnover time is usually expressed as the half-life of a protein in the tissues. Liver and plasma proteins have turnover times (a half-life) of 2 to 10 days, in contrast to 180 days for muscle protein and 1000 days for some collagen proteins. Muscle and connective tissue proteins appear to have a very prolonged turnover compared to liver and plasma proteins, which are rapidly synthesized from the amino acids in the amino acid pool. The concept of the dynamic state of body protein requires modification in view of the individuality of specific proteins.

THE SYNTHESIS OF PROTEIN

The process of synthesis of protein is always occurring in the body, especially in those tissues with a rapid turnover rate. A growing child or animal is continually building new tissue and therefore makes the greatest demand on the amino acid pool. The individual amino acids required for protein synthesis are apparently sorted out by the body and used to construct specific protein molecules. Although the tissues, particularly, the liver, are able to synthesize some amino acids, others must be present in the diet to assure a complete and proper assortment for synthetic purposes.

Essential Amino Acids

The amino acids that cannot be synthesized by the body and must therefore be supplied by the dietary protein are called **essential amino acids.** If an essential amino acid is lacking in the diet, the body is unable to synthesize tissue protein. If this condition occurs for any length of time, a negative nitrogen balance will exist, and there will be a loss of weight, a lowered level of serum protein, and a marked edema. Extensive feeding experiments on laboratory rats have established the following amino acids as essential for growth:

Histidine	Isoleucine
Methionine	Leucine
Arginine	Lysine
Tryptophan	Valine
Threonine	Phenylalanine

From the studies of Rose and his associates at the University of Illinois on the amino acid requirements of man, it has been suggested that all of the above ten except histidine and possibly arginine are essential to maintain nitrogen balance. An individual is in **nitrogen balance** when the nitrogen excreted equals the nitrogen intake in a given period of time. A growing child or a patient recovering from a prolonged illness is in **positive nitrogen balance.** Starvation, a wasting disease, or a diet lacking sufficient amounts of essential amino acids can result in a **negative nitrogen balance.**

Many common dietary proteins are deficient in one or more of these essential amino acids. Gelatin, for example, lacks tryptophan and is therefore an **incomplete protein.** If gelatin were the sole source of protein in the diet, a growing child could eat large quantities of this protein every day without building new body tissues. Zein and gliadin, the prolamines of corn and wheat respectively, are deficient in lysine, and zein is also low in tryptophan. Although an incomplete protein will not support growth when it is the only protein in the diet, we seldom confine ourselves to the consumption of a single protein. In an ordinary mixed diet the essential amino acids are best supplied by proteins of animal origin, such as meat, eggs, milk, cheese, and fish. A daily intake of protein of 1 to 1.5 grams per kilogram of body weight is recommended as adequate for body needs.

Mechanism of Protein Synthesis

For many years the synthesis of proteins in the body was explained as a reversal of the action of hydrolyzing enzymes, causing the formation of peptide bonds instead of the splitting of the bonds. The mechanism by which the tissues assured the proper assortment of amino acids in a newly synthesized protein was not explained. Although complete details of the synthesis of tissue proteins with the proper concentration and sequence of amino acids have not been settled, knowledge is growing rapidly in this field. In fact, a very active field of research at the present time involves the mechanism of protein synthesis, the sequence of amino acids in the protein being synthesized, and the nature of the genetic code responsible for this sequence.

Protein synthesis is initiated by the activation of amino acids. This process occurs by the combination of the amino acid with ATP and an enzyme specific for the amino acid, with the splitting off of two molecules of phosphoric acid.

$$\underset{\text{Amino acid (AA)}}{\text{R—}\underset{\text{NH}_2}{\underset{|}{\text{CH}}}\text{—COOH}} \xrightarrow[\text{ATP} \quad \text{PP}_i]{\text{aminoacyl synthetase}} \underset{\text{(E-AMP-AA)}}{\underset{\text{Amino acyl AMP enzyme complex}}{\text{Enz-Adenine-Ribose—O—}\overset{\text{O}}{\overset{\|}{\text{P}}}\text{—O—}\overset{\text{O}}{\overset{\|}{\text{C}}}\text{—}\underset{\text{NH}_2}{\underset{|}{\text{CH}}}\text{—R}}}$$

The second step of the process involves the transfer of the activated amino acid to a **transfer RNA molecule.** The t-RNA molecules are small (mol. wt. 30,000) nucleic acid molecules with a terminal grouping of cytidylic-cytidylic-adenylic acid represented as s-RNA-C-C-A. The transfer of the activated amino acid is under the influence of the same enzyme that produced the activation and CTP, or cytidine triphosphate.

$$\underset{\substack{\text{(activated}\\\text{amino acid)}}}{\text{E-AMP-AA}} + \underset{\substack{\text{transfer}\\\text{RNA}}}{\text{t-RNA}} \longrightarrow \underset{\substack{\text{transfer RNA-}\\\text{amino acid}\\\text{complex}}}{\text{t-RNA-AA}} + \text{AMP} + \text{E}$$

The third step involves the transfer of t-RNA-bound amino acids to the ribosomes of the cellular cytoplasm. Ribosomes are nucleoprotein particles composed of approximately 60 per cent basic proteins and 40 per cent ribosomal RNA (r-RNA). There are two subunits, a 30 S and a 50 S ribosome, that are joined to make a complete 70 S ribosome. In the cytoplasm ribosomes are joined by strands of messenger RNA to form polysomes, which consist of 4 to 6 ribosomes. Both the t-RNA and m-RNA molecules are bound by the ribosomes to achieve a synthesis of polypeptide from the activated amino acids carried by the t-RNA.

The active template that controls protein synthesis on the ribosomes is called messenger RNA (m-RNA). It is synthesized by RNA polymerase under the direction of DNA in the nucleus, and it carries information to the ribosomes to direct the sequence of alignment of t-RNA-bound amino acids. This first process of overall information transfer from DNA to m-RNA is called **transcription** (transcribing the message). The specific programming of the amino acids on the polysomes or m-RNA molecules to synthesize a protein containing a definite sequence of amino acids is called **translation** (translating the code).

The attachment of the activated amino acids bound to t-RNA to a specific area or site on the m-RNA is governed by the genetic code. There is a specific site on m-RNA consisting of three consecutive bases that binds a particular amino acid; this site is called a **codon.** The t-RNA for that particular amino acid (see alanyl t-RNA, p. 274) has a complementary triplet of bases called an **anticodon** that binds the t-RNA to the site on the m-RNA on the ribosome. Since there are four different nucleotide residues or bases in m-RNA, and a sequence of three bases is involved in coding for each amino acid, a total of 4^3 or 64 different combinations of the three bases would be available for coding. The genetic code is *nonoverlapping* in that it requires the action of a specific group of three bases on the m-RNA chain, and it is also said to be *degenerate* in that more than one codon may be employed by m-RNA to insert a specific amino acid into the peptide chain. The codons currently proposed for specific amino acids are shown in Table 22–1. Note that there is a specific codon for the initiation of a protein chain and three chain termination codons.

Each 70 S ribosome has two binding sites. One binds the t-RNA with a growing polypeptide chain, the other binds a t-RNA with a single activated amino acid. A sequence in the synthesis of a polypeptide chain is shown in Figure 22–1.

The scheme outlined in Figure 22–1 is capable of synthesizing polypeptide molecules with a specific amino acid sequence governed by the m-RNA that was coded by the DNA in the nucleus of the cell. The initiation and termination of the synthesis is governed by specific codons (Table 22–1), and the number and types of proteins synthesized is controlled by the supply of specific m-RNA molecules. It has been estimated that it requires about two minutes for a ribosome to synthesize a small protein molecule, and since there may be 4 to 6 ribosomes in a polysome or even larger numbers of ribosomes involved in the synthesis of a high-molecular weight protein, the time of synthesis may be less than 30 seconds.

Metabolic reactions obviously require an adequate supply of essential enzyme

TABLE 22-1 CODONS FOR SPECIFIC AMINO ACIDS

AMINO ACIDS	CODONS FOR m-RNA
Alanine	GCA, GCC, GCG, GCU
Arginine	AGA, AGG, CGA, CGG, CGC, CGU
Asparagine	AAC, AAU
Aspartic acid	GAC, GAU
Cysteine	UGC, UGU
Glutamic acid	GAA, GAG
Glutamine	CAG, CAA
Glycine	GGA, GGC, GGG, GGU
Histidine	CAC, CAU
Isoleucine	AUA, AUC, AUU
Leucine	CUA, CUC, CUG, CUU, UUA, UUG
Lysine	AAA, AAG
Methionine	AUG—Chain initiation
Phenylalanine	UUU, UUC
Proline	CCA, CCC, CCG, CCU
Serine	AGC, AGU, UCA, UCG, UCC, UCU
Threonine	ACA, ACG, ACC, ACU
Tryptophan	UGG
Tyrosine	UAC, UAU
Valine	GUA, GUG, GUC, GUU
Chain termination	UAA, UAG, UGA

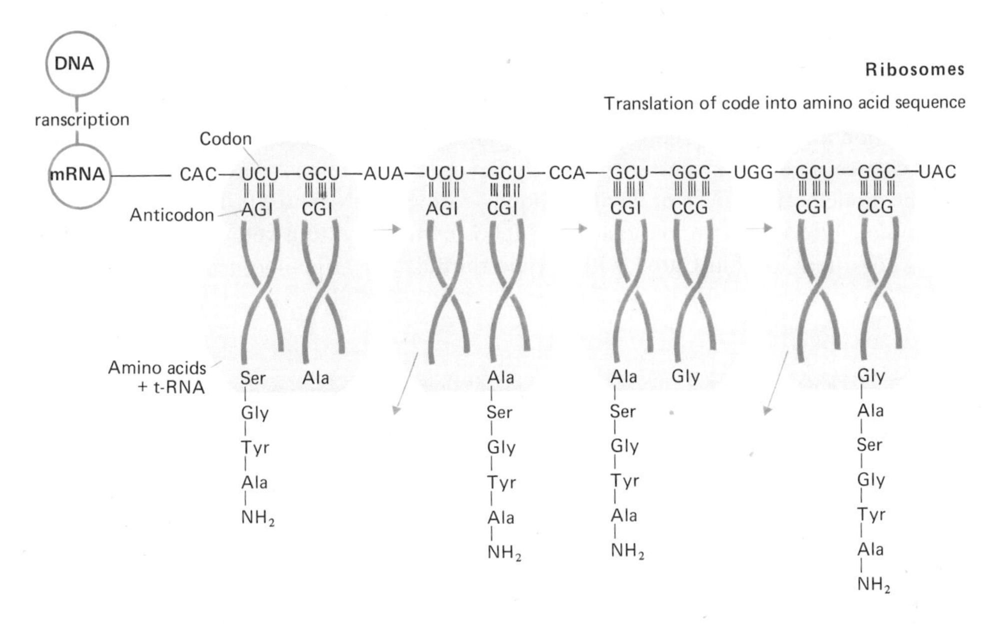

FIGURE 22-1 The figure demonstrates the two binding sites on the ribosome and the movement of the ribosomes on m-RNA (to the right) or movement of m-RNA (to the left). As the chain grows by the addition of one amino acid, a t-RNA molecule moves off the ribosome and a new single t-RNA amino acid moves onto the ribosome next to the polypeptide carrying the t-RNA.

molecules. If the cells can control the rate of m-RNA synthesis they can control the production of specific enzymes for metabolic cycles, thus controlling metabolism in general.

A mechanism of genetic control for the synthesis of enzyme proteins postulates that the chromosomes carry two types of genes, both **structural genes** and **operator genes.** The structural gene DNA directs the synthesis of protein molecules as described previously. The operator gene controls the action of adjacent structural genes in the synthesis of specific m-RNA molecules. The structural genes and their operator gene are designated an **operon.** The operator gene itself is controlled by a **regulator gene** which directs the synthesis of protein molecules called **repressors.** When the repressor combines with its operator gene, the structural genes cannot function in protein synthesis, and the operon is said to be repressed. If a situation in the cell inactivates the repressor and permits the operon to function, the operon is said to be derepressed. As suggested in Chapter 19, the end products of enzyme reactions, which are often metabolites, can affect the activity of enzyme systems. They may carry out this function by control of the repressor and the enzyme synthesis mechanism described previously.

METABOLIC REACTIONS OF AMINO ACIDS

The amino acids in the metabolic pool that are not immediately used for synthesis can undergo several metabolic reactions. They may follow the path of catabolism through deamination, urea formation, and energy production, or they may assist in the synthesis of new amino acids by the process of reamination and transamination. **Deamination, reamination, transamination,** and **urea formation** are processes common to all amino acids and are therefore very important to protein metabolism.

Deamination

A general reaction of catabolism is the splitting off of the amino group of an amino acid, with the formation of ammonia and a keto acid. This process is called **oxidative deamination** and is catalyzed by enzymes found in liver and kidney tissue called **amino acid oxidases.** These enzymes are generally flavoprotein enzymes containing either flavin adenine dinucleotide, FAD, or flavin mononucleotide, FMN. The enzyme dehydrogenates the amino acid to form an imino acid, which is hydrolyzed to a keto acid and ammonia. The process may be illustrated with a type formula for an amino acid.

$$\underset{\text{Amino acid}}{\mathrm{R{-}\underset{\underset{NH_2}{|}}{C}H{-}COOH}} \xrightarrow[\text{FAD} \rightarrow \text{FADH}_2]{\text{amino acid oxidase}} \underset{\text{Imino acid}}{\mathrm{R{-}\underset{\underset{NH}{\|}}{C}{-}COOH}} \xrightarrow[\mathrm{H_2O}]{\text{hydrolysis}} \underset{\text{Keto acid}}{\mathrm{R{-}\underset{\underset{O}{\|}}{C}{-}COOH}} + \mathrm{NH_3}$$

The fate of the keto acid depends on the amino acid from which it is derived. In general the catabolism of each amino acid must be studied separately. Glycine, for example, is the simplest amino acid, yet it can be transformed metabolically to formate, acetate, ethanolamine, serine, aspartic acid, fatty acids, ribose, purines, pyrimidines, and protoporphyrin. This amino acid may therefore play a role in carbohydrate, lipid, protein, nucleic acid, and hemoglobin metabolism, and it admirably illustrates the interrelationships that exist among the different types of metabolism in the body. Other amino acids undergo complex metabolic reactions that are beyond the scope of this book. In general, amino acids are classed as **glycogenic** or **ketogenic,** indicating their capacity to form glucose or glycogen and to follow the path of carbohydrates, or to enter the metabolic reactions of lipids and to form ketone bodies.

Transamination

The process of deamination results in the formation of many keto acids that are capable of accepting an amino group to form a new amino acid. That this process of **reamination** occurs was readily apparent from the isotope-labeling experiments of Schoenheimer. He observed a ready exchange of amino groups of dietary amino acids and tissue amino acids. A major mechanism for the conversion of keto acids to amino acids in the body is known as **transamination.** The original transamination reactions involved glutamic and aspartic acids. Glutamic acid, for example, could react with oxalacetic acid in the presence of a transaminase to form a new keto acid, α-ketoglutaric, and the new amino acid, aspartic acid.

$$\underset{\substack{\text{Glutamic}\\\text{acid}}}{\begin{array}{c}\mathrm{COOH}\\|\\\mathrm{CH_2}\\|\\\mathrm{CH_2}\\|\\\mathrm{CHNH_2}\\|\\\mathrm{COOH}\end{array}} + \underset{\substack{\text{Oxalacetic}\\\text{acid}}}{\begin{array}{c}\mathrm{COOH}\\|\\\mathrm{C{=}O}\\|\\\mathrm{CH_2}\\|\\\mathrm{COOH}\end{array}} \underset{\text{pyridoxal phosphate}}{\overset{\text{transaminase}}{\rightleftharpoons}} \underset{\substack{\alpha\text{-Ketoglutaric}\\\text{acid}}}{\begin{array}{c}\mathrm{COOH}\\|\\\mathrm{CH_2}\\|\\\mathrm{CH_2}\\|\\\mathrm{C{=}O}\\|\\\mathrm{COOH}\end{array}} + \underset{\substack{\text{Aspartic}\\\text{acid}}}{\begin{array}{c}\mathrm{COOH}\\|\\\mathrm{CH_2}\\|\\\mathrm{CHNH_2}\\|\\\mathrm{COOH}\end{array}}$$

The coenzyme **pyridoxal phosphate** is required in the reaction and **pyridoxamine phosphate** is formed.

$$R_1\text{—CH(NH}_2\text{)—COOH} + \text{Pyridoxal phosphate} \dashrightarrow R_1\text{—C(=N—CH}_2\text{—Py)—COOH} \longrightarrow R_1\text{—CO—COOH} + \text{Pyridoxamine phosphate}$$

Pyridoxal phosphate

Pyridoxamine phosphate

$$R_2\text{—CO—COOH} + \text{Pyridoxamine phosphate} \dashrightarrow R_2\text{—CH(N=CH—Py)—COOH} \longrightarrow R_2\text{—CH(NH}_2\text{)—COOH} + \text{Pyridoxal phosphate}$$

Pyridoxamine phosphate

Pyridoxal phosphate

The enzyme that catalyzes the reaction in the serum is called SGOT, serum glutamic oxalacetic transaminase, and a sharp rise in its concentration in the serum is indicative of myocardial infarction, a heart condition involving the cardiac muscle. There are many specific transaminases that serve as catalysts in the transfer of amino groups from amino acids to a variety of keto acids. Transamination reactions serve as important links joining carbohydrate, fat, and protein metabolism. A keto acid from any source can be used for the synthesis of an amino acid to be incorporated in tissue protein. For example, α-ketoglutaric acid is a constituent in the Krebs cycle and serves as a direct link from protein to carbohydrate and fat metabolism.

Formation of Urea

The ammonia, carbon dioxide, and water that result from the deamination and oxidation of the amino acids are combined to form urea. Urea formation takes place in the liver by a fairly complicated series of reactions, first described by Krebs and his coworkers. The ammonia and carbon dioxide combine with the amino acid ornithine to form another amino acid, citrulline. Another molecule of ammonia then combines with the citrulline to form the amino acid arginine, which is then hydrolyzed by means of the enzyme arginase, present in the liver, to form urea and ornithine. The ornithine may then enter the beginning of the cycle and combine with more ammonia and carbon dioxide from protein catabolism.

In recent years the detailed mechanism of the cycle has been worked out. Apparently ornithine does not react directly with CO_2 and NH_3 to form citrulline, but reacts with a compound called **carbamyl phosphate.** This compound is synthesized from ATP, CO_2, and NH_3 in the presence of the specific enzyme carbamyl phosphate synthetase and the cofactors N-acetylglutamate and Mg^{+2}.

$$CO_2 + NH_2 + 2ATP \xrightarrow[Mg^{+2},\ \text{N-acetylglutamate}]{\text{carbamyl phosphate synthetase}} NH_2\text{—C(=O)—O—P(=O)(O}^-\text{)—OH} + 2ADP + P_i$$

Carbamyl phosphate

The formation of arginine is also not a simple reaction of citrulline and NH_3 but involves a combination with aspartic acid to form argininosuccinic acid, which then splits into arginine and fumaric acid. The currently accepted **urea cycle** can be represented as follows:

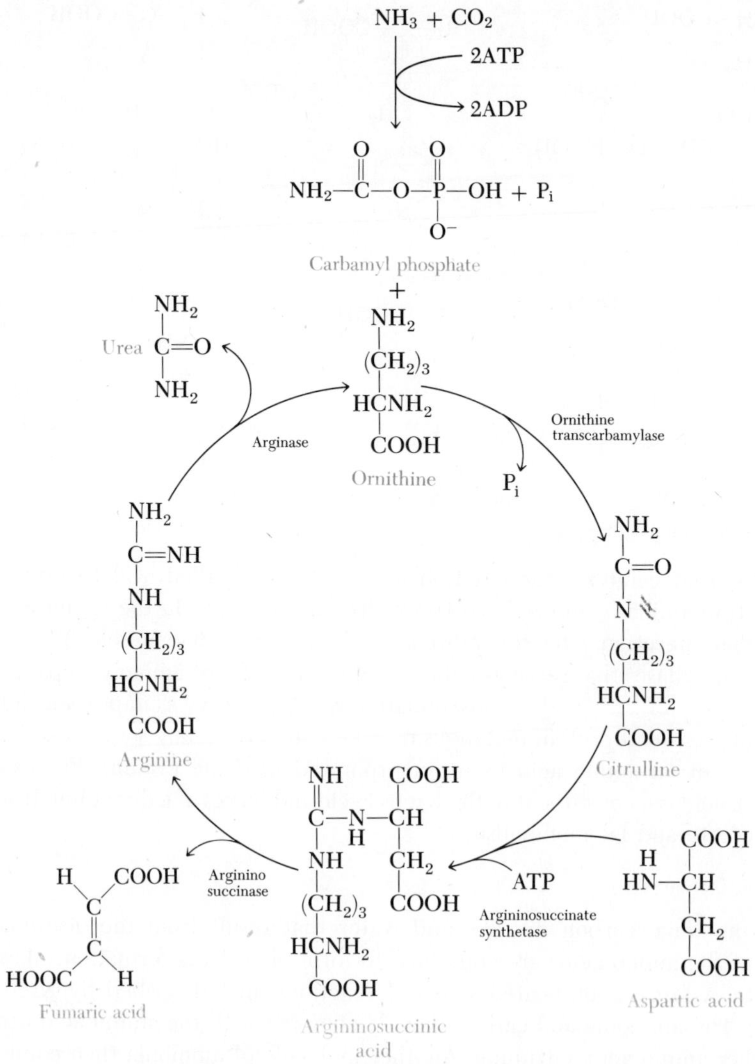

$NH_3 + CO_2$

2ATP → 2ADP

$NH_2—C(=O)—O—P(=O)(O^-)—OH + P_i$

Carbamyl phosphate

\+

Ornithine: $NH_2—(CH_2)_3—HCNH_2—COOH$

Ornithine transcarbamylase

P_i

Citrulline: $NH_2—C=O—NH—(CH_2)_3—HCNH_2—COOH$

Aspartic acid: $HOOC—CH_2—CH(NH_2)—COOH$

ATP

Argininosuccinate synthetase

Argininosuccinic acid: $NH=C(—NH—CH(COOH)—CH_2—COOH)—NH—(CH_2)_3—HCNH_2—COOH$

Arginino succinase

Fumaric acid: $HOOC—CH=CH—COOH$

Arginine: $NH_2—C=NH—NH—(CH_2)_3—HCNH_2—COOH$

Arginase

Urea: $NH_2—C=O—NH_2$

As urea is formed in the liver it is removed by the blood stream, carried to the kidneys, and excreted in the urine. Urea is the main end-product of protein catabolism and accounts for 80 to 90 per cent of the nitrogen that is excreted in the urine.

NUCLEOPROTEIN METABOLISM

In Chapter 16 nucleoproteins were shown to be constituents of nuclear tissue composed of a protein conjugated with nucleic acids. The important nucleic acids DNA and RNA are essential constituents of the cell nucleus, the chromosomes, and viruses and are involved in the synthesis of protein. During the process of digestion the protein is split from the nucleic acids and is broken down to amino acids. The nucleic acids are first attacked by ribonuclease and deoxyribonuclease to form nucleotides that are further hydrolyzed by nucleotidases to form phosphates and nucleosides. The nucleosides are absorbed through the intestinal mucosa and split by nucleosidases of the tissues into D-ribose, deoxyribose, purines, and pyrimidines. In metabolism the amino acids and sugar follow the ordinary process of protein and carbohydrate utilization. The phosphoric acid is used to form other phosphorus compounds in the body or may be excreted in the urine as phosphates.

Purine Metabolism

The synthesis of purines has been elaborated in recent years. This is a very complex process involving several steps, specific enzymes, and cofactors. Inosinic acid in the form of a mononucleotide is formed first and serves as an intermediate in the synthesis of adenylic acid and guanylic acid. With the help of tagged molecules, the precursors of the purine nucleus were established. These may be represented in the following diagram:

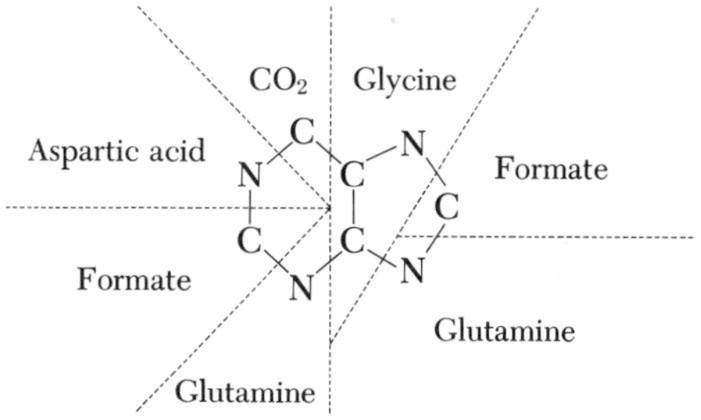

CO_2 Glycine
Aspartic acid Formate
Formate Glutamine
Glutamine

The purines that are formed by the hydrolysis of nucleosides in the tissues undergo catabolic changes, forming uric acid, which is excreted in the urine. The nucleosides **adenosine, inosine, guanosine,** and **xanthosine** are split into ribose plus adenine, hypoxanthine, guanine, and xanthine, respectively. These purines are not completely broken down to NH_3, CO_2, H_2O, and energy, but are progressively oxidized with the assistance of specific enzymes.

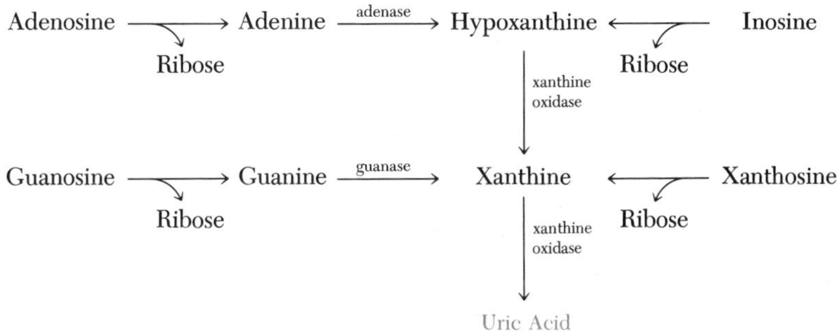

Adenosine → Adenine + Ribose; Adenine —adenase→ Hypoxanthine ← Inosine (− Ribose)

Hypoxanthine —xanthine oxidase→ Xanthine

Guanosine → Guanine + Ribose; Guanine —guanase→ Xanthine ← Xanthosine (− Ribose)

Xanthine —xanthine oxidase→ Uric Acid

In most mammals other than man and apes the uric acid is converted into **allantoin,** a more soluble substance.

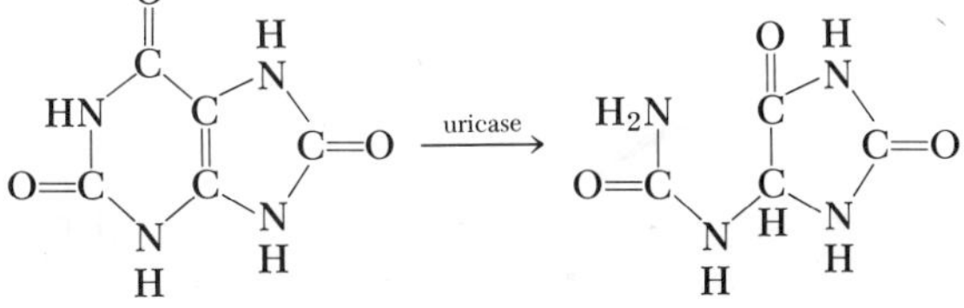

Uric acid —uricase→ Allantoin

Pyrimidine Metabolism

Pyrimidine synthesis starts with ammonia and carbon dioxide reacting with ATP to form **carbamyl phosphate** (see p. 351). This compound combines with aspartic acid to form carbamyl aspartic acid, which is converted to dihydroorotic acid, which is reduced to orotic acid. Orotic acid then reacts with 5-phosphoribosyl-1-pyrophosphate to form the nucleotide orotidine-5-phosphate. Decarboxylation of this nucleotide produces the primary pyrimidine, uridine-5-phosphate, or uridylic acid. The mononucleotide of uridylic acid apparently serves as the starting material for the synthesis of other pyrimidine nucleotides.

The pyrimidines that result from nucleoside hydrolysis in the tissues can be broken into small molecules in catabolism. Cytosine loses ammonia to form uracil, which is reduced to dihydrouracil utilizing the coenzyme $NADPH_2$. The ring is then opened to form a ureidopropionic acid, which is further split to β-alanine plus ammonia and carbon dioxide. These end products are eventually converted to urea for excretion.

Synthesis of Nucleic Acids

From the composition of nucleic acids described in Chapter 16, it is apparent that the synthesis would involve the copolymerization of four ribonucleotides for RNA and four deoxyribonucleotides for DNA. The DNA molecule has been synthesized by Kornberg and his coworkers by treating a mixture of deoxyribonucleotides with an enzyme isolated from bacteria. In the presence of all four deoxyribonucleotides, a DNA primer, and Mg^{+2}, the enzyme **DNA polymerase** can synthesize DNA. Enzymes called **RNA polymerases** or **transcriptases** found in plants, animals, and bacteria catalyze the synthesis of RNA from ribonucleoside triphosphates, ATP, GTP, CTP, and UTP. The synthesis is DNA-dependent, and requires Mg^{+2}, in addition to all four ribonucleoside triphosphates. The DNA serves as a template, and the base sequence in the DNA is transcribed into a corresponding sequence in the RNA. These nucleic acid molecules are similar to those from natural sources and are used to study the properties, compositions, and reactions of these large molecules.

CREATINE AND CREATININE

Creatine and creatinine are two nitrogen-containing compounds that are usually associated with protein metabolism in the body. **Creatine** is widely distributed in all tissues but is especially abundant in muscle tissue, where it is combined with phosphoric acid as **phosphocreatine,** or **creatine phosphate.** In the contraction of muscles, phosphocreatine apparently plays an important role as a reservoir of high-energy phosphate bonds readily convertible to ATP. The energy for the initial stages of muscular contraction probably

$$HN{=}C(NH_2)\text{—}NH\text{—}(CH_2)_3\text{—}HCNH_2\text{—}COOH + H_2N\text{—}CH_2COOH \xrightarrow[\text{transamidinase}]{\text{Arginine-Glycine}} NH_2\text{—}(CH_2)_3\text{—}HCNH_2\text{—}COOH + HN{=}C(NH_2)\text{—}HNCH_2COOH$$

Arginine + Glycine → Ornithine + Guanidoacetic acid

Guanidoacetic acid → (ATP → ADP; CH_3 from methionine; guanidino acetate methyl transferase) → Creatine phosphate

$$HN{=}C(NH\text{—}P(=O)(OH)O^-)\text{—}N(CH_3)CH_2COOH \xrightarrow{} H_2O + P_i + \text{Creatinine}$$

Creatinine: $HN{=}C$ in a ring with $N\text{—}C{=}O$ and $N(CH_3)\text{—}CH_2$

Creatinine

Creatine phosphate

comes from the hydrolysis of this compound to form creatine and phosphoric acid. These two substances are later combined during the recovery period of the muscle (see p. 330). Creatine is synthesized from the amino acids glycine, arginine, and methionine.

Creatinine is also present in the tissues but is found in much larger amounts in the urine. It is formed from either creatine phosphate or creatine and is an end product of creatine metabolism in muscle tissue.

INBORN ERRORS OF AMINO ACID METABOLISM

As early as 1906, Garrod, an English physician, described several abnormal patterns of metabolism. He called these abnormalities **inborn errors of metabolism,** since he recognized that the conditions were inherited. To the present time, about one hundred of these genetic diseases have been reported (p. 276). Many of them are extremely rare. The diseases due to a metabolic block in amino acid catabolism are of special interest since they occur somewhat more frequently. A common clinical manifestation of such conditions is mental deficiency. Three of these inborn errors, **phenylketonuria, alkaptonuria,** and **tyrosinemia,** will be considered in more detail as they are concerned with the metabolism of phenylalanine, an essential amino acid. Each involves a deficiency of a different single, specific enzyme whose synthesis is genetically controlled.

Phenylketonuria

Phenylketonuria, or PKU, was first recognized by Fölling in 1934 when he detected large amounts of **phenylpyruvic acid** in the urine of several mentally retarded patients. All individuals who have so far been found to excrete phenylpyruvic acid in their urine daily have shown some degree of mental deficiency; about 1 per cent of the patients in institutions for the mentally retarded excrete phenylpyruvic acid in their urine. Since the condition was recognized as a genetic disease, attention has been focused on parents as carriers of the defect and on detection and treatment of newborn infants with the disease.

The normal metabolism of phenylalanine involves its transformation into tyrosine with the aid of **phenylalanine hydroxylase.** If sufficient quantities of this enzyme are not synthesized, as in phenylketonuria, the concentration of phenylalanine in the blood, spinal fluid, and urine increases. In the tissues, the phenylalanine is converted into phenylpyruvic, phenyllactic, and phenylacetic acids and phenylacetyl glutamine as shown in the scheme on the next page. These metabolites of phenylalanine are excreted in the urine in large amounts in phenylketonuria.

As the disease itself is rather well understood today, the practical problem consists of identifying infants with the disease and initiating adequate treatment. Simple tests for the detection of phenylpyruvic acid in the urine or, preferably, tests for the increased concentrations of phenylalanine in the blood are available and are required in several states. The most successful treatment consists of restricting the amount of phenylalanine in the diet of the PKU infant, providing only the minimum quantity essential for normal growth and development. Fortunately, this regime allows the child to develop normally; the diet may apparently be relaxed somewhat after the age of six without serious effects on the child.

Alkaptonuria

Alkaptonuria is a rare genetic abnormality of tyrosine metabolism that is readily recognized by the characteristic changes in color which occur in the urine. When freshly passed the color is normal, but on standing it begins to darken. Alkalinity speeds up the change, and the urine passes through shades of brown to a final black color. Diapers

$C_6H_5-CH_2CH(NH_2)COOH \xrightarrow[\text{hydroxylase}]{\text{phenylalanine}} HO-C_6H_4-CH_2CH(NH_2)COOH$

Phenylalanine ⇅ Tyrosine

$C_6H_5-CH_2C(=O)COOH$

Phenylpyruvic acid

$C_6H_5-CH_2CH(OH)COOH$ $\quad C_6H_5-CH_2COOH \rightarrow C_6H_5-CH_2C(=O)-NH-CH(COOH)-CH_2-CH_2-COOH$

Phenyllactic acid — Phenylacetic acid — Phenylacetyl glutamine

wet with urine become darkly stained and as a result the condition is frequently recognized in early infancy. In adults, alkaptonuria may first be detected in a life insurance examination or routine medical check. The urine has strong reducing properties, which gives rise to a positive Benedict's test. These characteristics persist throughout life. As patients grow older their ligaments and cartilages tend to become dark blue in color, and the patients are prone to develop osteoarthritis.

The normal metabolism of tyrosine, which may be formed from phenylalanine as previously discussed, includes the formation of **homogentisic acid,** which is oxidized to fumaric acid and finally to acetoacetic acid. This oxidation is catalyzed by the enzyme **homogentisic acid oxidase.** The metabolic error is a block in the breakdown of the homogentisic acid, caused by a decrease in the concentration of the oxidase in the tissues; normal individuals can readily oxidize increased quantities of this acid without producing pigmented urine specimens. The formation of homogentisic acid and its metabolites is shown in the scheme on page 357. At present there is no specific treatment for this rare genetic disease.

Tyrosinemia

Tyrosinemia is a rare inborn error of tyrosine metabolism seen primarily in premature infants. In the scheme on page 357, the metabolic block associated with this defect occurs between *p*-hydroxyphenylpyruvic acid and homogentisic acid. The enzyme required for this step, ***p*-hydroxyphenylpyruvate oxidase,** has been found to be in low concentration in the liver of premature infants. In infantile tyrosinemia, kidney defects and a nodular cirrhosis of the liver occur, causing rickets, thrombocytopenia, darkening of the skin, and slight mental retardation. The urine contains large amounts of tyrosine and metabolites, including *p*-hydroxyphenylpyruvic, *p*-hydroxyphenylacetic, and *p*-hydroxyphenyllactic acids. The disease may be acute or chronic; the acute cases are characterized by diarrhea, failure to thrive, and death from liver failure in the first seven months of life.

Ascorbic acid is the coenzyme of *p*-hydroxyphenylpyruvate oxidase, and the disease may be alleviated in premature infants by feeding excess ascorbic acid. Longer term

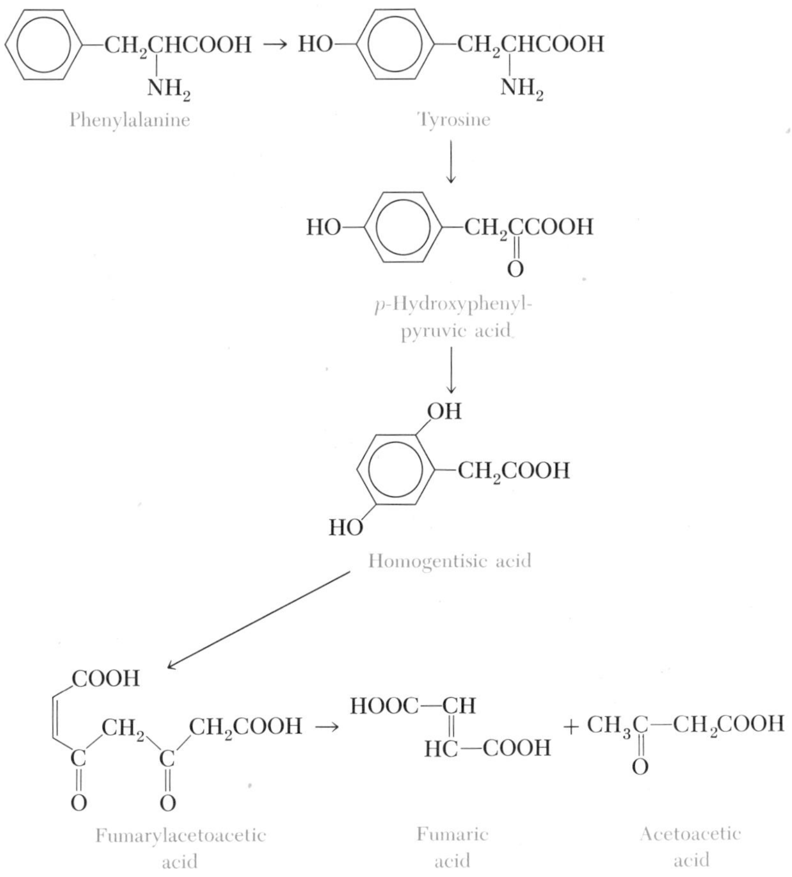

$C_6H_5{-}CH_2CH(NH_2)COOH \rightarrow HO{-}C_6H_4{-}CH_2CH(NH_2)COOH$

Phenylalanine → Tyrosine

↓

$HO{-}C_6H_4{-}CH_2C(=O)COOH$

p-Hydroxyphenyl-pyruvic acid

↓

$(HO)_2C_6H_3{-}CH_2COOH$

Homogentisic acid

↓

$HOOC{-}CH{=}CH{-}C(=O){-}CH_2{-}C(=O){-}CH_2COOH \rightarrow HOOC{-}CH{=}HC{-}COOH + CH_3C(=O){-}CH_2COOH$

Fumarylacetoacetic acid → Fumaric acid + Acetoacetic acid

treatment for tyrosinemia consists of a diet low in tyrosine and phenylalanine. If the dietary treatment is started early in life, both liver and kidney damage may be prevented.

CORRELATION OF CARBOHYDRATE, LIPID, AND PROTEIN METABOLISM

The correlation between carbohydrate and lipid metabolism has already been discussed. Since the catabolism of amino acids results in keto acids such as pyruvic acid, it can readily be seen that these products could enter the metabolic scheme of the carbohydrates. Furthermore, the glycogenic or glucose-forming amino acids and the ketogenic amino acids could enter the carbohydrate and lipid metabolism schemes. The over-all correlation of the three major types of metabolism is represented in the diagram on page 358.

IMPORTANT TERMS AND CONCEPTS

absorption
alkaptonuria
amino acid pool
anticodon
codon
creatinine
deamination
digestion
essential amino acids
inborn errors of metabolism
operon
phenylketonuria
ribosomes
m-RNA
t-RNA
transamination
transcription
translation
tyrosinemia
urea cycle
uric acid

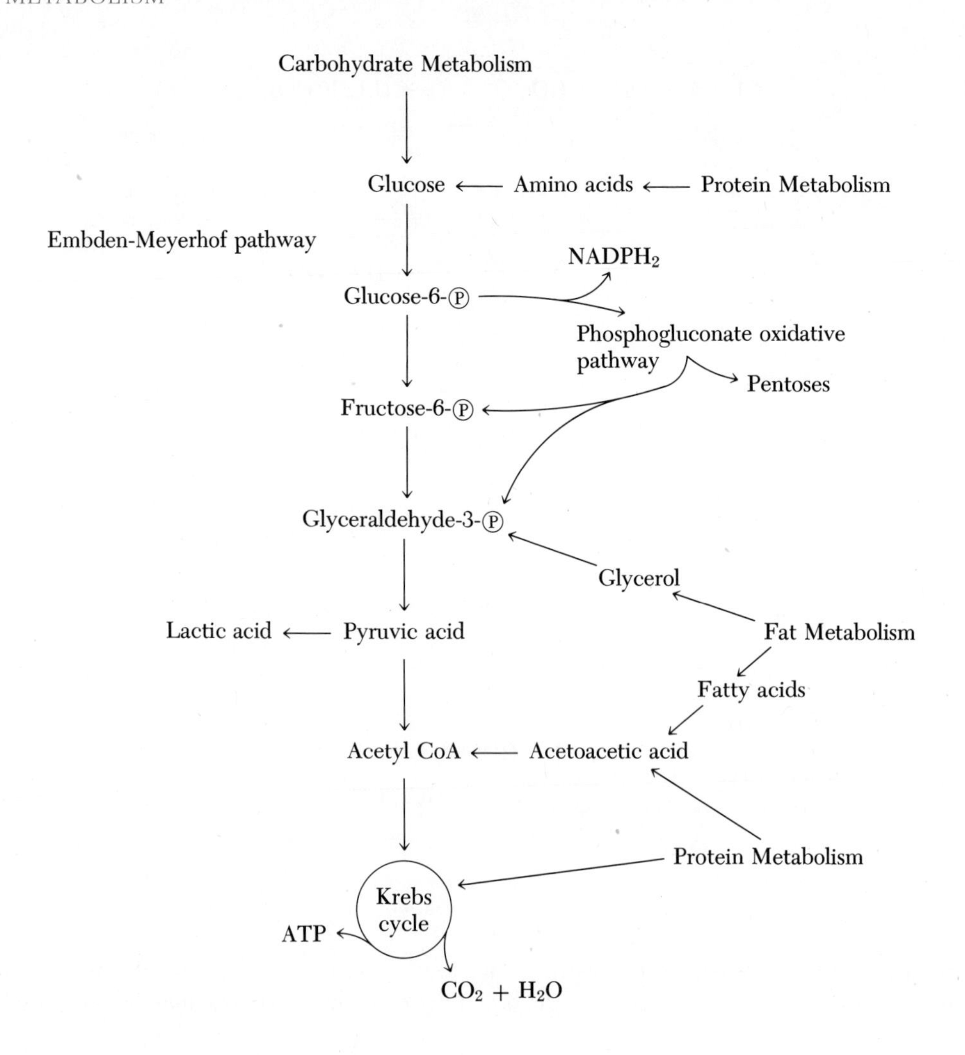

QUESTIONS

1. Outline the process of protein digestion and absorption.

2. Explain the concept of the amino acid pool.

3. What is meant by the dynamic state of tissue proteins in the body?

4. What is an essential amino acid? What would happen to a growing child that was deprived of adequate amounts of these amino acids? Why?

5. Describe the process of activation of amino acids and their combination with t-RNA prior to the synthesis of protein.

6. Illustrate the process of translation in protein synthesis.

7. In protein synthesis, explain the relationship between codons and anticodons.

8. What roles do ribosomes and m-RNA play in protein synthesis? Explain.

9. Explain the function of structural and operator genes and the operon in the synthesis of enzyme proteins.

10. What are the processes of deamination and transamination? Illustrate one process with equations.

11. How does pyridoxal phosphate function as a coenzyme in the process of transamination?

12. Outline the essential reactions in the urea cycle.

13. What products would result from the complete hydrolysis of RNA? Briefly describe the metabolic fate of each of the products.

14. What compounds are involved in the synthesis of the purine nucleus?

15. Briefly outline the process for the synthesis of creatine.

16. Explain how the process of transamination can serve as a common link between carbohydrate, fat, and protein metabolism?

17. What clinical manifestation and biochemical deficiency is characteristic of the three inborn errors of phenylalanine metabolism?

18. Why is it important to test the urine of newborn infants for phenylpyruvic acid or their blood for increased concentrations of phenylalanine? Explain.

SUGGESTED READING

Bergen: Tracer Isotopes in Biochemistry. Journal of Chemical Education, Vol. 29, p. 84, 1952.
The Chemistry of Life 1. How Cells Synthesize Proteins. Chemical & Engineering News, May 8, 1961, p. 81.
Clark and Marcker: How Proteins Start. Scientific American, Vol. 218, No. 1, p. 36, 1968.
Garrod: Inborn Errors of Metabolism. Oxford University Press, London, 1923.
Gorini: Antibiotics and the Genetic Code. Scientific American, Vol. 214, No. 4, p. 102, 1966.
Grünewald: The Evolution of Proteins. Chemistry, Vol. 41, No. 1, p. 11, 1968.
Howe: Amino Acids in Nutrition. Chemical & Engineering News, July 23, 1962, p. 74.
Kornberg: Biologic Synthesis of Deoxyribonucleic Acid. Science, Vol. 131, p. 1503, 1960.
Ptashne and Gilbert: Genetic Repressors. Scientific American, Vol. 222, No. 6, p. 36, 1970.
Roth: Ribonucleic Acid and Protein Synthesis. Journal of Chemical Education, Vol. 38, p. 217, 1961.
Scrimshaw and Behar: Protein Malnutrition in Young Children. Science, Vol. 133, p. 2039, 1961.
Yanotsky: Gene Structure and Protein Structure. Scientific American, Vol. 216, No. 5, p. 80, 1967.

BODY FLUIDS

CHAPTER 23

The *objectives* of this chapter are to enable the student to:

1. Describe the process of normal distribution of fluids between the circulatory system and the interstitial spaces.
2. Recognize the factors involved in normal water balance.
3. Describe the essential components of plasma or serum and the separation of proteins by electrophoresis.
4. Illustrate the normal composition of plasma electrolytes.
5. Describe the factors involved in the acid-base balance of the blood.
6. Describe the chemistry of hemoglobin and its function in respiration.
7. Define the isohydric cycle and explain its importance in respiration.
8. Recognize the essential functions of the kidney and the process of urine formation.
9. Describe the regulation of water, electrolyte, and acid-base balance by the kidney.
10. Describe the process of H^+ formation and ammonia synthesis and their importance in kidney function.

The complex function of a single cell requires that the intracellular fluid and the extracellular fluid surrounding the cell maintain a fairly constant chemical composition. This may be accomplished by the simple diffusion of cellular material back and forth between the intra- and extracellular fluids. As cells combine to form tissues and organs and finally the entire body, the process of diffusion is not sufficient to maintain the proper fluid balance. A circulatory, or transportation, system capable of carrying nutrient material, enzymes, hormones, and other regulatory substances to the tissues, and waste products to the organs of excretion, is required.

The **blood** is the most active transport system and consists of cellular elements suspended in plasma, the liquid medium. The **tissue fluid,** which surrounds the tissues, and the **lymph,** a slow-moving fluid that is similar to plasma and is carried in a system of vessels called the **lymphatics,** also assist the transportation system of the body. Lymph and tissue fluid are known together as **interstitial fluid** and make up about 15 per cent of the body's weight. Important interchange of material occurs from the blood to the tissue through the medium of the interstitial fluid. Plasma and interstitial fluid are collectively considered as **extracellular fluids.** A 160-pound man would have about 6.0 liters of blood, 3.5 liters of plasma, 10.5 liters of interstitial fluid, and 35 liters of intracellular fluid.

The intracellular fluids are separated from the extracellular fluids by the cellular membranes. Water, electrolytes, nutrient material, and waste products must pass through the membrane to maintain cellular function. The cell membrane is freely permeable to water, O_2, CO_2, urea, and glucose, but exhibits a selective permeability toward electrolytes, such as Na^+, K^+, Cl^-, Ca^{+2}, Mg^{+2}, and HCO_3^-. In experiments during which heavy water, D_2O, was injected into the extracellular fluid, it was found that the body required about 120 minutes to establish equilibrium between the intra- and extracellular fluids.

Water is rapidly transported throughout the body by the circulating plasma of the blood. The movement of water between the plasma and the interstitial spaces is dependent on the proteins in the plasma, which constitute a colloidal-osmotic machine that can draw water into the blood capillaries. This osmotic pressure, called the **oncotic pressure,** is equivalent to 23 torr and is opposed by the hydrostatic pressure transmitted to the capillaries from the heart. This pressure averages 32 torr at the arterial end and 12 torr at the venous end of the capillary. At the arterial end of the capillary system the hydrostatic pressure forces fluid into the interstitial space, whereas at the venous end the colloid osmotic pressure pulls it back into the capillaries. This mechanism is responsible for the normal distribution of fluids between the blood stream and the interstitial space, and may be diagramed as shown in Figure 23–1.

A decrease in plasma protein, which may occur in serious malnutrition or in kidney disease, would decrease the oncotic pressure and result in an increased flow of fluid to the interstitial space. This condition is known as **edema,** and it can also occur in any heart disease that results in an increase in blood pressure.

If water balance is to be maintained in the body, it is obvious that fluid intake and fluid excretion must be equal. The intake and output of a normal adult of average size can be represented as in the following tabulation:

Intake	*ml/Day*	*Output*	*ml/Day*
		Insensible perspiration	
		Lungs	700
		Skin	300
Water in beverages	1200	Sweat	300
Water in food	1500	Feces	200
Water of oxidation	300	Urine	1500
	3000		3000

Some concept of the magnitude of the daily fluid turnover by the body can be gained from the following data:

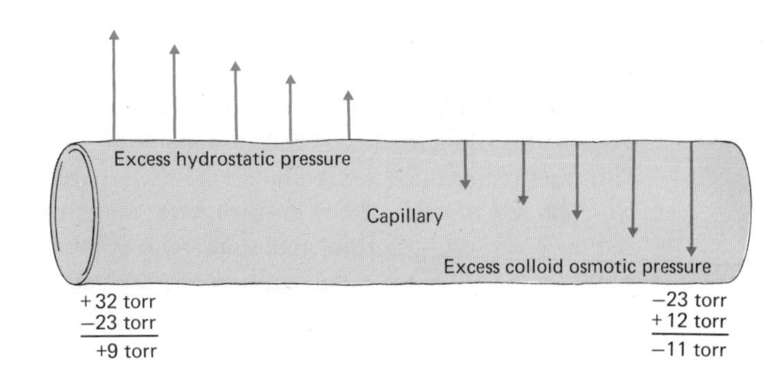

FIGURE 23–1 Correlation of colloid osmotic pressure and hydrostatic pressure in the maintenance of fluid balance between the plasma and the interstitial space.

Daily Volume of Digestive Fluids

	ml/Day
Saliva	1,500
Gastric secretion	2,500
Intestinal secretion	3,000
Pancreatic secretion	700
Bile	500
	8,200
Fluids lost from the body	3,000
Total fluid turnover	11,200
Compared to:	
Plasma volume	3,500
Total extracellular fluid	14,000

In general, if the fluid intake exceeds the output for any length of time, the tissue fluid will increase in volume and **edema** will result. The excessive loss of body fluids that may occur from vomiting, diarrhea, or copious sweating causes **dehydration** of the body.

BLOOD

Some idea of the importance of blood to the body can be gained from a consideration of its major functions:

1. The blood transports nutrient material to the tissues and waste products of metabolism to the organs of excretion.
2. It functions in respiration by carrying oxygen to the tissues and carbon dioxide back to the lungs.
3. It distributes regulatory substances, such as hormones, vitamins, and certain enzymes, to the tissues in which they exert their action.
4. The blood contains white corpuscles, antitoxins, precipitins, and so on, which serve to protect the body against microorganisms.
5. It plays an important role in the maintenance of a fairly constant body temperature.
6. It aids in the maintenance of acid-base balance and water balance.
7. It contains a clotting mechanism that protects against hemorrhage.

Blood comprises approximately 8 per cent of the body weight, which means that an average person has 4000 to 6000 ml of blood. Loss of blood by bleeding or by donating for transfusion has no serious effect on the body, since the blood volume is rapidly regenerated.

BLOOD CELLS

The two major portions of blood are the blood cells and the plasma. When separated by centrifugation, the blood cells occupy from 40 to 45 per cent by volume of the blood. This fraction contains the red blood cells, white blood cells, and thrombocytes (platelets).

The red blood cells, or **erythrocytes,** contain the respiratory pigment hemoglobin, and have several important functions. The number of red blood cells in men is approximately 5,000,000 per cubic millimeter, and in women approximately 4,500,000. The determination of the number of red cells in the blood is often carried out in the laboratory, since the value changes markedly in diseases such as anemia.

White blood cells, or **leukocytes,** are larger than red cells and have nuclei, which red cells do not. Normally, there are from 5000 to 10,000 white cells per cubic millimeter of blood. There are several types of white blood cells, and they all function to combat infectious bacteria. White cell counts are routinely determined in the laboratory since they increase in acute infections such as acute appendicitis.

Thrombocytes (platelets) are even smaller than red blood cells and do not contain a nucleus. There are from 250,000 to 400,000 thrombocytes per cubic millimeter in the blood of a normal person. Their major function is in the process of blood clotting, since they contain cephalin, a phospholipid involved in the early stages of clotting.

SERUM AND PLASMA

When freshly drawn blood is allowed to stand, it clots, and a pale yellow fluid soon separates from the clotted material. This fluid is called **serum** and is blood minus the formed elements and fibrinogen, which is used in the clotting process. If, on the other hand, blood collected in the presence of an anticoagulant is centrifuged, the fluid portion that separates from the cells is called **plasma.**

Plasma Proteins

The proteins of the plasma are present in a concentration of about 7 per cent. The most important of these are **albumin,** the **globulins,** and **fibrinogen.** The globulins have been separated into several fractions of different molecular size and properties by the technique of **electrophoresis** (Fig. 23–2). More recent techniques, including column, starch gel and cyanogum electrophoresis, and immunoelectrophoresis, are capable of separating plasma into more than 20 different proteins.

The plasma proteins have several functions in the body. One of the most essential is the maintenance of the effective osmotic pressure of the blood, which controls the water balance of the body. The globulins, IgG, IgA, and IgM, contain immunologically

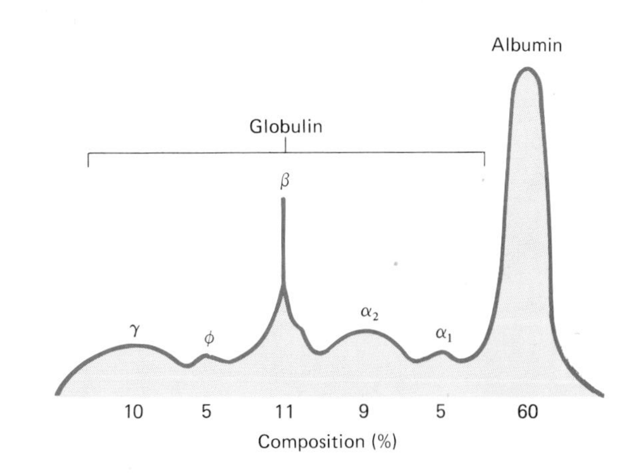

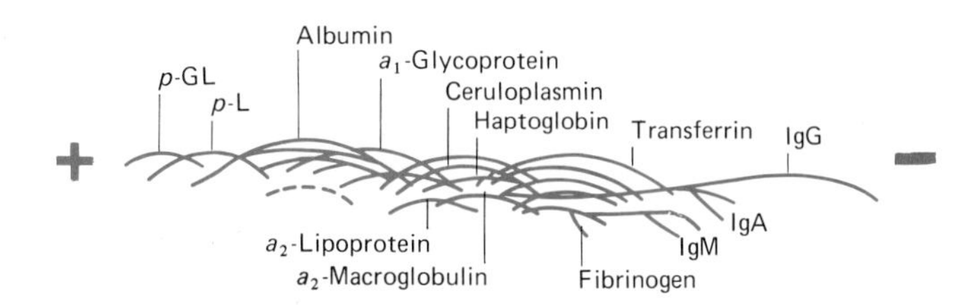

FIGURE 23–2 *Upper:* An electrophoretic pattern of normal plasma. *Lower:* Composite diagram of immunoelectrophoretic analysis of human plasma proteins. The anode is to the left. "Prealbumins" are labeled $_p$-GL (for glycoprotein) and $_p$-L (for lipoprotein). The pattern as drawn suggests the approximate relative positions of the more frequently observed precipitation arcs.

active antibodies against such diseases as diphtheria, influenza, mumps, and measles. The lipoproteins are combinations of lipids and α or β globulins, and function in the transport of lipids. Transferrin binds iron and ceruloplasmin binds copper, and these proteins function in the transport of these metals. Fibrinogen is an essential component in the blood clotting process.

SERUM ENZYMES

There are several enzymes in the plasma or serum that are released into the blood by the breakdown of body tissues. Since marked changes from the normal concentration often occur in disease, assays for specific enzymes are currently a powerful diagnostic tool in the laboratory. Clinically significant enzymes are amylase, lipase, acid and alkaline phosphatase, lactic dehydrogenase (LDH), creatine phosphokinase (CPK), aldolase, glutamic-oxalacetic transaminase (SGOT), and glutamic-pyruvic transaminase (SGPT). Amylase, for example, increases in acute pancreatitis, alkaline phosphatase in obstructive jaundice of the liver, and SGOT in myocardial infarction.

Plasma Electrolytes

The electrolytes that are present in the body fluids consist of positively charged ions, or **cations,** and negatively charged ions, or **anions.** They are mainly responsible for the osmotic pressure of the fluids and are involved in maintenance of the acid-base and water balance of the body. The major cations in the body fluids are Na^+, K^+, Ca^{+2}, and Mg^{+2}, whereas the major anions are HCO_3^-, Cl^-, HPO_4^{-2}, SO_4^{-2}, organic acids, and protein. In a consideration of electrolyte balance, the concentration of these ions is expressed in milliequivalents per liter of body fluid. Milliequivalents per liter (mEq/l) equals the atomic weight expressed in milligrams per liter divided by the valence. The concentration of plasma electrolytes as determined in the laboratory is sometimes expressed in mg/100 ml plasma. Calcium is often reported as 10.0 mg/100 ml for a normal individual, and the normal sodium concentration is about 327 mg/100 ml. To express these concentrations in mEq/l, we would proceed as follows:

$$\frac{\text{mg/100 ml} \times 10}{\dfrac{\text{atomic wt.}}{\text{valence}}} = \frac{\text{mg/l}}{\text{equiv. wt.}} = \text{mEq/l}$$

$$\frac{10\ \text{mg/100 ml} \times 10}{\dfrac{40}{2}} = \frac{100}{20} = 5\ \text{mEq/l of Ca}$$

$$\frac{327\ \text{mg/100 ml} \times 10}{\dfrac{23}{1}} = \frac{3270}{23} = 142\ \text{mEq/l Na}$$

The electrolyte balance of the ions in the plasma is shown in the following tabulation:

Cations	*mEq/l*	*Anions*	*mEq/l*
Na^+	142	HCO_3^-	27
K^+	5	Cl^-	103
Ca^{+2}	5	HPO_4^{-2}	2
Mg^{+2}	3	SO_4^{-2}	1
		Organic acids	6
		Protein	16
Total	155	Total	155

Comparisons of the electrolyte composition of body fluids and changes in disease are often illustrated in vertical bar graphs devised by Gamble. Variations in the electrolyte content of normal plasma, interstitial fluid, and intracellular fluid are shown in Figure 23–3.

In the discussion of water balance we listed the volumes of the daily secretions of digestive fluids. Their electrolyte composition must also be taken into account, especially when they are lost from the body by vomiting or diarrhea. In comparing the electrolytes of these fluids with those of plasma, only the major components that are involved in acid-base or electrolyte balance are shown in the bar graphs in Figure 23–4. It is readily apparent that the digestive secretions vary in their composition of electrolytes and in their contribution to the acid-base balance of the body.

ACID-BASE BALANCE

The normal metabolic processes of the body result in the continuous production of acids, such as carbonic, sulfuric, phosphoric, lactic, and pyruvic. In cellular oxidations

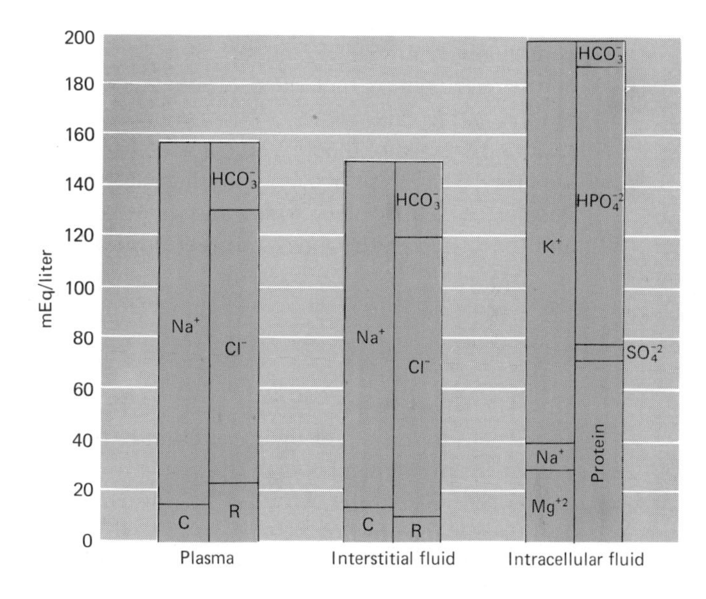

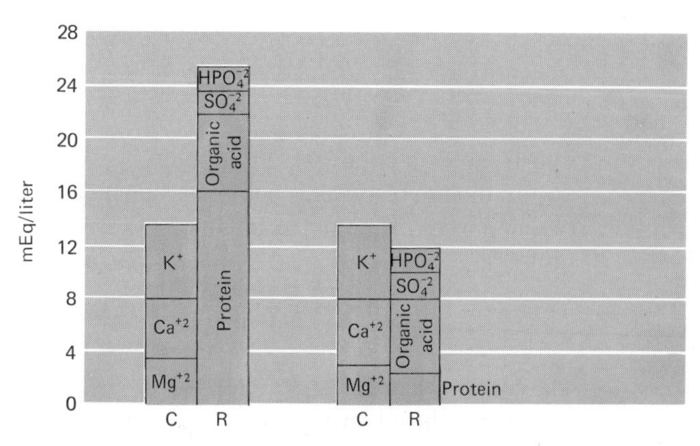

FIGURE 23–3 Electrolyte composition of normal body fluids. (The expanded scales below show the individual electrolytes in the C and R spaces.)

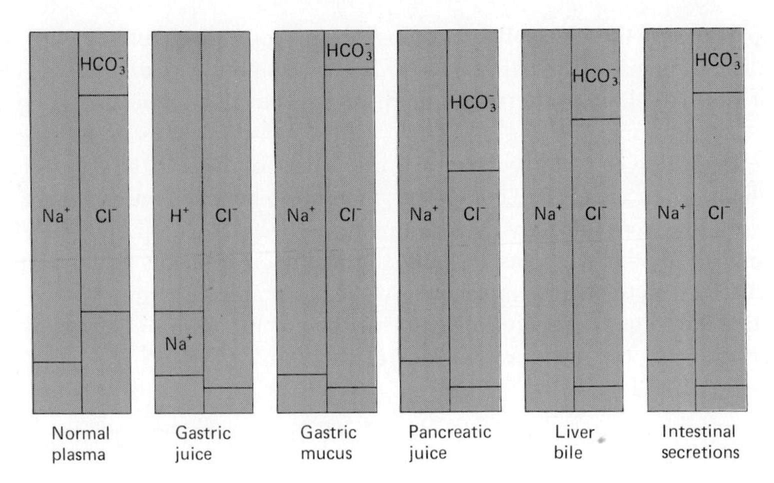

FIGURE 23-4 A comparison of the major electrolyte components in normal digestive secretions and normal plasma.

the main acid end product is H_2CO_3 with 10 to 20 moles formed per day, which is equivalent to one to two liters of concentrated HCl. Although some foods yield alkaline end products, the acid type predominates, and the body is faced with the necessity of continually removing the large quantities of acids that are formed within the cells. An added restriction is that these products must be transported to the organs of excretion via the extracellular fluids without a great change in their H^+ concentration. The 7.35 to 7.45 pH range of blood is one of the most rigidly controlled features of the electrolyte structure. The means of accomplishing this is the mechanism of the regulation of acid-base balance, which involves water and electrolyte balance, hemoglobin and blood buffers, and the action of the lungs and the kidneys.

BODY BUFFER SYSTEMS

The ability of extracellular fluids to transport acids from the site of their formation in the cells to the site of their excretion in the lungs and kidneys without an appreciable change in pH depends on the presence of effective buffer systems in these fluids and in the red blood cells. A **buffer** has already been defined as a mixture of a weak acid and its salt that resists changes in pH when small amounts of acid or base are added to the system (Chapter 1). The buffers in the plasma and extracellular fluid include the bicarbonate, phosphate, and plasma protein systems, which are represented as follows:

$$\frac{H_2CO_3}{BHCO_3} \qquad \frac{BH_2PO_4}{B_2HPO_4} \qquad \frac{\text{H Plasma Protein}}{\text{B Plasma Protein}}$$

The B stands for the base or cation.

In the red blood cells both bicarbonate and phosphate buffers are present along with two important hemoglobin buffers, as shown below:

$$\frac{\text{H Hemoglobin}}{\text{B Hemoglobin}} \qquad \frac{\text{H Oxyhemoglobin}}{\text{B Oxyhemoglobin}}$$

The buffers that are most effective in the regulation of acid-base balance are the bicarbonate, the plasma protein, and the hemoglobin buffers. Later in the chapter the important role of hemoglobin and its derivatives in the transportation of CO_2 from the tissues to the lungs without a change in pH will be considered.

The Bicarbonate Buffer

The bicarbonate buffer is by far the most important single buffer in acid-base balance. It is closely related to the constant production of CO_2, H_2CO_3, and $BHCO_3$; to the reactions of hemoglobin and oxyhemoglobin in the red cells; to the respiratory control of the H_2CO_3 concentration; and to the effect of the kidneys on $BHCO_3$ concentration. To illustrate the action of the buffer, reactions for the addition of a small amount of a strong acid and a strong base may be written as follows:

$$HCl + NaHCO_3 \rightarrow NaCl + H_2CO_3 \qquad (1)$$

$$NaOH + H_2CO_3 \rightarrow H_2O + NaHCO_3 \qquad (2)$$

In reaction (1), the basic member of the buffer pair reacts with the acid to form neutral NaCl and the acid member of the buffer pair. In reaction (2), the acid partner buffers the base to form water and the basic partner.

The buffering capacity of a buffer pair is related to its effectiveness in limiting changes of pH when acid or alkali is added to the system. The **Henderson-Hasselbalch equation** expresses the relation between the pH of the system, the pK of the acidic form of the buffer pair, and the concentration of each member of the buffer pair. The equation for the bicarbonate buffer is as follows:

$$pH = pK + \log \frac{[BHCO_3]}{[H_2CO_3]}$$

The pK is different for each buffer and is determined by measuring the pH of a solution that contains equal concentrations of the two components that make up the buffer pair. For example, if equal concentrations of $NaHCO_3$ and H_2CO_3 are present in a solution the pH equals 6.1, or

$$pH = pK = 6.1 \text{ when } \frac{[BHCO_3]}{[H_2CO_3]} = \frac{1}{1}, \text{ since } \log 1 = 0$$

Under normal conditions at the pH of blood, 7.4, the equation for the bicarbonate buffer is

$$7.4 = 6.1 + \log \frac{[BHCO_3]}{[H_2CO_3]} = 6.1 + \log \frac{27 \text{ mEq/l}}{1.35 \text{ mEq/l}} = 6.1 + \log \frac{20}{1} = 6.1 + 1.3$$

This unequal concentration of the buffer pairs would seem to indicate that the bicarbonate buffer is very ineffective at the pH of the blood. Because of the respiratory control of the H_2CO_3 concentration, however, buffering is remarkably effective.

The 1:20 ratio for H_2CO_3 and $BHCO_3$ in the plasma at pH 7.4 includes the normal values of 1.35 mEq/l for H_2CO_3 and 27 mEq/l for $BHCO_3$. In disease conditions, this ratio may be changed by increases or decreases in H_2CO_3 or $BHCO_3$, producing an acidosis or alkalosis. The most common changes involve the $BHCO_3$ concentration and are described as **metabolic acidosis** or **alkalosis.** Diseases that alter respiratory function affect the H_2CO_3 concentration and produce what is called **respiratory acidosis** or **alkalosis.** These are the four major abnormalities in acid-base balance.

HEMOGLOBIN

Hemoglobin is the respiratory protein of the red blood cell that has been described in detail by Perutz and his coworkers. This protein has a molecular weight of about 64,500, and is composed of four polypeptide chains and four heme molecules. Heme is a protoporphyrin derivative (p. 104), with one iron atom coordinated with each of the four pyrrole nitrogen atoms.

Heme

The four polypeptide chains, two α and two β chains, exist in the form of α helices which are folded and bent into three-dimensional structures. In the hemoglobin molecule, the four chains are grouped in a tetrahedron-shaped structure with the four heme molecules embedded in hollows in the folded chains. The relationship of the iron atom, the heme molecule, and one of the polypeptide chains may be seen in the model of one-fourth of a hemoglobin molecule (Fig. 23–5).

The normal concentration of hemoglobin in the blood varies from 14 to 16g per 100 ml. This means that a 150-pound person would have a total of approximately 900g of hemoglobin. Since the red blood cells that contain the pigment are constantly being broken down, there is a continuous degradation of hemoglobin into other pigments in

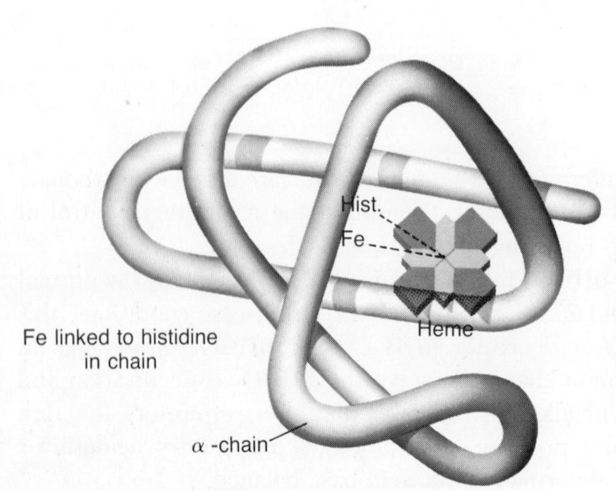

FIGURE 23–5 Artist's conception of the α-chain of the hemoglobin molecule with its associated heme group, illustrating the linkage of iron to histidine in the chain. (After Steiner, R. F.: The Chemical Foundations of Molecular Biology. Princeton, N. J., D. Van Nostrand, 1965.)

the body; for example, **bilirubin,** which is converted into pigments responsible for the characteristic color of bile, urine, and feces.

Normal and abnormal hemoglobin molecules may be separated by the electrophoresis of a drop of whole blood on a strip of special filter paper or cellulose acetate. Adult hemoglobin, fetal hemoglobin, and hemoglobin in sickle cell anemia are called hemoglobin a, f, and s, respectively. In addition, there are hemoglobins c, d, e, h, i, and j, which are present in various blood abnormalities. The majority of these hemoglobins differ by the substitution of only one amino acid in the amino acid sequence of their β-chains. It is surprising that this minor change in the molecule accounts for different rates of migration in electrophoresis and variations in physiological properties.

Respiration

Hemoglobin is often called the **respiratory pigment** of the blood and has the property of combining with gases such as oxygen and carbon dioxide. The transportation of oxygen by the blood depends on the reversible reaction between hemoglobin and oxygen to form **oxyhemoglobin.**

$$\underset{\text{Hemoglobin}}{Hb} + O_2 \rightleftharpoons \underset{\text{Oxyhemoglobin}}{HbO_2}$$

The oxygen capacity of the blood, about 1000 ml, is sufficient for normal tissue requirements. Some conception of the role of hemoglobin may be gained by a comparison of the oxygen capacity of plasma and whole blood. One liter of plasma can carry only 3 ml of oxygen in solution. In the absence of hemoglobin, the body's circulatory system would have to contain over 300 liters of fluid to supply oxygen to the tissues. This would represent a system four to five times our body weight.

In the process of respiration, hemoglobin comes into contact with a relatively rich oxygen atmosphere (partial pressure of 100 torr) in the alveoli of the lungs to form oxyhemoglobin. The oxyhemoglobin is carried by the arterial circulation to the tissues where a low oxygen concentration (partial pressure of 40 torr) and a high carbon dioxide concentration (partial pressure of 60 torr) combine to release the oxygen to the tissues. The carbon dioxide is then carried back to the lungs for excretion and the cycle is repeated.

URINE

In the previous chapters on metabolism several mechanisms that operate to maintain the constituents of the blood within fairly narrow limits of concentration have been considered. The utilization and transformation of nutrient material in the blood was emphasized in the chapters on metabolism. The removal of the waste products of metabolism, such as drugs, toxic substances, excess water, inorganic salts, and excess acid or basic substances, is essential to maintain the normal composition of the blood. The kidneys play a major role in the regular excretion of these substances from the blood and tissue fluids. The kidneys, therefore, are essential for the maintenance of blood and tissue fluid volume, the electrolyte and acid-base balance of the body, and the maintenance of normal osmotic pressure relationships of the blood and body fluids.

Water, carbon dioxide, and other volatile substances are eliminated from the body by the lungs. The skin excretes small amounts of water, inorganic salts, nitrogenous material, and lipids. Some inorganic salts are eliminated by the intestine, and the liver is involved in the excretion of cholesterol, bile salts, and bile pigments. Compared to the kidney, the other organs of excretion play a minor role.

The Formation of Urine

The kidney may be regarded as a filter through which the waste products of metabolism are passed to remove them from the blood. The blood enters the kidney by means of the renal arteries, which break up into smaller branches leading to the small filtration units called **malpighian corpuscles.** Each human kidney contains approximately 1,000,000 of these units. A malpighian corpuscle consists of a mass of capillaries from the renal artery which form the **glomerulus.** The glomerulus is enclosed within a capsule called **Bowman's capsule** which opens into a long tubule. Several of these tubules are connected to larger **collecting tubules** which carry the urine to the bladder. These anatomical structures of the kidney are illustrated in Figure 23–6.

The most generally accepted theory for the formation of urine can be outlined as follows: As the blood passes through the glomerulus, the constituents other than protein filter through the capillary walls and enter the tubules. As this filtrate passes down the tubules, a large proportion of the water and any substances which are of value to the body, such as glucose, certain inorganic salts, and amino acids, are reabsorbed into the blood stream. These substances are called **threshold substances.** Waste products of metabolism, such as urea and uric acid, are not completely reabsorbed, and therefore pass into the collecting tubules for excretion. A few substances, such as creatinine and potassium, are partially removed from the blood through excretion by the tubules in addition to filtration by the glomerulus. The function of the renal glomeruli may be regarded as that of ultrafiltration producing a protein-free filtrate of the plasma, followed by a process

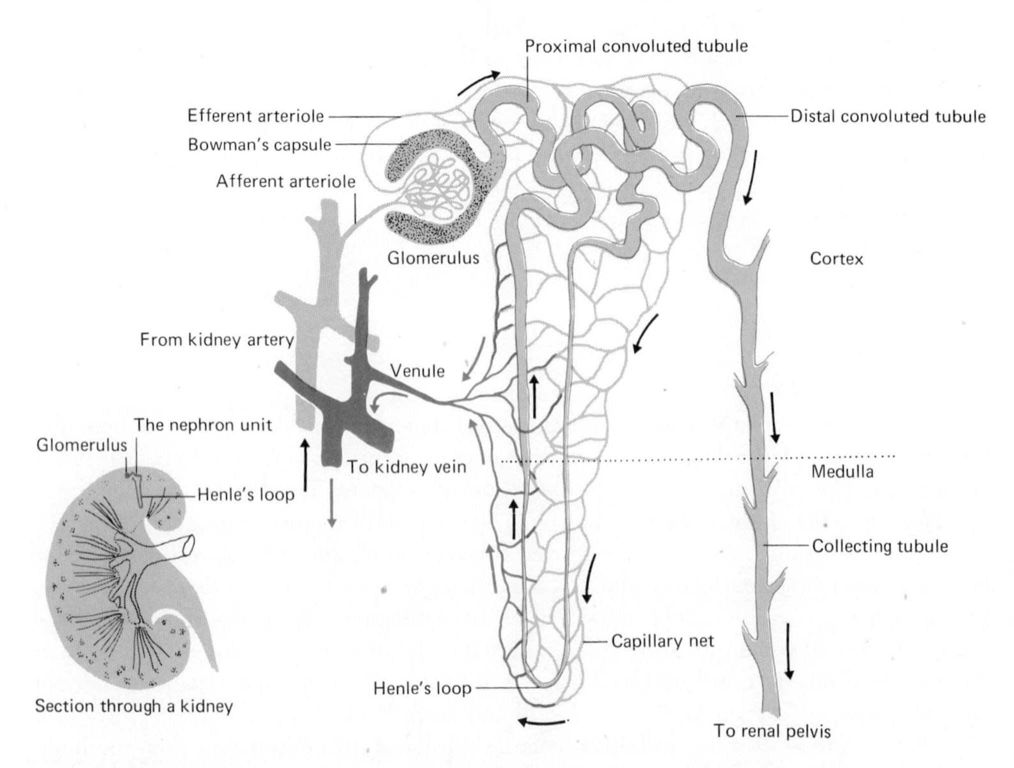

FIGURE 23–6 Diagram of a single kidney tubule and its blood vessels. (After Villee: Biology. 5th ed. Philadelphia, W. B. Saunders Company, 1967.)

of selective reabsorption by the renal tubules. Some concept of the magnitude of this process can be gained from the daily values for filtration and reabsorption. In a normal individual, 170 to 180 liters of water, 1000g of NaCl, 360g of $NaHCO_3$, and 170g of glucose are filtered through the glomeruli, and 168.5 to 178.5 liters of water, 988g of NaCl, 360g of $NaHCO_3$, and 170g of glucose are reabsorbed by the tubules in order to excrete about 30g of urea, 12g of NaCl, and other waste products in about 1500 ml of urine.

Regulatory Power of the Kidney

Although the kidney is considered mainly as an organ of excretion, it plays a regulatory role in water, electrolyte, and acid-base balance.

Water Balance

The control of the water content of blood, interstitial fluid, cell fluid, and digestive fluids in the body depends on normally functioning kidneys. Water in excess of body requirements is readily excreted by the kidney. When, however, water is needed to maintain the normal concentration of body fluids, the hypothalamus, in conjunction with the pituitary gland, secretes an antidiuretic hormone called **vasopressin.** This hormone causes an antidiuresis that results in less water being excreted in the urine and more being reabsorbed into the body fluids. Apparently changes in the osmotic pressure of the plasma regulate the secretion of vasopressin, which in effect is the "fine control" of urine volume.

Electrolyte Balance

To a certain extent the excretion or retention of electrolytes such as Na^+, K^+, Cl^-, HCO_3^-, and HPO_4^{-2} depends on the water balance of the body fluids. Electrolytes are excreted or reabsorbed along with the movement of water in the tubules. The steroid hormones of the adrenal cortex, such as **aldosterone** which is called a **mineralocorticoid,** exert more specific control over the excretion of electrolytes by the tubules. When the plasma has too much water, its osmotic pressure decreases, and it is said to be hypotonic. Under these conditions there is an increased secretion of adrenal cortex hormones that increase the retention or reabsorption of electrolytes into the plasma. If the plasma electrolytes were too concentrated and the osmotic pressure increased, the secretion of the cortical hormones would be depressed and the secretion of vasopressin would increase to assist in the excretion of electrolytes and the retention of water.

Acid-Base Balance

In the tubules Na reabsorption, in part at least, is involved in the regulation of acid-base balance through the action of the Na^+—H^+ exchange. This phase of Na^+ excretion is regulated to some extent by the pH of the blood plasma and the capacity of the tubular cells to acidify the urine by H^+ and NH_3 formation.

The glomerular filtrate contains the electrolytes and acids and bases present in the plasma. The chief cation is Na^+, and the chief anions are Cl^-, HCO_3^-, and HPO_4^{-2}, as electrolytes of NaCl, $NaHCO_3$, and Na_2HPO_4. At a pH of 7.4, about 95 per cent of the CO_2 is present as $NaHCO_3$ and about 83 per cent of the PO_4^{-3} as Na_2HPO_4. Normally, as the glomerular filtrate passes into the tubules most of the water is reabsorbed, and the greater proportion of the Na^+ is taken up by the tubular cells in exchange for H^+ formed in the tubular cells from H_2CO_3. This Na^+ is returned to the plasma in association with the HCO_3^- formed from H_2CO_3 in the tubular cells. The formation of H^+ and HCO_3^- from H_2CO_3 in both the proximal and distal tubular cells is catalyzed by the enzyme carbonic anhydrase, as follows:

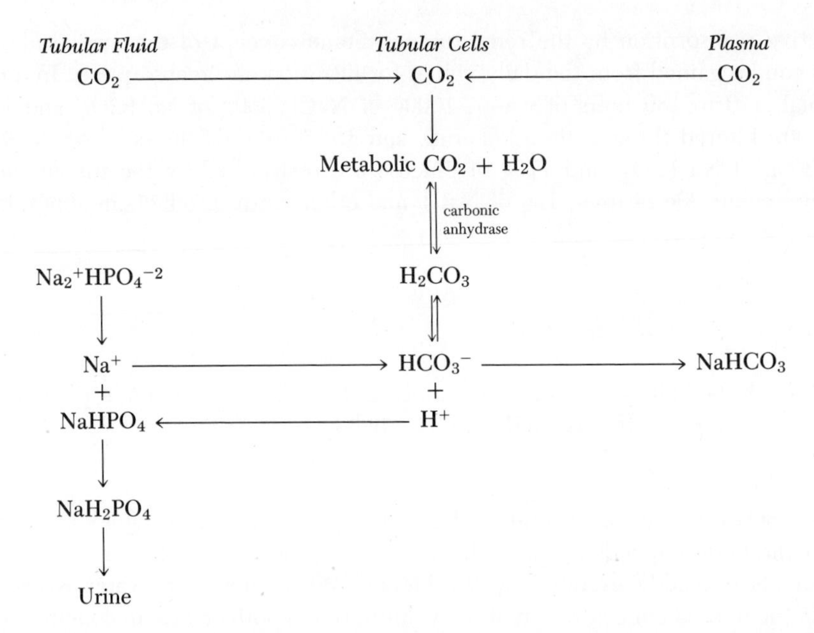

By varying the rate of ventilation, the lungs control the H_2CO_3 concentration in the plasma.

The task of the kidneys is to stabilize the concentration of the bicarbonate. This problem involves two aspects: first, the salvaging of all, or nearly all, of the bicarbonate contained in the glomerular filtrate (equivalent to approximately one pound of $NaHCO_3$ per 24 hours); and second, the neutralization of nonvolatile acids (H_2SO_4 and H_3PO_4). The kidney may conserve base in two ways: by conversion of neutral or basic salts to acid salts for excretion, as shown previously for Na_2HPO_4, and by the synthesis of ammonia.

The mechanism for the synthesis of ammonia to further conserve Na^+ and fixed base may be represented as follows:

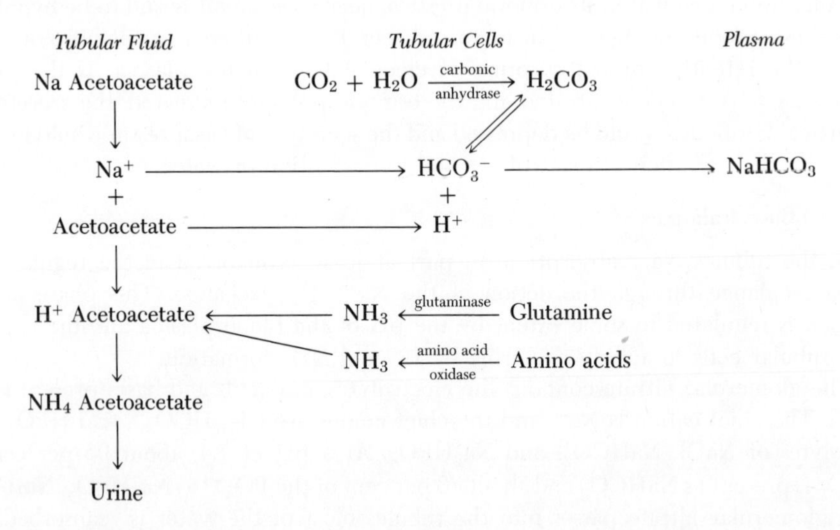

The urinary excretion of the ammonium ion in a person on a normal average diet is 30 to 70 mEq per day. In conditions of acidosis that might result from uncontrolled diabetes mellitus, acetone bodies, such as acetoacetic acid, are formed. The sodium salt of this acid must be excreted by the kidney, and to conserve the Na^+ for plasma buffers H^+ and NH_3 are formed in the tubular cells of the kidney.

IMPORTANT TERMS AND CONCEPTS

acid-base balance
bicarbonate buffer
blood
electrolyte balance
electrolytes
extracellular fluids
hemoglobin
isohydric cycle
oxyhemoglobin
plasma
plasma proteins
regulatory power of kidney
respiration
urine formation
water balance

QUESTIONS

1. List the important functional body fluids and their volumes in a normal adult.
2. What is oncotic pressure, and how does it assist in the normal distribution of fluids between the circulatory system and the interstitial space?
3. Outline the parameters involved in the balance between fluid intake and fluid output in the body.
4. Explain the difference between whole blood, plasma, and serum.
5. Compare an electrophoretic pattern of plasma proteins with an immunoelectrophoretic pattern. What information may be gained from each pattern?
6. The normal values for the concentration of calcium and chloride in the plasma, expressed as mEq/l, are 5 and 103, respectively. Show how you would calculate these concentrations as mg/l and mg/100ml.
7. Illustrate with a bar graph the electrolyte composition of normal plasma.
8. What are the most important buffers in the whole blood and plasma?
9. How is the pK of the acidic form of the buffer related to its buffering capacity? Explain.
10. Calculate the pH of a bicarbonate buffer solution that contains 20mEq/l of $BHCO_3$ and 2mEq/l of H_2CO_3.
11. What is the chemical nature of heme, and what is its relation to hemoglobin?
12. List the essential functions of the kidney and the process of formation of the urine.
13. What hormones are involved in the control of water and electrolyte balance by the kidney? Explain how they function.
14. Explain how the process of H^+ formation in the kidney tubules helps conserve Na^+ for the plasma.
15. Why would the kidney tubules increase the synthesis of ammonia in the condition of diabetes mellitus? Explain.

SUGGESTED READING

Edelman: The Structure and Function of Antibodies. Scientific American, Vol. 223, No. 2, p. 34, 1970.
Herron: Simplified Apparatus for Electrophoresis on Paper. Journal of Chemical Education, Vol. 46, p. 527, 1969.
Merrill: The Artificial Kidney. Scientific American, Vol. 205, No. 1, p. 56, 1961.
Smith: The Kidney. Scientific American, Vol. 188, No. 1, p. 40, 1953.
Surgenor: Blood. Scientific American, Vol. 190, No. 2, p. 54, 1954.

BIOCHEMISTRY OF DRUGS

CHAPTER 24

The *objectives* of this chapter are to enable the student to:

1. Recognize the important divisions of the peripheral nervous system.
2. Illustrate the passage of a nerve impulse across a synapse to the target organ.
3. Explain the use of aspirin in so many drug preparations.
4. Compare the action of cholinergic and anticholinergic drugs.
5. Recognize the importance of epinephrine and its functions in the body.
6. Explain the increased use of amphetamine-type drugs in today's society.
7. Recognize the similarities and differences between the many drugs used as tranquilizers or sedatives.
8. Recognize the difference in action and potency of marijuana, LSD, and heroin.
9. Describe the mechanism of action of the antifertility drugs.
10. Explain the use of cytotoxic drugs in the treatment of cancer.

If we accept a single definition of disease as *dis ease,* headaches, minor aches and pains, malnutrition, dietary deficiencies, metabolic abnormalities, endocrine disturbances, infections, and malignant cancer all qualify as diseases. Volumes have been written on diseases of the cell, tissues and organs of the body, and the therapeutic agents or drugs used to combat the disease process. It is becoming more and more apparent that all disease has a biochemical basis. The biochemistry of normal and abnormal heredity, deficiency diseases, errors of metabolism, and the process of infection is receiving considerable attention, study, and research.

ANALGESIC DRUGS

To aid in the understanding of the recent emphasis on health-related research, examples of different types of diseases and their treatment will be discussed. So many people are occasionally inconvenienced with headaches and minor aches and pains that their cause and treatment with analgesic drugs should be of interest. In general, these pains are caused by swelling of tissue, resulting in pressure on peripheral nerves, and also minor inflammation of tissues, accompanied by an increase in body temperature which affects nerve endings. A common analgesic drug that serves as the basis for a multitude of headache, cold, and flu remedies is acetylsalicylic acid or **aspirin** (p. 179). When combined with **phenacetin** and **caffeine,** the resultant preparation is the common APC tablets. The compound **N-acetyl-p-amino phenol** is a metabolic product of phenacetin and has replaced this drug in several preparations. Phenacetin is an organic amine derived from acetanilide, a compound originally used as an antipyretic drug (p. 193). Aspirin is an analgesic drug in that it reduces inflammation and swelling of tissues, exerts an antipyretic action in reducing fever, and probably exerts a chemical action on the peripheral nerves. Phenacetin and N-acetyl-p-amino phenol exhibit some of the effects of aspirin and are synergistic with respect to its action. The antipyretic effect of aspirin is thought to be related to the salicylic acid portion of the structure, whereas the antipyresis exhibited by phenacetin and N-acetyl-p-aminophenol depends on the amino-

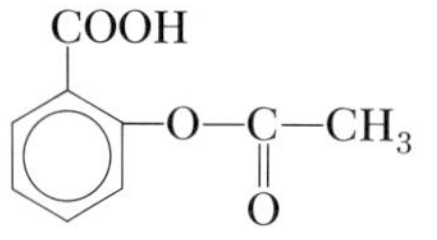

Aspirin 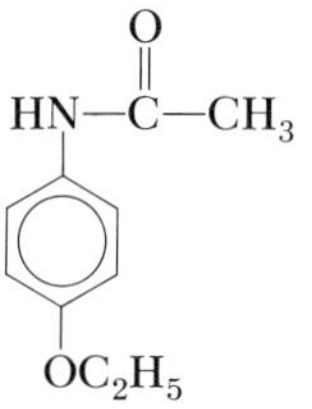 | Phenacetin 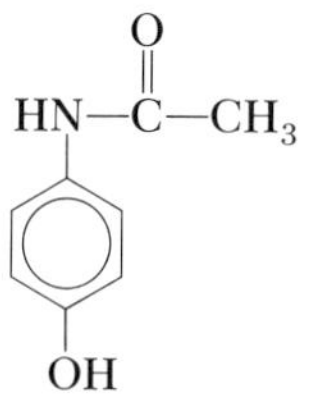 | N-acetyl-p amino phenol | Caffeine

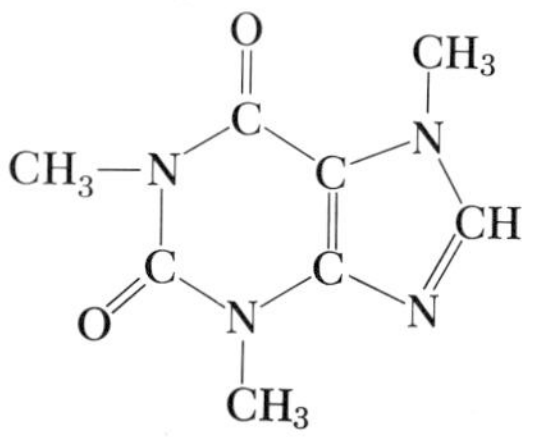

benzene portion of the compound. Caffeine is a diuretic and assists the kidney in excretion of the drug and the circulatory system in the transport of the drugs (p. 108). The addition of buffering agents to aspirin has been found to speed the absorption of the drug into the blood and tissues and has resulted in products such as Bufferin.

THE NERVOUS SYSTEM

The biochemical action of several drugs involves their effect on the nervous system. The central nervous system and the peripheral nervous system are the two main divisions of the body's nervous system; they are often affected in a different manner by specific drugs. The **central nervous system** includes the brain and the spinal cord. The pituitary gland which lies at the base of the brain is the master gland of the endocrine system; thus the central nervous system in association with the pituitary gland effectively coordinates the control of all body mechanisms and functions.

The peripheral nervous system serves as a relay between the central nervous system and the action that occurs in specialized tissues. The **somatic system,** for example, is the part of the peripheral nervous system that relays impulses to the voluntary muscles that function in the purposeful muscular movement involved in working, walking, and talking. Functions of the body that are regulated without conscious thought rely on another part of the peripheral nervous system called the **autonomic system.** This system is further divided into the **parasympathetic system,** which regulates the rate of our heartbeat and respiration and our digestive processes, and the **sympathetic system,** which functions in time of stress. Impulses from the central nervous system to the end organs or target organs in the tissues or from these organs back to the central nervous system are passed through an integrating control organ called a **ganglion.** The ganglia are collections of nerve cells that provide a measure of control over the two-way passage of nervous impulses. Nerve impulses involve the passage of electrical potential down the body of the nerve cell. The gap between a nerve cell carrying a message from the central nervous system and another nerve cell that may transmit the impulse to a receptor on a target organ is called a **synapse.** Compounds called chemical mediators are required to bridge the gaps and complete the transmission of the nervous impulse. Acetylcholine is the chemical mediator for ganglionic synapses in both parasympathetic and sympathetic systems and is also the chemical mediator between the nerve endings and target organs in the parasympathetic system. Norepinephrine is the chemical mediator between the nerve endings and the target organs of the sympathetic system.

When an impulse travels down the body of a nerve and reaches the end of the nerve cell at a synapse, acetylcholine is released. In the parasympathetic system this compound migrates across the synapse and contacts the receptor mechanism of the next nerve cell. The mechanism is stimulated to create an electrical potential in that nerve cell, which allows the nervous impulse to pass across the synapse. The acetylcholine is then removed by hydrolysis, which is catalyzed by the enzyme acetylcholinesterase, and the synapse is ready to pass the next nervous impulse. The processes that have just been described may be represented as follows:

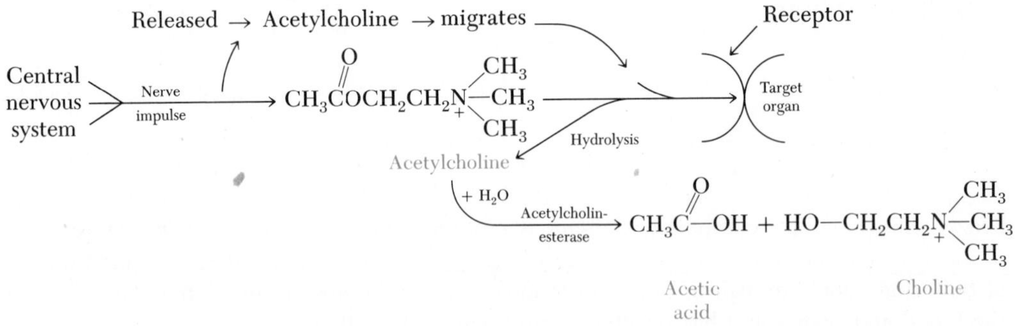

Released → Acetylcholine → migrates → Receptor

Central nervous system —(Nerve impulse)→ $CH_3\overset{O}{\overset{\|}{C}}OCH_2CH_2\overset{+}{N}(CH_3)_3$ → Target organ

Acetylcholine —(Hydrolysis)

$$\text{Acetylcholine} + H_2O \xrightarrow{\text{Acetylcholinesterase}} \underset{\text{Acetic acid}}{CH_3\overset{O}{\overset{\|}{C}}{-}OH} + \underset{\text{Choline}}{HO{-}CH_2CH_2\overset{+}{N}(CH_3)_3}$$

In addition to the action of mild analgesic drugs such as aspirin on the peripheral nervous system and narcotics such as morphine on the central nervous system, the mechanism of nerve impulse conduction presents several sites for drug action. The majority of these effects involve alterations in the reactions that occur in the area of the synapse. Drugs that affect the parasympathetic system will interfere with the normal regulatory processes in the body. Inhibitors of acetylcholinesterase such as diisopropylfluorophosphate (p. 145 and 312) can cause death by the irreversible blockage of nerve impulses across synapses. Atropine, an alkaloid present in the root of the deadly nightshade or belladonna plant (p. 377), specifically blocks synaptic transmission by the acetylcholine mechanism which results in an interference in the muscular control of the iris of the eye and a dilation of the pupils.

The sympathetic nervous system is affected by drugs related to norepinephrine and epinephrine and by enzyme inhibitors such as the monoamine oxidase inhibitors that affect the enzyme that catalyzes the oxidation and removal of norepinephrine from the receptor sites. Drugs that stimulate the sympathetic system are called adrenergic agents, as contrasted to adrenergic blocking agents that depress the system.

CHOLINERGIC AND ANTICHOLINERGIC DRUGS

These drugs serve to stimulate or depress the parasympathetic system. Although cholinergic drugs have no essential therapeutic uses, they do interfere with regulatory mechanisms of the body controlled by the parasympathetic system. Nicotine, an alkaloid found in tobacco leaves (p. 377), when administered in small doses (as in smoking tobacco), will stimulate the parasympathetic system, the central nervous system, and the cardiovascular system. Tremors, nausea and vomiting, increased blood pressure, abnormal heart beats, and convulsions may result from the administration of the drug.

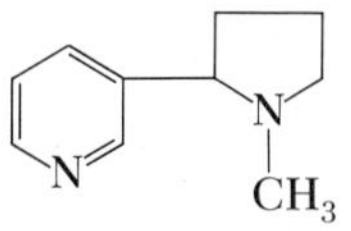

Nicotine

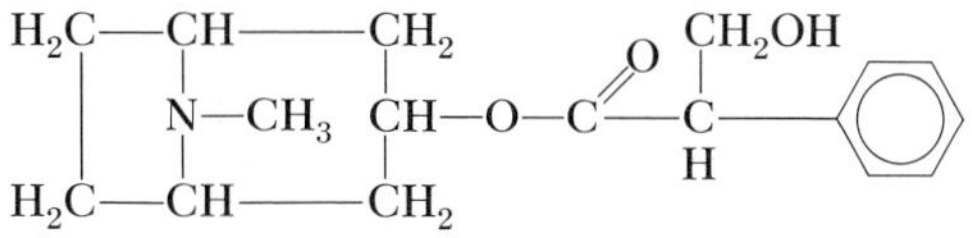

Atropine

Anticholinergic drugs include compounds like the belladonna alkaloids, atropine and scopolamine, which block the receptor sites in the nerve cells, and the acetylcholinesterase inhibitors. Atropine is used by ophthalmologists to dilate the pupils of the eye for examination of its interior, as discussed earlier. Scopolamine is chemically very similar to atropine (contains an oxygen atom bridged across the two carbon atoms on the left of the above structure) but has a stronger depressive action on the parasympathetic system. It is sometimes used to prepare patients for anesthesia, since its administration normally produces drowsiness, euphoria, amnesia, and dreamless sleep. The same dosage of the drug will occasionally cause excitement and restlessness, so the drug must be used with care.

As discussed earlier, acetylcholine serves as a transmitter of nerve impulses across the synapse, and must be immediately removed by hydrolysis to permit the passage of the next impulse. Acetylcholinesterase catalyzes this hydrolysis, which often occurs in a period of milliseconds. Any drug that inhibits the activity of the enzyme will act as an anticholinergic agent. The mechanism of action of acetylcholinesterase on acetylcholine and the inhibition of the enzyme has been the subject of many research investigations. The enzyme is believed to have two sites of interaction with the substrate, an anionic site and an esteratic site. When the enzyme hydrolyzes acetylcholine, the acetyl group splits off and acetylates the enzyme, which is later regenerated by hydrolysis to form acetic acid; the enzyme molecule inhibitors form complexes with the enzyme by attachment at either or both anionic and esteratic sites. Organic derivatives of phosphoric acid are potent inhibitors of acetylcholinesterase which phosphorylate the enzyme by complexing at the esteratic site. These compounds include the nerve gases, sarin, soman, and tagun, insecticides such as parathion and malathion, and diisopropylfluorophosphate (p. 145 and 312). Very small amounts of these inhibitors may produce a fatal toxicity when absorbed from the lungs or the skin. The lethal dose for a man may be as low as 0.01 mg/kg.

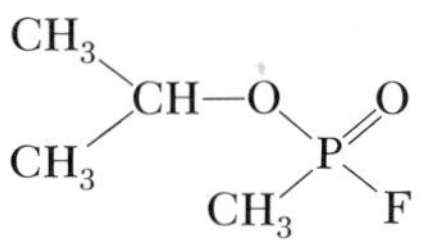

Sarin

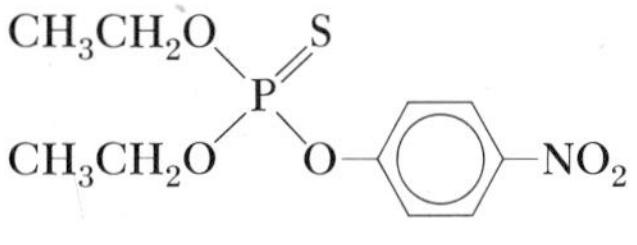

Parathion

ADRENERGIC AND ANTIADRENERGIC DRUGS

In addition to norepinephrine and epinephrine, the compound dopamine is also involved in the transmission of nerve impulses in the sympathetic system. These three compounds are synthesized in the nervous system and adrenal glands from the amino acids phenylalanine and tyrosine (p. 248), which form dihydroxyphenylalanine as a precursor to dopamine, norepinephrine, and epinephrine.

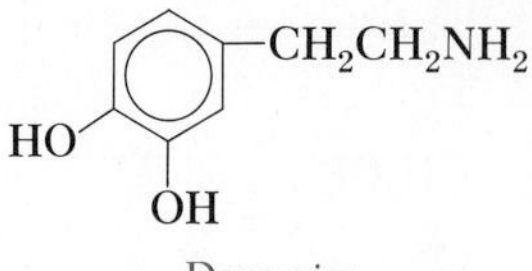

HO, OH — $CH_2CH_2NH_2$

Dopamine

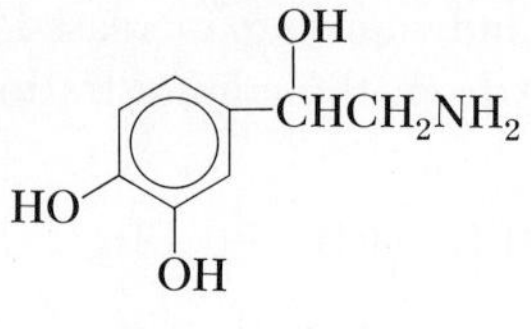

HO, OH — $CH(OH)CH_2NH_2$

Norepinephrine

HO, OH — $CH(OH)CH_2N(H)CH_3$

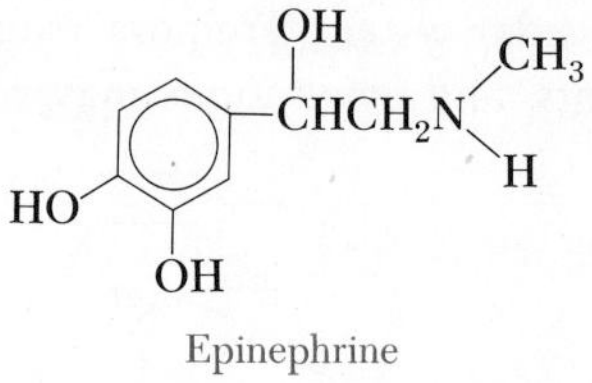

Epinephrine

These drugs are called catecholamines, a term related to their structures (organic amines of dihydroxy benzene or catechol). Dopamine transmits adrenergic impulses in areas of the brain and nervous system and serves as an intermediate precursor for norepinephrine in the granules of nerve cells involved in the mediation of impulses from the nerve endings to the target organs or receptor sites.

The essential nature of dopamine as a precursor for norepinephrine and epinephrine and as a transmitter of adrenergic impulses in the brain has been related to the treatment of **Parkinson's disease.** This is a disease of the brain affecting the metabolism of the catecholamines; it is also known as paralysis agitans, since it involves both a progressive paralytic rigidity and tremors of the extremities. Recently it was found that the dopamine concentration of certain areas of the brain was markedly deficient in chronic patients who had died of the disease. When administered as a drug, the dopamine would not pass into the brain tissue from the blood; however, its precursor, L-dihydroxyphenylalanine or L-dopa (p. 112), readily passed from the blood into the brain tissue, where it was converted into dopamine. The administration of the drug **L-dopa** results in considerable relief and improvement of symptoms for these patients and is being hailed as a breakthrough in Parkinson's disease therapy.

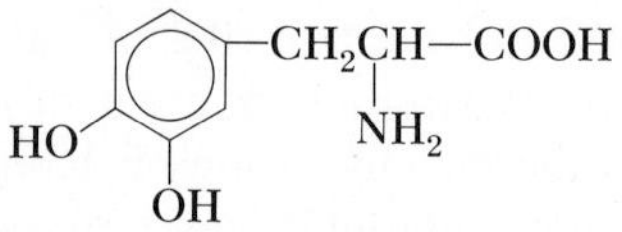

HO, OH — $CH_2CH(NH_2){-}COOH$

L-Dopa

Epinephrine is a hormone produced by the adrenal glands which stimulate the target organs in the sympathetic system similar to the action of the other catecholamines. In addition, since the adrenal gland releases epinephrine into the bloodstream, this compound exhibits many effects that are not directly related to the action of the nervous system. Epinephrine released by the adrenals or administered by injection stimulates the heart, causing an increase in rate and often an alteration of the rhythm of the beat; it also produces a constriction of blood vessels with an increase in blood pressure and a subsequent increase in blood flow to the muscles, affects respiration by relaxing the bronchial muscles (especially in asthma), and increases blood sugar formation from glycogen breakdown to provide additional energy for stressful situations.

There are several compounds not produced by the body that mimic the action of the catecholamines and are therefore called sympathomimetic amines. Isoproterenol, amphetamine, methamphetamine, and ephedrine are examples of this type of drug. Isoproterenol is used clinically as a bronchodilator in respiratory disorders and as a cardiac stimulant in any type of heart block. Drugs related to amphetamine, which is a racemic mixture, include dextroamphetamine, methamphetamine, and ephedrine. In addition to many of the effects described previously for catecholamine administration, the amphetamine-type drugs exert powerful central nervous system stimulating actions. These include a decreased sense of fatigue, increased initiative and ability to concentrate, often an elevation of mood, elation, and euphoria, and increased motor activity. Unfortunately, prolonged use of these drugs is almost always followed by mental depression and fatigue.

The drugs are commonly used as appetite depressants in obesity to counteract depressive syndromes and behavioral disorders and as supportive therapy for overdoses of central nervous system depressant drugs. Ephedrine exhibits many of the effects of epinephrine but has a lower potency and a much longer duration of action. It is used in mild cases of asthma or in chronic cases that require continued medication.

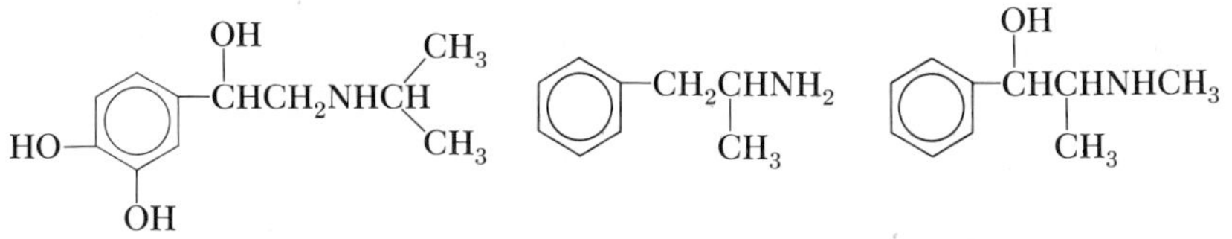

Isoproterenol

Amphetamine

Ephedrine

Another type of drug that is closely related to the action of adrenergic agents is the inhibitor of the enzyme monoamide oxidase. This enzyme is involved in the removal of norepinephrine from the receptor sites after the passage of a sympathetic nerve impulse. Compounds such as isocarboxazid and nialamide inhibit the activity of the enzyme and prevent the destruction of the norepinephrine, thus causing a stimulating effect. These drugs are used to elevate the mood of depressed patients and are called psychic energizers, although the relationship between their ability to inhibit the enzyme and their therapeutic action is not understood.

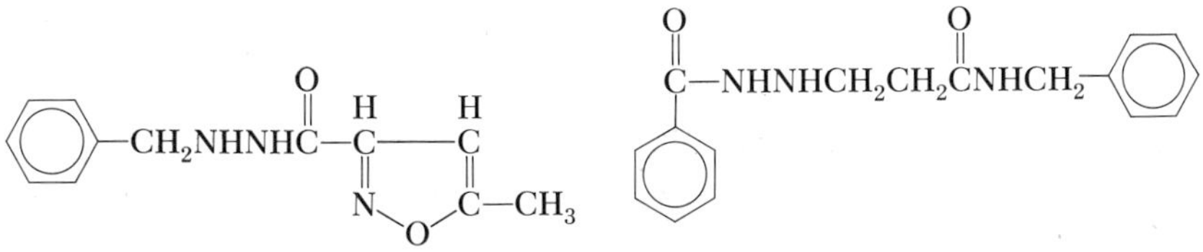

Isocarboxazid

Nialamide

TRANQUILIZERS, SEDATIVES, AND HALLUCINOGENIC DRUGS

Drugs that depress the sympathetic system by blocking the adrenergic impulses are often used as tranquilizers. Phenothiazine, which has a three-ringed structure in which two benzene rings are linked by a sulfur atom and a nitrogen atom, forms the basis for a group of drugs that are potent adrenergic blockers. The phenothiazines are widely used in the treatment of pyschiatric patients and in the treatment of nausea and vomiting. There is a close relationship between the chemical structure and the activity of the drug. Substitution of a chlorine or methoxy group in position R′ in the basic structure increases the potency of the drug for depressing motor activity and altering psychotic behavior in patients. A CF_3 substitution in this position increases antiemetic and antipsychotic potency. One of the most potent phenothiazines has a CF_3 on position R′ and a piperazine group on position R″. Chlorpromazine is the most frequently prescribed phenothiazine, while fluphenazine is one of the most potent members of this group. Their relationship to the basic structure is shown as follows:

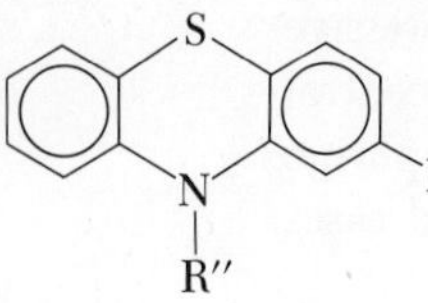

Phenothiazine (Basic structure)

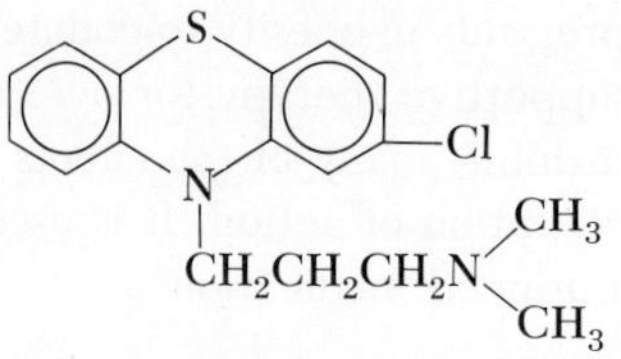

Chlorpromazine

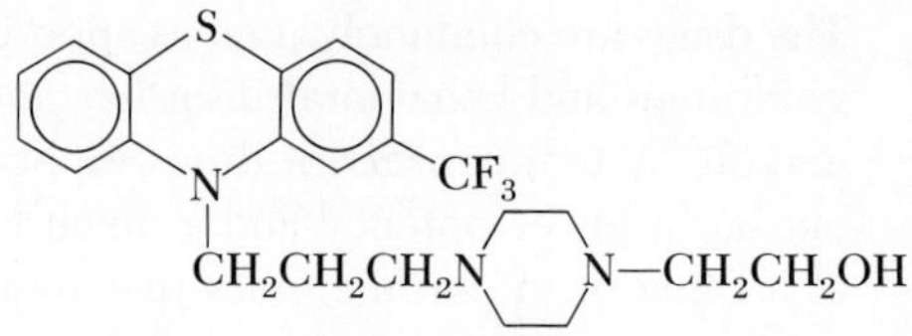

Fluphenazine

Chlorpromazine also has a sedative effect and reduces the blood pressure.

Alkaloids obtained from the roots of the *Rauwolfia serpentina*, a plant found mainly in India, are represented by the tranquilizer reserpine. This drug depletes the stores of catecholamines and serotonin in the brain and other tissues and therefore interferes with the normal conduction of adrenergic impulses. Reserpine also depresses the central nervous system and acts on the cardiovascular system to lower the blood pressure. In addition to its use as a tranquilizer it is commonly employed in the treatment of hypertension.

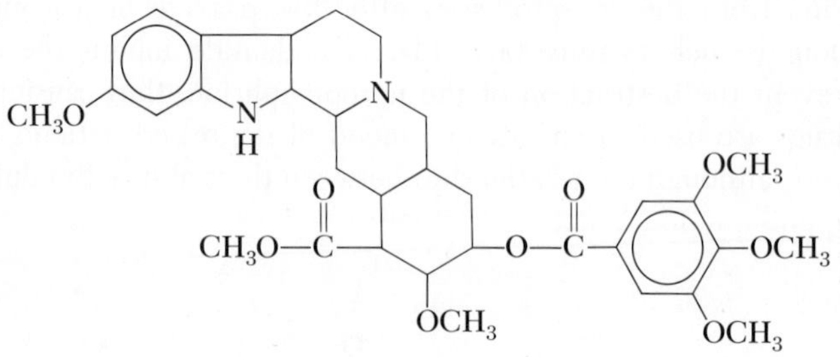

Reserpine

Other compounds used as mild tranquilizers or sedatives, such as barbiturates and meprobamate, are related to the pyrimidines and to urea (p. 107 and 180). These drugs depress the central nervous system. In the case of barbiturates their speed of action and duration of effect depends on an increased lipid solubility and an increase in the length of the side chain, as typified by the structure of Seconal versus that of Barbital. The pharmacological effects of meprobamate are very similar to those of the barbiturates.

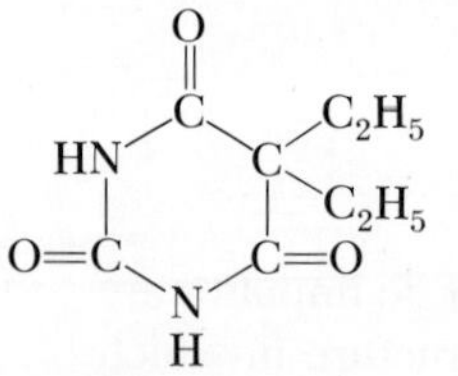

Barbital

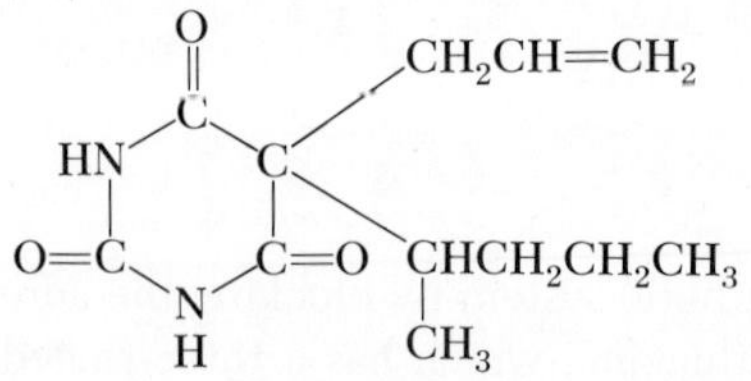

Seconal

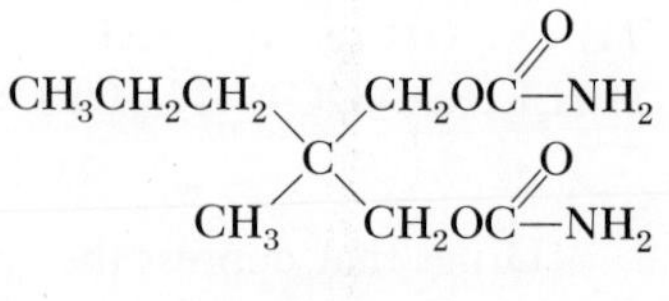

Meprobamate

Continued use of mild sedatives or tranquilizers, like barbiturates, leads to a dependency on the drug (p. 107). Overdosage or combinations of alcoholic beverages and barbiturates may lead to coma and accidental death. The more potent tranquilizers are subjected to strict control by physicians and are not as widely prescribed. When drugs with strong analgesic and sedative properties are required, narcotic drugs, such as **morphine** and **demerol,** are employed. Morphine, first isolated from the opium poppy, has a complex chemical structure containing a phenanthrene nucleus and a piperidine nucleus. **Codeine,** which is also commonly used as a weaker narcotic, is a methyl ester of morphine. **Heroin** is the diethyl ester of morphine and is used by drug addicts because it is more lipid-soluble and faster-acting than morphine. Demerol was synthesized as a substitute for morphine but proved to be no less habit-forming.

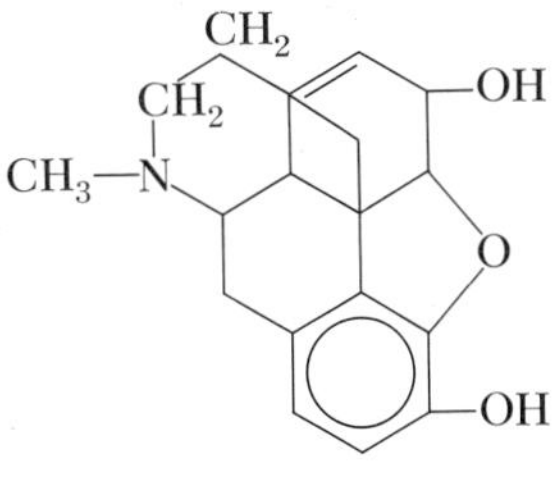

Morphine

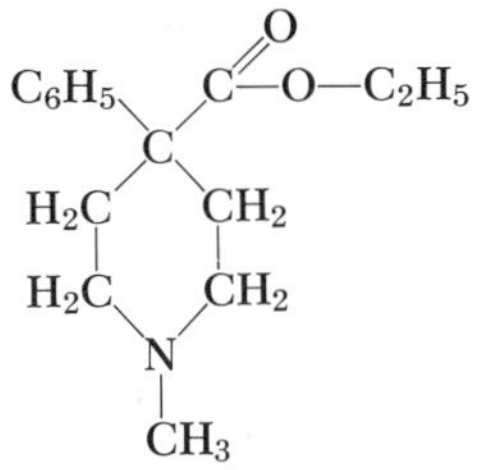

Demerol

A long step further in the use of mind-influencing drugs is represented by the **psychedelic** or **hallucinogenic drugs.** It is difficult to characterize the medical effect of these drugs, although many nonmedical experiments are being conducted. In view of the change in personality and mental state of an individual taking this type of drug, the results of ingestion are often unpredictable. **Marijuana** represents a mild type of hallucinogen and is obtained from the *Cannabis sativa* or hemp plant. The most potent marijuana is obtained from the yellow resin produced from the flowers of the ripe plant and is called **hashish.** Chemically the drug is a derivative of an alcohol, **cannabinol,** and the active constituent is believed to be a delta-L form, which has recently been synthesized. **Mescaline** and **lysergic acid diethylamide** (**LSD**) (p. 28) are examples of more potent hallucinogens.

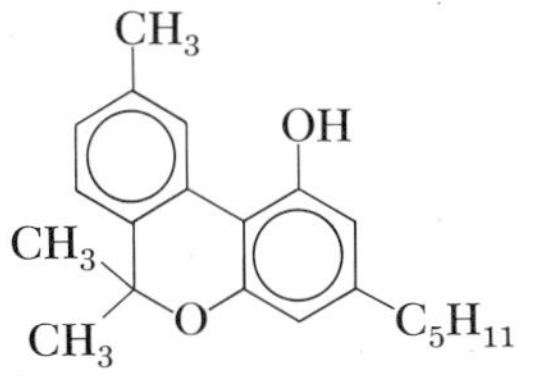

Cannabinol

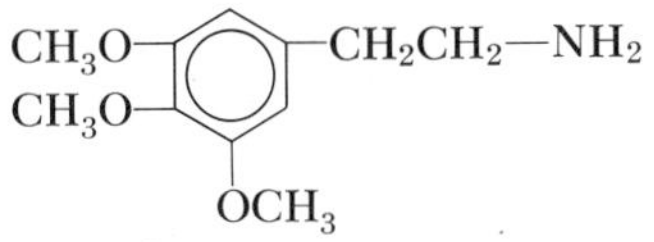

Mescaline

ANTIHISTAMINES

Histamine, which is formed by the decarboxylation of the amino acid histidine (see p. 248), is a powerful pharmacological agent with effects on the vascular system, smooth muscles, and exocrine glands, especially the gastric glands. The administration or release of histamine causes dilatation of capillaries and small blood vessels with a subsequent drop in systemic blood pressure; the dilatation of cerebral vessels results in a histamine headache which may be very severe. Smooth muscles, especially the bronchioles, are stimulated by histamine and may cause respiratory problems in persons suffering from bronchial asthma and other pulmonary diseases. Histamine is a powerful gastric secretogogue and produces a copious secretion of gastric juice of high acidity. It also stimulates nerve endings and causes itching when introduced into the superficial layers of the skin.

Antihistamines are drugs that antagonize the pharmacological actions of histamine and also reduce the intensity of allergic reactions. A portion of the chemical structure of various antihistamines is similar to that in histamine, and these drugs act as competitive antagonists to histamine. Apparently they occupy the receptor sites on the effector cells and exclude histamine from these sites. The common core of the chemical structure in both histamine and antihistamines is a substituted amine, i.e., ethylamine. It is believed that it is this portion of the molecule that competes with histamine for the receptors.

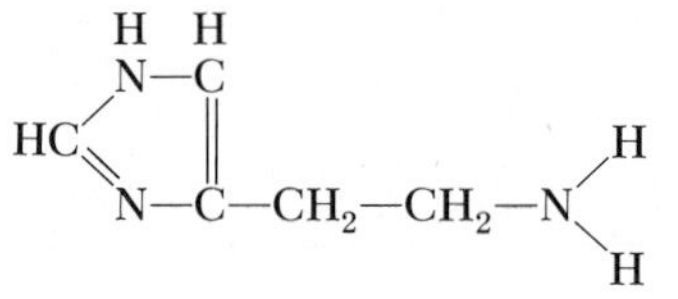

Histamine

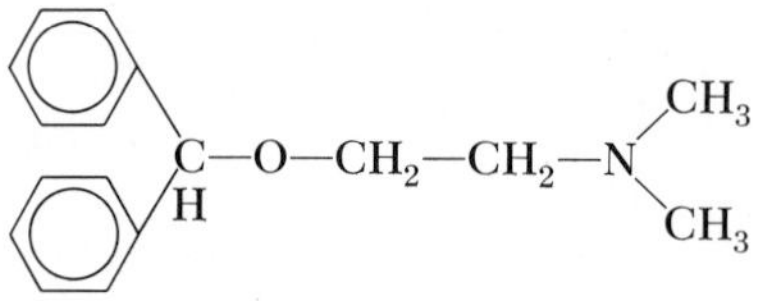

Benadryl (Diphenhydramine)

Therapeutically the antihistamines are most commonly used in the symptomatic treatment of various allergic diseases. Patients with bronchial asthma, hay fever, allergic rhinitis, and chronic rhinitis with superimposed acute colds gain considerable relief by the use of antihistamines. Various types of allergic dermatitis, contact dermatitis, insect bites, and poison ivy are benefited by the topical application of antihistamine-containing lotions. Some of these drugs, especially **Dramamine,** which relies on diphenhydramine as the active agent, are very effective against motion sickness. One common side effect of most antihistamines is their tendency to induce sedation, which restricts their daytime use when it is necessary to operate motor vehicles. **Chlor-trimeton,** shown below, is less prone to produce drowsiness than most other preparations. The prominent hypnotic effect of antihistamines related to **Benadryl** and **Pyribenzamine** has resulted in their use in various proprietary remedies for insomnia, such as Sominex and Nytol.

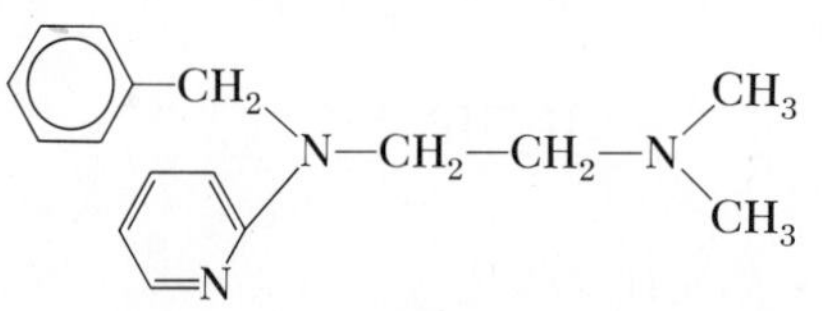

Pyribenzamine (Tripelennamine)

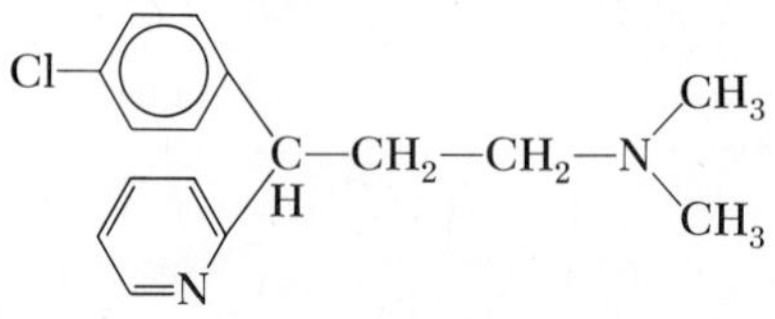

Chlor-trimeton (Chlorpheniramine)

ANTIBACTERIAL AND ANTIBIOTIC DRUGS

The body has to guard against infection by bacteria or viruses from birth to death. The newborn infant is fortified with antibodies against disease upon receiving the gamma globulins in the mother's milk. The layer of skin covering the body, the hydrochloric acid of the gastric juice, the digestive enzymes, and the various phagocytic cells in the circulation, all serve as a first-line defense against infection. Vaccination during childhood stimulates the production of antibodies to certain diseases. Until the mid-thirties, a serious infection or infectious disease was viewed with alarm by physicians and laymen alike. Recovery depended mainly on the body's natural defense and supportive general therapy. The first major group of chemotherapeutic agents were the sulfonamides or **sulfa drugs,** prepared by the acylation of sulfanilamide (p. 97). These drugs were antibacterial agents that inhibited the synthesis of a compound like folic acid that was essential for the continued growth of the invading bacteria. Sulfanilamide, for example, acts as a competitive inhibitor of the enzyme that is involved in the utilization of p-aminobenzoic acid in the synthesis of tetrahydrofolic acid by the bacteria (see p. 310). **Sulfanilamide** was found to be effective in the treatment of streptococcus infections, pneumonia, puerperal

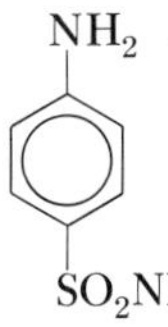

Sulfanilamide 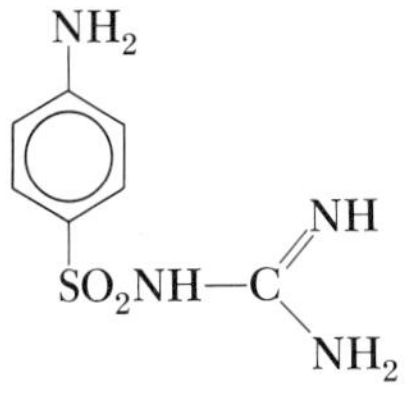   Sulfaguanidine 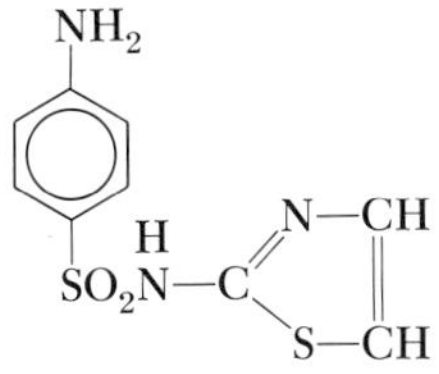   Sulfathiazole 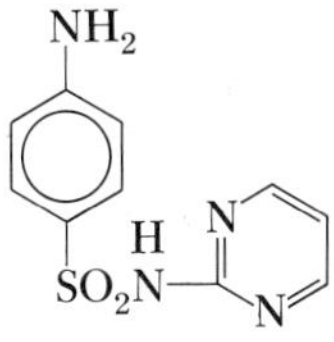   Sulfadiazine

fever, gonorrhea, and gas gangrene. The drug is only slightly soluble in water and may damage the kidney by accumulation in that organ during excretion. Other toxic reactions, including methemoglobinemia, resulted in the development of other derivatives, such as **sulfaguanidine, sulfathiazole,** and **sulfadiazine.** A thorough study of the therapeutic properties of each sulfa drug resulted in better treatment and control of various infectious diseases. Sulfadiazine, for example, is less toxic than the other sulfa drugs, yet is one of the most effective in the treatment of pneumonia and staphylococcus infections. The antibacterial action of the sulfonamides is dependent on the free p-amino group in the structure of the drug.

A few years after the development of the sulfa drugs, a new type of antimicrobial agent was accidently discovered by Fleming. He observed that a staphylococcus culture on a bacterial plate did not grow around the periphery of a blue-green mold that had contaminated the culture plate. **Penicillin** was isolated from the secretion of the mold and was termed an antibiotic, since it interfered with the growth of the bacteria. A large number of antibiotic agents have been isolated from similar experiments with other molds. It required nine years of intensive research to synthesize penicillin. Other antibiotics, including **streptomycin, tetracycline,** and **prostaphlin,** have been synthesized, and prostaphlin shows considerable promise as an effective control of infections caused by staphylococcal bacteria. As in the case of the sulfa drugs, a family of antibiotics with specific antimicrobial properties has now been developed.

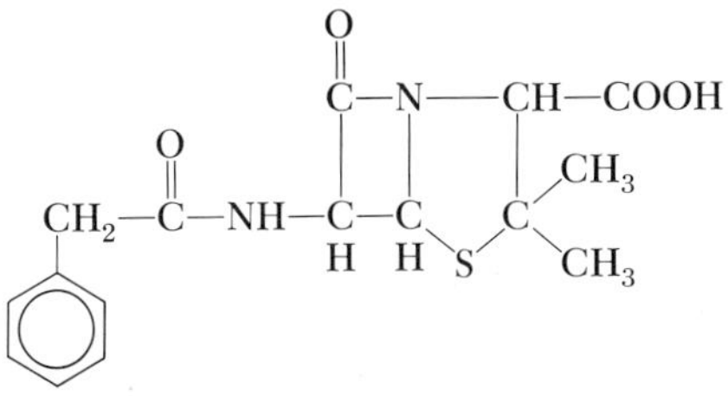

Penicillin G 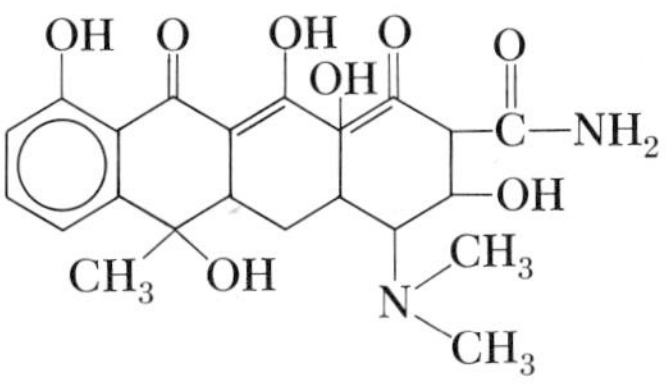   Tetracycline

STEROID DRUGS IN RHEUMATOID ARTHRITIS

There are several types of collagen diseases associated with inflammatory changes in connective tissue which affect mainly the joints, skin, heart, and muscle. **Rheumatoid**

arthritis and **lupus erythematosus** are examples of collagen disease. Arthritis is most common and affects men and especially women in their forties and fifties. For many years, aspirin has been used as a mild antiinflammatory agent in arthritis and continues as the maintenance therapy. More recently **cortisone** has been found to reduce inflammation of the joints in arthritis and to reverse the course of this and other collagen diseases. The dosage level required to produce these desirable effects also produced several unwanted side effects. Steroid derivatives of cortisone were developed to decrease the incidence of side effects and increase the therapeutic potency of the drugs. The 9-fluoro-16-methyl derivative of **prednisolone,** a steroid closely related to cortisol, is 100 to 250 times as potent as cortisone in the treatment of rheumatoid arthritis. At present, arthritic patients are maintained with aspirin and given small doses of potent steroid drugs whenever an acute inflammatory process flares up in their joints. The relationship between the structure and the function of cortisone and related steroids has already been discussed in Chapter 14 (see p. 236).

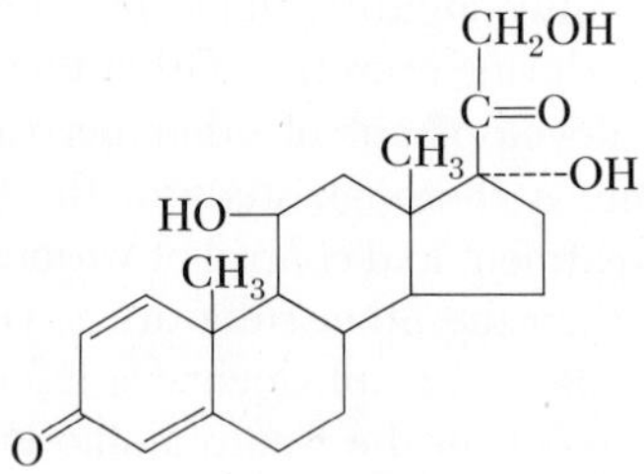

Prednisolone

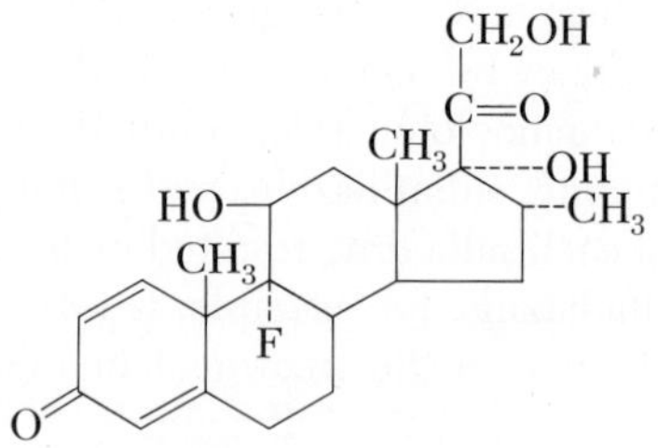

9-Fluoro-16-methyl prednisolone

ANTIFERTILITY DRUGS

Another type of steroid drug related to the sex hormones is the **antifertility drugs.** These drugs are unique in that they are given to inhibit a normal physiological process, whereas the great majority of drugs is used to treat a disease process or to alleviate the symptoms of a disease. As early as 1937 it was shown that the hormone progesterone would inhibit ovulation in rabbits. In the 1950's, several laboratories attempted to develop orally active steroids that possessed the properties of progesterone. **Norethindrone, 17α-ethynyl-19-nortestosterone,** and **norethynodrel,** the progestational component of Enovid, resulted from these studies. These drugs were related to testosterone (see p. 238) and contained a 17α-ethynyl group.

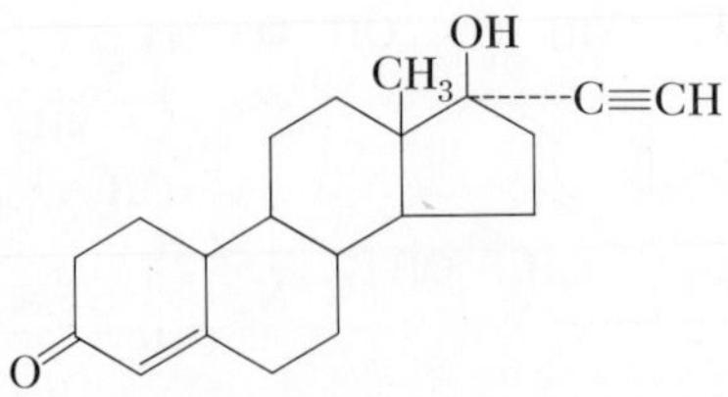

Norethindrone

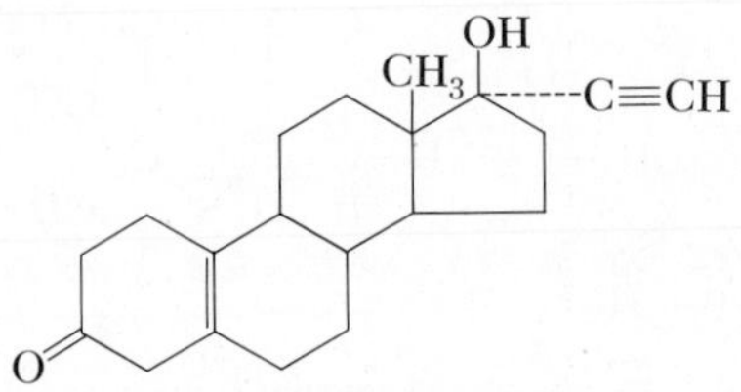

Norethynodrel

Since progesterone is relatively inactive when given orally, derivatives of this compound were studied for oral potency. It was found that 17-acetoxy-progesterone was active orally and that the addition of an α-methyl group to produce **medroxyprogesterone acetate** further enhanced this activity.

Mestranol

Medroxyprogesterone acetate

In clinical trials in Puerto Rico, Haiti, and the United States, the testosterone and progesterone derivatives described previously were found to effectively suppress ovulation in women. Also, it was discovered that estrogen enhanced the suppressive effect of the progesterone and that the 3-methyl ether of ethynyl estradiol or **mestranol** served as a potent estrogen. A combination of orally effective progesterone and estrogen active drugs are therefore commonly used in "the pill." The original dose was 10 mg of the progesterone and 0.15 mg mestranol, but this is being reduced toward 1 mg progesterone and 0.05 mg mestranol to decrease the occurrence of such side effects as nausea, headaches, dizziness, and thrombosis. The drug is usually taken on days five to 25 of the menstrual cycle and then withdrawn to permit normal menstruation. It may be possible in the near future that these drugs may be injected intramuscularly with effective action from one to six months.

From a study of the mode of action of oral contraceptives it was concluded that the primary effect is inhibition of follicular maturation (see p. 237), which prevents the occurrence of ovulation.

This mechanism is outlined in Figure 24-1, which illustrates the blocking action of the progesterone derivatives on the hormones of the pituitary gland responsible for development of an ovarian follicle and the subsequent release of an unfertilized ovum by the process of ovulation.

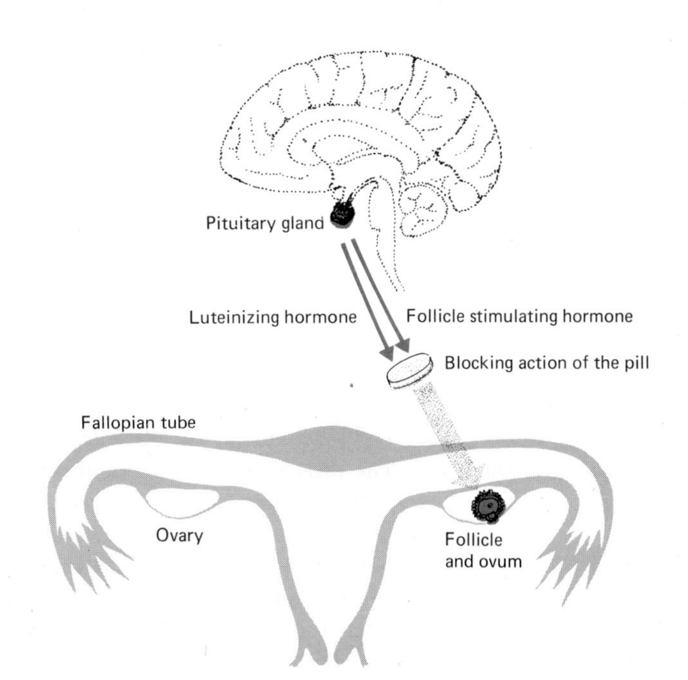

FIGURE 24-1 The effect of oral contraceptives of the process of ovulation.

HYPOGLYCEMIC DRUGS IN DIABETES MELLITUS

Diabetes mellitus, because of its frequency, is probably the most important metabolic disease. The fundamental difficulty in the disease is a relative or complete lack of insulin, which is necessary for normal carbohydrate metabolism. Since the metabolic pathways of carbohydrates, fats, and proteins are known to be closely interwoven, any essential fault in carbohydrate metabolism also involves the metabolism of fat and protein, as well as water, electrolyte, and acid-base balance. There is evidence that diabetes is a hereditary disease and that the genetic tendency toward the disease results in the common condition known as prediabetes, which exists in many relatives of diabetics. The chemistry of **insulin** (p. 256) and its function in diabetes (p. 317) have already been discussed. The beta cells of the islet tissue in the pancreas produce insulin, and any condition resulting in hyperglycemia stimulates the pancreas to secrete greater quantities of this hormone. To control diabetes, insulin must be injected into the muscle tissue daily. In view of the discomfort and inconvenience to the patient, many efforts have been made to prolong the action of insulin and to develop drugs with insulin-like activity that can be taken orally. The combination of insulin with other proteins has resulted in preparations with prolonged activity. **Protamine zinc insulin** is the best example of such a combination. It has been known for several years that salicylates produce a hypoglycemic effect, but the potency was not sufficient to warrant use of the drug as an insulin substitute. In 1942 it was observed that a sulfamidothiazole compound exhibited a potent hypoglycemic effect. Related compounds were tested, and by 1955 sulfonylurea derivatives such as **tolbutamide** (Orinase) were available as antidiabetic agents. Guanidine derivatives also produced hypoglycemia, and phenethylbiguanide (Phenformin) represents another type of antidiabetic drug. When administered orally these two drugs exhibit different properties.

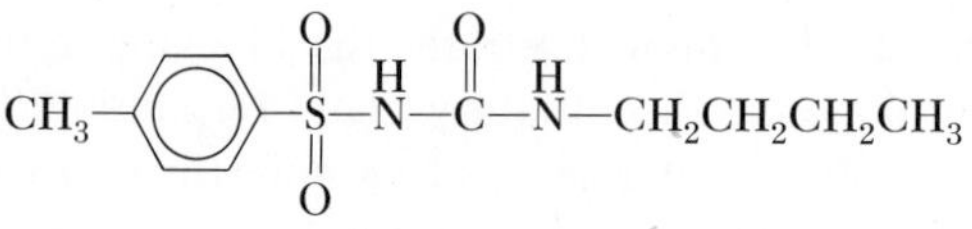

$$CH_3-C_6H_4-S(=O)_2-NH-C(=O)-NH-CH_2CH_2CH_2CH_3$$

Orinase (Tolbutamide)

$$C_6H_5-CH_2CH_2-NH-C(=NH)-NH-C(=NH)-NH_2$$

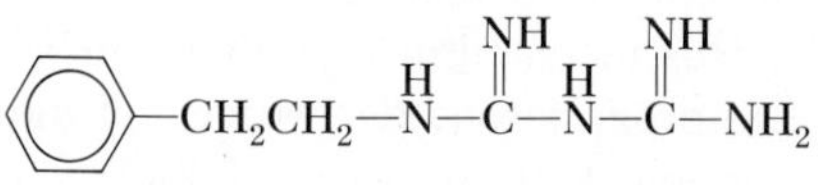

Phenformin (Phenethylbiguanide)

Orinase stimulates the secretion of insulin from the beta cells, while Phenformin stimulates the oxidation of glucose by the peripheral tissues. Both of the drugs are effective in the treatment of diabetic patients over 40 years of age with stable and mild diabetes. Their use is ineffective in young unstable diabetics, especially those prone to ketoacidosis. In patients that do not respond to Orinase alone, the concurrent administration of both drugs is often effective. It has also been found that diabetics with congestive heart failure or severe renal insufficiency should not be treated with either of these drugs.

URICOSURIC DRUGS IN GOUT

Gout is a chronic disease that is characterized by an increased level of uric acid in the blood, hyperuricemia, acute episodes of gouty arthritis, and degenerative changes in the joints. The disease is seen predominantly in men and the initial stages occur in their forties. In the early days, gout appeared to be more prevalent among the aristocracy, and at present, professors and clinicians exhibit the greatest incidence of the disease; however, even vegetarians and lower economic groups show a predilection for the disease. A diet rich in protein and glandular meats containing nucleoproteins and more frequent visits to a physician or clinic may explain the observed increase in occurrence in the upper income groups.

In the treatment of the disease an attempt is made to decrease the level of uric acid in the blood and in the body stores. Salicylates in the form of aspirin have long been used as a uricosuric agent, but fairly high doses and prolonged treatment are required. Drugs with a more potent action often produced toxic hepatitis. **Probenecid** (Benemid), a derivative of benzoic acid, was found to inhibit the enzymes responsible for the reabsorption of uric acid by the kidney. It is a safe drug that lowers serum uric acid by about 50 per cent within two to four days and maintains the reduced level as long as therapy is continued. A more recent therapeutic agent is **allopurinol** that inhibits the enzyme xanthine oxidase and thus reduces the formation of uric acid from its immediate precursors, hypoxanthine and xanthine. Treatment with allopurinol maintains the

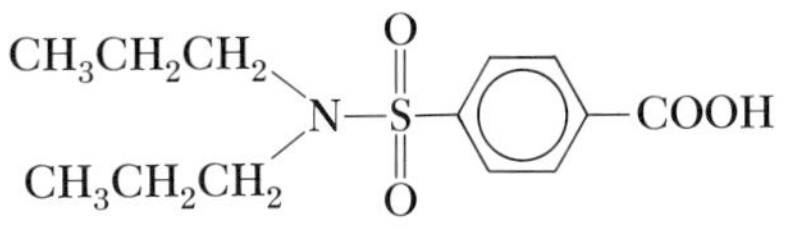

$(CH_3CH_2CH_2)_2N-SO_2-C_6H_4-COOH$

Benemid

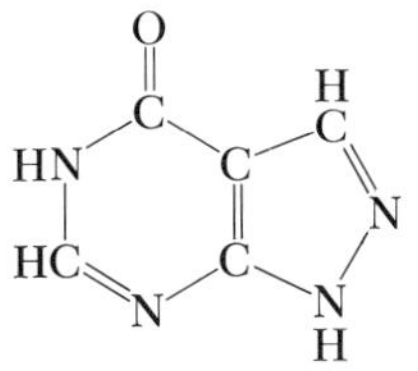

Allopurinol

blood uric acid level at a normal value and lowers the body pool of uric acid by constant excretion through the kidney. Both of these drugs were first "tailor-made" by pharmacologists for a specific biochemical purpose. Probenecid was developed to decrease the renal tubular secretion of penicillin, and was found to be very effective in decreasing the reabsorption of uric acid by the tubules. Allopurinol was synthesized to serve as a potent inhibitor of xanthine oxidase to prevent the oxidation of 6-mercaptopurine and thus extend its effective action in the treatment of leukemia.

CYTOTOXIC DRUGS IN CANCER

Cancer is a general term used to describe rapid multiplication of cells and increased growth of certain tissues in the body. Cancers are also called *tumors* and *neoplasms* and are classed as benign when they may be removed by surgery without reoccurrence, and as malignant when they spread to other parts of the body and exhibit reoccurrence of growth after surgery. *Leukemia* is a cancer of leucocytes or white blood cells, *carcinoma* involves epithelial cells, and *sarcoma* is a tumor of muscle or connective tissue. Considerable research effort and money has been expended in extensive studies on cancer in recent years. Tumors have been transplanted in experimental animals and produced by *carcinogenic agents*, such as dimethylbenzanthracene, to aid in the study of the metabolism of cancer cells and possible therapeutic agents. Although several classes of compounds including alkylating agents, antimetabolites, and purine and pyrimidine analogs show promise in the treatment of leukemia, surgical removal of tumors and irradiation with x-rays and radioactive isotopes remain the major mode of attack on cancer. The drug **6-mercaptopurine** is thought to be converted first to hypoxanthine and then to ribonucleotides (see p. 269). It also is believed to suppress the biosynthesis of purines within the cell. These actions interfere with the production of RNA and DNA in the tumor cells and exert a cytotoxic action, especially on bone marrow and intestinal epithelial cells. **Cyclophosphamide** is an example of an alkylating agent that was synthesized to overcome the undesirable side effects of other similar drugs. It is a cytotoxic agent that has shown good initial results in the treatment of Hodgkin's disease and lymphosarcoma. A folic acid antagonist which has demonstrated dramatic results in the treatment of acute leukemia in children is **methotrexate.** An analog of pyrimidine, **fluorouracil,** has been particularly effective in the treatment of advanced carcinoma, especially of the breast and the gastrointestinal tract. The structure of some of the cytotoxic drugs used in the treatment of cancer are shown as follows:

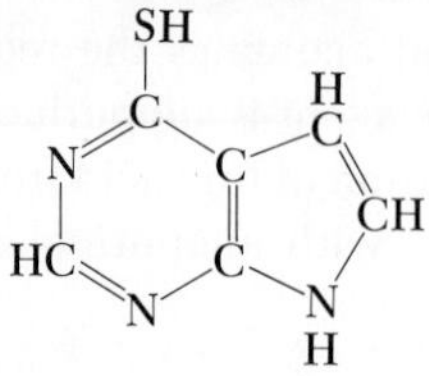

6-Mercaptopurine

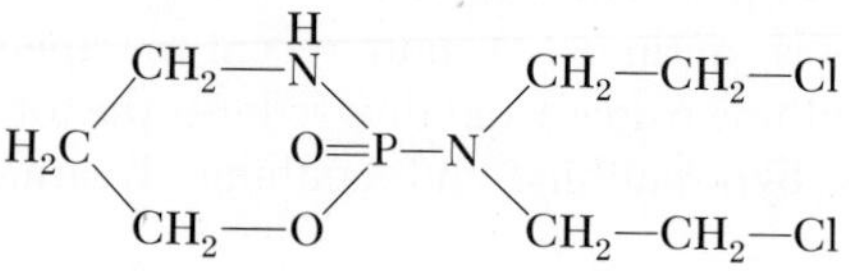

Cyclophosphamide

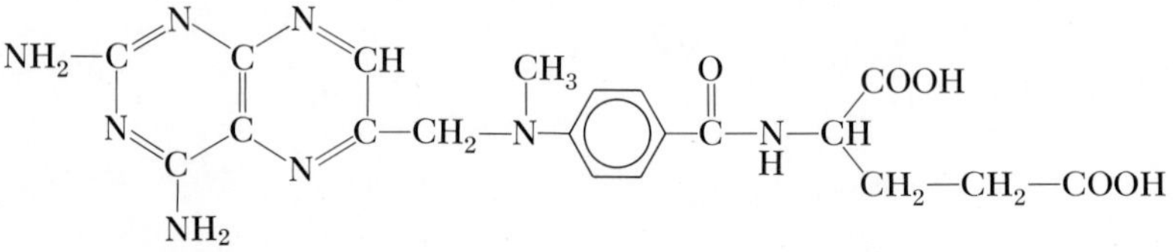

Methotrexate

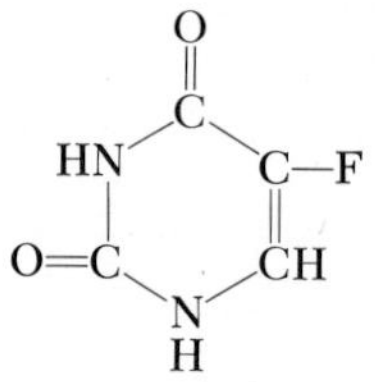

Fluorouracil

IMPORTANT TERMS AND CONCEPTS

- acetylcholinesterase
- adrenergic
- analgesic
- antibiotics
- antifertility drugs
- antihistamine
- cytotoxic drugs
- epinephrine
- ganglion
- heroin
- hypoglycemic drugs
- marijuana
- nervous system
- sulfa drugs
- sympathetic system
- synapse
- tranquilizers

QUESTIONS

1. Why can it be stated that all disease has a biochemical basis? Explain.
2. Why is aspirin used in so many proprietary drug preparations? Explain.
3. Outline the essential divisions of the peripheral nervous system.
4. Draw a diagram that represents the passage of a nerve impulse across a synapse to the target organ.
5. Give an example of a cholinergic drug and describe its action in the body.
6. Explain the action of parathion as an insecticide.
7. What type of drug is epinephrine, and why does it affect so many systems in the body?
8. Discuss the chemical nature and action of L-dopa in Parkinson's disease.
9. Why are amphetamine-type drugs used in such large quantities today? Explain.
10. Why are modifications of the phenothiazine structure important therapeutically? Explain.

11. Isocarboxazid is an example of a monoamine oxidase inhibitor. Describe the usefulness of the drug and compare its action with that of reserpine.

12. Compare meprobamate and the barbiturates on the basis of their chemical structure, potency, and mode of action.

13. Give two examples of a hallucinogen drug. Should the use of these drugs be controlled by law? Explain.

14. Outline the main uses of antihistaminic drugs.

15. Explain the differences and similarities of sulfa drugs and antibiotics such as penicillin.

16. Give an example of a steroid drug used in the treatment of arthritis and explain its action.

17. Describe the mechanism of action of the antifertility drugs.

18. Explain the action of one oral insulin substitute and its advantages.

19. Compare the mechanism of action of benemid and allopurinol in the treatment of gout.

20. Why is 6-mercaptopurine used in the treatment of leukemia?

21. Compare the chemical structure of allopurinol and 6-mercaptopurine. Should allopurinol be effective in the treatment of leukemia? Explain.

SUGGESTED READING

Barron, Jarvick, and Bunnell: The Hallucinogenic Drugs. Scientific American, Vol. 210, No. 4, p. 29, 1964.
Berelson, and Freedman: A Study in Fertility Control. Scientific American, Vol. 210, No. 5, p. 29, 1964.
Bogue: Drugs of the Future. Journal of Chemical Education, Vol. 46, p. 468, 1969.
Braun: The Reversal of Tumor Growth. Scientific American, Vol. 213, No. 5, p. 75, 1965.
Collier: Aspirin. Scientific American, Vol. 209, No. 5, p. 96, 1963.
Drugs and the Caltech Student. Chemistry, Vol. 42, p. 8, 1969.
Frei, and Frereich: Leukemia. Scientific American, Vol. 210, No. 5, p. 88, 1964.
Gates: Analgesic Drugs. Scientific American, Vol. 215, No. 5, p. 131, 1966.
Grinspoon: Marihuana. Scientific American, Vol. 221, No. 6, p. 17, 1969.
Linder: The Health of the American People. Scientific American, Vol. 214, No. 6, p. 21, 1966.
Marijuana Program Advances at NIMH. Chemical & Engineering News, July 6, 1970, p. 30.
Nichols: How Opiates Change Behavior. Scientific American, Vol. 212, No. 2, p. 80, 1965.
Scientific Methods of Crime Investigation. Chemistry, Vol. 42, p. 12, 1969.
Smith: Death from Staphylococci. Scientific American, Vol. 218, No. 2, p. 84, 1968.
Weeks: Experimental Narcotic Addiction. Scientific American, Vol. 210, No. 3, p. 46, 1964.

APPENDIX 1

Table of Logarithms

	0	1	2	3	4	5	6	7	8	9
1.0	.0000	.0043	.0086	.0128	.0170	.0212	.0253	.0294	.0334	.0374
1.1	.0414	.0453	.0492	.0531	.0569	.0607	.0645	.0682	.0719	.0755
1.2	.0792	.0828	.0864	.0899	.0934	.0969	.1004	.1038	.1072	.1106
1.3	.1139	.1173	.1206	.1239	.1271	.1303	.1335	.1367	.1399	.1430
1.4	.1461	.1492	.1523	.1553	.1584	.1614	.1644	.1673	.1703	.1732
1.5	.1761	.1790	.1818	.1847	.1875	.1903	.1931	.1959	.1987	.2014
1.6	.2041	.2068	.2095	.2122	.2148	.2175	.2201	.2227	.2253	.2279
1.7	.2304	.2330	.2355	.2380	.2405	.2430	.2455	.2480	.2504	.2529
1.8	.2553	.2577	.2601	.2625	.2648	.2672	.2695	.2718	.2742	.2765
1.9	.2788	.2810	.2833	.2856	.2878	.2900	.2923	.2945	.2967	.2989
2.0	.3010	.3032	.3054	.3075	.3096	.3118	.3139	.3160	.3181	.3201
2.1	.3222	.3243	.3263	.3284	.3304	.3324	.3345	.3365	.3385	.3404
2.2	.3424	.3444	.3464	.3483	.3502	.3522	.3541	.3560	.3579	.3598
2.3	.3617	.3636	.3655	.3674	.3692	.3711	.3729	.3747	.3766	.3784
2.4	.3802	.3820	.3838	.3856	.3874	.3892	.3909	.3927	.3945	.3962
2.5	.3979	.3997	.4014	.4031	.4048	.4065	.4082	.4099	.4116	.4133
2.6	.4150	.4166	.4183	.4200	.4216	.4232	.4249	.4265	.4281	.4298
2.7	.4314	.4330	.4346	.4362	.4378	.4393	.4409	.4425	.4440	.4456
2.8	.4472	.4487	.4502	.4518	.4533	.4548	.4564	.4579	.4594	.4609
2.9	.4624	.4639	.4654	.4669	.4683	.4698	.4713	.4728	.4742	.4757
3.0	.4771	.4786	.4800	.4814	.4829	.4843	.4857	.4871	.4886	.4900
3.1	.4914	.4928	.4942	.4955	.4969	.4983	.4997	.5011	.5024	.5038
3.2	.5051	.5065	.5079	.5092	.5105	.5119	.5132	.5145	.5159	.5172
3.3	.5185	.5198	.5211	.5224	.5237	.5250	.5263	.5276	.5289	.5302
3.4	.5315	.5328	.5340	.5353	.5366	.5378	.5391	.5403	.5416	.5428
3.5	.5441	.5453	.5465	.5478	.5490	.5502	.5514	.5527	.5539	.5551
3.6	.5563	.5575	.5587	.5599	.5611	.5623	.5635	.5647	.5658	.5670
3.7	.5682	.5694	.5705	.5717	.5729	.5740	.5752	.5763	.5775	.5786
3.8	.5798	.5809	.5821	.5832	.5843	.5855	.5866	.5877	.5888	.5899
3.9	.5911	.5922	.5933	.5944	.5955	.5966	.5977	.5988	.5999	.6010
4.0	.6021	.6031	.6042	.6053	.6064	.6075	.6085	.6096	.6107	.6117
4.1	.6128	.6138	.6149	.6160	.6170	.6180	.6191	.6201	.6212	.6222
4.2	.6232	.6243	.6253	.6263	.6274	.6284	.6294	.6304	.6314	.6325
4.3	.6335	.6345	.6355	.6365	.6375	.6385	.6395	.6405	.6415	.6425
4.4	.6435	.6444	.6454	.6464	.6474	.6484	.6493	.6503	.6513	.6522
4.5	.6532	.6542	.6551	.6561	.6571	.6580	.6590	.6599	.6609	.6618
4.6	.6628	.6637	.6646	.6656	.6665	.6675	.6684	.6693	.6702	.6712
4.7	.6721	.6730	.6739	.6749	.6758	.6767	.6776	.6785	.6794	.6803
4.8	.6812	.6821	.6830	.6839	.6848	.6857	.6866	.6875	.6884	.6893
4.9	.6902	.6911	.6920	.6928	.6937	.6946	.6955	.6964	.6972	.6981
5.0	.6990	.6998	.7007	.7016	.7024	.7033	.7042	.7050	.7059	.7067
5.1	.7076	.7084	.7093	.7101	.7110	.7118	.7126	.7135	.7143	.7152
5.2	.7160	.7168	.7177	.7185	.7193	.7202	.7210	.7218	.7226	.7235
5.3	.7243	.7251	.7259	.7267	.7275	.7284	.7292	.7300	.7308	.7316
5.4	.7324	.7332	.7340	.7348	.7356	.7364	.7372	.7380	.7388	.7396
5.5	.7404	.7412	.7419	.7427	.7435	.7443	.7451	.7459	.7466	.7474
5.6	.7482	.7490	.7497	.7505	.7513	.7520	.7528	.7536	.7543	.7551
5.7	.7559	.7566	.7574	.7582	.7589	.7597	.7604	.7612	.7619	.7627
5.8	.7634	.7642	.7649	.7657	.7664	.7672	.7679	.7686	.7694	.7701
5.9	.7709	.7716	.7723	.7731	.7738	.7745	.7752	.7760	.7767	.7774

Table of Logarithms Continued

	0	1	2	3	4	5	6	7	8	9
6.0	.7782	.7789	.7796	.7803	.7810	.7818	.7825	.7832	.7839	.7846
6.1	.7853	.7860	.7868	.7875	.7882	.7889	.7896	.7903	.7910	.7917
6.2	.7924	.7931	.7938	.7945	.7952	.7959	.7966	.7973	.7980	.7987
6.3	.7993	.8000	.8007	.8014	.8021	.8028	.8035	.8041	.8048	.8055
6.4	.8062	.8069	.8075	.8082	.8089	.8096	.8102	.8109	.8116	.8122
6.5	.8129	.8136	.8142	.8149	.8156	.8162	.8169	.8176	.8182	.8189
6.6	.8195	.8202	.8209	.8215	.8222	.8228	.8235	.8241	.8248	.8254
6.7	.8261	.8267	.8274	.8280	.8287	.8293	.8299	.8306	.8312	.8319
6.8	.8325	.8331	.8338	.8344	.8351	.8357	.8363	.8370	.8376	.8382
6.9	.8388	.8395	.8401	.8407	.8414	.8420	.8426	.8432	.8439	.8445
7.0	.8451	.8457	.8463	.8470	.8476	.8482	.8488	.8494	.8500	.8506
7.1	.8513	.8519	.8525	.8531	.8537	.8543	.8549	.8555	.8561	.8567
7.2	.8573	.8579	.8585	.8591	.8597	.8603	.8609	.8615	.8621	.8627
7.3	.8633	.8639	.8645	.8651	8657	.8663	.8669	.8675	.8681	.8686
7.4	.8692	.8698	.8704	.8710	.8716	.8722	.8727	.8733	.8739	.8745
7.5	.8751	.8756	.8762	.8768	.8774	.8779	.8785	.8791	.8797	.8802
7.6	.8808	.8814	.8820	.8825	.8831	.8837	.8842	.8848	.8854	.8859
7.7	.8865	.8871	.8876	.8882	.8887	.8893	.8899	.8904	.8910	.8915
7.8	.8921	.8927	.8932	.8938	.8943	.8949	.8954	.8960	.8965	.8971
7.9	.8976	.8982	.8987	.8993	.8998	.9004	.9009	.9015	.9020	.9026
8.0	.9031	.9036	.9042	.9047	.9053	.9058	.9063	.9069	.9074	.9079
8.1	.9085	.9090	.9096	.9101	.9106	.9112	.9117	.9122	.9128	.9133
8.2	.9138	.9143	.9149	.9154	.9159	.9165	.9170	.9175	.9180	.9186
8.3	.9191	.9196	.9201	.9206	.9212	.9217	.9222	.9227	.9232	.9238
8.4	.9243	.9248	.9253	.9258	.9263	.9269	.9274	.9279	.9284	.9289
8.5	.9294	.9299	.9304	.9309	.9315	.9320	.9325	.9330	.9335	.9340
8.6	.9345	.9350	.9355	.9360	.9365	.9370	.9375	.9380	.9385	.9390
8.7	.9395	.9400	.9405	.9410	.9415	.9420	.9425	.9430	.9435	.9440
8.8	.9445	.9450	.9455	.9460	.9465	.9469	.9474	.9479	.9484	.9489
8.9	.9494	.9499	.9504	.9509	.9513	.9518	.9523	.9528	.9533	.9538
9.0	.9542	.9547	.9552	.9557	.9562	.9566	.9571	.9576	.9581	.9586
9.1	.9590	.9595	.9600	.9605	.9609	.9614	.9619	.9624	.9628	.9633
9.2	.9638	.9643	.9647	.9652	.9657	.9661	.9666	.9671	.9675	.9680
9.3	.9685	.9689	.9694	.9699	.9703	.9708	.9713	.9717	.9722	.9727
9.4	.9731	.9736	.9741	.9745	.9750	.9754	.9759	.9763	.9768	.9773
9.5	.9777	.9782	.9786	.9791	.9795	.9800	.9805	.9809	.9814	.9818
9.6	.9823	.9827	.9832	.9836	.9841	.9845	.9850	.9854	.9859	.9863
9.7	.9868	.9872	.9877	.9881	.9886	.9890	.9894	.9899	.9903	.9908
9.8	.9912	.9917	.9921	.9926	.9930	.9934	.9939	.9943	.9948	.9952
9.9	.9956	.9961	.9965	.9969	.9974	.9978	.9983	.9987	.9991	.9996

INDEX

H

N

O

Q